The Goddard Guide to Arthropods of Medical Importance, Seventh Edition

T0134061

The Goddard Guide to Arthropods of Medical Importance, Seventh Edition

By
Gail Miriam Moraru
Jerome Goddard II

CRC Press
Taylor & Francis Group
Boca Raton London New York

CRC Press is an imprint of the
Taylor & Francis Group, an **informa** business

Tick (non-blue background). Dr. Lorenza Beati, National Tick Collection
Tick (blue background). Dr. Blake Layton, Mississippi State University Extension Service
Mosquito. Centers for Disease Control, Photo by James Gathany
Millipede. Copyright 2011 by Jerome Goddard, Ph.D.
Sand fly. Armed Forces Pest Management Board
Kissing bug. Copyright 2008 by Jerome Goddard, Ph.D.
Human bot larva. Copyright 2007 by Jerome Goddard, Ph.D.
Scorpion. Copyright 2008 by Jerome Goddard, Ph.D.
Tarantula. Copyright 2011 by Jerome Goddard, Ph.D.

CRC Press
Taylor & Francis Group
6000 Broken Sound Parkway NW, Suite 300
Boca Raton, FL 33487-2742

First issued in paperback 2022

© 2019 by Taylor & Francis Group, LLC
CRC Press is an imprint of Taylor & Francis Group, an Informa business

No claim to original U.S. Government works

ISBN-13: 978-1-138-06943-5 (hbk)
ISBN-13: 978-1-03-233852-1 (pbk)
DOI: 10.1201/b22250

This book contains information obtained from authentic and highly regarded sources. Reasonable efforts have been made to publish reliable data and information, but the author and publisher cannot assume responsibility for the validity of all materials or the consequences of their use. The authors and publishers have attempted to trace the copyright holders of all material reproduced in this publication and apologize to copyright holders if permission to publish in this form has not been obtained. If any copyright material has not been acknowledged please write and let us know so we may rectify in any future reprint.

Except as permitted under U.S. Copyright Law, no part of this book may be reprinted, reproduced, transmitted, or utilized in any form by any electronic, mechanical, or other means, now known or hereafter invented, including photocopying, microfilming, and recording, or in any information storage or retrieval system, without written permission from the publishers.

For permission to photocopy or use material electronically from this work, please access www.copyright.com (http://www.copyright.com/) or contact the Copyright Clearance Center, Inc. (CCC), 222 Rosewood Drive, Danvers, MA 01923, 978-750-8400. CCC is a not-for-profit organization that provides licenses and registration for a variety of users. For organizations that have been granted a photocopy license by the CCC, a separate system of payment has been arranged.

Trademark Notice: Product or corporate names may be trademarks or registered trademarks, and are used only for identification and explanation without intent to infringe.

Publisher's Note

The publisher has gone to great lengths to ensure the quality of this reprint but points out that some imperfections in the original copies may be apparent.

Library of Congress Cataloging-in-Publication Data

Names: Goddard, Jerome, author. | Moraru, Gail Miriam, 1986- author.
Title: The Goddard guide to arthropods of medical importance / by Gail Miriam
Moraru, Jerome Goddard.
Other titles: Physician's guide to arthropods of medical importance |
Arthropods of medical importance
Description: 7th edition. | Boca Raton, Florida : CRC Press, [2019] |
Preceded by Physician's guide to arthropods of medical importance / Jerome
Goddard. 6th ed. 2013. | Includes bibliographical references and index.
Identifiers: LCCN 2018047453| ISBN 9781138069435 (hardback : alk. paper) |
ISBN 9781315115283 (e-book)
Subjects: | MESH: Parasitic Diseases | Arthropod Vectors |
Arthropods--pathogenicity | Bites and Stings--therapy
Classification: LCC RA641.A7 | NLM WC 695 | DDC 614.4/32--dc23
LC record available at https://lccn.loc.gov/2018047453

Visit the Taylor & Francis Web site at
http://www.taylorandfrancis.com

and the CRC Press Web site at
http://www.crcpress.com

Dr. Gail M. Moraru
To the memory of my mother

Dr. Jerome Goddard, II
To my wife, Lindsey, the love of my life and my constant support

Supplemental online materials are available through the Taylor & Francis website available at www.crcpress.com/9781138069435. These include PowerPoint presentations, photos, and brief educational videos.

CONTENTS

Part I Pathological Conditions Caused by Arthropods and Principles of Their Treatment

PREFACE

We are living in a digital information world unlike anything ever seen before, with almost every single idea, theory, and fact known to humankind readily accessible on one's smartphone. How are people supposed to evaluate the reliability of all this information? How can they synthesize these "facts" into meaningful, useful paradigms? As a millennial, I have seen in me and in my peers intense efforts to quickly process and categorize information/facts/data in ways that make sense. Take the discipline of medicine as a specific example. I have trained students in a variety of health fields (pre-nursing students and veterinary students, as examples), and what has become abundantly clear to me is that there is so much information available for health professionals that it is absolutely impossible to remember it all. Therefore, *reliable* references are needed — those in which the reader doesn't have to be afraid the authors are trying to sell them a particular product or service. This guide, originally written by Jerome Goddard Sr., and now modified by me and his son, is just such a guide. It is written for physicians, public health professionals, medical entomologists, and other interested persons who need a readily accessible source of up-to-date information on the myriad ways insects and other arthropods affect human health. Medicine benefits from the input and collaboration of academic scientists, and vice versa. This is especially important in a quickly changing world.

Climate change and international travel mean entomology is not a static field. While for most people living in the industrialized nations, the threat from insects, spiders, mites, and scorpions lies primarily with stings and bites of various species and the reactions, both allergic and nonallergic, to them, vector-borne diseases continue to afflict millions in other parts of the world. Because many arthropods are very adaptable and can now be transported long distances by human travel or may change their distribution due to environmental change, these diseases will not remain isolated to those parts of the world where they currently are.

This book was written to provide a reference of the insects, mites, scorpions, and spiders of public health importance as well as various ancillary topics related to these organisms. Voluminous works could be developed on many of these topics; however, a deliberate effort has been made to keep extraneous information to a minimum. Also, as in all areas of science, entomology includes controversies over certain points and "facts." In many cases, these facts are constantly changing and being revised. Accordingly, we have tried to limit the references in this book and to provide views that represent a general consensus of the current status of each subject.

The primary focus of this arrangement is to provide easy, almost instant access to essential information concerning these topics. It is not the intent of this reference to make entomologists out of the readership. Specialists should be consulted whenever possible for definitive identification of an arthropod. Extensive technical jargon has been avoided as much as possible in the "General Description" sections, and a glossary is included to aid the reader in defining and locating descriptive terms and characters.

The volume begins with several chapters on the pathologic conditions caused by arthropods, and the principles of treating those conditions. These are provided because a physician may first have to identify the nature of an arthropod-caused problem (sting, bite, blistering, etc.). Chapter 1, "Principles of Treatment for Arthropod Bites, Stings, and Other Exposure," includes the rationale behind the various treatment regimes. This should be helpful because, although specific recommendations may change through time, the underlying principles of controlling the immune response will not. When arthropods are mentioned in Part I, there is a parenthetical reference indicating where in Part III the reader can go for more detailed information. Part II consists of a chapter on identification principles of arthropods and a new chapter for physicians on clues to recognizing and treating arthropod bites and stings.

The third major part of the book is an alphabetical arrangement of the arthropods of medical importance with clearly marked subheadings for easy information access. To find a topic or insect section, the reader should look for that name or topic alphabetically. Keep in mind that all flies are grouped together, as are all lice, all mites, etc. A person wishing to find the topic "screwworm fly" would consult the flies chapters, in this case, "Flies that May Cause Myiasis." Also, it is important to remember that common names vary with locality. A "blue-tailed darner" may mean one thing to the authors and something totally different to someone else. Part IV of the book contains discussion of the various personal protection measures that may be employed against insects and other arthropods, including repellents and bed nets. There is an appendix containing a list of common signs and symptoms of arthropod-borne diseases. The index includes the various pathologic conditions and as many of the common names as possible to aid the reader in finding a particular topic or insect. Also, it is important to remember that, if a patient brings in to the clinic an insect, mite, or spider associated with a particular health problem, it is prudent not only to deal with the problem at hand (with the aid of this reference) but also to submit the specimen to a university extension service or health department entomologist for definitive identification. This might be important for later follow-up, consultation, or legal matters.

The seventh edition of Physician's Guide to Arthropods of Medical Importance marks 25 years of this book being continuously in print, a testament to its usefulness. As mentioned, in this edition the authors changed (see the section on biography of Jerome Goddard) and all chapters have been updated with much of the latest information and current references. In addition, we have added supplemental online materials, PowerPoint presentations, photos, and brief educational videos accessible through the Taylor & Francis website available at www.crcpress.com/9781138069435.

Finally, every effort has been made to ensure that the treatment recommendations herein are current and widely recognized as appropriate; however, it must be emphasized that treatment recommendations may change over time and should not be construed to be the sole specific treatment guidelines for any one case. Physicians should consult appropriate medical literature (Conn's Current Therapy, for example) and/or drug package inserts for the most up-to-date treatment recommendations.

Gail Miriam Moraru
Albany, OR

ACKNOWLEDGMENTS

First Edition

This book would not have been possible without the help and advice of numerous individuals. My special thanks are due to two medical entomologists. Chad McHugh, a uniquely insightful civilian United States Air Force (USAF) entomologist (Brooks Air Force Base, Texas), most generously read every chapter (sometimes more than once) and offered invaluable advice and comments. LTC Harold Harlan (an outstanding Army entomologist at the Uniformed Services University of the Health Sciences, Bethesda, Maryland) also read the entire book, giving helpful advice and additional information. Both of these individuals were more than willing to take time out of their busy schedules to work through a quite voluminous manuscript.

A few physicians with whom I work directly or indirectly reviewed portions of the manuscript and/or offered much-needed comments: Drs. Mary Currier and Tom Brooks (Mississippi Department of Health), and Drs. David Conwill and John Moffitt (University of Mississippi Medical Center).

During the formative stages of the manuscript the following individuals reviewed specific chapters or subject areas: Dr. Hans Klompen (Institute of Arthropodology and Parasitology, Georgia Southern University), Dr. Paul Lago (Biology Department, University of Mississippi), Dr. Robert Lewis (Department of Biology/Zoology, Iowa State University), Maj. Tom Lillie (USAF entomologist), Tim Lockley (Imported Fire Ant Lab, Animal and Plant Health Inspection Service [APHIS], U.S. Department of Agriculture [USDA]), and Dr. Hal Reed (Biology Department, Oral Roberts University).

Information on specific arthropods and/or photographs were provided by Dr. Virginia Allen (Geisinger Medical Center, Pennsylvania); Steve Bloemmer (Tennessee Valley Authority, Land Between the Lakes); Dr. Tom Brooks (University of Mississippi Medical Center and Mississippi Department of Health); Drs. Richard Brown, Clarence Collison, and Bob Combs (Entomology Department, Mississippi State University); Ian Dick (Environmental Health Department, The London Borough of Islington), Sandra Evans (U.S. Army Environmental Hygiene Agency, Aberdeen Proving Ground), Harry Fulton (Bureau of Plant Industry, Mississippi Department of Agriculture), LTC Harold Harlan (U.S. Army entomologist), Dr. James Jarratt (Entomology Department, Mississippi State University), Dr. Hans Klompen (Institute of Arthropodology and Parasitology, Georgia Southern University), John Kucharski (Agricultural Research Service, USDA), Dr. Paul Lago (Biology Department, University of Mississippi), Maj. Tom Lillie (USAF entomologist), Tim Lockley (Fire Ant Lab, APHIS, USDA), Dr. Chad McHugh (USAF civilian entomologist), Dr. Hal Reed (Biology Department, Oral Roberts University), Dr. Richard Robbins (Armed Forces Pest Management Board, Defense Pest Management Information Analysis Center), Dr. John Schneider (Entomology Department, Mississippi State University), and Sue Zuhlke (Mississippi Gulf Coast Mosquito Control Commission).

The USAF medical entomology fact sheets (from the Epidemiology Division of the USAF School of Aerospace Medicine) were instrumental in writing some chapters, as were some written sections and illustrations from the Mississippi Department of Health publication *The Mosquito Book*, by Ed Bowles. Les Fortenberry (Mississippi Department of Health) did most of the original artwork. Much of Chapter

28 (Ticks) was taken from a previous military manual written by the author entitled *Ticks and Tick-Borne Diseases Affecting Military Personnel*. Artwork in that publication was originally done by Ray Blancarte (USAF School of Aerospace Medicine), and some photography was done by Bobby G. Burnes (also of the USAF School of Aerospace Medicine). The Centers for Disease Control's *Key to Arthropods of Medical Importance*, which is revised and included as a figure in Chapter 8, was originally written by H.D. Pratt, C.J. Stojanovich, and K.S. Littig.

My wife, Rosella M. Goddard, did much of the typing and encouraged me during the three years of manuscript preparation. I owe a great deal of gratitude to her.

Second Edition

Chad McHugh (USAF civilian entomologist), Dr. David Conwill (University of Mississippi Medical Center), and Dr. Mary Currier (Mississippi Department of Health) provided helpful comments.

The following persons provided photographs or permission to use their material: Dr. Mary Armstrong, Ralph Turnbo, and Tom Kilpatrick (all at the Mississippi Department of Health); Dr. Alan Causey (University of Mississippi Medical Center); Mike and Kathy Khayat (Pascagoula, Mississippi); and Dr. Gary Groff (Pascagoula).

As always, my wife, Rosella, and my sons, Jeremy and Joseph, helped me immensely. Many of the sting or bite lesions were photographed by my boys as we spent time in the field collecting specimens.

Third Edition

As scientific knowledge continues to expand at an unprecedented rate, it is obvious that no one person can hope to keep up. Accordingly, I continue to utilize several scientists/physicians as resource persons. Their help is critical; I could not keep this book up to date without their help. They are Dr. Chad McHugh (USAF civilian entomologist), Dr. Hans Klompen (currently at Ohio State University), Drs. David Conwill and William Lushbaugh (University of Mississippi Medical Center), Dr. Fernando de Castro (Dermatology Associates, Lexington, Kentucky), and Drs. Mary Currier and Risa Webb (Mississippi Department of Health). Phyllis Givens (Jackson, Mississippi) and George Allen (Jackson, Mississippi) provided photographs or specimens.

Again, my wife, Rosella, and my sons, Jeremy and Joseph, helped me immensely. Fourteen of the pictures in this book are of Jeremy or Joseph, either to illustrate lesions or to demonstrate a particular activity.

Fourth Edition

Dr. Chad McHugh (USAF civilian entomologist), Dr. Chris Paddock (CDC), Dr. Mary Currier (Mississippi Department of Health), and Drs. John Moffitt and Richard deShazo (University of Mississippi Medical Center) provided helpful comments. Dr. deShazo was invaluable in helping me update the allergy sections and graciously allowed me to use a brief portion of his writing in Chapter 1 under "Mechanisms of Allergic Reactions."

The following persons provided photographs, specimens, or permission to use their material: Dr. Mike Brooks (Laurel, Mississippi), Dr. Barry Engber (North Carolina Department of Health), Dr. James Jarratt (Mississippi State University), Dr. Richard Russell and Stephen Doggett (Westmead Hospital, Westmead, Australia), and Sheryl Hand and Dr. Sally Slavinski (Mississippi Department of Health). I am especially indebted to James Jarratt, a long-time friend who has helped me through the years photograph specimens and allowed me to use his (much better) photographs.

Fifth Edition

Dr. Bruce Harrison (North Carolina Department of Environment and Natural Resources) graciously helped me revise the mosquito chapter and Dr. Chris Paddock (CDC Pathology Activity) provided comments on insect bite pathology. David Notton and Nigel Wyatt (Department of Entomology, The Natural History Museum, London) allowed me to examine/study several specimens of African Diptera for this book. Several of the best photographs included in this edition could not have been possible without the help of my assistant, Wendy Varnado (Mississippi Department of Health). My son, Jerome Goddard II, single-handedly developed the CD-ROM for inclusion in this book. I never cease to be amazed at his computer programming abilities. Many of the pictures on the CD are from the CDC, Armed Forces Pest Management Board, Armed Forces Institute of Pathology, or the USDA.

The following persons provided photographs, specimens, or permission to use their material: Kailen Austin (student, Mississippi College), Dr. Michael Brooks (Laurel [Mississippi] Ear, Nose, and Throat Surgical Clinic), Mallory Carter (student, Mississippi College), Rachel Freeman (student, Mississippi College), Dr. Blake Layton (Mississippi State University), Margaret Morton (Mississippi Department of Health), Joe MacGown (Mississippi Entomological Museum, Mississippi State University), Wendy Varnado (Mississippi Department of Health), and Gretchen Waggy (Grand Bay National Wildlife Refuge).

Sixth Edition

Dr. Bruce Harrison (North Carolina Department of Environment and Natural Resources) graciously helped with the mosquito chapter, and Dr. Andrea Varela-Stokes (College of Veterinary Medicine, Mississippi State University) and Dr. Richard Robbins (Armed Forces Pest Management Board) helped with the tick chapter. Dr. Chris Paddock (CDC Pathology Branch) provided comments on various arthropod groups and infectious disease pathology. Several of the best photographs included in this edition would not have been possible without the help of Wendy Varnado (Mississippi Department of Health), Dr. Blake Layton and Joe MacGown (both at the Entomology Department, Mississippi State University). My son, Jerome Goddard II, updated the CD-ROM for this new edition. Many of the pictures on the CD-ROM are from the CDC, Armed Forces Pest Management Board, Armed Forces Institute of Pathology, or the USDA. Videos on the CD-ROM include a bed bug informational clip courtesy of the American Medical Association and used with permission, as well as two short videos about tick and bed bug biology and control, used with permission from the Department of Agricultural Communications, Mississippi State University Extension Service. The other videos included on the CD-ROM concerning general entomology and pest control are copyright 2007 by the author.

The following persons provided photographs, specimens, or permission to use their material in the sixth edition: Dr. Gerald T. Baker (Entomology Department, Mississippi State University), Dr. Richard deShazo (Department of Medicine, The University of Mississippi Medical Center), Dr. Mark Feldlaufer (USDA, Beltsville, MD), Dr. Lane Foil (Entomology Department, Louisiana State University), Dr. Andrea Varela-Stokes (College of Veterinary Medicine, Mississippi State University), Dr. Brian Fisher (California Academy of Sciences), Mona Goodin (Starkville, Mississippi), Dr. Vidal Haddad (Botucatu School of Medicine, UNESP, São Paulo, Brazil), Dr. Ashley Lovell (Clanton, Alabama), Tom Mann (Mississippi Museum of Natural Sciences), Dr. Jerome Goddard II (Auburn University, Montgomery Campus), Brook Burton (London, UK), Mallory Carter Pickering (Brandon, Mississippi), DeAnne Seigler (Oahu, Hawaii), Elmer Gray (University of Georgia), Officer Joseph D. Goddard (Camden, Alabama), Lindsey Carpenter Goddard (Montgomery, Alabama), Lauren Little Goddard (Camden, Alabama), Rachel Freeman Ford (Jackson, Mississippi), Audrey Sheridan (Starkville, Mississippi), and Eugene Skiles (Vilonia, Arkansas).

Seventh Edition

The Seventh Edition was a significant departure from previous editions and would not have been possible without the help and guidance of many people. The following persons read portions of the manuscript, supplied specimens, photographs, or portions of text, and/or offered helpful advice: Michelle Allerdice (Centers for Disease Control, Atlanta, GA), Dr. Lorenza Beati (U.S. National Tick Collection, Statesboro, GA), Lawrence Bircham (Brandon, MS), Tom and Pat Baker (Oakdale, CT), Dr. Patrick Carrington (Shreveport, LA), Dr. Claudia Cuervo (Pontificia Universidad Javeriana, Bogota, Colombia), Dr. Trisha Dubie (Oklahoma State University), Camellia Rose Goddard (Montgomery, AL), Emilia Kooienga (graduate student, Mississippi State University), Dr. Blake Layton (Mississippi State University), Joe MacGown (Mississippi State University), Sarah McInnis (graduate student, Mississippi State University), Dr. Bruce Noden (Oklahoma State University), Dr. Richard de Shazo (University of Mississippi Medical Center, Jackson, MS), Dr. Patricia H. Stewart (University of Mississippi Medical Center, Jackson, MS), Dr. Wendy C. Varnado (Mississippi Department of Health, Jackson, MS), and Dr. Julie Porter Wyatt (University of Mississippi Medical Center, Jackson, MS).

Finally, Rosella M. Goddard digitally created the worldwide distribution maps for the various arthropod vectors and diseases. We are especially grateful for that.

BIOGRAPHY

The first six editions of the *Physician's Guide to Arthropods of Medical Importance* were written by Jerome Goddard over a 21-year period from 1993 to 2013. Unlike most medical and entomology textbooks, which may have dozens of authors, the first six editions were written entirely by Dr. Goddard. Jerome Goddard had a gift for communicating entomology in a clear and interesting manner to non-entomologists. The seventh edition transferred authorship to two well-qualified individuals, and from this point onward, the book will be titled *The Goddard Guide to Arthropods of Medical Importance.* In light of that change, the following biography is included for the original author of this book, Dr. Jerome Goddard.

Jerome Goddard received his bachelor's and master's degrees in biological science from the University of Mississippi in 1979 and 1981 and his doctorate in medical entomology from Mississippi State University in 1984. In December of 1985, he was commissioned as an officer in the U.S. Air Force and served as a medical entomologist in the Epidemiology Division of the USAF School of Aerospace Medicine, Brooks Air Force Base, Texas, for three and a half years. In 1988, he was named Best Academic Instructor in the Residents in Aerospace Medicine Course and Company Grade Officer of the Year. From 1989 to 2008, Dr. Goddard served in the capacity of State Medical Entomologist at the Mississippi Department of Health, Jackson, Mississippi, where he designed, implemented, and supervised all vector control programs relating to public health throughout the state of Mississippi. Since 2008 he has been an extension professor of medical and veterinary entomology in the Department of Biochemistry, Molecular Biology, Entomology, and Plant Pathology at Mississippi State University. He is also an affiliate faculty member in the School of Medicine, The University of Mississippi Medical Center, in Jackson.

Dr. Goddard has authored or coauthored over 200 scientific publications and 6 book chapters in the field of medical entomology and is the author of *Ticks and Tick-Borne Diseases Affecting Military Personnel,* published by the USAF; *Infectious Diseases and Arthropods, Third Edition,* published by Springer International (Humana Press Imprint); and *Public Health Entomology,* published by Taylor & Francis (CRC Press). In 2001, Dr. Goddard published a novel about a mosquito-borne disease outbreak entitled *The Well of Destiny,* and between 2006 and 2018 he published three more novels, two of which were released by Livingston Press, a university press. He has been featured in *Reader's Digest and Time Magazine,* on the

National Geographic Channel, and on a series entitled "Living with Bugs" on the Learning Channel. In 2010, he appeared on the "Colbert Report" to discuss bed bugs. He is a member of the American Association for the Advancement of Science, the Entomological Society of America, the Mississippi Mosquito and Vector Control Association, and the Mississippi Entomological Association. His main research interests are the ecology and epidemiology of tick-borne diseases, the health and medical impacts of bed bugs, and mosquitoes and mosquito-borne diseases. He and his wife, Rosella, live in Starkville, Mississippi, have two grown children, and four grandchildren.

AUTHORS

Gail M. Moraru is currently an instructor of biology at Linn-Benton Community College in Albany, Oregon. She obtained her Bachelor's degree in biological science from Cornell University and her Ph.D. in Veterinary Medical Science from Mississippi State University where she studied the natural history of a novel spotted fever rickettsiosis. Subsequently, she conducted research as a post-doctoral fellow at Haifa University in Israel, looking at how natural and anthropogenic disturbances impact mosquito populations and other aquatic communities. Dr. Moraru's second post-doctoral position was at Mississippi State University where she investigated rickettsial interference phenomena in ticks. Before coming to Linn-Benton Community College, she worked as a Senior Extension Research Associate at Mississippi State University studying Zika virus vectors throughout the state.

Dr. Moraru won the Couvillion endowed graduate scholarship (parasitology), has team-taught veterinary parasitology, wildlife diseases, and several different biology and microbiology classes, and has authored or coauthored over 15 scientific publications. She is a member of the Entomological Society of America, the Southeastern Society of Parasitologists, Mississippi Entomological Association, and the Mississippi Mosquito and Vector Control Association. Her main research interests are disease ecology and parasitology. Dr. Moraru is the daughter of Jewish Romanian immigrants and grew up with her sister in California. She currently resides in Corvallis, OR with her husband and five animals.

Jerome Goddard II is currently an associate professor of mathematics at Auburn University Montgomery (AUM), located in Montgomery, Alabama. He received a Bachelor of Science degree in Mathematics with a minor in Computer Science, and then a Master of Science in Mathematics (both) from Mississippi College in Clinton, MS. He subsequently earned a Ph. D. in Mathematical Sciences from Mississippi State University, specializing in Partial Differential Equations under the supervision of Prof. Ratnasingham Shivaji, currently a W. L. Giles Distinguished Professor Emeritus (Mississippi State University) and H. Barton Excellence Professor & Head of the Department of Mathematics (University of North Carolina Greensboro).

Jerome's research interests lie in study of nonlinear boundary value problems with nonlinear boundary conditions. These types of PDEs arise in models from population dynamics and combustion theory. Recently, his research has been externally funded by the National Science Foundation. That collaborative project between mathematicians and an ecologist is an integration of modeling of population dynamics via reaction diffusion models, mathematical analysis, and experimental analysis of an invertebrate (insect) system to explore the effects of habitat fragmentation, conditional dispersal, predation, and interspecific competition on herbivore population dynamics from the patch level to the landscape level. He has published 18 articles in peer-reviewed journals and has given 60+ presentations at regional, national, and international conferences. Dr. Goddard is married to Lindsey Carpenter Goddard and, along with Millie and Braxton, their children, they reside in Montgomery, AL. Jerome enjoys spending time with his family and outdoor hobbies including fishing, hunting, camping, hiking, and backpacking. He has an enthusiasm for teaching and research, but his true passion lies in directing student research which combines the two—mathematics and biology.

Part I

**Pathological Conditions Caused by Arthropods
and Principles of Their Treatment**

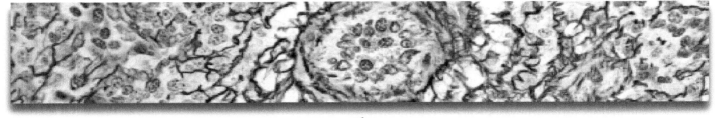

1

PRINCIPLES OF TREATMENT FOR ARTHROPOD BITES, STINGS, AND OTHER EXPOSURE

I. Introduction

Arthropods adversely affect humans in a number of ways. There are direct effects, such as tissue damage due to stings and bites, as well as vesicating fluid exposure (e.g., from blister beetles) and tissue infestation by the bugs themselves (e.g., myiasis). Additionally, some venoms produce necrosis in human tissues, and others produce neurological effects. Indirect effects on human health include disease transmission and allergic reactions to bites and stings, as well as to arthropod skins or emanations.

Because different underlying mechanisms produce the pathological reactions associated with arthropods, it is imperative that attending physicians properly categorize a reaction in order to counteract those ill effects. This chapter is designed to present an overview of the underlying principles of treating arthropod exposure, focusing on the different types of pathological conditions produced by arthropods. No effort is made to explain in detail the immunological and physiological bases underlying these types of pathology; instead, an overview of the mechanisms involved and ways to counteract or control them is given. In addition, only brief discussion is offered concerning arthropod-borne parasitic infections. We will leave that to textbooks on medical parasitology.

II. Direct Effects of Arthropod Exposure

Certainly, bees, wasps, or ants can sting and produce toxic effects in humans by their venom alone, regardless of hypersensitivity. Venoms in many social ants, wasps, and bees contain substances that produce pain and release histamine (directly, not mediated by immunoglobulin E (IgE); see Chapter 3

for a more detailed discussion). Stings or bites cause some tissue damage and inflammation. Inflammation is a result of at least three events: (1) an increase in blood supply to the affected area; (2) an increase in capillary permeability, allowing larger molecules to cross through the endothelium; and (3) leukocytes, mostly neutrophils and macrophages, migrating into the affected tissues.

It has often been estimated that between 500 and 800 honey bee stings could cause death in humans due to the toxic effects of the venom alone. One author calculated that 1,500 honey bee stings would constitute the median lethal dose for a 75-kg person based on extrapolation from the LD_{50} of bee venom for mice.[1] These direct toxic effects (from honey bees or other social Hymenoptera) would include release of histamine, contraction of smooth muscle, increase in capillary permeability, vasodilation with a resulting drop in blood pressure, destruction of normal tissue barriers, destruction of red blood cells, and pain. Severe cases would probably result in renal failure. Treatment strategies would include symptomatic treatment until the venom effects were diminished. As histamine is a component of bee, wasp, and hornet venoms, and as melittin (found in honey bee venom) causes histamine to be released from cells, administration of antihistamines would be indicated. In addition, therapeutic agents to counteract the ill effects of histamine release (e.g., bronchodilators) would also be helpful.

Biting insects produce direct effects on humans as well. Mosquitoes are a nuisance because of their biting behavior, and they may produce tiny punctate hemorrhages (with or without a halo) or persistent papular lesions.[2] Sometimes, large wheals with gross surrounding edema are produced owing to sensitization. Significant blood loss may also occur from mosquito bites. Snowpool mosquitoes in the northern

United States and Canada emerge by the trillions each spring, and landing counts on the human forearm have been reported as high as 300 per minute. This rate of biting could reduce the total blood volume in a human body by half in 90 minutes.[3] Black flies, attacking by the thousands, may cause severe annoyance and small itchy papules and swelling. Ceratopogonid midges also bite in vast numbers, causing irritation and numerous minute papular lesions that may persist for several days. Other biting insects that affect human health directly include bed bugs, kissing bugs, horse and deer flies, stable flies, fleas, and lice. Treatment principles for the direct effects of biting insects generally involve palliative antipruritic lotions or creams and a brief course of systemic corticosteroids if necessary.[2]

Some caterpillars possess venom-filled spines that break off in human skin upon handling or other contact, a condition called eucerism (see Chapters 5 and 14). These spines release venom into the skin upon contact, and pathology similar to a sting may develop. Except in systemic reactions, treatment generally involves topical application of palliatives. In addition, the embedded broken-off spines themselves may have to be removed. This may be done with clear adhesive tape in a repeated "stripping" action.

Myiasis, the invasion of human tissue by living fly maggots, is also a direct effect of arthropods on human health (see Chapters 6 and 21). Although inflammation and secondary infection may be involved, the primary treatment is to remove the maggots. Most pathology associated with myiasis resolves fairly readily after removal.

Certain beetles may cause blistering of human skin. Blister beetles contain the vesicating agent cantharidin, which produces water-filled blisters on human skin a few hours after exposure. Blisters resulting from exposure are generally not serious but may require efforts to prevent secondary infection.

Millipedes and certain true bugs may cause staining of human skin upon exposure. Fluids from millipedes (see Chapter 23) cause burns and dark mahogany brown discoloration,[4] while stink bugs and related insects may stain or cause inflammatory plaques on skin. One report described brown macules on the feet of a child due to burrowing bugs (Family Cydnidae).[5]

III. Hypersensitivity Reactions to Arthropod Venom or Saliva

Sometimes the human immune system produces undesirable results when trying to protect the body. In a hypersensitive or "allergic" person, a relatively innocuous antigen elicits an out-of-proportion immune reaction. Thus, the tissue damage resulting from hypersensitivity is worse than the actual damage produced by the salivary secretion, venom, or other antigen itself.

A. Hypersensitivity Reactions

Hypersensitivity reactions fall into two principal categories, reflecting the two major subdivisions of the immune system. The first category includes reactions initiated by an antibody (for instance, immediate hypersensitivity reactions), in which symptoms are manifest almost immediately after exposure to the antigen by a sensitized person. The second category includes reactions initiated by T lymphocytes (delayed hypersensitivity), and symptoms are usually not obvious for a number of hours or days.

Some authors break hypersensitivity down into four types: I, II, III, and IV. Types I to III involve antibody-mediated reactions. Type I reactions are IgE-mediated immediate hypersensitivity reactions. A systemic reaction to a honey bee sting is a good example of type I hypersensitivity. Because the allergen is directly introduced into the blood or tissue, a severe reaction can occur, such as anaphylactic shock. Type II reactions are antibody-mediated "cytotic" reactions similar to those occurring with some hemolytic reactions. Type III reactions are mediated by circulating antibody–antigen complexes and cause clinical syndromes such as serum sickness. Type IV reactions are mediated by T lymphocytes and macrophages and occur independently of any antibody.

Allergens and specific antibodies produced in response to allergens do not by themselves cause the pathological symptoms associated with immediate hypersensitivity. Instead, the chemical substances (called mediators) released or activated in the host's tissues, resulting from the antigen–antibody binding in solution or on the cell membranes, cause the characteristic tissue damage associated with hypersensitivity.

In atopic persons, the initial exposure to an allergen stimulates an immunoglobulin E (IgE) response. IgE is a minor component of normal blood serum, having a concentration of approximately 1 µg/mL. IgE levels are generally higher in atopic persons than in normal individuals of the same age; however, a normal IgE level does not exclude atopy. IgE levels are also elevated in persons with parasitic worm infections, which indicates its beneficial role in humans. IgE-sensitized mast cells in the gut mucosa provide a good defense against the worms attempting to traverse the gut wall. The IgE produced in atopic individuals in response to allergens sensitizes mast cells, which degranulate upon exposure to the allergen.

Mast cells are similar structurally and functionally to basophils. They are found in association with mucosal epithelial cells as well as in connective tissue. Mast cells characteristically contain approximately 1,000 granules, which upon degranulation release pharmacological mediators causing the allergic symptoms.

Mechanisms of Allergic Reactions. Having "allergies" reflects an autosomal dominant pattern of inheritance with incomplete penetrance. This pattern of inheritance shows up as a propensity to respond to allergen exposure by producing high levels of allergen-specific IgE. Excess production of IgE appears to be controlled by various immune response genes located in the major histocompatibility complex (MHC) on chromosome 6.

The IgE response is dependent on prior sensitization to the allergen. The allergen must first be internalized by antigen-presenting cells, including macrophages, dendritic cells, activated T lymphocytes, and B lymphocytes. After allergen processing, peptide fragments of the allergen are presented with class II (MHC) molecules of host antigen-presenting cells to CD4+ T lymphocytes. These lymphocytes have receptors for the particular MHC–peptide complex. This interaction results in release of cytokines by the CD4+ cell. T-helper lymphocytes (CD4+) are apparently of two classes: T_H1 and T_H2. If the CD4+ cells that recognize the allergen are of the T_H2 class, a specific group of mediators is released, including interleukin-4 (IL-4), IL-5, and IL-9. Other cytokines such as IL-2, IL-3, IL-10, IL-13, and granulocyte–macrophage colony-stimulating factor (GM-CSF) are also released in the process of antigen recognition but are not specific to the T_H2 class. Cytokines such as IL-4, IL-5, and IL-6 are involved in B-cell proliferation and differentiation. Activated B lymphocytes (with bound allergen) are stimulated by these cytokines to multiply and secrete the IgM antibodies. IL-4, IL-6, IL-10, and IL-13 from T_H2 cells promote B-cell isotype switching to IgE production. Thus, atopy appears to be a result of predisposition toward T_H2-type responses, resulting in production of large quantities of allergen-specific IgE.

IgE antibodies specific to a certain allergen bind to mast cells and basophils. When these "sensitized" cells are re-exposed to the offending allergen, IgE molecules attached to the surface of mast cells and basophils become cross-linked by the allergen, leading to a distortion of the IgE molecules and a subsequent series of enzymatic reactions and cell degranulation that releases mediators into the bloodstream and local tissues. The most important preformed mediator is histamine, which causes vasodilation, increased vascular permeability (leading to edema), and mucous secretion (respiratory tract). Other mediators are formed during degranulation such as prostaglandin D_2 (PGD_2); the sulfidopeptide leukotrienes LTC_4, LTD_4, and LTE_4 (slow-reacting substance of anaphylaxis); platelet-activating factor (PAF); and bradykinin. PAF is a potent chemotactic factor, and the sulfidopeptide leukotrienes and bradykinin are vasoactive compounds. Cross-linking of IgE on mast cells also activates phospholipase A_2 and releases arachidonic acid from the A_2 position of cell membrane phospholipids. Mast cells then metabolize arachidonic acid through the cyclooxygenase pathway to form prostaglandin and thromboxane mediators or through the lipoxygenase pathway to form leukotrienes.

Once the allergic reaction begins, mast cells amplify it by releasing vasoactive agents and cytokines such as GM-CSF, tumor necrosis factor α (TNF-α), transforming growth factor β, IL-1 to IL-6, and IL-13. These cytokines lead to further IgE production, mast cell growth, and eosinophil growth,

chemotaxis, and survival. For example, IL-5, TNF-α, and IL-1 promote eosinophil movement by increasing their expression of adhesion receptors on endothelium. Arriving eosinophils then secrete IL-1, which favors T_H2 cell proliferation and mast cell growth factor IL-3. Eosinophils release oxygen radicals and proteins that are toxic to affected tissues.

B. Local Hypersensitivity Reactions

Local allergic reactions involve the nose, lung, and, occasionally, the skin. These areas where the allergen makes contact with sensitized (IgE-"loaded") tissues are usually the only ones affected in these reactions. Allergic and perennial rhinitis, as well as asthma, may be due to arthropods or their emanations. A good example of this is house dust mite allergy (see Chapter 24).

C. Systemic Hypersensitivity Reactions

Systemic allergic reactions are more likely to occur when the allergen reaches the blood or lymph circulations and involve several organ systems. Anaphylaxis is the term often used to describe the rapid, sometimes lethal sequence of events occurring in certain cases upon subsequent exposure to a particular allergen. Initial signs of anaphylaxis are often cutaneous, such as generalized pruritus, urticaria, and angioedema. If the reaction continues, excessive vasodilation and increased vascular permeability caused by histamine and the other mediators may lead to irreversible shock. When angioedema affects the larynx, oropharynx, or tongue, the upper airway can become occluded. Pulmonary edema and bronchial constriction may lead to respiratory failure.

D. Late Hypersensitivity Reactions

A cutaneous late-phase IgE-mediated response in allergic individuals may appear 2–48 hours after challenge and is characterized by a second wave of inflammatory mediators and dramatic influx of immune and inflammatory cells to the site of antigen exposure. These reactions, also called large local reactions, are pruritic, painful, erythematous, and edematous, and they often peak within 12 hours after stings. The edema from large local reactions can, in extreme cases, be severe enough to cause compression of nerves or blood vessels to an extremity.[6] Late-phase asthma and anaphylaxis occur via similar mechanisms.

E. Delayed Hypersensitivity Reactions

An allergic dermatitis, characterized by eczema-like eruptions on the skin, may develop in response to insect or mite body parts, saliva, or feces secondary to the immediate reaction. Delayed-type hypersensitivity reactions typically appear over a period of several days, perhaps not maximal until 48 or 72 hours after antigen exposure. This is cell-mediated immunity wherein CD4-positive T lymphocytes react with the antigen and release lymphokines into tissues. These lymphokines may serve as attractants for monocytes.

F. Treatment Principles for Hypersensitivity Reactions

Antihistamines block most, if not all, of the effects of histamine release. This is accomplished by competing for histamine at its receptor sites, thus preventing histamine from attaching to these receptor sites and producing an effect on body tissues. Oral administration of antihistamines is often recommended for local reactions. In treating generalized systemic or anaphylactic reactions, epinephrine remains the most important treatment and can be life-saving. Antihistamines such as diphenhydramine hydrochloride are given parenterally.

Localized wheal-and-flare reactions to mosquito bites may be prevented by use of oral antihistamines. One study demonstrated that persons who had previously had dramatic cutaneous reactions to mosquito bites had, when taking cetirizine (Zyrtec®), a 40% decrease in the size of the wheal response at 15 minutes and in the size of the bite papule at 24 hours.[7]

Corticosteroids have an anti-inflammatory effect. They act by various mechanisms, including vasoconstriction, decreasing membrane permeability, decreasing mitotic activity of epidermal cells, and lysosomal membrane stabilization within leukocytes and monocytes. In antigen-dependent T cell activation reactions (delayed hypersensitivity), steroids inhibit antigen-specific lymphocyte activation and proliferation. Also, inhibition of the influx of inflammatory cells by glucocorticoids leads to inhibition of the appearance of inflammatory mediators during the late phase. Applied topically to the skin, steroids deplete Langerhans cells of CD1 and HLA-DR molecules, blocking their antigen-presenting function.

In certain arthropod-related allergies (including asthma), such as dust mite or cockroach allergies, inhaled steroids, leukotriene antagonists, or cromolyn sodium may sometimes be used. For severe asthma, research has shown that anti-IgE products such as omalizumab can significantly reduce the number of days with asthma and allow reduction of inhaled steroid use.[8,9] Other monoclonal antibodies such as benralizumab which targets an interleukin-5 receptor have shown benefit for persons with severe asthma.[10] Cromolyn stabilizes mast cells against degranulation, thus preventing release of histamine, leukotrienes, and other pharmacological mediators. The use of epinephrine in severe or systemic hypersensitivity reactions acts to suppress (stabilize) mediator release from mast cells and basophils and reverses many of the end-organ responses to the pharmacological mediators of anaphylaxis, resulting in bronchodilation and relaxation of smooth muscle. The prompt use of epinephrine can often lead to complete resolution of the clinical manifestations of anaphylaxis within minutes.[11]

Other specific interventions may be needed to manage anaphylaxis (see Chapter 2 for more detail). These include actions such as supplemental inspired oxygen, endotracheal intubation, cricothyrotomy, adrenergic stimulants (such as isoproterenol, dopamine, norepinephrine, nebulized β_2 agonists), glucagon (for β-blocked patients), H_1 and H_2 antihistamines, and glucocorticoids.[11,12] Careful monitoring of each individual case, with particular attention to the intensity and relative progression of the anaphylaxis, should enable the attending physician to decide which of these additional measures are indicated.

IV. Neurotoxic Venoms

A. Mechanisms of Toxicity

Widow spiders and some scorpions produce ill effects in humans by neurotoxic venoms. The primary toxin in widow spider (Latrodectus spp.) venom is α-latrotoxin which binds to specific presynaptic receptors (neurexin 1a and CIRL), precipitating neurotransmitter release, particularly norepinephrine and acetylcholine.[13] This leads to sweating, piloerection, muscular spasm, weakness, tremor, and sometimes paralysis, stupor, and convulsions. This type of venom may not produce obvious skin lesions but will primarily produce these systemic reactions.

Scorpion venom is also neurotoxic. It contains multiple low-molecular-weight basic proteins (the neurotoxins), mucus (5–10%), salts, and various organic compounds such as oligopeptides, nucleotides, and amino acids. Unlike most spider and snake venoms, scorpion venom contains few or no enzymes. The low-molecular-weight proteins increase permeability through the neuronal sodium channels. These toxins directly affect the neuronal portion of the neuromuscular junction, causing depolarization of the nerve and myocyte. They may also increase permeability of neuronal sodium channels in the autonomic nervous system. Systemic symptoms of scorpion envenomation include blurred vision, sweating, spreading partial paralysis, muscle twitching, abnormal eye movements, excessive salivation, hypertension, and, sometimes, convulsions. Death (if it occurs) is usually a result of respiratory paralysis, peripheral vascular failure, and myocarditis.

B. Treatment Principles for Neurotoxic Venoms

Strategies for treating an arthropod bite or sting that is neurotoxic in nature involve counteracting the effects of the venom and supportive treatment. Antivenins are commercially available for many of the widow spider venoms and the venoms of some scorpion species. Muscle relaxants, opioid analgesics, and/or antivenin are used for widow spider bites (see Chapter 29).

Antivenin is sometimes used in treating scorpion stings along with anticonvulsants, vasodilators, assisted ventilation, and other supportive measures as needed (see Chapter 28).

V. Necrotic Venoms

A. Mechanisms of Toxicity

In contrast to the widow spiders, violin spiders (brown recluse being one of the most notable) have venom that is necrotic in activity coupled with hyaluronidase which acts as a spreading factor. Brown recluse spider venom contains a lipase enzyme, sphingomyelinase D, which is significantly different from phospholipase A in bee and wasp venoms. This specific lipase is the primary necrotic agent involved in the formation of the typical lesions (see Chapter 29). It is possible that neutrophil chemotaxis is induced by sphingomyelinase D.[14] The subsequent influx of neutrophils into the area is critical in the formation of the necrotic lesion.

B. Treatment Principles for Necrotic Venoms

Treatment of a necrotic arthropod bite (e.g., brown recluse) is controversial because controlled studies are lacking and the severity of the bite is variable.[15] Treatment is mostly supportive, and more serious interventions such as early surgical excision, dapsone, and hyperbaric oxygen are not well supported by evidence (see Chapter 29).[16] King[17] suggested that application of ice packs may be very important in limiting necrosis because activity of the necrotic enzyme in brown recluse venom is related to temperature.

References

1. Camazine S. Hymenopteran stings: Reactions, mechanisms, and medical treatment. *Bull. Entomol. Soc. Am.* 1987;Spring 1988 Issue:17–21.
2. Alexander JO. *Arthropods and Human Skin.* Berlin: Springer-Verlag; 1984.
3. Foster WA, Walker ED. Mosquitoes. In: Mullen GR, Durden LA, eds. *Medical and Veterinary Entomology.* New York: Academic Press; 2002:203–262.
4. Shpall S, Freiden I. Mahogany discoloration of the skin due to the defensive secretion of a millipede. *Pediatr Dermatol.* 1991;8:25–26.
5. Malhotra AK, Lis JA, Ramam M. Cydnidae (burrowing bug) pigmentation. *JAMA Dermatol.* 2015;151:232–233.
6. Moffitt JE, de Shazo RD Allergic and other reactions to insects. In: Rich RR, Fleisher WT, Kotzin BL, Schroeser HW, eds. *Rich's Clinical Immunology: Principles and Practice.* 2nd ed. New York: Mosby; 2001.
7. Reunala T, Brummer-Korvenkotio H, Karppinen A, Coulie P, Palosuo T. Treatment of mosquito bites with cetirizine. *Clin. Exp. Allergy.* 1993;23:72–75.
8. Busse WW, Morgan WJ, Gergen PJ, et al. Randomized trial of omalizumab (anti-IgE) for asthma in inner-city children. *N. Engl. J. Med.* 2011;364(11):1005–1015.
9. Walker S, Monteil M, Phelan K, Lasserson TJ, Walters EH. Anti-IgE for chronic asthma in adults and children. *Cochrane Database Syst. Rev.* 2006(2):CD003559.
10. Bleeker ER, Fitzgerald JM, Chanez P, et al. Efficacy and safety of benralizumab for patients with severe asthma uncontrolled with high-dosage inhaled corticosteroids and long-acting B_2 agonists (SIROCCO): A randomized, multicenter, placebo-controlled phase 3 trial. *Lancet.* 2016;388:2115–2126.
11. Kemp SF, de Shazo RD. Prevention and treatment of anaphylaxis, chap 40. In: Lockey RF, Bukantz SC, Bousquet J, eds. *Allergens and Allergen Immunotherapy.* 3rd ed. New York: Marcel Dekker; 2004.
12. Wasserman SI. Anaphylaxis. In: Rich RR, Fleisher WT, Kotzin BL, Schroeser HW, eds. *Rich's Clinical Immunology: Principles and Practice.* 2nd ed. New York: Mosby; 2001.
13. Offerman SR, Daubert GP, Clark RF. The treatment of black widow spider envenomation with antivenin *Latrodectus mactans*: A case series. *Permanente J.* 2011;15:76–81.
14. King LE, Jr. Spider bites. *Arch. Dermatol.* 1987;123:41–43.
15. Buescher LS. Spider bites and scorpion stings. In: Rakel RE, Bope ET, eds. *Conn's Current Therapy.* Philadelphia: Elsevier; 2005:1302–1304.
16. Bope ET, Kellerman R. *Conn's Current Therapy.* Philadelphia: Elsevier Saunders; 2017.
17. King LE. Brown recluse bites: Stay cool. *J. Amer. Med. Assoc.* 1986;254:2895–2896.

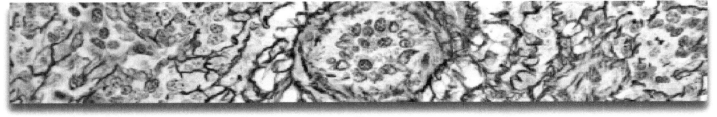

2

ALLERGY TO ARTHROPODS

I. Allergy to Stings or Bites

A. Introduction and Medical Significance

People encounter insects almost everywhere. Inevitably, thousands of persons are stung or bitten daily. For most people, local pain, swelling, and itching are the only effects, and they gradually abate. For others, life-threatening allergic reactions occur. More people die each year in the United States from bee and wasp stings than from snake bites.[1] Why? Probably because more people are exposed to stinging insects than to venomous snakes; therefore, some individuals become hypersensitive to such stings. Consider fire ants. They are so numerous and widespread in the southern United States that >50% of persons may be stung each year,[2] and as much as 17% of the population is sensitized.[3] Freeman[4] reported that fire ants were responsible for 42% of visits to an allergy clinic in San Antonio, Texas.

Stinging insects in the order Hymenoptera such as bees, wasps, and ants can kill people in two ways: by the sheer numbers of stings producing toxic effects and by the allergic reactions in susceptible individuals. It generally takes 500 or more bee stings to kill an individual by the toxic effects of the venom alone (see Chapter 1 for a discussion of direct effects), but just one sting may prove fatal for a person with a bee sting allergy.

Numerous arthropods can cause allergic reactions in persons by their stings, including various wasps, bees, ants, scorpions, and even caterpillars; however, the ones most commonly involved are paper wasps and yellowjackets, honey bees, and fire ants (see Chapter 31, Chapter 11, and Chapter 10, respectively, for discussions of each of these groups). In addition to stings, bites from some arthropods may produce allergic reactions, including anaphylaxis and other systemic effects (Figure 2.1); however, systemic hypersensitivity reactions to arthropod bites are much less common than those resulting from stings. The groups most often involved in producing systemic effects by their bites are the kissing bugs (genus *Triatoma*) (Figure 2.2), black flies, horse flies, and deer flies.[5,6] Mosquitoes, to a lesser extent, are involved, with several reports in the literature of large local reactions, urticaria, angioedema, headache, dizziness, lethargy, and even asthma.[7] Tick bites may sometimes cause extensive swelling and rash. Ticks reported to do so are the hard ticks *Ixodes holocyclus* and *Amblyomma triguttatum*, and the soft tick *Ornithodoros gurneyi*. Arthropod saliva from biting insects contains anticoagulants, enzymes, agglutinins, and mucopolysaccharides. Presumably, these components of saliva serve as sensitizing allergens.

Normal Reaction to Stings or Bites. A normal reaction to one or a few stings involves only the immediate area of the sting and appears within 2–3 minutes. Usually, it consists of redness, itching, swelling, pain, and formation of a wheal at the site. The reaction usually abates within a day or so. If a person is stung by numerous hymenopterans, the acute toxic reaction (nonallergic) resulting from large amounts of venom can be severe. Murray[8] described a man who was stung over 2000 times by bees and exhibited signs of histamine overdosage—severe headache, vomiting, diarrhea, and shock.

Severe Local (or Large Local) Reaction to Stings or Bites. Large local reactions (Figure 2.3) are characterized by painful, pruritic swelling often up to 10cm in diameter (but still contiguous with the sting site) and may involve an entire extremity. Large local reactions increase in size for 24–48 hours and take 3–10 days to resolve. Most patients with large local reactions have detectable venom-specific immunoglobulin E (IgE) antibodies. Large local reactions have not been shown to significantly increase the risk for anaphylaxis upon subsequent stings. In fact, the risk of a systemic reaction in patients who experience large local reactions is no more than 4–10%.[9] Venom immunotherapy has been shown to be effective in preventing large local reactions to some hymenopteran venoms but is rarely required.

CASE HISTORY

ALLERGIC REACTION TO FIRE ANT STING?

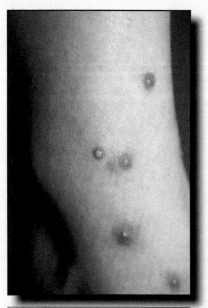

Typical fire ant lesions approximately 24 hours after sting.

A woman called my office saying she was having an allergic reaction to an ant sting. From her description of the event and the mound the specimen was likely an imported fire ant. She quickly described her lesion—a small pustule—and how she had felt since the sting. The sting had happened the day before. What should she do? Go to the hospital? Before I heard anything more about the case, I told her that if she thought she was having an allergic reaction to a sting she should go to the doctor immediately. She persisted in telling the story. It seemed obvious that she was not having an allergic reaction. It had happened the day before, and there was neither swelling nor any systemic effect. Wheal and flare are common initial signs of fire ant stings; pustular lesions are normal 24 hours post-sting (see figure).

Comment: Fire ants are responsible for thousands of human stings in the southern United States each year. Whenever their mound is disturbed, they spill out aggressively looking for the intruder. There are generally three types of reactions to stings: normal, large local, and systemic. Large local reactions can occur for several days after a sting and are characterized by extensive swelling over a large area. For example, if a person is stung on the hand, he or she may swell past the elbow. A systemic reaction—generalized urticaria, angioedema, anaphylaxis—usually begins 10–30 minutes after the sting; however, very rarely, symptoms may not start for several hours. In the case under discussion, the woman may have been confused about terminology. Sometimes people trying to describe a bite or sting site use words that have totally different meanings to a healthcare provider.

Source: Adapted from Goddard, J., *Lab. Med.,* 25, 366, 1994.

Figure 2.1 Hypersensitivity reaction to numerous mosquito bites. (Photograph courtesy of Dr. Elton Hansens.)

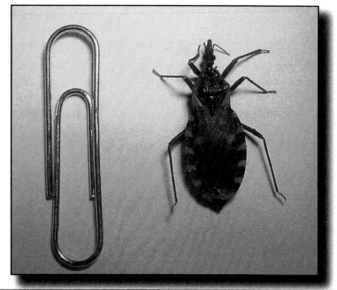

Figure 2.2 Kissing bug. (Photograph © 2005 by Jerome Goddard, Ph.D.)

Systemic Reaction to Stings or Bites. Systemic allergic reactions produce symptoms in areas other than the sting site. Thus, the allergic person may have both the local pain, wheal, and itching from the sting, as well as generalized pruritus, urticaria, angioedema, respiratory difficulty, syncope, stridor, gastrointestinal distress, and hypotension (Table 2.1). Systemic reactions usually begin with widespread

cutaneous symptoms such as angioedema or urticaria. These skin manifestations may be the extent of the systemic reaction, or there may be progression to a generalized pruritus, widespread edema, and upper respiratory distress.[10] In severe reactions, shock begins to develop with a rapid pulse and

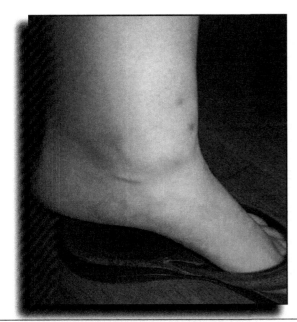

Figure 2.3 Large local reaction to fire ant stings. (Photograph courtesy of Lauren E. Goddard and used with permission.)

Table 2.1 Signs and Symptoms of Anaphylaxis

General	Apprehension, uneasiness, weakness, malaise, metallic taste, paresthesia in extremities In children—sudden behavior changes, cessation of play
Cutaneous	Skin—erythema, pruritus, urticaria, angioedema, flushing Eyes—periorbital swelling and erythema
Gastrointestinal	Abdominal cramps, vomiting, diarrhea
Genitourinary	Urinary or fecal incontinence, uterine cramps
Respiratory	Chest tightness, cough, dyspnea, stridor, wheezing, cyanosis, respiratory collapse/arrest
Cardiovascular	Dizziness, lightheadedness, hypotension, syncope
Neurologic	Headache, confusion, tunnel vision

Source: Adapted from Levine, M.I. and Lockey, R.F., Eds., *Monograph on Insect Allergy*, 2nd ed., American Academy of Allergy, Asthma, and Immunology, Milwaukee, WI, 1986; Arnold, J.J. and Williams, P.M., *Am. Fam. Phys.*, 84(10), 1111–1117, 2011.

low blood pressure. The victim may feel a constriction in his throat and chest, and breathing continues to become difficult. Sometimes, a severe allergic reaction results in anaphylactic shock and death within 10–15 minutes, although 20–30 minutes is more common. One report of 641 deaths from Hymenoptera stings in the United States found that respiratory conditions accounted for 53% of the deaths,[11] and another more recent study showed that laryngeal edema and circulatory failure were the most common causes of death from anaphylaxis.[12]

Cross-Reactivity among Venoms. There is considerable cross-reactivity among the vespid venoms (yellowjacket and hornet), meaning that a person sensitized to one vespid venom could have a serious reaction to a sting from other members of the group.[9] There is only infrequent cross-reactivity between honey bee and vespid venoms. (*Note:* There are exceptions, as some individual patients have shown cross-reactivity between honey bee and yellowjacket venoms.) Immunologic cross-reactivity among fire ant venom and other stinging insect venoms is very limited.

B. Management and Treatment

Normal Reaction. An extremely useful algorithm for the management of insect sting reactions is provided by Golden et al.[9] Treatment for a normal or mild local reaction involves the use of ice packs or pain relievers to minimize pain and washing the wound to lessen the chances of secondary infection. Oral antihistamines may help counteract the effects of histamine (IgE-mediated or not) in the affected tissues resulting from certain venom components. Topical antihistamines and corticosteroids may also be helpful.[13] The use of meat tenderizer containing the enzyme papain is of no therapeutic value.[14] The theory behind the use of papain is valid because *in vitro* incubation of papain and venom leads to destruction of the venom activity. However, in a laboratory experiment with mice, there was no marked inhibition of lesion development in mice receiving papain or Adolph's® meat tenderizer by intradermal injection or topical application.[15]

Severe Local (or Large Local) Reaction. In the case of a large local reaction characterized by considerable swelling and tenderness around the sting site rest and elevation of the affected limb may be needed. The patient should avoid exercise because it may exacerbate the swelling. If the sting site is on or near the throat, nose, or eye area, or if there is widespread swelling, patients should definitely seek medical care. Treatment involves analgesics, topical high-potency steroids, oral antihistamines to relieve itching, and perhaps systemic steroids (such as prednisone) if swelling is severe.[9,16] Superimposed infections such as cellulitis or septicemia, unusual with Hymenoptera envenomation, could require aggressive treatment that may include hospitalization and systemic antibiotics. If the offending arthropod is a biting fly such as a mosquito, cutaneous reactions may even be prevented by use of high-potency topical or oral antihistamines. One study demonstrated that persons who had previously had dramatic cutaneous reactions to mosquito bites, when taking cetirizine (Zyrtec®), had a 40% decrease in the size of the wheal response at 15 minutes and the size of the bite papule at 24 hours.[17]

Systemic Reaction. Persons who experience a generalized allergic reaction (even mild) could be at risk of a more serious reaction upon the next sting (days, weeks, or months later). In the event of a systemic reaction, the most important aspect of care is for that person to get to an emergency facility for immediate treatment. If the individual has an epinephrine autoinjector, it should be used. An ice pack on the sting site may delay absorption of venom, and removal of a honey bee stinger may also reduce venom absorption. People should be reminded, however, that these measures should not delay seeking emergency treatment in any way.

Physicians often do several things to treat a severe allergic reaction. There may be some minor differences in procedures used (depending on the reference consulted), but the immediate goal is the same—maintain an adequate airway and maintain blood pressure. The American Academy of Allergy and Immunology has published steps for the management of anaphylaxis (Table 2.2).[18] The following is a modification (see Wasserman[19]) of suggestions made by Stafford et al.[20] for treatment of severe reactions to fire ant stings, which is fairly typical for the management of similar reactions to all Hymenoptera stings:

An immediate subcutaneous injection of 0.3 to 0.5 mL of a 1:1000 solution of epinephrine (preferably intramuscularly in the lateral thigh) should be administered, and repeated, with blood pressure monitoring, at 10-min intervals if necessary. Intravenous epinephrine may be administered at a rate of 2 µg/min for treatment of severe shock or cardiac arrest, but bolus administration should be avoided. The airway must be established and maintained by using endotracheal intubation or cricothyrotomy, if necessary. Intravenous fluids should be given to replenish depleted intravascular volume in the treatment of anaphylactic shock. Norepinephrine, H_1 and H_2 blocking agents may be required. Systemic corticosteroids and both types of antihistamines may prevent recurrent or biphasic anaphylaxis. Glucagon is appropriate for patients on beta-blockers.

Administration of oxygen (see Table 2.2) may be needed to minimize development of hypoxia, which by itself may contribute to vascular collapse and cerebral edema. Also, wheezing that is refractory to repeated doses of epinephrine can be treated with continuously nebulized beta agonists such as albuterol.[21] However, the administration of epinephrine is the most important element of treatment. It acts to suppress mediator release from mast cells and basophils and reverses many of the end-organ responses to mediators of anaphylaxis. Complete resolution of the clinical manifestations of anaphylaxis often occurs within minutes. The critical and immediate use of epinephrine is the reason why people who are allergic to insect stings carry autoinjector syringes loaded with the drug (Figure 2.4). There are both adult and pediatric versions of autoinjectors. In case of a sting, the allergic person can give himself an injection that may very well save his life. Alexander[14] recommended that one or two close and reliable relatives of the allergic person should also be carefully instructed in the correct use of the epinephrine autoinjector.

It is important to note that just because an individual uses the epinephrine injection in case of a sting does not mean prompt

Table 2.2 Management of Anaphylaxis

General therapeutic measures
  Assessment
  Epinephrine (intramuscularly in the lateral thigh)
  Glucagon if on beta blockers
Specific interventions
  Airway obstruction
    Upper airway obstruction
      Supplemental inspired oxygen
      Extension of the neck
      Oropharyngeal airway
      Endotracheal intubation
      Cricothyrotomy
    Lower airway obstruction
      Supplemental inspired oxygen
      Inhaled beta agonists
      Conventional treatment for status asthmaticus
  Hypotension
    Peripheral vascular defects
      Trendelenberg position
      Intravenous isotonic sodium chloride
      Vasopressors if required (dopamine, intravenous epinephrine, norepinephrine)
        Diphenhydramine plus cimetidine
    Cardiac dysfunction
      Conventional therapy of dysrhythmias
      Diphenhydramine plus cimetidine
Suppression of persistent or recurrent reactions
  Direct observation after anaphylaxis event (this time period can vary due to several factors and may be extended considerably)
    Systemic glucocorticoids
Formulate plan to minimize future reactions
Educate in insect avoidance techniques
Medical identification tag
Self-injectable epinephrine
Venom immunotherapy

Source: Originally adapted from Levine, M.I. and Lockey, R.F., Eds., *Monograph on Insect Allergy*, 2nd ed., American Academy of Allergy, Asthma, and Immunology, Milwaukee, WI, 1986. Subsequently modified with information from Yates et al.17 and personal communication with Patricia H. Stewart, M.D.

medical treatment is not necessary. It is still vital to get to a hospital or physician as quickly as possible. The loaded syringes are meant only to stave off fulminating symptoms long enough for the victim to get to a hospital. This is especially important in light of the fact that sometimes there is a second phase of anaphylaxis 4–10 hours after the initial reaction.[22] Epinephrine autoinjector syringes must be prescribed by a physician, so any person who has suffered even mild symptoms of an allergic reaction should be seen by an allergist for evaluation. It might also be a good idea for all insect-allergic persons to wear a medical identification tag or card to alert medical personnel of their allergy in case they lose consciousness.

Interestingly, compared to other causes of anaphylaxis, such as from food or medicines, the prevalence of mast cell disorders is higher in patients with anaphylactic reactions to insect stings. Elevated basal serum tryptase has been shown to be correlated with risk of anaphylaxis to stings, and especially those reactions characterized by hypotensive shock. Therefore, physicians should consider measuring basal serum tryptase levels in patients with anaphylaxis to a sting.[9]

Long-Term Management of Insect Sting Allergy. Sting-allergic patients and their physicians should also think of long-term management of the problem. There is always the possibility of being stung again. Venom immunotherapy (VIT) is a procedure used by allergists to increase the allergic person's tolerance to insect venom. The process works by stimulating serum-venom-specific IgG and decreasing titers of serum-venom-specific IgE. It is accomplished by numerous injections of venom from offending insects (or from extracts from whole bodies in the case of fire ants). Initially, the injections are very weak. The dosages are gradually increased over time until the patient can tolerate approximately the same amount of venom as is in a sting. The patient is then kept on a maintenance dose to keep up that tolerance. Once

initiated, VIT should usually be continued for at least 3–5 years.[9] Graft[23] showed that immunotherapy is highly effective and safe for the prevention of future systemic reactions to Hymenoptera stings. In a study[24] of 65 patients on a maintenance dose of fire ant whole body extract, only 1 patient (2%) of 47 who were subsequently stung by fire ants had an anaphylactic reaction. Physicians deciding whether or not to initiate VIT base their decision on clinical history and results of venom skin tests and venom-specific radioallergosorbent tests (RAST, or more recently ImmunoCap RAST) (see also Table 2.3). Adults with a history of systemic reaction (especially hypotensive reactions) and a positive intradermal skin test or RAST should see an allergist about receiving immunotherapy.[9,25] Physicians should avoid VIT based solely upon *in vivo* and *in vitro* testing for venom IgE without a history of systemic reaction to a sting. Lastly, some patients may think it is too expensive and inconvenient to undergo immunotherapy, but it is a way for persons with insect allergy to lead a relatively normal life.

C. Avoidance of Offending Insects

Here are some ways for both allergic and nonallergic individuals to avoid stinging insects:

1. Each year, have someone eliminate bee, wasp, and fire ant nests around the home—preferably early in the summer before the nests get large. Pest control operators will usually do this for a fee. Homeowners who wish to accomplish nest elimination should wait until night or a very cool morning to minimize the threat of stings (persons who are allergic to insect stings should not attempt this). If nest elimination is accomplished at night, a flashlight should not be used unless a red filter is used. Bees and wasps will zero in on the beam.
2. Wear light- or khaki-colored clothing when outside during warm weather. These colors are less attractive

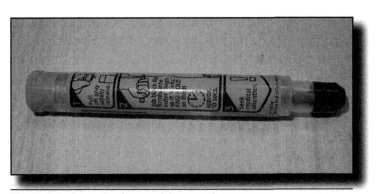

Figure 2.4   Example of an autoinjector syringe

Table 2.3   Selection of Patients for Venom Immunotherapy

Sting Reaction	ST/RAST[a]	Venom Immunotherapy
Systemic, non-life-threatening (child) immediate, generalized, confined to skin (urticaria, angioedema, erythema, pruritus)	+/–	No
Systemic, life-threatening (child) immediate, generalized, may involve cutaneous symptoms but also has respiratory (laryngeal edema or bronchospasm) or cardiovascular (hypotension/shock) symptoms	+	Yes
Systemic (adult)	+	Yes
Systemic	–	No

Source: Golden, D.B.K. et al. Stinging insect hypersensitivity: a practice parameter update, 2016. *Ann. Allergy Asthma Immunol.* 2017, 118: 28–54. Levine, M.I. and Lockey, R.F., Eds., *Monograph on Insect Allergy*, 2nd ed., American Academy of Allergy, Asthma, and Immunology, Milwaukee, WI, 1986. With permission.
Note: Avoid VIT based solely on *in vivo* and *in vitro* testing for venom IgE without a history of systemic reaction to a sting.
[a] Venom skin test or venom-specific radioallergosorbent test (RAST), more recently ImmunoCap RAST.

than dark ones. Be careful to avoid brightly colored floral-patterned clothes.

3. While driving a car during the warm weather months, keep car windows closed and use the air-conditioning.

4. Do not walk barefoot during the warm weather months. Bees are often found feeding on flowers at ground level, and fire ants have numerous feeding trails (even long distances from their mounds).

5. Wear long pants and long sleeves when working outdoors.

6. Wear gloves when gardening. A lot of people are stung on the hand while picking flowers or vegetables.

7. Avoid the use of scented sprays, perfumes, shampoos, suntan lotions, and soaps when working outdoors.

8. Avoid clover patches, gardens full of blossoms, blossoming trees, fields of goldenrod, and other areas with concentrations of bees, wasps, or ants.

9. Be cautious around rotting fruit, garbage cans, and littered picnic areas, especially in the late summer and early fall. Yellowjackets often feed in those areas.

10. Avoid drinking sodas or eating popsicles, ice-cream cones, watermelons, and other sweets outdoors. This may attract bees and yellowjackets.

11. If you see a bee or wasp nest or encounter a nest before the insects become agitated, retreat slowly. Do not panic. Once the nest is disturbed, however, it is best to run immediately, even though hymenopterans are attracted to movement.

OFTEN-ASKED QUESTION

MY DOCTOR SAID I WAS ALLERGIC TO KISSING BUGS. HOW CAN THAT BE, AS THEY DO NOT STING?

Most people are aware that you can become sensitized to venoms from many different stinging insects, leading to allergic reactions ranging from mild to severe (including anaphylactic shock); however, there is confusion when it comes to allergic reactions to bites. Arthropod bites may produce allergic reactions as well, though rare, presumably a result of hypersensitivity to salivary components secreted during the biting process. Arthropod saliva contains anticoagulants, enzymes, agglutinins, and mucopolysaccharides, which may serve as sensitizing allergens. Reactions have occurred following bites by many different types of arthropods but most commonly from bites by *Triatoma* (kissing bugs), horse and deer flies, and mosquitoes.

***Triatoma* allergy.** Kissing bugs—so named because of the nasty habit of taking a blood meal from the face—belong to the insect family Reduviidae (hence, the sometimes-used moniker reduvid bugs), specifically the subfamily Triatominae. Within this subfamily, some (not all) species fall under the genus *Triatoma*; triatomines may also be in other genera. There are at least ten *Triatoma* species found in the United States, but only about six of these are likely to be encountered.[1,2] Allergic reactions have been reported from bites by five species (*T. protracta, T. gerstaeckeri, T. sanguisuga, T. rubida,* and *T. rubrofasciata*[3]), although in the United States *T. protracta* is the species most often reported in allergic reactions.[4,5] Kissing bug bites may be painless, leaving a small punctum without surrounding erythema, or they may cause delayed local reactions appearing like cellulitis. Anaphylactic reactions include itchy, burning sensations, respiratory difficulty, and other typical symptoms of anaphylaxis.[2]

Triatoma bugs feed on vertebrate hosts such as bats, other small- and medium-sized mammals, birds, and humans. Accordingly, the pests are often found in association with their host nest or habitation (e.g., caves, bird nests, rodent burrows, houses); for example, *T. protracta* is found in woodrat nests. Bugs periodically fly away from the nests of their hosts (nocturnal cyclical flights) and may be attracted to lights at dwellings, subsequently gain entrance, and try to feed. Some species are able to colonize houses; they seem especially prolific in substandard structures with many cracks and crevices, mud walls, thatch roofs, etc.

Protection from *Triatoma* bites. Personal protection measures from kissing bugs involve avoidance (if possible), such as not sleeping in adobe or thatched-roof huts in endemic areas, and exclusion methods such as erecting bed nets.[6] Domestic or peridomestic kissing bug species (Mexico, Central America, and South America) may be controlled by proper construction of houses, sensible selection of building materials, sealing of cracks and crevices, and precision-targeting of insecticides within the home. In the United States, prevention of bug entry into homes may involve outdoor light management (i.e., lights placed away from the house, shining back toward it, instead of lights installed on the house) and efforts to find and seal entry points around the home.

REFERENCES

1. Schofield, C.J. and Dolling, W.R., Bed bugs and kissing bugs, in *Medical Insects and Arachnids,* Lane, R.P. and Crosskey, R.W., Eds., Chapman & Hall, London, 1993, pp. 483–516.
2. Rohr, A.S., Marshall, N.A., and Saxon, A., Successful immunotherapy for *Triatoma protracta* induced anaphylaxis, *J. Allerg. Clin. Immunol.,* 73, 369–375, 1984.
3. Ryckman, R.E., Host reactions to bug bites: a literature review and annotated bibliography, *Calif. Vector Views,* 26, 1–23, 1979.
4. Marshall, N., Liebhaber, M., Dyer, Z., and Saxon, A., The prevalence of allergic sensitization to *Triatoma protracta* in a southern California community, *J. Med. Entomol.,* 23, 117–124, 1986.
5. Marshall, N.A. and Street, D.H., Allergy to *Triatoma protracta.* I. Etiology, antigen preparation, diagnosis, and immunotherapy, *J. Med. Entomol.,* 19, 248–252, 1982.
6. Goddard, J., Kissing bugs and Chagas' disease, *Infect. Med.,* 16, 172–175, 1999.

II. Allergy and Asthma: Irritation Caused by Consuming or Inhaling Insect or Mite Parts

A. Introduction and Medical Significance

The prevalence of asthma has been steadily increasing over the last 30 years and is now the most frequent medical handicap among children.[26,27] Several insect or mite species (or their body parts) may cause irritation, asthma, or other allergic reactions when inhaled and, less commonly, when ingested. The major inhalant offenders are house dust mites, *Dermatophagoides farinae* (and *D. pteronyssinus*); cockroaches; several species of mayflies and caddisflies; some nonbiting chironomid midges; and ladybugs. Research has increasingly revealed that cockroach exposure results in severe asthma outcomes, particularly in children living in urban environments. As these arthropods die, their decaying cast skins become part of the environmental dust. In addition, insect emanations such as scales, antennae, feces, hemolymph from reflexive bleeding of the joints, and saliva are suspected as being sources of sensitizing antigens. Compounding the problem, the average child spends 95% of his or her time indoors, providing plenty of time for sensitization. As for the digestive route, cockroach vomit, feces, and pieces of body parts or shed skins contaminating food are most often the cause of insect allergy via ingestion.

Reactions via the Respiratory Route. Until the mid-1960s, physicians simply diagnosed certain people as being allergic to house dust; subsequently, Dutch researchers made the first link between house dust allergy and house dust mites[28,29] (see also Chapter 24). There are many species of house dust mites, but three species are considered the main source of mite allergen: *Dermatophagoides pteronyssinus, D. farinae,* and *Euroglyphus maynei.* The mites commonly infest homes throughout much of the world (especially the first two) and feed on shed human skin scales, mold, pollen, feathers, and animal dander (Figure 2.5). Keratin is their primary food source. They are barely visible to the naked eye and live most commonly in mattresses and other furniture where people spend a lot of time. Dust mites are not poisonous and do not bite or sting, but they contain powerful allergens in their excreta, exoskeleton, and scales. For the hypersensitive individual living in an infested home, this can mean perennial rhinitis, urticaria, eczema, and asthma, often severe. In fact, house dust mite allergens are crucial in the development of allergic rhinitis and asthma. House dust mites can also be triggers for atopic dermatitis.[30]

Recent evidence indicates that early and prolonged exposure to inhaled allergens (such as dust mites and cockroaches) plays an important role in the development of both bronchial hyperreactivity and acute attacks of asthma.[26] Accordingly, bronchial provocation with house dust mite or cockroach allergen can increase nonspecific reactivity for days or weeks.

Figure 2.5 House dust mites contained in a sample. (Photograph © 2009 by Jerome Goddard, Ph.D.)

So, the root cause of asthma onset is sometimes exposure to house dust mites or cockroaches. Asthma-related health problems are most severe among children in inner-city areas. It has been hypothesized that cockroach-infested housing is at least partly to blame. In one study of 476 asthmatic inner-city children, 50.2% of the children's bedrooms had high levels of cockroach allergen in the dust.[31] That study also found that children who were both allergic to cockroach allergen and exposed to high levels of this allergen had 0.37 hospitalizations a year, as compared to 0.11 for other children.[31]

The northern fowl mite (NFM), *Ornithonyssus sylvarium*, found in poultry operations in temperate and subtropical regions worldwide, has been found to cause allergic rhinitis and asthma in farmers.[32] The poultry workers reported respiratory illness while at work in the houses but significant reduction in symptoms when outside. Clinical and laboratory findings including skin tests, RAST, and direct bronchial challenge supported the link between poultry farmers and NFM.

Mayflies (Figure 2.6) and caddisflies are delicate insects that spend most of their lives underwater as immatures. They emerge as adults in the spring and summer in tremendous numbers, are active for a few days, and then die. They do not bite or sting, but body particles from mass emergence of these insects have been well documented as causing allergies.

Nonbiting midges in the family Chironomidae have also been implicated as causes of insect inhalant allergy. A greater prevalence of asthma has been demonstrated in African populations seasonally exposed to the "green nimitti" midge, *Cladotanytarsus lewisi*.[33,34] Kagen et al.[35] implicated *Chironomus plumosus* as a cause of respiratory allergy in Wisconsin.

The Asian lady beetle (Figure 2.7), or ladybug, *Harmonia axyridis*, has recently been described as a cause of seasonal inhalant allergy, causing allergic rhinitis, asthma, and urticaria.[36–38] The primary allergens are found in ladybug hemolymph (blood), which is released from leg joints in a process called "reflexive bleeding."[37] The bugs were imported into the United States from Asia between 1916 and 1990 to serve as a "natural" or biological form of pest control. Unfortunately, they seek winter hibernation sites within homes and other buildings, exposing people to hundreds or thousands of the insects. Ladybug allergy prevalence in endemic areas has been reported to be as high as 10%,[37] and self-reported hypersensitivity rates among people with home infestations has been estimated at 50%.[39]

In areas heavily infested with cockroaches, constant exposure to house dust contaminated with cockroach allergens is unavoidable. Accordingly, many people become sensitized and develop cockroach allergy. In a study in Thailand, 53.7% of 458 allergic patients reacted positively to cutaneous tests of cockroach body parts.[40] In a study in New York City, the figure was even higher—over 70% of almost 600 allergic patients routinely visiting seven hospitals reacted positively to cockroach antigen.[41]

The desert locust, *Schistocerca gregaria* (actually a species of grasshopper), emerges in huge swarms and may fly long distances to eat lush vegetation, including crops. This species has caused significant crop losses in Africa and Asia for thousands of years. One such "plague" of locusts in central Sudan in 2003 caused an epidemic of allergic attacks which Sudanese officials called "lung eczema."[42] One official in the affected area said the illness was linked to a pheromone released by the locusts during their mating season, which caused asthma in affected people.

Reactions via the Digestive Tract. Adult beetles and larval flies, moths, or beetles, as well as their cast skins, often contaminate food and may be responsible for irritation and allergic responses through ingestion. The confused

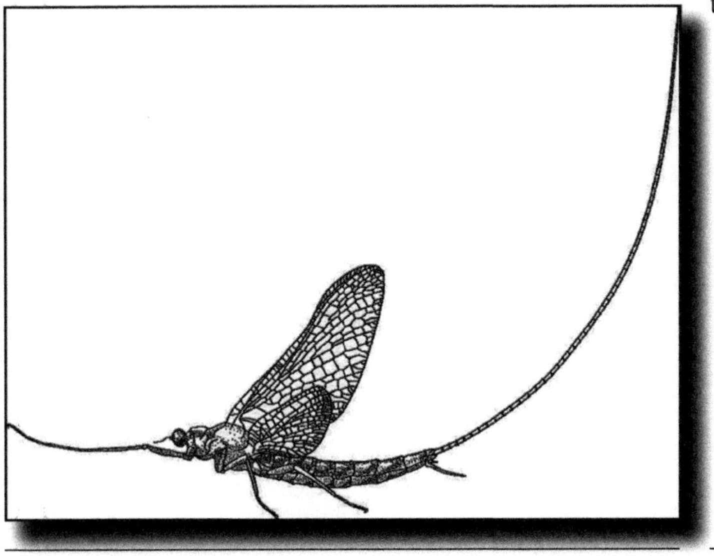

Figure 2.6 May fly adult. (From Steyskal, G.C. et al., Eds., *Insects and Mites: Techniques for Collection and Preservation*, USDA ARS Misc. Publication No. 1443, U.S. Agricultural Research Service, Washington, DC, 1986.)

Figure 2.7 Ladybug adult. (From Steyskal, G.C. et al., Eds., *Insects and Mites: Techniques for Collection and Preservation*, USDA ARS Misc. Publication No. 1443, U.S. Agricultural Research Service, Washington, DC, 1986.)

flour beetle, *Tribolium confusum*, and rice weevil, *Sitophilus granarius*, have been reported to cause allergic reactions in bakery workers.[43] In addition, physicians are often confronted with parents worried about their children who have inadvertently eaten a maggot in their cereal, candy bar, or other food product. These maggots may be moth, beetle, or fly larvae (Figure 2.8) and generally cause no problems upon ingestion; however, some beetle larvae (primarily the family Dermestidae) found in stored food products possess minute barbed hairs (hastisetae) and slender elongate hairs (spicisetae) that apparently can cause enteric problems[44] (Figures 2.9 and 2.10). The symptoms experienced after ingesting dermestid larvae have been attributed to mechanical action of the

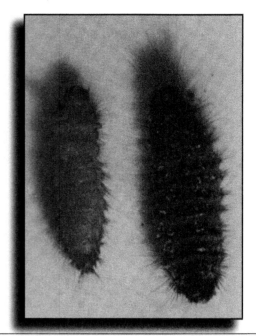

Figure 2.10 Warehouse beetle larvae (Coleoptera: Dermestidae). Left: dorsal view; right: ventral view. (From Lillie, T.H. and Pratt, G.K., *USAF Med. Serv. Dig.*, 31, 32–34, 1980. With permission.)

hastisetae and spicisetae resulting in tissue damage or irritation in the alimentary tract. Clinical symptoms include diarrhea, abdominal pain, and perianal itch.[45,46]

Cockroaches seem to be most often involved in allergic responses. Allergens are present in cockroach feces, which can be inadvertently ingested in heavily infested areas (Figure 2.11). Other allergens are present in cockroach saliva and exoskeletons, which can be introduced into foodstuffs.

B. Management and Treatment

House Dust Mites. For prevention, to decrease the likelihood of sensitization, house dust mite allergen levels should be maintained at less than 2 μg/g.[47] Once patients are sensitized, management of house dust mite allergy may be achieved by immunotherapy, which has recently included use of sublingual solutions or tablets.[48,49] Environmental sanitation efforts such as encasing mattresses with plastic, keeping mattresses free of dust, using a synthetic (or washable) pillow, keeping airborne dust levels low, use of tile or wood floors instead of carpet, and efficient and frequent housecleaning. Studies have demonstrated that a dust-free bedroom is a practical and effective method for decreasing childhood asthma in those with house dust mite allergy.[50] Even relatively simple and inexpensive interventions such as using mattress encasements and carpet exclusion have been associated with concentrations of house dust mite allergens falling below reported sensitization thresholds.[51] For vacuuming, double-thickness filters or high-efficiency particulate air (HEPA) filters are necessary for maximum results. Housecleaning tasks are best accomplished by a nonallergic person or family member. As the mites require

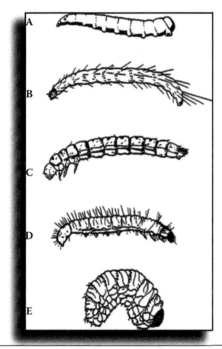

Figure 2.8 Insect larvae often collectively termed maggots: (A) fly larva, (B) flea larva, (C) mealworm larva, (D) moth larva, and (E) weevil larva. (From CDC, *Pictorial Keys to Arthropods, Reptiles, Birds, and Mammals of Public Health Significance*, U.S. Centers for Disease Control and Prevention, Atlanta, GA, 1963.)

Figure 2.9 Magnified view of hairs from dermestid beetle. Left: hastisetae (175×); right: spicisetae (400×). (From Lillie, T.H. and Pratt, G.K., *USAF Med. Serv. Dig.*, 31, 32–34, 1980. With permission.)

Figure 2.11   Cockroach found in bottle of creamer after top was left open.

Figure 2.12   Cockroach feces seen around a household pipe.

a relative humidity of 60% to flourish, humidity control—by increasing ventilation or using a dehumidifier—will limit mite numbers in a house. In fact, a relative humidity of less than 50% critically reduces house dust mite prevalence.[47] Some studies have demonstrated the effectiveness of benzyl benzoate or tannic acid for the treatment of mites in carpets; however, results are equivocal (sometimes failing[52]), and tannic acid may cause color changes in carpets. Other miticides for house dust mites are under development.[53,54] Even with the use of acaricides, however, airborne allergen loads may not be reduced below sensitization levels. One thing that does seem to show promise is the combined use of heat and steam treatment of home furnishings. One study showed that a single treatment of home furnishings reduced mite allergen loads below sensitization levels and improved the asthmatic patients' bronchial hyperreactivity fourfold.[55]

Cockroaches. Because exposure to indoor cockroach allergens and concomitant sensitization may lead to allergy and asthma, interventions aimed at the exposure interface should help mitigate the problem. Cockroach allergy may be managed by symptomatic therapy, immunotherapy with cockroach extracts, and intense sanitation and pest control measures to reduce roach populations and clean up their feces (Figure 2.12). Research has shown that allergen reduction below clinically relevant thresholds by cockroach control alone is possible and highly desirable.[56–58] A more recent study demonstrated that a single pest control intervention—strategic placement of cockroach baits—can result in eradication of cockroaches and improved asthma outcomes for

children.[59] Obviously, because the shelf life of cockroach allergen is several years, it is better to live in a dwelling that has never had a cockroach infestation, so preventive pest control is warranted. Medical treatment of insect inhalant allergy includes avoidance and symptomatic therapy. Where avoidance is impossible or impractical because of residential or occupational exposure, antihistamines, inhaled steroids, beta-2 agonists, or cromolyn sodium may be indicated. Newer therapies, including anti-IgE drugs such as omalizumab may also be helpful.[60]

Ladybugs. Although pesticide applications may occasionally be needed, prevention and management of ladybugs primarily involves exclusion and vacuuming. Exclusion includes sealing cracks and crevices where insects may seek to enter the house and making sure doors and windows are tight fitting and properly screened. Large concentrations of ladybugs inside a house can be removed by vacuuming, although allergic persons should not be doing the vacuuming. Medical treatment of seasonal allergies to ladybugs is similar to that for cockroaches and house dust mites.

Food Contamination. To prevent allergic reactions resulting from ingesting insect parts or fluids, sanitation is the key. Food preparation areas should be thoroughly cleaned and sanitized prior to cooking, even if the areas look clean. Cockroaches may have contaminated these surfaces during the night. Leftovers should be properly covered and/or refrigerated to prevent cockroach feeding. Cereal, nuts, candy, flour, and cornmeal should be examined before consumption for evidence of insect infestation such as small beetles or weevils. Staples such as flour, meal, and cereals can be placed in tight-fitting plastic containers to prevent insect infestation.

Both insect inhalant and sting allergy, similar to any other allergy, can usually be effectively managed and treated. By avoiding the offending types or species involved, practicing

CASE HISTORY

Bugs in Peanut Brittle. A woman brought in the remains of a peanut brittle bar purchased at a grocery store. Upon eating it, she said, she observed bugs in it. She claimed that she immediately threw up and was very ill. She asked that I identify the specimens and document the event for her lawyer. The food item was examined and found to contain larvae and adults of the Indian meal moth, a common food pest. I wrote a letter detailing the product brought in (lot number and other package details) and providing species identification of the insect and its habits. I also made statements to the effect that the insects in question would likely not cause immediate nausea and that I could not verify that the food product was infested at time of purchase.

Comment: When people bring in opened food products claiming that the product was infested with bugs upon purchase, medical personnel should not say or write things that confirm the allegation. Actual physical illness is usually not the issue—these cases almost always result in lawsuits. Medical personnel generally have no objective information about how long the product had been at the client's home or how it had been treated, intentionally or otherwise.

Source: Adapted from Goddard, J., Lab. Med., 25, 371, 1994.

good sanitation (in the case of inhalant or ingestant allergies), and applying appropriate immunotherapy or antihistamine therapy, allergic persons should be able to conduct their lives in a relatively normal manner.

References

1. Parrish HM. Analysis of 460 fatalities from venomous animals in the U.S. *Am. J. Med. Sci.* 1963;245:129–145.
2. de Shazo RD, Butcher BT, Banks WA. Reactions to the stings of the imported fire ant. *N. Engl. J. Med.* 1990;323:462–466.
3. Caplan EL, Ford JL, Young PF, Ownby DR. Fire ants represent an important risk for anaphylaxis among residents of an endemic region. *J. Allergy Clin. Immunol.* 2003;111:1274–1277.
4. Freeman T. Hymenoptera hypersensitivity in an imported fire ant endemic area. *Ann. Allergy Asthma Immunol.* 1997;78:369–372.
5. Hoffman DR. Allergic reactions to biting insects. In: Levine MI, Lockey RF, eds. *Monograph on Insect Allergy.* 2nd ed. Milwaukee, WI: American Academy of Allergy and Immunology; 1986.
6. Moffitt JE, Venarske D, Goddard J, Yates AB, deShazo RD. Allergic reactions to *Triatoma* bites. *Ann. Allergy Asthma Immunol.* 2003;91(2):122–128; quiz 128–130, 194.
7. Gluck JC, Pacin MP. Asthma from mosquito bites: A case report. *Ann. Allergy.* 1986;56:492–493.
8. Murray JA. A case of multiple bee stings. *Cent. Afr. J. Med.* 1964;10:249–250.
9. Golden DB, Demain J, Freeman T, et al. Stinging insect hypersensitivity: A practice parameter update 2016. *Ann. Allergy Asthma Immunol.* 2017;118(1):28–54.
10. Arnold JJ, Williams PM. Anaphylaxis: Recognition and management. *Am. Fam. Phys.* 2011;84:1111–1117.
11. Levine MI, Nall TM. Pathologic findings in Hymenoptera deaths. In: Levine MI, Lockey RF, eds. *Monograph on Insect Allergy.* 2nd ed. Milwaukee, WI: American Academy of Allergy and Immunology; 1986.
12. Pumphrey RS. Lessons for management of anaphylaxis from a study of fatal reactions. *Clin. Exp. Allergy.* 2000;30:1144–1150.
13. Freeman TM. Clinical practice. Hypersensitivity to hymenoptera stings. *N. Engl. J. Med.* 2004;351(19):1978–1984.
14. Ross EV, Badame AJ, Dale SE. Meat tenderizer in the acute treatment of imported fire ant stings. *J. Am. Acad. Dermatol.* 1987;16:1189–1192.
15. Agostinucci W, Cardoni AA, Rosenberg P. Effect of papain on bee venom toxicity. *Toxicon.* 1981;19:851–853.
16. Yates AB, Moffitt JE, de Shazo RD. Anaphylaxis to arthropod bites and stings: Consensus and controversies. *Immunol. Allergy Clin. NA.* 2001;21:635–651.
17. Reunala T, Brummer-Korvenkotio H, Karppinen A, Coulie P, Palosuo T. Treatment of mosquito bites with cetirizine. *Clin. Exp. Allergy.* 1993;23:72–75.
18. Levine MI, Lockey RF. *Monograph on Insect Allergy.* 2nd ed. Milwaukee, WI: American Academy of Allergy and Immunology; 1986.
19. Wasserman SI. Anaphylaxis. In: Rich RR, Fleisher WT, Kotzin BL, Schroeser HW, eds. *Rich's Clinical Immunology: Principles and Practice.* 2nd ed. New York: Mosby; 2001.
20. Stafford CT, Hoffman DR, Rhoades RB. Allergy to imported fire ants. *South. Med. J.* 1989;82:1520–1527.
21. Kemp SF, de Shazo RD. Prevention and treatment of anaphylaxis. In: Lockey RF, Bukantz SC, Bousquet J, eds. *Allergens and Allergen Immunotherapy.* 3rd ed. New York: Marcel Dekker; 2004:chap 40.
22. Sullivan TJ. Treatment of reactions to insect stings and bites. In: Levine MI, Lockey RF, eds. *Monograph on Insect Allergy.* 2nd ed. Milwaukee, WI: American Academy of Allergy and Immunology; 1986.
23. Graft DF. Venom immunotherapy for stinging insect allergy. *Clin. Rev. Allergy.* 1987;5:149–153.
24. Hylander RD, Ortiz AA, Freeman TM, Martin ME. Imported fire ant immunotherapy: Effectiveness of whole body extracts. *J. Allergy Clin. Immunol.* 1989;83:232–238.
25. Graft DF. Indications for venom immunotherapy. In: Levine MI, R LR, eds. *Monograph on Insect Allergy.* 2nd ed. Milwaukee, WI: American Academy of Allergy and Immunology; 1986:chap 8.
26. Gaffin JM, Phipatanakul W. The role of indoor allergens in the development of asthma. *Curr. Opin. Allergy Clin. Immunol.* 2009;9(2):128–135.

27. Dahlen SE, Dahlen B, Drazen JM. Asthma Treatment Guidelines Meet the Real World. *N. Engl. J. Med.* 2011;364(18):1769–1770.

28. Spieksma FTM. The mite fauna of house dust, with particular reference to the house dust mite. *Acarologia.* 1967;9:226–234.

29. Spieksma FTM. The house dust mite, *Dermatophagoides pteronyssinus*, producer of house dust allergen. Thesis, University of Leiden, Netherlands, 65 pp.; 1967.

30. Cameron MM. Can house dust mite-triggered atopic dermatitis be alleviated using acaricides. *Br. J. Dermatol.* 1997;137:1–8.

31. Rosenstreich DL, Eggleston P, Kattan M, et al. The role of cockroach allergy and exposure to cockroach allergen in causing morbidity among inner-city children with asthma. *N. Engl. J. Med.* 1997;336:1356–1360.

32. Lutsky I, Bar-Sela S. Northern fowl mite in occupational asthma of poultry workers. *Lancet.* 1982;320(8303):874–875.

33. Gad el Rab MO, Kay AB. Widespread immunoglobulin E-mediated hypersensitivity in the Sudan to the "green nimitti" midge, *Cladotanytarsus lewisi. J. Allergy Clin. Immunol.* 1980;66:190–193.

34. Kay AB, MacLean CM, Wilkinson AH, Gad El Rab MO. The prevalence of asthma and rhinitis in a Sudanese community seasonally exposed to a potent airborne allergen, the "green nimitti" midge, *Cladotanytarsus lewisi. J. Allergy Clin. Immunol.* 1983;71:345–352.

35. Kagen SL, Yuninger JW, Johnson R. Lake fly allergy: Incidence of chironomid sensitivity in an atopic population. *J. Allergy Clin. Immunol.* 1984;73:187.

36. Goetz DW. Harmonia axyridis ladybug invasion and allergy. *Allergy Asthma Proc.* 2008;29(2):123–129.

37. Goetz DW. Seasonal inhalant insect allergy: Harmonia axyridis ladybug. *Curr. Opin. Allergy Clin. Immunol.* 2009;9(4):329–333.

38. Nakazawa T, Satinover SM, Naccara L, et al. Asian ladybugs (*Harmonia axyridis*): A new seasonal indoor allergen. *J. Allergy Clin. Immunol.* 2007;119(2):421–427.

39. Sharma K, Muldoon SB, Potter MF, Pence HL. Ladybug hypersensitivity among residents of homes infested with ladybugs in Kentucky. *Ann. Allergy Asthma Immunol.* 2006;97(4):528–531.

40. Choovivathanavanich P. Insect allergy: Antigenicity of the cockroach and its excrement. *J. Med. Assoc. Thailand.* 1974;57:237–240.

41. Cornwell PB. *The Cockroach.* Vol 1. London: Hutchinson and Company; 1968.

42. Bhattacharya S. Plague of locusts causes mass allergy attack. New Scientist Magazine (U.K.), November 3, 2003.

43. Arlian LG. Arthropod allergens and human health. *Ann. Rev. Entomol.* 2002;47:395–433.

44. Lillie TH, Pratt GK. The hazards of ingesting beetle larvae. *USAF Med. Ser. Dig.* 1980;31:32.

45. Jupp WW. A carpet beetle larva from the digestive tract of a woman. *J. Parasitol.* 1956;42:172.

46. Okumura GT. A report of canthariasis and allergy caused by *Trogoderma. California Vect Views.* 1967;14:19–20.

47. Calderon MA, Linneberg A, Kleine-Tebbe J, et al. Respiratory allergy caused by house dust mites: What do we really know? *J. Allergy Clin. Immunol.* 2015;136(1):38–48.

48. Bergmann KC, Demoly P, Worm M, et al. Efficacy and safety of sublingual tablets of house dust mite allergen extracts in adults with allergic rhinitis. *J. Allergy Clin. Immunol.* 2014;133(6):1608–1614 e1606.

49. Radulovic S, Calderon MA, Wilson D, Durham S. Sublingual immunotherapy for allergic rhinitis. *Cochrane Database Syst. Rev.* 2010;12:CD002893.

50. Murray AB, Ferguson AC. Dust-free bedrooms in the treatment of asthmatic children with house dust or house dust mite allergy: A controlled trial. *Pediatrics.* 1983;71:418–422.

51. Hill DJ, Thompson PJ, Stewart GA, et al. The Melbourne house dust mite study: Eliminating house dust mites in the domestic environment. *J. Allergy Clin. Immunol.* 1997;99:323–329.

52. Huss RW, Huss K, Squire EN, Jr., et al. Mite allergen control with acaricide fails. *J. Allergy Clin. Immunol.* 1994;94:27–32.

53. Mori T, Takada Y, Hatakoshi M, Matsuo N. New trifluoromethanesulfonanilide compounds having high miticidal activity against house dust mites. *Biosci. Biotechnol. Biochem.* 2004;68(2):425–427.

54. Yu SJ. *The Toxicology and Biochemistry of Insecticides.* Boca Raton, FL: CRC Press; 2008.

55. Htut T, Higenbotta TW, Gill GW, Darwin R, Anderson PB, Syed N. Eradication of house dust mites from the homes of atopic asthmatic subjects: A double blind study. *J. Allergy Clin. Immunol.* 2001;107:55–59.

56. Gore JC, Schal C. Cockroach allergen biology and mitigation in the indoor environment. *Ann. Rev. Entomol.* 2007;52:439–463.

57. Sever ML, Arbes SJ, Jr., Gore JC, et al. Cockroach allergen reduction by cockroach control alone in low-income urban homes: A randomized control trial. *J. Allergy Clin. Immunol.* 2007;120(4):849–855.

58. Arbes SJ, Jr., Sever M, Mehta J, et al. Abatement of cockroach allergens (Bla g 1 and Bla g 2) in low-income, urban housing: Month 12 continuation results. *J. Allergy Clin. Immunol.* 2004;113(1):109–114.

59. Rabito FA, Carlson JC, He H, Werthmann D, Schal C. A single intervention for cockroach control reduces cockroach exposure and asthma morbidity in children. *J. Allergy Clin. Immunol.* 2017;140:565–570.

60. Busse WW, Morgan WJ, Gergen PJ, et al. Randomized trial of omalizumab (anti-IgE) for asthma in inner-city children. *N. Engl. J. Med.* 2011;364(11):1005–1015.

3

STINGS

I. Introduction and Medical Significance

As discussed in the first chapter, stings by venomous arthropods can produce direct effects in humans by the toxic action of the venom alone or indirect effects due to allergic reactions (Table 3.1). Direct toxic effects are very rare but may include cerebral infarction, neuropathies (even optic), and seizures.[1–3] In addition, secondary infection may arise from stings, especially if the lesion is scratched (Figure 3.1). The direct effects of a sting can be mild such as pain, itching, wheal, flare, etc., or serious when numerous stings are received and the large amount of venom injected produces toxic effects. Small children are at a higher risk of developing severe toxicity because of their smaller body weight. One account of a toxic reaction in a child from massive hornet stings described clinical features such as coma, respiratory failure, coagulopathy, renal failure, and liver dysfunction. But, for most individuals, the risk of a severe reaction resulting from either a toxic or allergic mechanism is quite low. Lightning claims more lives annually in the United States than stinging arthropods,[2] and adverse reactions to penicillin kill seven times as many.[3]

Many arthropods can sting, but several groups are particularly notorious offenders. Parrish[4] analyzed fatalities due to venomous animals in the United States from 1950 to 1959 and found that 229 of 460 recorded deaths were due to the stings of yellowjackets, other wasps, ants, and bees. Honey bees, being practically ubiquitous, accounted for most of those numbers. A more recent study reported 162,000 cases of bee stings treated in emergency departments annually in the United States.[5] In other countries, combined with these hymenopterans, scorpions cause significant mortality. One reference[6] listed 20,352 deaths from scorpion stings in Mexico alone during the periods 1940 to 1949 and 1957 to 1958. In Brazil, more than 5,000 reported stings and 48 deaths from scorpions occur each year.[7]

II. Pathology Produced by Arthropod Stings

Alexander[8] described a typical hymenopteran sting (excluding those from ants) as a central white spot marking the actual sting site surrounded by an erythematous halo (Figure 3.2). The entire lesion generally is a few square centimeters in area. He also reported an initial rapid dermal edema with neutrophil and lymphocyte infiltration. Plasma cells, eosinophils, and histiocytes appear later. Lesions produced by fire ant stings are characterized by a central wheal with surrounding erythema, followed by the development of a vesicle, and finally a pustule. According to Caro et al.,[9] the pustules are thin-roofed and contain polymorphonuclear cells and lymphocytes after 24 hours, and eosinophils, plasma cells, polymorphonuclear cells, and lymphocytes after 72 hours. Of course, the histopathology of arthropod stings varies with the insect and whether or not the victim has preexisting antibody to an insect venom. Most large local reactions reflect the presence of immunoglobulin E (IgE) antibody. DeShazo et al.[10] described such reactions in detail in studies of fire ant stings. Whereas the typical wheal-and-flare reactions followed by a sterile pustule were composed of a nonspecific cellular infiltrate, the erythematous, indurated, and pruritic large local reactions occurring in individuals with venom-specific IgE consisted of an eosinophil-rich mixed cellular infiltrate with densely polymerized fibrin. The fibrin gel structure is manifested by the edematous, indurated quality of these lesions, which take 3–5 days to resolve.

Table 3.1 Various Reactions to Insect Stings

Reactions	Response	Comments
Normal	Local pain; itching; swelling	Generally subsides in 2 hours
Large local	Extensive swelling	Subsides in several days
Anaphylaxis	Life-threatening shock; difficulty in breathing; angioedema	Immediate medical attention required
Toxic reactions	Headache; vomiting; diarrhea; shock	Nonallergic, caused by direct effect of many stings
Unusual syndromes	Serum sickness Vasculitis Neuritis Reversible renal diseases	May result from multiple stings or sting in or near a nerve

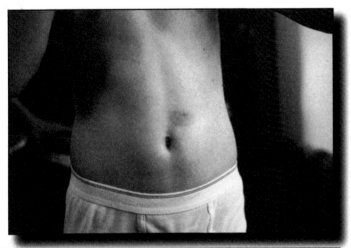

Figure 3.1 Secondary infection of the fire ant sting lesion due to scratching.

III. Stinging Behavior

Most stinging wasps and some bees are solitary or subsocial insects, and they use their stings primarily for subduing prey. This offensive use of stinging and venom by these species rarely leads to human envenomization, except in a few cases of inadvertent or deliberate handling of the specimens. These venoms generally cause slight and temporary pain to humans.

The social wasps, bees, and ants are a different story. They use the sting primarily as a defensive weapon, and their venom causes intense pain in vertebrates (Figure 3.3). Workers (sterile, female insects) of all these groups instinctively defend their nests. One study showed that 51% of people were stung by fire ants within 3 weeks of summertime exposure in an infested area.[11] Encountering a single bee, wasp, or ant out foraging for food is generally not dangerous and will not usually result in a sting; however, walking too near a nest will elicit rapid defensive stinging behavior by numerous guard bees, wasps, or ants. In yellowjackets,

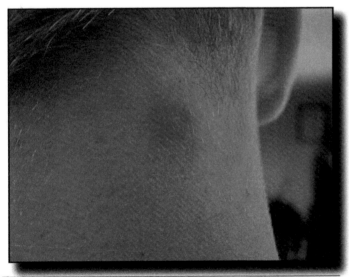

Figure 3.2 Normal sting reaction.

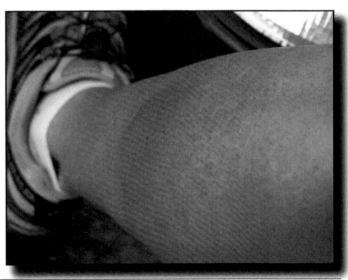

Figure 3.3 Painful wasp sting in a patient. (Photograph courtesy of Wendy C. Varando and used with permission.)

the numerical response is in proportion to the extent of the disturbance; the defensive flight is brief (1.5–5 minutes) and is usually confined to a radius of about 7 meters around the nest.[8]

IV. Morphology of the Sting Apparatus

In all stinging wasps, bees, and ants, the stinger is a modified ovipositor, or egg-laying device, that may no longer function in egg laying. Accordingly, in the highly social Hymenoptera only a queen or other reproductive caste member lays eggs; the workers gather food, conduct other tasks, and sting intruders.

A typical ovipositor (nonstinging device) consists of three pairs of elongate structures, called valves, which can insert the eggs into plant tissues, soil, etc. One pair of the valves makes up a sheath and is not a piercing structure, whereas the other two pairs form a hollow shaft that can pierce the substrate in order for the eggs to pass down through. Two accessory glands within the body of the female inject secretions through the ovipositor to coat the eggs with a gluelike substance.

For the stinging configuration, the ovipositor has several modifications to enable a stinging function (Figure 3.4). The genital opening from which the eggs pass is anterior to the sting apparatus, which is flexed up out of the way during egg laying. Also, the accessory glands have been modified. One now functions as a venom gland and the other (the Dufour's gland) may be important in production of pheromones. The venom gland leads to a venom reservoir or poison sac, which may contain up to 0.1 mL of venom in some of the larger hymenopterans.

The stinger itself is well adapted for piercing the skin of vertebrates. In the case of yellowjackets, there are two lancets and a median stylet that can be extended and thrust into a victim's skin (Figure 3.5). Penetration is not a matter of a single stroke, but instead by alternating forward strokes of the lancets sliding along the shaft of the stylet. The tips of the lancets are slightly barbed (and actually recurved like a fishhook in the case of honey bees) so they are essentially sawing their way through the victim's flesh. Contraction of the venom sac muscles injects venom through the channel formed by the lancets and shaft. The greatly barbed tip of the lancets in honey bees prevents the stinger from being withdrawn from vertebrate skin; thus, the entire sting apparatus is torn out as the bee flies away. Other hymenopterans, on the other hand, can generally sting repeatedly.

V. Venom Components and Activity

Table 3.2 lists some of the active constituents of the vespid wasp, honey bee, and fire ant venoms, adapted from Schmidt.[12,13] Table 3.3 shows the World Health Organization (WHO) allergen nomenclature. Venoms are highly complex mixtures of pharmacologically and biologically active agents.

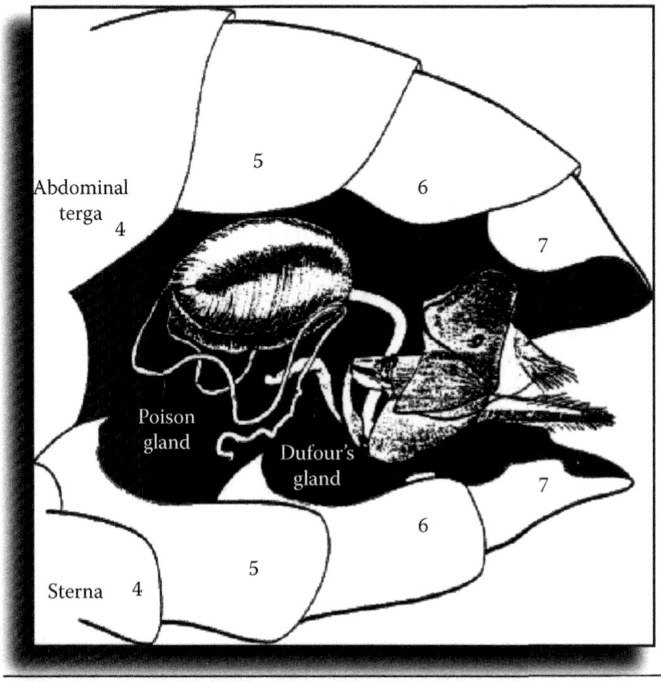

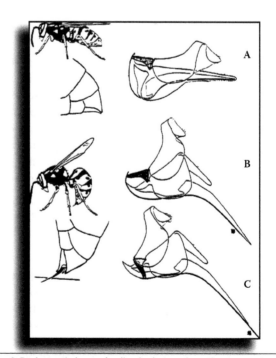

Figure 3.4 Cutaway view of yellowjacket sting apparatus. (From Akre, R.D. et al., *The Yellowjackets of America North of Mexico*, Agriculture Handbook No. 552, U.S. Department of Agriculture, Washington, DC, 1981.)

Figure 3.5 Lateral views of yellowjacket worker sting apparatus, retracted and extended during maximum thrust. (From Akre, R.D. et al., *The Yellowjackets of America North of Mexico*, Agriculture Handbook No. 552, U.S. Department of Agriculture, Washington, DC, 1981.)

Table 3.2 Some Proteins Found in Hymenopteran Venoms

	Honeybees	Vespid Wasps	Fire Ants
Enzymes			
Phospholipase A_1	–	+	+[a]
Phospholipase A_2	10–12%	–	—
Phospholipase B	1%	+	+
Hyaluronidase	1–2%	+	–
Acid phosphatase	15%	–	+
Alkaline phosphatase	+	–	—
Lipase	–	+	—
Esterase	+	+	—
Protease	–	+	–
Peptides			
Hemolysins	Melittin 40%	+	—
Mastolytic peptides	MCD peptide 2%	Mastoparans	—
Neurotoxins	Apamin 3%	+	—
Antigen 5	–	+	+
Kinins	–	+	—
Group specific allergens	+	+	+

Source: Adapted from Schmidt, J.O., *Clin. Exp. Allergy*, 24, 511, 1994. With permission.
Note: +, present, –, absent, —, not investigated.
[a] Specificity unknown.

Table 3.3 World Health Organization (WHO) Nomenclature for Some Common Fire Ant, Vespid Wasp, and Bee Allergens

Arthropod	Allergen	Common Name
Fire ants	Sol i I	Phospholipase
	Sol i II	—
	Sol i III	Antigen 5 group
	Sol i IV	—
Vespid wasps	Dol m I	Phospholipase A_1
	Dol m II	Hyaluronidase
	Dol m III	Acid phosphatase
	Dol m V	Antigen 5
Honey bees	Api m I	Phospholipase A_2
	Api m II	Hyaluronidase
	Api m III	Melittin
	Api m IV	Acid phosphatase

Source: Adapted from Moffitt, J.E. and deShazo, R.D., in *Rich's Clinical Immunology Principles and Practice*, 2nd ed., Rich, R.R. et al., eds., Mosby, New York, 2001, p. 47.6.

Because some venoms are similar, there may be cross-reactivity reactions in humans, but not always. Histamine is the most predominant low-molecular-weight component. Serotonin, dopamine, noradrenalin, and acetylcholine are also present. The amount of serotonin seems to be directly related to the painfulness of the sting. Melittin is a protein polypeptide toxin that is a primary constituent of honey bee venom. It is a direct agent of hemolysis. Apamin is one of the smallest polypeptides known (molecular weight: 2,038 g/mol). It is a neurotoxin, and its interaction with the spinal cord is well established. Mast cell degranulating (MCD) peptide, as a mastocytolytic agent, is very effective in releasing histamine. Interestingly, MCD peptide only comprises approximately 2% of bee venom but can produce the same effects as melittin, which comprises as much as 50% of bee venom. Kinins occupy an intermediary position between biogenic amines and high-molecular-weight compounds. The relative importance of kinins in envenomization is, however, yet to be clarified. Phospholipase A is an enzyme that can attack structural phospholipids, resulting in damage to biological membranes, mitochondria, and other cellular constituents. There are two types of phospholipase: A_1 and A_2 (see Table 3.3 for current names for these allergens). Phospholipase A_2 is present in honey bee venom, whereas vespid (wasps, yellowjackets, and hornets) venoms contain phospholipase A_1. Hyaluronidase is a spreading factor that opens the way for other venom components to move through host tissues. It works by hydrolyzing hyaluronic acid, which resists the spread of harmful substances through epithelial and connective tissue.

As for which of these venom components serve as allergens, honey bee allergens are phospholipase A_2, hyaluronidase, acid phosphatase, and melittin. Allergens contained in vespid venoms are phospholipase A_1, hyaluronidase, and a protein called antigen 5.[14]

Imported fire ants (IFAs) possess venom of an alkaloid nature, which is apparently the first venom of animal origin recognized to be of this type. IFA venom exhibits potent necrotoxic activity. About 95% of this venom consists of water-insoluble 6-n-alkyl, or alkenyl, 2-methyl piperidines, which do not produce allergic reactions in people but are responsible for pain and pustule formation (Figure 3.6). Sometimes other, nontypical lesions may occur from IFA stings such as nodules (Figure 3.7). The other portion of fire ant venom is an aqueous solution of proteins, peptides, and other small molecules that produces the allergic response in hypersensitive individuals. Of the proteins found, four allergens have been isolated (from *Solenopsis invicta*): Sol i I, Sol i II, Sol i III, and Sol i IV.[15,16] Some patients react to every combination of the four allergens. Sol i II, Sol i III, and Sol i IV are not immunologically related to bee or wasp venom proteins, so there is no cross-reactivity with bee or wasp venoms; however, people sensitized to other Hymenopteran venoms do show in vitro sensitivity to Sol i I. The clinical interpretation of this finding is unclear. This may explain how someone living outside an IFA area could have a systemic reaction to his or her first fire ant sting upon entering an area containing IFAs.

Harvester ant venom is made up of approximately 70% proteins, including phospholipase A_2 and phospholipase B, hyaluronidase, lipase, acid phosphatase, esterases, and others. In the harvester ant species *Pogonomyrmex badius*, high levels of lipase have been found.

Scorpion venom is primarily neurotoxic, containing multiple low-molecular-weight basic proteins (the neurotoxins), mucus (5–10%), salts, and various organic compounds, such as oligopeptides, nucleotides, and amino acids.[17] It contains few or no enzymes. Scorpion sting reactions may be either localized (and transitory) or systemic. Localized responses are characterized by pain and swelling, similar to that of a wasp or bee sting. In some localized responses, the sting site might develop into an indurated lesion, even from harmless scorpion species.[18] These local reactions may persist up to 72 hours and be followed by development of blood blisters at the sting site. Systemic reactions are highly variable and may not necessarily be life-threatening. Symptoms may range from aching, burning, and numbness in various tissues to slurred speech, tightness in the chest, and rapid heartbeat. More severe systemic reactions may include profuse sweating, respiratory problems, and convulsions. Death, when it does occur, is often due to cardiac or respiratory failure.

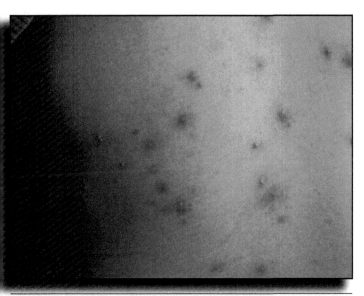

Figure 3.6 Pustules on patient's back resulting from fire ant stings. (Photograph courtesy of Mallory Carter Pickering.)

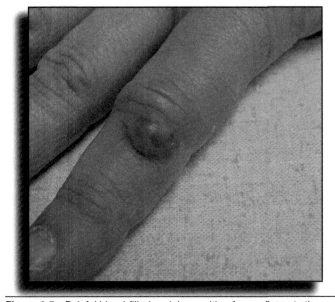

Figure 3.7 Painful blood-filled nodule resulting from a fire ant sting.

References

1. Watemberg N, Weizman Z, Shahak E, Aviram M, Maor E. Fatal multiple organ failure following massive hornet stings. *Clin. Toxicol.* 1995;33:471–474.
2. Camazine S. Hymenopteran stings: Reactions, mechanisms, and medical treatment. *Bull. Entomol. Soc. Am.* 1987;Spring 1988 Issue:17–21.
3. Idsoe O, Guthe T, Wilcox RR, De Weck AL. Nature and extent of penicillin side-reactions, with particular reference to fatalities from anaphylactic shock. *Bull. WHO.* 1968;38:129–145.
4. Parrish HM. Analysis of 460 fatalities from venomous animals in the U.S. *Am. J. Med. Sci.* 1963;245:129–145.
5. O'Neil ME, Mack KA, Gilchrist J. Epidemiology of non-canine bite and sting injuries treated in U.S. emergency departments, 2001–2004. *Public Health Rep.* 2007;122:764–775.
6. Mazzotti L, Bravo-Becherelle MA. Scorpionism in the Mexican republic. In: Keegan HL, MacFarlane WV, eds. *Venomous and Poisonous Animals and Noxious Plants in the Pacific Region.* Oxford: Pergamon Press; 1963:119.
7. Warrell DA. Venomous bites and stings in the tropical world. *Med. J. Aust.* 1993;159:773–779.
8. Alexander JO. *Arthropods and Human Skin.* Berlin: Springer-Verlag; 1984.
9. Caro MR, Derbes VJ, Jung RC. Skin responses to the sting of the imported fire ant. *Arch. Dermatol.* 1957;75:475–479.

10. deShazo RD, Griffing C, Kwan TH, Banks WA, Dvorak HF. Dermal hypersensitivity reactions to imported fire ants. *J. Allergy. Clin. Immunol.* 1984;74:841–845.

11. Tracy JM, Demain JG, Quinn JM, Hoffman DR, Goetz DW, Freeman T. The natural history of exposure to the imported fire ant. *J. Allergy Clin. Immunol.* 1995;95:824–828.

12. Schmidt JO. Chemistry, pharmacology, and chemical ecology of ant venoms. In: Piek T, ed. *Venoms of the Hymenoptera.* New York: Academic Press; 1986:425–509.

13. Schmidt JO. Hymenopteran venoms: Striving toward the ultimate defense against vertebrates. In: Evans DL, Schmidt JO, eds. *Insect Defenses, Adaptive Mechanisms, and Strategies of Prey and Predators.* Albany, NY: State University of New York Press; 1990:387–419.

14. Reisman RE. Insect stings. *N. Engl. J. Med.* 1994;331:523–527.

15. deShazo RD, Butcher BT, Banks WA. Reactions to the stings of the imported fire ant. *N. Engl. J. Med.* 1990;323:462–466.

16. Hoffman DR, Dove DE, Jacobson RS. Allergens in Hymenoptera venom. XX. Isolation of four allergens from imported fire ant venom. *J. Allergy Clin. Immunol.* 1988;82:818–821.

17. Polis GA. *The Biology of Scorpions.* Stanford, CA: Stanford University Press; 1990.

18. Mullen GR, Stockwell SA. Scorpions (Scorpiones). In: Mullen GR, Durden LA, eds. *Medical and Veterinary Entomology, Second Edition.* New York: Elsevier; 2009:397–409.

4

BITES

I. Introduction

Bites by arthropods can be as medically significant as stings, especially for hypersensitive individuals. In this chapter, I have lumped all bites together for discussion. However, the term bite probably should be restricted in meaning to purposeful biting by species for either catching prey or blood feeding and not to accidental or inadvertent biting by plant-feeding species. Phytophagous or predaceous insect species may "bite" in self-defense, piercing the skin with their proboscis, but the injury is actually a stab wound and not a true bite.

The method of obtaining blood differs among blood-feeding arthropods. Some species, such as mosquitoes, bed bugs, kissing bugs, and sucking lice, obtain blood directly from venules or small veins—a method termed solenophagy. Others, such as ticks, horse flies and deer flies, black flies, and biting midges, obtain blood by lacerating blood vessels and feeding from the pool of blood thus formed—a method termed telmophagy. The method of blood feeding likely plays a significant role in whether a species is able to acquire and transmit pathogenic microorganisms.

II. Mouthpart Types

Insect mouthparts, at least in the medically important species, can be generally divided into three broad categories: (1) biting and chewing (Figure 4.1A), (2) sponging (Figure 4.1B), and (3) piercing–sucking (Figure 4.1C). Within these categories there are numerous adaptations and specializations among the various insect orders. The biting and chewing mouthpart types, such as those in food pest insects, and sponging mouthpart types, found in the filth fly groups, are of little significance with regard to human bites, but the piercing–sucking mouthparts,

especially the bloodsucking types, are of considerable importance. Insect piercing–sucking mouthparts vary primarily in the number and arrangement of the stylets, which are needlelike blades, and the shape and position of the lower lip of insect mouthparts, the labium. Often, what is termed the proboscis of an insect with piercing–sucking mouthparts is an ensheathment of the labrum, stylets, and labium. These mouthparts are arranged in such a way that they form two tubes. One tube is usually narrow, being a hollow pathway along the hypopharynx, and the other is wider, formed from the relative positions of the mandibles or maxillae. Upon biting, saliva enters the wound via the narrow tube, and blood returns through the wider tube by action of the cibarial or pharyngeal pump.

The true bugs (order Hemiptera), such as bed bugs, kissing bugs, and assassin bugs, have a labium formed into a three- or four-segmented cylindrical proboscis (Figure 4.2). Four stylets are formed from the mandibles and maxillae. In some cases, as in bed bug feeding, the labium folds up above the skin surface, allowing the fascicle (the stylets linked together in a complete unit) to penetrate the skin for feeding.

Sucking lice (order Phthiraptera, suborder Anoplura) do not have an elongated proboscis, but they do have piercing mouthparts that contain several recurved hooks that serve to anchor the mouthparts during feeding. There are no palps. Three stylets make up the fascicle in anoplurans; they are pushed into host tissues by muscular action in the act of biting. Two of the stylets represent the mandible and maxilla, and the other forms the tube functioning as a salivary duct. Salivary secretions are released into the wound, and the pharyngeal pump begins to draw blood into the digestive tract via the food duct.

Fleas (order Siphonaptera) have a pair of stylets composed of maxillary laciniae that pierce the host's skin. There is also an unpaired stylet formed from the epipharynx. All three structures form the fascicle. The laciniae are bladelike

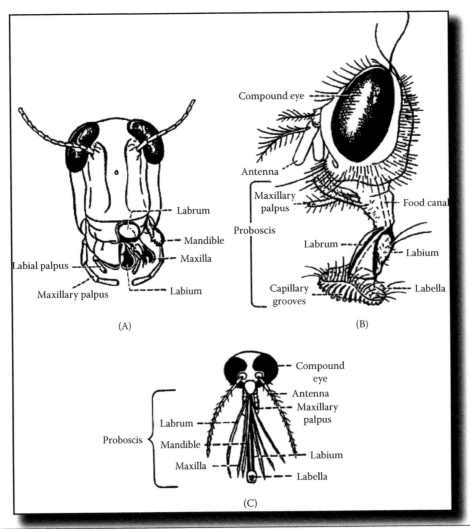

Figure 4.1 (A) Chewing, (B) sponging, and (C) piercing–sucking mouthparts. (Figure courtesy of U.S. Centers for Disease Control and Prevention, Atlanta, GA.)

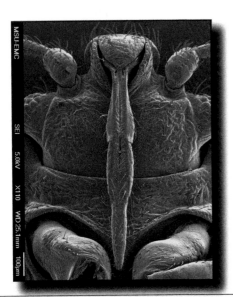

Figure 4.2 True bugs (order Hemiptera) often have a three- or four-segmented cylindrical proboscis. (Photograph copyright 2011 by Jerome Goddard, Ph.D., and courtesy of Dr. G.T. Baker.)

and produce the wound. Blood is then pumped into the pharynx by pharyngeal and cibarial pumps.

Not all flies have piercing–sucking mouthparts, but in those that do there is considerable variation. Mosquitoes have six stylets (two mandibles, two maxillae, the hypopharynx, and the labrum–epipharynx) ensheathed in an elongated, cylindrical labium. This combined structure forms the prominent proboscis of mosquitoes (Figure 4.3A, B and Figure 4.4). Upon biting, only the fascicle is inserted into the host's skin; the labium folds up above the skin surface (Figure 4.3C).

Although relatively short, sand fly mouthparts are efficient organs of penetration. The labrum is broad and tapering to a point, and it bears a number of small spiny projections. Only the female has mandibles, which take the form of blades bearing numerous small serrations. The maxillary stylets, formed by the galeae, are also bladelike and bear numerous fine denticles on their inner margins and a few additional teeth near the end on their outer margins. The labium is elongated and channeled to carry the stylet and hypopharynx. It has a pair of soft labellar lobes located distally. At the sides of

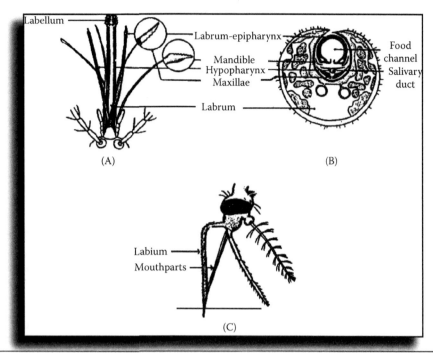

Figure 4.3 (A) Mosquito mouthparts, (B) cross-section of a fascicle, and (C) mouthparts' position upon biting. (From U.S. Naval Medical School, Laboratory Guide to Medical Entomology with Notes on Malaria Control, National Naval Medical Center, Bethesda, MD, 1945.)

Figure 4.4 Microscopic view of mosquito proboscis.

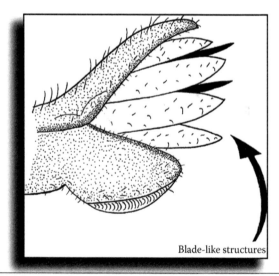

Figure 4.5 Horse fly mandibles are large and bladelike.

the proboscis are the two maxillary palps, which are longer than the proboscis itself.

Horse flies, deer flies, black flies, and biting midges basically have the same type of mouthparts—scissorlike (Figure 4.5). Most of the stylets are much flattened compared to those of mosquitoes. In addition, the mandibles are flattened and move in a transverse manner. Upon feeding, the mandibles move in a scissorlike fashion and the maxillae are thrust in and out of the wound causing pooled blood in the host's tissues. A more specific description of horse fly mouthparts is provided by Snow.[1] The horse fly labrum is relatively soft with a blunt tip and is not used for penetration; however,

the mandibles are large and bladelike. These and the styliform maxillary galeae are the piercing organs. Once these mouthparts have made a wound and fluids are flowing out of it, the labium, which does not enter the wound itself but kinks up on the surface of the skin, draws up blood by means of pseudotracheae on the labellar lobes.

Stable flies have a hardened, conspicuous proboscis (primarily the labium) that can clearly be seen held out in front of the head. The proboscis has the small labella at its tip composed of everted rasping teeth. With a hard thrust, aided by twisting of the labella, the anterior end of the labium is inserted into the host's skin. The hardened long proboscis enables stable flies to inflict painful bites even through socks, sleeves, and other tight-fitting clothing.

Tsetse flies apply pressure to the skin surface of their hosts with their proboscis. Rasps and teeth on the labellum aid the labium in penetrating the skin. Back-and-forth movements of the fly's head enable the labium to rupture capillaries in the skin. Blood is then rapidly sucked into the food canal of the labrum by the cibarial pump located in the fly's head. During the feeding process, saliva enters the wound via the salivary canal of the hypopharynx.

Some noninsect arthropods such as spiders, mites, and ticks have piercing–sucking mouthparts, although the structures are derived from different morphological features than those of insect mouthparts. Mites and ticks (subclass Acari, sometimes also called acarines) have a feeding structure, or gnathosoma, that is headlike and may be mistaken for a true head. The gnathosoma, consisting of mouthparts and palps, is arranged to form a tubular structure for obtaining food and passing it into the digestive tract. The cutting–piercing mouthpart structures of acarines are the chelicerae. Chelicerae may be used for tearing, as in the case of scabies mites, or piercing, as in the case of chiggers. In ticks, there is also an anchoring hypostome, which is a very prominent structure and bears teeth on its ventral surface.

III. Pathology of Arthropod Bites

Arthropod bites basically consist of punctures made by the mouthparts of bloodsucking insects; the actual mechanical injury to human skin is generally minimal; however, in the case of horse flies, there may be more tissue damage owing to their slashing–lapping feeding method and their large size. By far the most lesions on human skin are produced by the host's immune reactions to the offending arthropod salivary secretions or venom.[2] Sometimes hypersensitivity develops to antigens found in insect saliva.[3] Arthropod saliva is injected while feeding in order to (1) lubricate the mouthparts upon insertion, (2) increase blood flow to the bite site, (3) inhibit coagulation of host blood, (4) anesthetize the bite site, (5) suppress the host's immune and inflammatory responses, and (6) aid in digestion. Upon repeated exposure to a particular arthropod's salivary secretions, humans may become allergic to its bites. After becoming sensitized, it is not unusual to have large local reactions to bites by that particular arthropod. Interestingly, there seems to be a progression of sensitivity. Hypersensitivity to insect bites seems to be practically nonexistent during the first year of life, then reaction intensity and lesion size may increase rapidly and peak in the 4- to 7-year-old group (Figure 4.6), then taper off during adolescence.[4]

Arthropod bites are often characterized by urticarial wheals, papules, vesicles, and, less commonly, blisters. After a few days, or even weeks, secondary infection, discoloration, scarring, papules, or nodules may persist at the bite site.

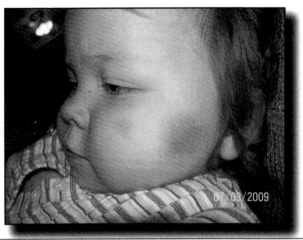

Figure 4.6 Insect bites on infants and toddlers are often prominent due to hypersensitivity (Photograph courtesy of Mona Goodin, Mississippi State University, and used with permission.)

WHY DO CHILDREN REACT TO MOSQUITO BITES MORE THAN ADULTS?

There is a natural cycle of sensitization and ultimate desensitization to mosquito salivary allergens. This process generally begins when humans are infants and may last months or years. Note: a person who has undergone this cycle may still develop cutaneous lesions when exposed to new or different species to which they are not accustomed. The cycle consists of five steps:

1. Persons never exposed to a particular mosquito species do not develop reactions to initial bites by that species.
2. Subsequent bites produce delayed local skin reactions.
3. After more bites, immediate wheals may develop.
4. With further exposure, delayed local reactions decrease and eventually cease.
5. Eventually, upon continued biting events, the person no longer has immediate reactions either.

Figure 4.7 shows secondary infection resulting from scratching a mosquito bite. Complicating the picture further is the development of late cutaneous allergic responses in some atopic individuals. Diagnosis may be especially difficult in the case of biopsies of papules or nodules. Biopsy specimens can reveal a dense infiltrate of a mixture of inflammatory cells composed variably of lymphocytes, plasma cells, macrophages, neutrophils, or eosinophils (Figure 4.8). Lesions containing a majority of uniform lymphocytes are occasionally mistaken for a lymphomatous infiltrate. If the infiltrate is predominantly perivascular and extending throughout the depths of the dermis, the lesion might be confused with lupus erythematosus. Eosinophils are frequently seen in papules or nodules resulting from arthropod bites. There may

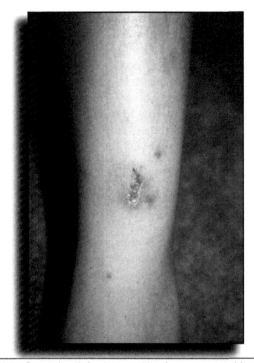

Figure 4.7 Secondary infection resulting from mosquito bites.

Figure 4.8 Striking vasculopathy resulting from an insect bite 6 hours earlier. (Photograph courtesy of Drs. R.D. deShazo and Mark Feldlaufer.)

be a dense infiltration of neutrophils, resembling an abscess. Occasionally remains of arthropod mouthparts elicit a foreign body granulomatous response with giant cells. Scabies mites occur in the stratum corneum and can usually be seen upon microscopic examination. New lesions from scabies such as papules or vesicles are covered by normal keratin, whereas older lesions have a parakeratotic surface; there may also be a perivascular infiltrate of lymphocytes, histiocytes, and eosinophils.[5]

Histopathological studies of late cutaneous allergic responses have revealed mixed cellular infiltrates, including lymphocytes, polymorphonuclear leukocytes, and some partially degranulated basophils. A prominent feature of late cutaneous allergic reactions is fibrin deposition interspersed between collagen bundles in the dermis and subcutaneous tissues.

Infectious Complications. Secondary infection with common bacterial pathogens can occur in any lesion in which the integrity of the dermis is disrupted, whether by necrosis or excoriation.[6] Infection may result in cellulitis, impetigo, ecthyma, folliculitis, furunculosis, and other manifestations. Three findings may be helpful in making the diagnosis of secondary bacterial infection:[6]

> Increasing erythema, edema, or tenderness beyond the anticipated pattern of response of an individual lesion suggests infection.
> Regional lymphadenopathy can be a useful sign of infection, but it may also be present in response to the primary lesion without infection.
> Lymphangitis is the most reliable sign and suggests streptococcal involvement.

IV. Recognizing Arthropod Bites

A. Diagnosis

If a patient recalls no insect or arachnid exposure, arthropod bites may pose frequent difficulty in diagnosis (see Chapter 9); however, bites should be considered in the differential diagnosis of any patient complaining of itching. Diagnosis of insect bites depends on (1) maintaining a proper index of suspicion in this direction (especially during the summer months), (2) a familiarity of the insect fauna in one's area, and (3) obtaining a good history.[7] It is very important to find out what the patient has been doing lately (e.g., hiking, fishing, gardening, cleaning out a shed). Even history can be misleading, however, because patients may present a lesion that they think is an insect bite, when in reality the correct diagnosis is something like urticaria, folliculitis, or delusions of parasitosis. Physicians need to be careful not to diagnose insect bites based on lesions alone and should call upon entomologists to examine samples.

B. Characteristics of Lesions

True Bugs. Members of the Hemiptera produce variable lesions upon biting. Bed bug bites are multiple, may be linear in distribution, and often are on the head, neck, and chest. The linear pattern of bites is not due to a bed bug feeding multiple times in a line but instead to numerous bugs feeding at the bed/human body interface. Bed bugs crawl out from hiding and line up to feed where the victim's arm(s), back, neck, or other body parts contact the mattress. The most common bed bug bite reactions for which medical attention is sought are 2- to 5-mm pruritic maculopapular, erythematous lesions at bed bug feeding sites, one per insect. These usually itch and if not abraded resolve within a week or so. The size and pruritis associated with common reactions may increase in some individuals upon repeated biting.[8-10] Some patients experience complex cutaneous reactions. Reports of these have included pruritic wheals (local urticaria) around a central punctum, papular urticaria, and diffuse urticaria at bite sites usually noted when the patient gets up in the morning.[5,11-14] Bullous rashes may occur upon subsequent biting events days later.[15-19] In some cases, these reactions evolve into pruritic papules or nodules that when scratched may become secondarily infected (impetiginous) and persist for weeks.[4,5,8,20-22] The timing of cutaneous reactions to bed bugs may change with multiple exposures. This appears to reflect host immunological responses to salivary proteins. Usinger[23] fed a colony of bed bugs on himself weekly for 7 years and noted that his reactions progressed from delayed to immediate, with no evidence of desensitization.

Assassin bugs (including wheel bugs) inflict a bite that is immediately and intensely painful. There may be a small red spot or vesicle at the bite site and considerable swelling; numbness may also occur. Kissing bugs generally have a painless bite, but they may produce four distinct reactions depending on sensitivity: (1) a papule with a central punctum, (2) small vesicles grouped around the bite site with swelling and a little redness, (3) a large urticarial lesion with a central punctum and surrounding edema, or (4) hemorrhagic nodular-to-bullous lesions.

Lice. Pubic lice bites produce intense pruritus and discoloration of the skin (bluish-gray maculae) if the infestation is longstanding. Other than that, individual lesions are usually not discernable. Head lice bites may be pustular in appearance. Secondary infection such as impetigo contagiosa on the scalp, head, or neck may indicate head lice infestation. The direct effect of body lice feeding is intense irritation, probably due to proteins in their saliva. This leads to widespread excoriation. The usual clinical presentation for body lice is pyoderma in covered areas. Uninfected bites present as small papules or puncta on an erythematous base.[24] In most clinical settings, the patient may already be undressed when the physician gets to see him or her, and only the ulcerations, infections, and excoriations are present—the actual lice are to be found in the seams of the patient's clothing. In general, primary sensitization to body lice is attained 3–8 months after the original infestation. From that point on, feeding by the lice will cause intense itching. Secondary sensitization may develop some 12–18 months after the initial infestation, resulting in a systemic reaction to louse bites characterized by a feeling of malaise and pessimistic frame of mind. The body-lice-infested person becomes apathetic if left alone and irritable if roused (this is the origin of the term *feeling lousy*). In addition, the skin of people who continually harbor body lice becomes hardened and darkly pigmented, a condition known as *vagabond's syndrome*.

Fleas. Flea bites often occur in irregular groups of several to a dozen or more. The typical flea bite on a human consists of a central spot surrounded by an erythematous ring. There is usually little swelling, but the center may be elevated into a papule, vesicle, or bulla. Papular urticaria is seen in persons with chronic exposure to flea bites. The lesions appear in crops, and all stages can be seen simultaneously—fresh wheals, persistent papules, vesicles, scratch marks, exudation, encrustation, and often secondary infection.

Mosquitoes and Biting Midges. Mosquitoes and biting midges (Ceratopogonidae) cause itching and welts by their bites. In many individuals, the typical wheal-and-flare reaction is evident approximately 30 minutes after the bite (Figures 4.9 and 4.10). The size and extent of the lesions are largely a result of the species of insect involved, the sensitivity status of the individual, and age of the individual. Some patients exhibit large urticarial plaques appearing hours after the bite. Delayed, small but pruritic papules are common after mosquito bites. These generally appear several hours after the bite and usually last 1–3 days, but occasionally persist for weeks. During the spring and summer months, children who play outdoors often exhibit numerous mosquito bite lesions of varying ages.

Horse Flies and Deer Flies. Female horse flies and deer flies (family Tabanidae) have broad, scissorlike mouthparts that can slash a deep painful wound in their host's skin. Bites by tabanids often become secondarily infected, leading to *cellulitis*; however, in the absence of allergy or secondary infection, there is generally no reddening or swelling.

Black Flies. Black flies often produce reddened, itching papules. Subsequently, multiple nodules may develop. It is

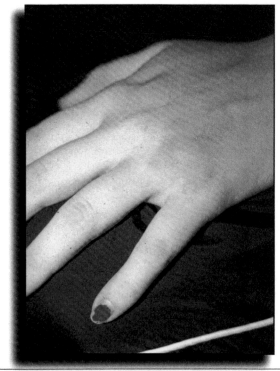

Figure 4.9 Wheal-and-flare reactions may occur from mosquito bites within 30 minutes of biting.

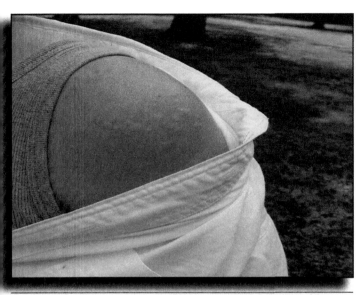

Figure 4.10 Numerous mosquito bites on arm after 10-minute exposure outdoors. (Photograph courtesy of Wendy C. Varnado and used with permission.)

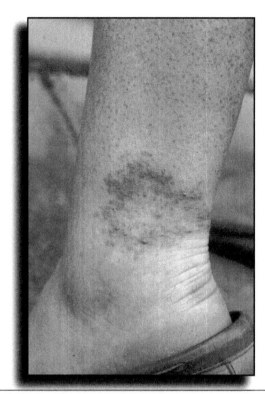

Figure 4.11 Localized reaction to tick bite.

not uncommon for the entire affected area to swell markedly. The arms, legs, and face are common sites of attack, but a favorite site is the posterior cervical region, particularly at the base of the hair. In the case of black fly-transmitted onchocerciasis, the condition is extremely pruritic and widespread papules and lichenification occur, particularly around the shoulders, upper arms, buttocks, and thighs. Microfilariae invade the dermis and destroy the elastic tissue so that the skin may hang in folds, especially in the groin area.

Sand Flies. Sand fly bites may produce itching papules, vesicles, wheals, and a condition closely resembling papular urticaria. Additionally, sand flies may transmit the parasite responsible for cutaneous leishmaniasis. In those cases, exposed skin areas are the most commonly bitten, resulting in papules and nodules that break down, ulcerate, and crust. There is usually healing within a year, but with considerable scarring.

Spiders and Centipedes. Widow spiders differ from violin spiders with regard to reactions to their bites. In the case of widow spiders, there is often a pinprick sensation initially. Two red puncture marks may be present at the bite site, and there may develop a dull, numbing pain around the site that builds to a peak 1–3 hours after the bite. Centipedes may also produce two puncture wounds at the bite site, but there is usually redness and swelling present with immediate intense pain. Violin spider bites are rarely, if ever, accompanied by pain initially; however, within 2–8 hours there is mild to severe pain at the site, erythema, blanching, itchiness, and vesiculation.

Ticks. Ticks will attach to human skin and remain attached for several days. The actual penetration is usually not painful, except in the case of some species with very long mouthparts; however, inflammation or even hypersensitivity reactions may occur after a few days of tick attachment

(Figures 4.11). Even after tick removal, a reddened nodule may persist at the bite site for weeks or even months.

Mites. Chigger bites usually lead to an intensely itchy dermatitis composed of pustules and wheals. Lesions occur within 3–6 hours after exposure to chigger-infested areas. Alexander[5] stated that an edematous wheal with a tiny central vesicle may persist for 8–15 days.

Bites by the straw itch mite produce itching a few hours after attack. Often, small urticarial papules with a tiny central vesicle (0.5 mm in diameter) mark the site of skin puncture. Other human-biting mites such as the tropical rat mite produce immediate sharp itching pain upon biting and cause small urticated papules.

Scabies mites have an affinity for the wrists and the areas between the fingers (although they certainly occur in other areas). Accordingly, scabies is often characterized by rashlike lesions that typically occur in tracks in the webbing between the fingers or on the wrists. Alexander[5] described the burrow as a grayish line resembling a pencil mark, about 5 mm long on average. Sometimes, however, scabies lesions may be singular (papules or vesicles) or even bullous. (Note: rashes may occur over much of the body as a result of scabies infestation, but the rash may not correspond to the location of mite burrows.)

Cheyletiella mites (such as *C. parasitivorax, C. yasguri, C. blakei,* and *C. furmani*) produce scattered or localized urticae, papules, or 2- to 6-mm papulovesicles in the areas of skin where direct contact with an infested pet occurs—the flexor sides of arms, breasts, and abdomen.[25]

References

1. Snow KR. *Insects and Disease.* New York: John Wiley and Sons; 1974.

2. Frazier CA. Diagnosis of bites and stings. *Cutis.* 1968;4:845–849.

3. Moffitt JE. Allergic reactions to insect stings and bites. *South. Med. J.* 2003;96(11):1073–1079.

4. McKiel JA, West AS. Nature and causation of insect bite reactions. *Ped. Clin. NA.* 1961;8:795–814.

5. Alexander JO. *Arthropods and Human Skin.* Berlin: Springer-Verlag; 1984.

6. Kemp ED. Bites and stings of the arthropod kind. *Postgrad Med.* 1998;103:88–94.

7. Allington HV, Allington RR. Insect bites. *J. Am. Med. Assoc.* 1954;155:240–247.

8. Liebold K, Schliemann-Willers S, Wollina U. Disseminated bullous eruption with systemic reaction caused by *Cimex lectularius. J. Eur. Acad. Dermatol. Venereol.* 2003;17:461–463.

9. Bartley JD, Harlan HJ. Bed bug infestation: Its control and management. *Mil. Med.* 1974;139:884–886.

10. Cestari TF, Martignago BF. Scabies, pediculosis, and stinkbugs: Uncommon presentations. *Clin. Dermatol.* 2005;23:545–554.

11. Brasch J, Schwarz T. 26-year-old male with urticarial papules. *J. Dtsch. Dermatol. Ges.* 2006;4(12):1077–1079.

12. Gbakima AA, Terry BC, Kanja F, Kortequee S, Dukuley I, Sahr F. High prevalence of bedbugs *Cimex hemipterus* and *Cimex lectularis* in camps for internally displaced persons in Freetown, Sierra Leone: A pilot humanitarian investigation. *West Afr. J. Med.* 2002;21(4):268–271.

13. Honig PJ. Arthropod bites, stings, and infestations: Their prevention and treatment. *Pediatr. Dermatol.* 1986;3:189–197.

14. Rook AJ. Papular urticaria. *Ped. Clin. NA.* 1961;8:817–820.

15. Cooper DL. Can bedbug bites cause bullous erythema? *J. Am. Med. Assoc.* 1948;138:1206.

16. Hamburger F, Dietrich A. Lichen urticatus exogenes. *Acta Paediat.* 1937;22:420.

17. Kemper H. Beobachtungen ueber den Stech-und Saugakt der Bettwanze und seine Wirkung auf die menschliche Haut. *Zeitschr f Desinfekt.* 1929;21:61–65.

18. Kinnear J. Epidemic of bullous erythema on legs due to bedbug. *Lancet.* 1948;255:55.

19. Patton WS, Evans A. *Insects, Ticks, and Venomous Animals of Medical and Veterinary Importance, Part I.* Croydon, UK: H.R. Grubb Ltd.; 1929.

20. Abdel-Naser MB, Lotfy RA, Al-Sherbiny MM, Sayed Ali NM. Patients with papular urticaria have IgG antibodies to bedbug (*Cimex lectularius*) antigens. *Parasitol. Res.* 2006;98(6):550–556.

21. Crissey JT. Bedbugs: An old problem with a new dimension. *Int. J. Dermatol.* 1981;20:411–414.

22. Masetti M, Bruschi F. Bedbug infestations recorded in central Italy. *Parasitol. Int.* 2007;56(1):81–83.

23. Usinger RL. *Monograph of Cimicidae.* Vol 7. College Park, MD: Entomological Society of America, Thomas Say Foundation; 1966.

24. Elgart ML. Pediculosis. *Dermatol. Clin.* 1990;8:219–228.

25. Van Bronswijk JEMH, de Kreek EJ. *Cheyletiella* of dog, cat, and domesticated rabbit. *J Med Entomol.* 1976;13:315–319.

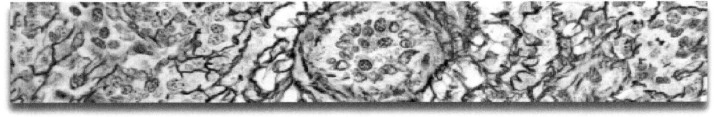

5

BLISTERING, DERMATITIS, AND URTICARIA FROM CONTACT WITH ARTHROPODS

I. Blistering from Exposure to Blister Beetles

A. Medical Significance

Most beetles are harmless and can be handled by humans with no ill effects; however, blister beetles, primarily species in the families Meloidae, Oedemeridae, and Staphylinidae, possess vesicating chemical substances in their body fluids (Figure 5.1; see also Chapter 12). The blistering fluid in meloids and oedemerids is cantharidin but is a somewhat different compound in staphylinids (some call it pederin). When the meloid beetles sense danger, they exude cantharidin by filling their breathing tubes with air, closing their breathing pores, and building up body fluid pressure until fluid is pushed out through the leg joint between the femur and tibia, a process called reflex bleeding.[1] This secretion readily penetrates human skin. Although there may be slight tingling and burning initially at the site of exposure, the fluid-filled blisters do not develop until a few hours later.

The blisters that result from exposure to live blister beetles or their dried pulverized body parts are generally few and self-limiting. Alexander[2] observed that generally there is no pain or itch associated with the blisters as long as they do not rupture. Extensive blister beetle exposure may irritate the kidneys owing to the penetrating activity of cantharidin.[3] If ingested, the substance can cause nausea, diarrhea, and vomiting; horses have died from ingesting numerous striped blister beetles.[4]

Frazier and Brown[3] reported that a good indication of a blister beetle attack is that the blisters are all in the same stage of development and there is no accompanying rash around them. In addition, there may be a line or track of blisters where the beetle crawled over the skin. Alexander[2] reported

that areas commonly affected are the face, neck, chest, thighs, and calves, as well as the buttocks in children.

Severe cases, including accidental or purposeful ingestion of cantharidin, are characterized by inflammation of the mucosa of the oral cavity with excessive salivation, nausea, vomiting (with blood), severe cramping and abdominal pain, bloody diarrhea, and frequent urination.[5,6] The lethal dose of cantharidin is estimated to range from 1.0 to 8.0 mg in adults, although there is at least one report of someone surviving a dose of >75 mg.[5,6]

B. Contributing Factors and Species

Blister beetle dermatoses are seasonal, with most cases reported in the late summer as the beetles emerge and feed in gardens and field crops. Three common blister beetles in the United States are the ash-gray and striped blister beetles in the central and southeastern parts of the country and oil beetles (several species) in the southwest. Contact with the beetles may occur intentionally, such as in the case of a child picking one up, or accidentally, such as the beetle crawling on an unsuspecting victim sitting on the ground, working in the garden, etc. In southern Texas in late summer, the author has observed hundreds of black oil beetles crawling over the ground in a random, solitary fashion.

C. Management and Treatment

Blisters resulting from blister beetle exposure are generally not serious, and reabsorption usually occurs in a few days if the blisters are unruptured. If blisters are ruptured, the skin may flake and there may be a mild reddening of the area before clearing in a week or so.[1,7] For treatment, exposed areas should be washed, antibiotic ointments applied to

Figure 5.1 Blister beetles can release fluids that cause development of blisters on human skin. (Photograph courtesy of Dr. Blake Layton, Mississippi State University Extension Service.)

prevent secondary infection, and the areas bandaged for protection until the blisters reabsorb.

II. Dermatitis and Urticaria from Exposure to Larval Lepidopterans

A. Medical Significance

Caterpillars of many species can produce mild to severe dermatitis, urticaria, nodular conjunctivitis, pain, headache, and even convulsions (rare) via tiny hairs that irritate or inject a venom (Figure 5.2; see also Chapter 14).[8] Visible lesions are usually preceded by a burning itch. The itch is often followed by the appearance of 2- to 5-mm tiny, rose-red maculopapules or papules. Keegan[9] has given an account of the direct injury produced by caterpillars and moths, and an in-depth discussion of human reactions to lepidopterans can be found in Alexander.[2] In the scientific literature, distinction is usually made between direct contact with urticating caterpillars—called erucism—and contact with scales or hairs of adult butterflies or moths—called lepidopterism (see next section).

Relatively speaking, only a few lepidopteran species are involved in erucism; however, urticating caterpillars can become numerous, resulting in a significant health hazard. In the 1920s, the San Antonio schools had to be closed due to a large outbreak of puss caterpillars and the resulting numerous stings among the school-age children.[10] The caterpillars were so numerous on posts, walls, and bushes that virtually any outdoor activity resulted in a sting.

Even a few urticating caterpillars can be a serious problem if a person has an eye lesion or exhibits acute systemic symptoms when exposed. The puss caterpillar is especially dangerous, sometimes causing cardiovascular and neurological manifestations. Frazier and Brown[3] reported a case in which a young woman stepped barefoot on a puss caterpillar.

Figure 5.2 Many caterpillars with thick or prominent spines can cause a sting similar to that of wasps.

Within minutes she experienced symptoms of shock with respiratory distress; upon arrival at the hospital, she had hypotension and bradycardia. She was successfully treated for shock and pain but reported numbness in the foot and leg for several days. Eye lesions may be relatively benign when single pointed hairs are found on the palpebral conjunctiva (they can generally be removed by an ophthalmologist), or they can be extremely serious, with loss of vision in the case of numerous deeply embedded hairs or spines that progressively penetrate deeper and deeper owing to eye movement and barbs on the hairs.[2]

B. Contributing Factors and Species

Both mechanical irritation and injection of venom by the caterpillar spines or hairs contribute to the urticarial response in humans. Many of the hairs, setae, or spines of the offending species are hollow and connected to venom glands. Upon exposure, these structures may be broken, allowing venom to either be injected into the skin or ooze out onto abraded skin. In addition, in a study of gypsy moth caterpillars, whole first instar larvae contained an average of 17.3 ng of histamine, and the hairlike setae of the fifth instar larvae contained 80 ng of histamine per organism.[11] Accordingly, histamine release, as well as a hypersensitivity reaction, likely plays a role in urticaria and dermatitis associated with these caterpillars. Four common species of Lepidoptera that have caterpillars that can sting upon exposure to human skin are the saddleback caterpillar, the io moth caterpillar, the brown-tail moth caterpillar, and the puss caterpillar. Although not true "stinging" caterpillars, the gypsy moth caterpillar and related groups have been known to cause rashes and dermatitis on human skin upon exposure.[12] Detailed discussions of the biologies of these species are provided in Chapter 14.

C. Management and Treatment

Treatment of erucism mainly involves personal protection and avoiding the offending species. Embedded spines or hairs can sometimes be removed by "stripping" the lesion with cellophane tape. Acute urticarial lesions may respond satisfactorily to topical corticosteroid lotions and creams (desoximetasone gel has been frequently used), which reduce the intensity of the inflammatory reaction. Oral antihistamines may relieve itching and burning sensations. Systemic administration of corticosteroids has also been used for itching and pain. Pain relievers may also be needed, especially in the case of puss caterpillar stings. Systemic reactions require close monitoring and aggressive treatment (see Chapter 2).

III. Inhalant Irritation or Contact Dermatitis from Moth Hairs or Scales

A. Medical Significance

The term lepidopterism refers to urticaria and irritation caused by adult moths and their hairs or scales (see Chapter 26 for further discussion of the species involved). Reports of particularly severe lepidopterism have come from the South American countries of Peru, Brazil, Venezuela, and Argentina. In that area, female moths in the genus *Hylesia* have barbed urticating setae, which are broken off and become airborne as the moths flutter around lights. This results in urticating rashes and upper respiratory irritation. In the United States, the tussock moths and their relatives (family Lymantriidae) cause rashes, upper respiratory irritation, and eye irritation to forest workers in the Pacific Northwest.[13]

B. Contributing Factors and Species

In most cases of lepidopterism, huge outbreaks of an offending moth species lead to environmental air becoming laden with broken hairs or scales. In the United States, the Douglas fir tussock moth (*Orgyia pseudotsugata*) female covers her egg masses with froth and body hairs. These hairs, along with other hairs from the tips of the female abdomen, become airborne and cause irritation to humans present in and near infested forests. Outbreaks of the Douglas fir tussock moth appear to develop almost explosively and then subside abruptly after a year or two. During an outbreak, there are literally billions of caterpillars crawling on the ground, trees, brush, and buildings. Accordingly, airborne levels of the offending hairs are quite high. A related species, the white-marked tussock moth (*O. leucostigma*), occurs throughout most of North America and may be involved in cases of lepidopterism, especially in the east. Rothschild et al.[14] also listed moths in the genera *Epanaphe*, *Anaphe*, *Gazalina*, and *Epicoma* (family Notodontidae) as being causes of occasional lepidopterism.

C. Management and Treatment

Treatment mainly involves personal protection and avoiding the offending species. During the season of *Hylesia* moth emergence (South America), it may help to turn off outdoor lights. In affected areas of the United States, air-conditioning and frequent changes of filters may help reduce airborne levels of spines or hairs. Sensitive persons should avoid walking in forests heavily infested with tussock or gypsy moths. Physicians should be aware that cases diagnosed as simple conjunctivitis or keratoconjunctivitis occurring in areas with extreme moth infestation (tussock moths in the Pacific Northwest and gypsy moths in the east) may be due to contact with airborne moth hairs.

Acute urticarial lesions may respond satisfactorily to topical corticosteroid lotions and creams (desoximetasone gel has frequently been used), which reduce the intensity of the inflammatory reaction. Oral antihistamines may relieve itching and burning sensations. In more serious cases, oral prednisone may be indicated. Rosen[15] reported that systemic administration of corticosteroids in the form of intramuscular triamcinolone acetonide has been remarkably effective in relieving severe itching due to gypsy moth dermatitis.

References

1. Lehman CF, Pipkin JL, Ressmann AC. Blister beetle dermatitis. *Arch. Dermatol.* 1955;71:36–41.
2. Alexander JO. *Arthropods and Human Skin*. Berlin: Springer-Verlag; 1984.
3. Frazier CA, Brown FK. *Insects and Allergy*. Norman, OK: University of Oklahoma Press; 1980.
4. Bahme AJ. Cantharides toxicosis in the equine. *Southwest Vet.* 1968;21:147–150.
5. Presto AJ, Muecke EC. A dose of Spanish fly. *J. Am. Med. Assoc.* 1970;214:591–592.
6. Wertelecki W, Vietti TJ, Kulapongs P. Cantharidin poisoning from ingestion of a "blister beetle". *Pediatrics.* 1967;39:287–289.
7. Swarts WB, Wanamaker JF. Skin blisters caused by vesicant beetles. *J. Am. Med. Assoc.* 1946;131:594–594.
8. Diaz JH. The epidemiology, diagnosis, and management of caterpillar envenoming in the southern US. *J. Louisiana State Med. Soc.* 2005;157(3):153–157.
9. Keegan HL. Some medical problems from direct injury by arthropods. *Int. Pathol.* 1969;10:35–45.

10. Foot NC. Pathology of the dermatitis caused by *megalopyge opercularis,* a Texas caterpillar. *J. Exp. Med.* 1922;35:737–741.

11. Shama SK, Etkind PH, Odell TM, Canada AT, Soter NA. Gypsy moth dermatitis. *N. Engl. J. Med.* 1982;306:1300–1303.

12. CDC. Caterpillar-associated rashes in children – Hillsborough Coubty, Florida, 2011. *MMWR.* 2012;61:209–211.

13. Perlman F, Press E, Googins GA, Malley A, Poarea H. Tussockosis: Reactions to Douglas fir tussock moth. *Ann. Allergy.* 1976;36:302–306.

14. Rothschild M, Reichstein T, Von Euw J, Alpin R, Harman RRM. Toxic Lepidoptera. *Toxicon.* 1970;8:293–297.

15. Rosen T. Caterpillar dermatitis. *Dermatol. Clin.* 1990;8:245–252.

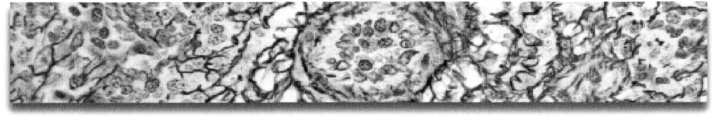

6

MYIASIS (INVASION OF HUMAN TISSUES BY FLY LARVAE)

I. Introduction and Medical Significance

Fly larvae infesting the organs and tissues of people or animals is referred to as myiasis. The condition occurs in several forms that can be primarily classified—at least from their evolutionary roots—as saprophagous and sanguinivorous.[1,2] Not all myiasis involves long-term tissue infestation; for example, the Congo floor maggot in Africa does not embed itself in tissue but only sucks blood from its host for a short time (see Chapter 21). Specific cases of myiasis are clinically defined by the areas affected; for example, there may be traumatic (wound), gastric, rectal, auricular, and urogenital myiasis, among others. Myiasis can be accidental, when fly larvae occasionally find their way into the human body, or facultative, when fly larvae enter living tissue opportunistically after feeding on decaying tissue in neglected, malodorous wounds. Myiasis can also be obligate, in which the fly larvae must spend part of their developmental stages in living tissue. Obligate myiasis is the most serious form of the condition from a pathogenic standpoint and constitutes true parasitism.

Fly larvae are not capable of reproduction; therefore, myiasis under normal circumstances should not be considered contagious from patient to patient. Transmission of myiasis occurs only via an adult female fly. See Chapter 21 for discussions of the most common species involved and their biologies.

A. Accidental Myiasis

Accidental enteric myiasis (sometimes referred to as pseudomyiasis) is mostly a benign event, but the larvae could possibly survive temporarily, causing stomach pains, nausea, or vomiting. There is some question as to whether or not these cases should be classified as true myiasis, given the lack of fly development after the ingested eggs hatch. Numerous fly species in the families Muscidae, Calliphoridae, Syrphidae, and Sarcophagidae may produce accidental enteric myiasis. Some notorious offenders are pomace flies and fruit flies (*Drosophila* spp.); the cheese skipper, *Piophilia casei*; the black soldier fly, *Hermetia illucens* (Figure 6.1); and the rat-tailed maggot, *Eristalis tenax*. Other instances of accidental myiasis occur when fly larvae enter the urinary passages or other body openings. Flies in the genera *Musca*, *Muscina*, *Fannia*, *Megaselia*, and *Sarcophaga* have often been implicated in such cases.

B. Facultative Myiasis

Facultative or "opportunistic" myiasis occurs when flies lay their eggs on living tissue instead of dead animals. This may cause pain and tissue damage as fly larvae leave necrotic tissues and invade healthy ones. Cases of facultative myiasis frequently occur in medical and long-term care facilities where incapacitated patients are housed.[3-6] Numerous species of Muscidae, Calliphoridae, and Sarcophagidae have been implicated in facultative myiasis (Figure 6.2). In the United States, the calliphorid *Phaenicia sericata* has been reported as causing facultative myiasis on several occasions.[4,7] Another calliphorid, *Chrysomya rufifacies*, has been introduced into the United States from the Australasian region and is also known to be regularly involved in facultative myiasis.[8] Other muscoid fly species that may be involved in this type of myiasis include *Calliphora vicina*, *Phormia regina*, *Cochliomyia macellaria*, and *Sarcophaga haemorrhoidalis*.

C. Obligate Myiasis

As opposed to facultative myiasis, some types of flies must develop in the living tissues of a host. This is termed obligate

CASE HISTORY

HUMAN PARASITES OR FLY LARVAE?

A woman sent me several wormlike specimens collected from inside her toilet and on the floor around the toilet. "How are these things getting there?" she asked. "Does this mean these are intestinal parasites?" Upon examination, the specimens were identified as soldier fly larvae, *Hermetia illucians*, common pests inhabiting decaying or putrefying organic matter. This species has been reported as a cause of intestinal myiasis, but—as in this case—when these larvae are found in or near toilets, it is difficult to tell if they have been excreted into the toilet or were deposited as eggs by the adult fly in the toilet area due to decaying wood and build-up of organic matter at the toilet/floor interface. I recommended that the woman have the toilet taken up, thoroughly cleaned, and a new wax seal installed. She called back later to say that her husband found many more specimens under the toilet where the flange meets the sewer pipe and even in the wax seal itself.

Comment: Soldier fly larvae are often found in toilets. I have encountered at least ten such events. Often, they are mistaken at first for parasites coming from inside the people using the toilet, when in reality they are merely larvae feeding in and around a toilet. Very frequently toilets leak a little around the connection to the sewer, allowing water and organic debris to accumulate—perfect conditions for soldier fly larvae.

Figure 6.1 Accidental myiasis. Soldier fly larva found in a child's vomit. (Photograph copyright 2014 by Jerome Goddard, Ph.D.)

Figure 6.2 Facultative myiasis. Maggots and adult blow flies on a dead deer. Ordinarily, these flies are attracted to dead animals but may mistake a foul-smelling or neglected wound for carrion.

myiasis and is caused by species affecting sheep, cattle, horses, and many wild animals. In people, obligate myiasis is primarily due to the screwworm flies (Old and New World) and the human bot fly (Figure 6.3). Obligate myiasis is rarely fatal in the case of the human bot fly of Central and South America, but it has led to considerable pathology and death in the case of screwworm flies. Screwworm flies use wildlife and domestic livestock as primary hosts, but they do attack humans. If, for example, a female screwworm fly oviposits just inside the nostril of a sleeping human, hundreds of developing maggots may migrate throughout the turbinal mucous membranes, sinuses, and other tissues. Surgically, it would be extremely difficult to remove all of the larvae. Fortunately, on account of the sterile male release program, screwworm flies have been virtually eliminated from the United States and Mexico. A recent outbreak of the screwworm fly in the Florida Keys resulted in the loss of many Key deer (Figure 6.4) (see also Chapter 21). After repeated releases of sterile screwworm flies, the problem was abated.[9]

D. Intentional Myiasis (Maggot Debridement Therapy)

Physicians occasionally utilize fly maggots (primarily blow fly larvae) for the debridement of wounds and ulcers both in the United States and internationally.[10–12] Historically, maggot therapy was commonly used in medicine until the advent of antibiotics in the 1940s, but lately the practice is increasingly being used again, especially in cases where antibiotics are ineffective or surgery is not possible.[11] When raised on sterile media and properly handled, blow fly maggots are capable of debriding a wound without spreading further infection or feeding on living tissues. The most commonly used species of larvae for maggot debridement therapy (MDT) is *Lucilia sericata* in the family Calliphoridae.[13,14]

These sterile, live, medical-grade "biosurgery" maggots can be applied using a special "maggot cage" dressing that confines larvae to the wound site and prevents escape. The maggots are able to dissolve necrotic tissue and bacterial biofilms.[15] Their excretions kill bacteria, dissolve old tissue, and stimulate generation of granulation tissue—a type of new, healthy tissue that forms in healing wounds.[16] Once the maggots have been applied (dosing range for maggots per square centimeter varies from 4–10 maggots depending on healthcare provider and maggot type being used). Once they are sealed against the wound with dressings, they must be checked every 2–4 hours to ensure that the outer dressing is dry, the seal is strong, and that there are no openings from which the maggots can escape. Maggots can be left in the wound until the area has been debrided. Once necrotic tissue has been dissolved and ingested, the maggots will cease feeding, leaving only healthy tissue; they are not intended to harm the patient. After treatment is completed, the maggots can be disposed of with normal medical waste after flushing the wound thoroughly with sterile fluid to remove any residual maggots or excretions. Besides chronic wounds, MDT can be used for patients with malignant wounds, venous wounds, and burns.[15] MDT may also be prescribed for patients whose condition is not stable enough to undergo surgical debridement. Patients

Figure 6.4 Obligate myiasis. Screwworm fly larvae in a Key deer. (Photograph courtesy of Dr. Samantha Gibbs, U.S. Fish and Wildlife Service, with permission.)

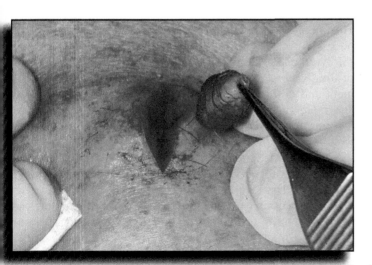

Figure 6.3 Obligate myiasis. Human bot fly larva removed from a patient. (Photograph courtesy of Dr. Mark Mehrany, Department of Dermatology, Mayo Clinic, Rochester, MN.)

with rapidly advancing infections or a deteriorating condition should not undergo MDT. The entire treatment can take anywhere from 4 days to several weeks to completely debride a wound, depending on the extent of the injury. Wounds must be open to the outside of the body, not completely dry, not near major blood vessels, and cannot include bone or tendon tissues.

II. Contributing Factors

A. Accidental Myiasis

Accidental enteric myiasis generally occurs from ingesting fly eggs or young maggots on uncooked foods or previously cooked foods that have been subsequently infested. Cured meats, dried fruits, cheese, and smoked fish are the most commonly infested foods. Other cases of accidental myiasis may occur from the use of contaminated catheters, douching syringes, or other invasive medical equipment or from sleeping with the body exposed.

B. Facultative Myiasis

Several fly species lay eggs on dead animals or rotting flesh (Figure 6.2). Accordingly, the flies may mistakenly oviposit in a foul-smelling wound of a living animal. The developing maggots may subsequently invade healthy tissue. Facultative myiasis most often is initiated when flies oviposit in necrotic, hemorrhaging, or pus-filled lesions. Wounds with watery alkaline discharges (pH 7.1–7.5) have been reported to be especially attractive to blow flies. Facultative myiasis frequently occurs in helpless semi-invalids who have poor (if any) medical care. Often, in the case of the very elderly, their eyesight is so weak that victims do not detect the myiasis. In clinical settings, facultative myiasis is most likely to occur in incapacitated patients who have recently had major surgery or those having large or multiple uncovered or partially covered festering wounds. Not all human cases of facultative myiasis occur in or near a wound, however. In the United States, larvae of the blow fly, *Phaenicia sericata*, have been reported from the ears and nose of healthy patients with no other signs of trauma in those areas.[3]

C. Obligate Myiasis

Obligate myiasis is essentially a zoonosis; humans are not the ordinary host but may become infested. Human infestation by the human bot fly is very often via a mosquito bite—the eggs are attached to mosquitoes and other biting flies (see Part III for more details); however, human screwworm fly myiasis is a result of direct egg laying on a person, most often in or near a wound or natural orifice. Egg-laying activity of screwworm flies occurs during daytime.

III. Prevention, Management, and Treatment

A. Prevention

Prevention and good sanitation can avert much of the accidental and facultative myiasis occurring in the industrialized world. To prevent flies from ovipositing on exposed foodstuffs, these should not be unattended for any length of time. Covering and preferably refrigerating leftovers should be done immediately after meals. Washing fruits and vegetables prior to consumption should help remove developing maggots, although visual examination should also be done while slicing or preparing these items. Other forms of accidental myiasis may be prevented by protecting invasive medical equipment from flies and avoiding sleeping nude, especially during daytime. To prevent facultative myiasis, extra care should be taken to keep wounds clean and dressed, especially on elderly or helpless individuals. Daily or weekly visits by a home health nurse can go a long way in preventing facultative myiasis in patients who stay at home. In institutions containing invalids or otherwise helpless patients, every effort should be made to control entry of flies into the facility. This might involve taking such precautions as keeping doors and windows screened and in good repair, thoroughly sealing all cracks and crevices, installing air curtains over doors used for loading and unloading supplies, and installing ultraviolet electrocuter (or the newer electronic types) fly traps in areas accessible to the flies but inaccessible to patients. Prevention of obligate myiasis involves avoiding sleeping outdoors during daytime in screwworm-infested areas and using insect repellents in Central and South America to prevent bites by bot fly egg-bearing mosquitoes.

B. Management and Treatment

Treatment of accidental enteric myiasis is probably not necessary (although there may be rare instances of clinical symptoms), as in most cases there is no development of the fly larvae within the highly acidic stomach environment and other parts of the digestive tract. They are killed and merely carried through the digestive tract in a passive manner. Treatment of other forms of accidental myiasis as well as facultative or obligate myiasis involves removal of the larvae. Alexander[17] recommended debridement with irrigation. Others have suggested surgical exploration and removal of larvae under local anesthesia.[3] Care should be taken not to burst the maggots upon removal. Human bot fly larvae have been successfully removed using "bacon therapy," a treatment method involving covering the punctum (breathing hole in the skin) with raw meat or pork.[18] In a few hours, the larvae migrate into the meat and are then easily extracted (see Chapter 21 for more discussion).

Maggot infestation of the nose, eyes, ears, and other areas may require surgery if larvae cannot be removed via natural orifices. Alternatively, infested tissues may be flushed with nitrofurazone.[19] Because blow flies and other myiasis-causing flies lay eggs in batches, there could be tens or even hundreds of maggots in a wound.

References

1. Stevens JR, Wallman JF. The evolution of myiasis in humans and other animals in the Old and New Worlds (Part I): Phylogenetic analysis. *Trends Parasitol.* 2006;22:129–143.
2. Stevens JR, Wallman JF. The evolution of myiasis in humans and other animals in the Old and New Worlds (Part II): Biological and life-history studies. *Trends Parasitol.* 2006;22:181–188.
3. Anderson JF, Magnarelli LA. Hospital acquired myiasis. *Asepsis.* 1984;6:15.
4. Greenberg B. Two cases of human myiasis caused by Phaenicia sericata in Chicago area hospitals. *J. Med. Entomol.* 1984;21:615.
5. Daniel M, Sramova H, Zalabska E. Lucilia sericata (Diptera: Calliphoridae) causing hospital-acquired myiasis of a traumatic wound. *J. Hosp. Infect.* 1994;28(2):149–152.
6. Ahadizadeh EN, Ketchum HR, Wheeler R. Human cutaneous myiasis by the Australian sheep blow fly, Lucilia cuprina (Diptera: Calliphoridae), in Oklahoma. *J. Forensic. Sci.* 2015;60(4):1099–1100.
7. Merritt RW. A severe case of human cutaneous myiasis caused by Phaenicia sericata. *California Vect. Views.* 1969;16:24–26.
8. Richard RD, Ahrens EH. New distribution record for the recently introduced blow fly Chrysomya rufifaces in North America. *Southwest Entomol.* 1983;8:216–218.
9. Skoda SR, Phillips PL, Welch JB. Screwworm in the United States: Response to and elimination of the 2016–2017 outbreak in Florida. *J. Med. Entomol.* 2018;55:doi:10.1093/jme/tjy1049.
10. Schwarck L. Maggot debridement therapy. *J. Contin. Educ. Nurs.* 2009;40(1):14–15.
11. Sherman RA. Maggot therapy takes us back to the future of wound care: New and improved maggot therapy for the 21st century. *J. Diabetes Sci. Technol.* 2009;3(2):336–344.
12. Sherman RA, Pechter EA. Maggot therapy: A review of therapeutic applications of fly larvae in human medicine, especially for treating osteomyelitis. *Med. Vet. Entomol.* 1988;2:225–230.
13. Sherman RA, Hall MJR. Medicinal maggots: An ancient remedy for some contemporary afflictions. *Ann. Rev. Entomol.* 2000;45:55–81.
14. Williams KA, Cronje FJ, Avenant L, Villet MH. Identifying flies used for maggot debridement therapy. *S. Afr. Med. J.* 2008;98:196–198.
15. Anonymous. Maggot debridement therapy (MDT) in adults and children. British Columbia, Canada, Provincial Nursing Skin and Wound Committee, www.clwk.ca/buddydrive/file/procedure-maggot-debridement-therapy/; 2012.
16. Grey JE, Enoch S, Harding KG. Wound assessment. *BMJ.* 2006;332:285–288.
17. Alexander JO. *Arthropods and Human Skin.* Berlin: Springer-Verlag; 1984.
18. Brewer TF, Wilson ME, Gonzalez E, Felsenstein D. Bacon therapy and furuncular myiasis. *J. Am. Med. Assoc.* 1993;270:2087–2088.
19. Lima-Junior SM, Asprino L, Prado AP, Moreira RW, de Moraes M. Oral myiasis caused by Cochliomyia hominivorax treated nonsurgically with nitrofurazone: Report of two cases. *Oral Surg. Oral Med. Oral Pathol. Oral Radiol. Endod.* 2010;109(3):70–73.

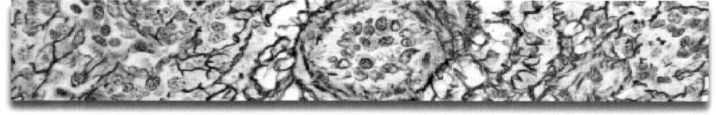

7

DELUSIONS OF PARASITOSIS (IMAGINARY INSECT OR MITE INFESTATIONS)

I. Introduction

Arthropods have historically affected humans psychologically, and there have even been "psychological epidemics." In 14th-century Europe, for example, there were reports of a strange dancing mania in which affected persons danced wildly in the streets screaming and foaming at the mouth. In Italy, some of these victims claimed that they had been bitten by a wolf spider, or tarantula (not the same as the American tarantula), and the dance mania was therefore called tarantism. Apparently, those bitten by the spiders felt compelled to dance until the poison was out of their system.

Some people today have delusions of parasitosis (DOP) in which they think they are infested with arthropod parasites. The condition is not uncommon. Any public health or extension service entomologist who has worked with the public has encountered this problem in one form or another. Schrut and Waldron[1] reported seeing over 100 cases in 5 years. Alexander[2], Slaughter et al.,[3] and Hinkle[4] have provided excellent overviews of this subject. Before discussing this phenomenon, some clarifying definitions are in order.

Entomophobia—Entomophobia is often erroneously used as a catchall term to explain abnormal reactions in some people to insects or mites. This word has been loosely used to include several distinct psychological phenomena such as abnormal fear of insects or arachnids, delusions of parasitic mite infestations, and dermatitis caused by "computer mites," "paper mites," or "cable mites" in the workplace. A more precise definition of entomophobia is a terrible fear or dread of insects, mites, or spiders. It may be characterized by hysterical reactions at the sight of the feared arthropod. Entomophobia may be related to,

but is distinctly different from, delusions of parasitosis and computer or cable mite dermatitis.

Delusions of parasitosis—DOP is an emotional disorder in which the patient is convinced that tiny, almost invisible parasites are present on or in his or her body. These parasites could be insects, mites, or various helminths (worms). Some researchers prefer to label the disorder pyschogenic parasitosis because not all patients are delusional in the classic sense.[3,5] In most cases, DOP is a monosymptomatic hypochondriacal psychosis in which there are no other thought disorders and the delusions are not a result of another psychiatric illness; however, DOP cases can be associated with paranoia or schizophrenia, which can quite possibly endanger the healthcare provider. One DOP patient attempted to murder her general practitioner.[6] The original author of this book (JG) also encountered patients who subsequently began to think that health department officials (including the entomologist!) were "out to get him." The mental and emotional stress from DOP may be severe, leading to destructive behaviors such as quitting jobs, burning furniture, abandoning homes, using pesticides dangerously and repeatedly, etc.

Workplace infestations—Sometimes there is an imaginary insect or mite infestation at the workplace. Workers may complain of cable mites, paper mites, or computer mites biting them and producing lesions. Strictly speaking, this situation may not be DOP in that often there is a real substance in the work environment causing the problem (e.g., fiberglass or other insulation, chemicals, paper or cardboard splinters).

Prior to categorizing persons into any one of these three groups, it is imperative that a thorough entomological inspection be carried out at the residence or workplace. This should be done with an objective and open mind because it is possible that there is a real insect or mite problem. A pest management professional (exterminator) could do this, but a university, health department, or extension entomologist would be better suited as they will generally have had more entomological training. Provided in the following text is an overview of the insect or mite species that could be the cause of a real infestation (please refer to the corresponding sections of Part III for further details).

WHAT ABOUT MORGELLONS DISEASE?

The term *Morgellons disease*, referring to a medical case described in 1674, is often used synonymously with delusions of parasitosis. Patients with Morgellons have an unshakable belief that fibers are emerging from their skin and that the fibers are causing biting and stinging sensations. Along with their wide range of skin conditions, people with this disorder also often exhibit severe fatigue, arthralgia, cognitive decline, and mood disorders. In fact, one recent study showed that 70% of 115 patients complained of chronic fatigue.[1] Morgellons may be psychologically "infective," leading to other family members and co-workers (seemingly) getting the disease. At this time, there is no evidence incriminating insects or mites as the cause of Morgellons.[1] Many, if not most, physicians do not recognize Morgellons as a valid clinical entity and generally view the disorder in one of three ways:[2]

1. Some health professionals believe that Morgellons disease is a specific condition that needs to be confirmed by future research.
2. Some health professionals believe that the signs and symptoms of Morgellons disease are caused by another condition, often mental illness.
3. Other health professionals do not acknowledge Morgellons disease or are reserving judgment until more is known about the condition.

REFERENCES

1. Pearson, M.L. et al., Clinical, epidemiologic, histopathologic and molecular features of an unexplained dermopathy, *PLoS ONE*, 7(1), e29908, 2011.
2. Anon., *Morgellons Disease: Managing a Mysterious Skin Condition*, Mayo Clinic, Rochester, NY, 2012 (www.mayoclinic.com/health/morgellons-disease/sn00043).

II. Actual Arthropod Causes of Dermatitis

There are two human-specific mites: the human scabies mite, *Sarcoptes scabei*, and the human follicle mite, *Demodex folliculorum*. The human scabies mite burrows beneath the outer layer of skin, leaving crusted papules and/or a sinuous red trail along the skin for a few centimeters. The infestation causes itching and commonly occurs in the webbing between the fingers and on the wrists and elbows. Skin scrapings of affected areas will usually contain the mites. The human follicle mite occurs, in most cases, in hair follicles around the nose and eyelids but generally causes no pathology or discomfort. Physicians should be aware of the follicle mite so that, when detected during a skin scraping seeking the cause of a mysterious biting, the mites will not be inadvertently reported as the cause.

There are a few other mite species (not host specific for humans) that will readily bite people. Usually, these are only opportunistic infestations that are self-limiting. They do not take up residence and reproduce on or in the skin. The tropical fowl mite, *Ornithonyssus bursa* (a parasite of sparrows); the northern fowl mite, *O. sylviarum* (a parasite of chickens); and the tropical rat mite, *O. bacoti* (a parasite of rats) will aggressively bite people, causing pain and itching. In a lot of cases of "mysterious" mite infestation, a rat or bird nest can be found someplace in the dwelling, and the patient is being bitten by one of these mite species. These mites can be seen (they are about 1 mm long) and are often captured while biting or crawling on the skin.

Other mites may occasionally be involved in cases of dermatitis, such as the grocer's itch mite, *Glycyphagus domesticus*; the spiny rat mite, *Laelaps echidnina*; the straw itch mite, *Pyemotes tritici*; the baker's mite, *Acarus siro*; and the dried fruit mite, *Carpoglyphus lactis*. These cases are usually limited to persons whose occupations bring them into prolonged, close association with these mites. Again, none of these species can live on or in human skin.

Body and head lice, *Pediculus humanus* corporis and *P. humanus capitis*, respectively, can cause discomfort and itching. Pubic lice, *Pthirus pubis*, live on humans in the pubic and perianal regions. Lice, however, are large enough to see and thus are usually found during a physical examination.

If a patient is elderly or has poor eyesight, fleas may be the cause of mysterious bites in the home. Slowly pulling a white towel or part of a sheet around in the patient's house should yield several fleas if the house is infested. Incidentally, homeowner flea problems have increased in the last year or two.

Bed bugs also could be the cause of mysterious bites, and their incidence is currently on the rise in the United States. Contrary to popular opinion, they are not only found where unsanitary conditions exist, but may also occur in affluent homes, hotels, and institutions. One sign of bed bug infestation is the presence of small specks of blood or dark feces on bedding. Upon examination of the premises, bed bugs can be found (they are plenty big enough to see) beneath loosened wallpaper, in the seams of bedding, or in cracks and crevices around the bed and furniture. Adults are about the size of an apple seed, while nymphs are much smaller but still visible with the naked eye.

There is one report of collembolans (springtails) (Figure 7.1) found in skin scrapings from 18 individuals diagnosed with DOP.[7] However, identification of the insects was made from photographs alone, using special imaging software, and no specimens were recovered and retained as voucher specimens. Currently, results of this paper are controversial and not accepted by a majority of entomologists. More research is required to ascertain the true prevalence, if any, of Collembola in humans.

Again it must be emphasized that cases of imaginary insect or mite infestations must be carefully investigated to rule out actual arthropod causes. Scrapings of skin lesions should be taken for examination for scabies mites. A competent entomologist should examine the patient's workplace and residence for evidence of insects or mites. Careful attention should be given to rat- or bird-infested dwellings, as they may harbor parasitic mites. If repeated collection attempts, skin scrapings, and insecticidal treatments are unsuccessful, and if the patient exhibits symptoms consistent with delusions of parasitosis, then a diagnosis of DOP should be strongly considered.

III. Delusions of Parasitosis

Dr. Jerome Goddard at the Mississippi Department of Health reported encountering 5 to 10 cases of DOP each year (Table 7.1 is a partial listing of delusory parasitosis cases and workplace infestations Dr. Goddard investigated). The disorder shows patterns of typical behavior.[4,8]

Sometimes an initial and real insect infestation precedes and triggers the delusion. In the majority of cases, the victim is an elderly white female, and the "bugs" may appear and disappear while they are being watched; they enter the skin and reappear and invade the hair, nose, and ears. (In one case an entomologist was investigating, the woman suddenly cried out loudly saying that one of the mites had suddenly run up her nose.) Oddly, DOP patients often claim that the bugs are in their rectum or vagina, only coming out at night. The patients claim that the "bugs" are able to survive repeated insecticidal sprays and the use of medicated shampoos and lotions. Frequently there is a history of numerous visits to medical doctors and dermatologists. Lesions may be present, although neurotic excoriation may be the cause.[9] The physician or entomologist is often presented with tissue paper, small bags, pill bottles (Figure 7.2), or other containers with the presumed pests. Tape samples are

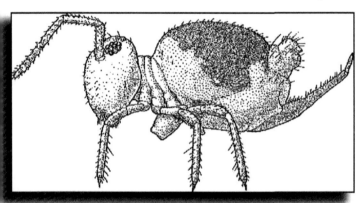

Figure 7.1 A springtail or collembolan. (From Steyskal, G.C. et al., eds., *Insects and Mites: Techniques for Collection and Preservation*, USDA Misc. Publ. No. 1443, Agricultural Research Service, U.S. Department of Agriculture, Washington, DC, 1986.)

Table 7.1 Breakdown of Selected Imaginary Insect or Mite Infestations Reported to the Mississippi State Department of Health, 1989–1995

Case No.	Entity Affected	Age	Sex	Race	Complaint
1.	Individual	~60	M	B	"Parasites in rectum and penis; come out at night to bite"
2.	Individual	Elderly	F	W	"Mites on the skin"
3.	Individual	40	M	B	"Bugs going in and out of skin"
4.	Individual	~60	F	W	"Bugs in my rectum"
5.	Individual	65	F	W	"Bugs biting like fire; making life miserable"
6.	Individual	70	F	W	"Long-standing mite infestation on skin"
7.	Individual	33	M	W	"Mite infestation; appears as white smoke or flakes"
8.	Individual	Elderly	F	W	"Being taken over by bugs"
9.	Individual	75	F	W	"Must be ticks and fleas in my house"
10.	Individual	50	M	W	"Mysterious bug bites for years"
11.	Individual	~60	F	B	"Tiny bugs gliding over my body"
12.	Individual	~60	F	W	"Tiny insects in my ear for 9 months"
13.	Individual	~60	F	W	"Tiny mites biting and making life miserable"
14.	Workplace	25–55	F	W	"Paper mites in medical records"
15.	Workplace	18–50	F	W, B	"Mites in boxes and shipping crates"
16.	Workplace	21–48	F	W, B	"Paper and computer mites in medical records"
17.	Workplace	21–50	F	W, B	"Paper and computer mites in medical records"

CASE HISTORY

"MITES" IN A WOMAN'S EAR

A 60-year-old white female came to my office after she had been to a family practitioner and three dermatologists to no avail with the problem of "tiny insects living on her skin, but especially inside the ears." Upon being seen, she presented samples she had collected—several coffee cups, small pill bottles, and vials, all containing skin scrapings, ear wax, and/ or dried blood. During the interview she stated that the problem began 9 months earlier as an infestation on the skin but, due to repeated pesticide treatments (on her skin), was now in her ears. She also revealed that she regularly forced Kwell® (1% lindane) into her ears via a syringe. Other extreme measures to rid her and her house of the "bugs" included intensive scrubbing of floors and walls, and "rubbing her skin raw."

Careful examination of the patient (by a health department physician) and the samples she had brought revealed no evidence of insect or mite infestation. Previous examinations by dermatologists had ruled out scabies. Based on the characteristic presentation of "imaginary bugs" in this case, results of the interview and exam, and history of numerous visits to various physicians, a tentative diagnosis of delusions of parasitosis (DOP) was made. The patient was referred back to her family physician with instructions to have him contact the author for further discussion.

Comment: DOP is more common than one might think. Over a span of 3 years, 14 cases were reported to the Mississippi Department of Health. Certainly, many more cases never get reported to a state public health agency.

Source: Adapted from Goddard, J. and Currier, M.M., *J. Agromed.*, 2, 53, 1995. Copyright 1995 Haworth Medical Press, Binghamton, NY. With permission.

also commonly submitted (Figure 7.3). However, these usually contain dust, specks of dirt, dried blood, pieces of skin, and occasionally common (nonharmful) household insects or their body parts (Figure 7.4). Alexander said that presentation of such "evidence," together with the intensity of the patient's belief, is almost pathognomonic of the disorder. DOP sufferers are often confident of the cause of their problem and may even say it is something "new to science." Further, some patients, out of their paranoid inclinations, may say their infestation is part of a government conspiracy. Out of desperation, the victims may move out of their home, only to report later that the "bugs" have followed them there, too. An affected person may be so positive of his or her infestation and give such a detailed description that other family members may agree with the patient. They may even be "infected" themselves; thus, the delusion has been transferred. Sometimes affected persons become excessively preoccupied with cleanliness and spend all available money on cleansers, soaps, disinfectants, insecticides, etc., in order to thoroughly clean, scrub, and sterilize their homes.

The following are some excerpts from letters received by Dr. Goddard from patients apparently suffering from DOP:

> **Patient 1: Elderly Female.** The attached package of specimens were [sic] taken from objects in my house when they could be seen—the beds, bathtub, and

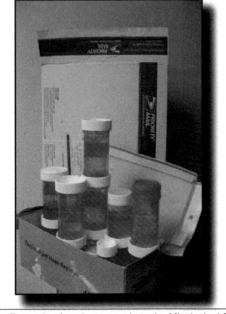

Figure 7.2 Example of package sent in to the Mississippi Department of Health for possible "insect infestation."

Figure 7.3 Pieces of tape with samples sent in by a DOP patient.

from myself when I could both see and feel them. I picked them up with a Lysol solution cloth or paper towel. …My house has been sprayed and fumigated many times. …I have been told that if I have mites, I could not see them. …Please let me know how to capture something so small, and mites which cannot be seen. These are making life miserable for me. …Is there a possibility that my trouble could be some type of fly larva?

Patient 2: Elderly Female. I couldn't get everything in the [other] containers into this one, and left those that had fallen to the bottom, here. In transferring from 3 containers to this one I'm sure some specimens were left clinging. This is a 3-day accumulation. There were definitely 6–8 live, jumpy specimens. Thank you! I hope you will inspect everything, even scraping the sidewalls [of the container], and top and those that are almost invisible to the naked eye.

From reading these two letters it seems that affected patients obsessively feel there are mites, invisible to the naked eye, that

Figure 7.4 Pieces of skin found in samples submitted by a DOP patient.

readily infest humans. It may be a fear of the unknown—the existence of invisible insects or mites that nobody knows much about (even "doctors")—that contributes to this delusion. Thus, whenever these patients feel a tingling sensation or have an itchy place on their skin, they are afraid and later become convinced that mites are on or under their skin.

On the other hand, there may be internal physiological causes of the pruritus, mistakenly believed to be arthropod-produced, such as diabetes, icterus, atopic dermatitis, and lymphoblastomas. At times, the mental disturbance of pellagra takes the form of delusion of parasitosis and disappears with appropriate therapy.[9] Hinkle[10] listed 50 common prescription drugs whose side effects include parathesias, itching, urticaria, and rash. Cases of DOP have also been attributed to cocaine or methamphetamine abuse.[11,12]

A. Diagnosis by a Physician

Delusions of parasitosis is diagnosed as a delusional disorder, somatic type, if symptoms persist for more than 1 month.[13] Medical causes may be ruled out by thorough laboratory and neurological evaluation. Schizophrenia and schizophreniform disorders may be eliminated with a detailed patient history and cognitive testing. Physicians should also check for a comorbid psychiatric disorder that may be perpetuating the delusion, because DOP often co-occurs with axis I disorders, including major depressive disorder, substance abuse, dementia, and mental retardation. After ruling out actual arthropod causes (see previous section) and underlying medical conditions, physicians may wish to refer the patient to a psychiatrist, but most DOP patients will not see a psychiatrist. Instead, they will seek out another physician, thus starting the whole process over. An interdisciplinary approach, mainly involving family practice physicians, dermatologists, psychiatrists, and entomologists (Figure 7.5), is needed to help

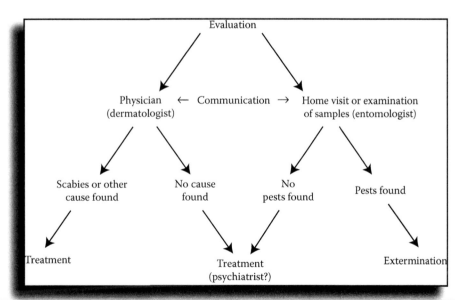

Figure 7.5 An interdisciplinary approach to treatment of DOP. (From Goddard, J., *Infect. Med.*, 15, 169, 1998. With permission.)

DOP patients. Family practice physicians or general practitioners are usually the first providers who see DOP patients. Physicians need to be careful not to diagnose "insect bites" based on lesions alone and should call upon entomologists to examine samples. Entomologists need to understand the medical complexity of delusions—that there are intensive worries, true delusions, and a host of abnormal personality traits associated with DOP—and avoid any hint of medical evaluation of the patient. Treatment strategies, including antipsychotic drugs, have been proposed.[12–17] Previously, dermatologists reported good success in treating DOP patients with pimozide; however, the drug has several adverse effects and should only be used with careful supervision. Risperidone and olanzapine have been proposed instead of pimozide.[12]

IV. Workplace Infestations

Computer, paper, or cable mite dermatitis in a workplace may not be strictly a delusion of parasitosis. Often, these people are actually experiencing a prickling, tingling, or creeping sensation caused by a real substance in their environment (rock wool or fiberglass insulation, paper fragments, dust, or other debris).[18] This problem is exacerbated by mass hysteria (e.g., "Every time I come to work here I get bitten by tiny mites!") Imaginary mite infestations in the workplace typically involve females who work with old, dusty records or elaborate electronic equipment. This has been seen several times in hospital medical record offices. Contributing factors include inadvertent confirmation by physicians, bites from fleas or mosquitoes at home being scratched at work, and unscrupulous pest control operators willing to spray for computer or cable mites in the absence of a real pest.

Dealing with an imaginary workplace infestation is difficult. First of all, a competent pest control operator should inspect the premises, looking for possible causes of the mysterious biting. He or she should set glue boards (sticky traps) in affected areas and check them at regular intervals. If there are actual arthropod pests in sufficient numbers to bite employees, they will likely be captured on the monitoring devices. If no actual insect or mite infestation can be identified in the affected area, then environmental factors such as carpets, paper splinters and particles, static electricity, ventilation systems, and indoor air pollutants must be considered. An excellent handout about invisible itches is available from the University of Kentucky Cooperative Extension Service.[19] Sometimes, hiring an "outside" cleaning crew to thoroughly clean the area will eliminate the supposed infestation. If problems continue, an industrial hygienist should be consulted for inspection for these types of irritants.

References

1. Schrut AH, Waldron WG. Psychiatric and entomological aspects of delusionary parasitosis. *J. Am. Med. Assoc.* 1963;186:213–216.
2. Alexander JO. *Arthropods and Human Skin.* Berlin: Springer-Verlag; 1984.
3. Slaughter JR, Zanol K, Rezvani H, Flax JF. Psychogenic parasitosis: A case series and literature review. *Psychosomatics.* 1998;39:491–500.
4. Hinkle N. Ekbom syndrome: The challenge of "invisible bug" infestations. *Ann. Rev. Entomol.* 2010;55:77–94.
5. Zanol K, Slaughter J, Hall R. An approach to the treatment of psychogenic parasitosis. *Int. J. Dermatol.* 1998;37:56–63.
6. Bourgeois ML, Duhamel P, Verdoux H. Delusional parasitosis: Folie a deux and attempted murder of a family doctor. *Br. J. Psychiatry.* 1992;161:709–711.
7. Altschuler DZ, Crutcher M, Dulceanu N, Cervantes BA, Terinte C, Sorkin LN. Collembola found in scrapings from individuals diagnosed with delusory parasitosis. *J. New York Entomol. Soc.* 2004;112:87–95.
8. Goddard J. Analysis of 11 cases of delusions of parasitosis reported to the Mississippi Department of Health. *South. Med. J.* 1995;88:837–839.
9. Obermayer ME. Dynamics and management of self-induced eruptions. *Calif. Med.* 1961;94:61–71.
10. Hinkle N. Delusory parasitosis. *Am. Entomol.* 2000;46:17–25.
11. Elpern DJ. Cocaine abuse and delusions of parasitosis. *Cutis.* 1988;42:273–274.
12. Scheinfeld N. Delusions of parasitosis: A case with a review of its course and treatment. *Skinmed.* 2003;2(6):376–378.
13. Matthews AM, Hauser P. A creepy-crawly disorder. *Curr. Psychiatry.* 2005;4:88–93.
14. Driscoll MS, Rothe MJ, Grant-Kels JM, Hale MS. Delusional parasitosis: A dermatologic, psychiatric, and pharmacologic approach. *J. Am. Acad. Dermatol.* 1993;29:1023–1033.
15. Gould WM, Gragg TM. Delusions of parasitosis – An approach to the problem. *Arch. Dermatol.* 1976;112:1745–1748.
16. Koblenzer CS. *Psychocutaneous Disease.* Orlando, FL: Grune and Stratton; 1987.
17. Torch EM, Bishop ER. Delusions of parasitosis – Psychotherapeutic engagement. *Amer. J. Psychother.* 1981;35:101–109.
18. Possick PA, Gellin GA, Key MM. Fibrous glass dermatitis. *Am. Ind. Hyg. Assoc.* 1970;Jan-Feb Issue:12.
19. Potter MF. Invisible Itches: Insect and Non-Insect Causes. University of Kentucky, Cooperative Extension Service Publ. No. ENT-58. 4 pp.; 1997.

Part II

**Identification of Arthropods and
the Diseases They Cause**

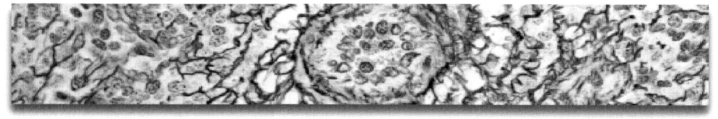

8

IDENTIFICATION OF MEDICALLY IMPORTANT ARTHROPODS

I. Principles of Identification and Naming

Precise identification of an offending arthropod is extremely important in medical settings and may have far-reaching implications; for example, a person stung by a non-dangerous scorpion does not need the expensive and intense treatments that a person stung by a more deadly species needs. Also, treatment recommendations for pubic lice differ radically from those for nymphal ticks in the pubic region (yet the two are often confused). Obviously, accurate identification is crucial from the outset.

Entomologists have their own jargon and terminology, as do all specialists, which readers will be able to decipher upon understanding the basis for arthropod naming. Generally, arthropods are grouped into classes, orders, families, genera, and species according to shared common morphological structures. Specimens with eight legs and two body regions are placed in one category, specimens with six legs and three body regions are placed in another, etc. Making things easier, the names assigned to these categories are often descriptive of the arthropods in that category; for example, insects with two wings are in the order Diptera, meaning "two wings" (other common insect orders are provided in Table 8.1). Species names may further describe the arthropod—*Buggus erythrocephala* would mean having a red head, the term melanogaster would mean dark or black belly, and the name *Calliphora vomitoria* or *Cynomya cadaverina* for a blow fly would give the reader some idea of what kind of things these flies are attracted to or breed in.

Not all names given to arthropods are descriptive of the organism; some memorialize a specialist in that field. *B. parkeri*, *B. bacoti*, *B. kochi*, *B. blakei*, *B. walkeri*, and *B. smithi* were named after Drs. Parker, Bacot, Koch, Blake, Walker, and Smith, respectively. The first person to describe a new species (and publish the name) becomes the author of that scientific name, and that person's name is often cited with the arthropod—*Phlebotomus diabolicus* Hall. If subsequently the species is moved to another genus through some kind of taxonomic revision, the author's name is retained but in parentheses—*Lutzomyia diabolica* (Hall). If two species are subsequently determined to be the same species (this happens frequently), the oldest described species—the first published one—becomes the official name. The other name is then considered a taxonomic synonym and is discarded. Governing all this is the standard code of rules of nomenclature that has been laid down by the International Commission on Zoological Nomenclature.

A. Morphological Identification

The traditional method for identifying arthropods is morphological, i.e., basing the identification on size, shape, or number of certain morphological characters. This process often involves using published diagnostic keys similar to a flow chart to arrive at the proper identification. Good written keys include numerous high-quality photos or line drawings of the structures mentioned (sometimes line drawings are actually better at showing detail than photographs). Picture keys generally do not offer detailed morphological information, but instead provide actual whole-organism pictures in pairs (Figure 8.1). The reader compares the pictures and makes a choice each time throughout the flow chart, ultimately arriving at an identification.

A written diagnostic key may look something like this:

1. Second antennal segment without a laterodorsal longitudinal seam. . . . . Acalypteratae
1' Second antennal segment with a complete laterodorsal longitudinal seam . . . go to 2

2. Hypopleuron (Meron) usually without hairs or bristles . . . . Anthomyiaria

2' Hypopleuron (Meron) usually with hairs or bristles in one or more rows . . . . go to 3

A picture key may look something like the one on the following page (Figure 8.1).

Identification keys may continue on like this for many pages. In fact, some diagnostic keys comprise entire books, volumes, or monographs. Complicating matters, keys can use complex and unfamiliar terminology, making identification difficult. Well-written keys avoid jargon specific to any one group of arthropods and attempt to avoid couplets with vague phrases such as "such-and such structure is longer." If the person making the identification has never seen a specimen with the second option, they cannot make a decision. How long is longer? Therefore, morphological characteristics in a key should be quantitative whenever possible. Lastly, making a morphological identification does not harm the specimen, so voucher specimens can be retained in a museum for future examination if there is ever a question.

B. Molecular Identification

Molecular identification of arthropods involves a genetic analysis obtained by grinding up the arthropod in question (or pieces thereof), running a polymerase chain reaction (PCR) procedure to amplify portions of one or more key genes of that specimen, and then comparing those sequences with known (published) sequences found in the National Center for Biotechnology Information, National Library of Medicine, "GenBank." Searching GenBank can show which specimens have sequences that most closely match the unknown sequence. For example, if you did a PCR on an unidentified tick specimen using published primers for identifying ticks and then submitted your sequence to GenBank, you would get a result showing the closest matches. Your result may say, "100% match with *Amblyomma maculatum*." If so, then that is the precise identification. If your result says "closest match is *Amblyomma americanum* 86%," then you

can assume something went wrong with the analysis, or perhaps you have a specimen with no sequences in GenBank, or maybe you have found a new species. To improve accuracy of molecular identifications, more than one gene should be used, the more the better.

In the last decade, there has been an effort to barcode the world's animal life using the cytochrome *c* oxidase subunit I gene (*COXI*). Use of DNA barcodes has supposedly been validated,[1] although not all scientists agree.[2] According to Will,[2] people seeking a panacea of molecular identification of species will encounter all sorts of constraints and inconsistencies in their work: most importantly, judgments about species boundaries. This problem is addressed in a paper by Sperling,[3] wherein he discusses what DNA sequence or allozyme divergence number is the cutoff above which populations can be considered separate species. Based on data from *Paplio* butterflies, Sperling says it is unreasonable to expect any kind of simple relationship between percent sequence divergence and maintenance of genetic integrity.[3]

C. Choosing Morphological versus Molecular Identification

Choosing an identification strategy is more complicated than it seems. Some scientists might say that morphological identification methodology is outdated and should be abandoned. However, some questions cannot be answered with a molecular analysis of an arthropod. The best example is the developmental stage: whereas in morphological identification one can almost instantly determine the life stage of an arthropod, this determination is more difficult using molecular methods. Certain arthropods contain different concentrations of mRNA at different points in their development, suggesting that you could identify the developmental stage by analyzing the specimen for mRNA concentration; however, this often does not work. If identification of the developmental stage is required, morphological analysis is the best (and quickest) method to use. If developmental stage is not a point of concern, but identification of the species is, then maybe molecular analysis is sufficient. However, genetic material is not always available, especially from historic collections. Additionally, molecular analysis is time-consuming, and a laboratory needs access to DNA extraction kits, reagents for amplification, a thermal cycler to perform the amplification, access to sequencing capacity or electrophoresis, and software to analyze completed sequences. In addition to the expense, a complex understanding of the organism's genome is necessary to successfully use molecular analysis. For example, many hard ticks are closely related genetically, and some loci differentiate between species by only a few base pairs. If the targeted region of the genome is not informative between related species, this differentiation can be convoluted and inconclusive.

As mentioned, while molecular analysis can potentially provide clear distinctions between species, these analyses are

Table 8.1 Names of Some Common Insect Orders

Insect Order	Common Names
Blattaria	Cockroaches
Coleoptera	Beetles
Diptera	Flies (includes mosquitoes)
Ephemeroptera	Mayflies
Hemiptera	True bugs, includes kissing bugs and bed bugs
Hymenoptera	Wasps, ants, bees
Isoptera	Termites
Lepidoptera	Butterflies, moths
Odonata	Dragonflies, damselflies
Orthoptera	Grasshoppers, crickets
Phthiraptera	Lice
Siphonaptera	Fleas

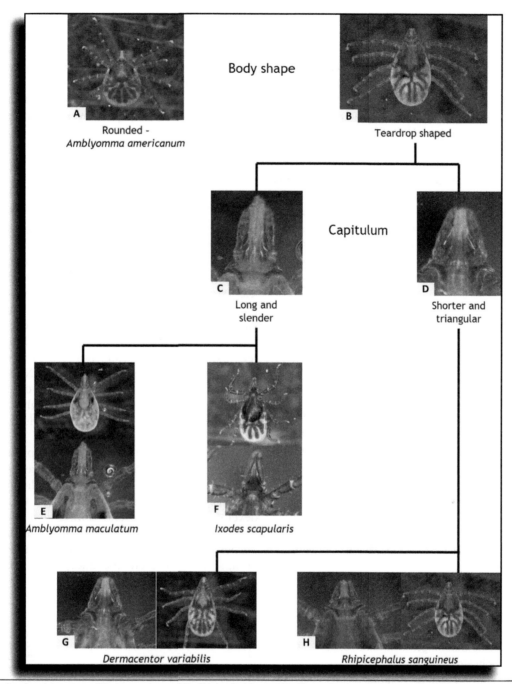

Body shape

A Rounded -
Amblyomma americanum

B Teardrop shaped

Capitulum

C Long and
slender

D Shorter and
triangular

E *Amblyomma maculatum*

F *Ixodes scapularis*

G *Dermacentor variabilis*

H *Rhipicephalus sanguineus*

Figure 8.1 Example of a picture key. (Photo courtesy of Dr. Trisha Dubie and Dr. Bruce Noden, Oklahoma State University, used with permission.)

dependent upon the availability of genetic data. GenBank has a huge cache of genetic data with which to compare results, but there is no formal review process to submit genetic data to GenBank, allowing for at least a proportion of the data to be inaccurate. How to determine what is and is not accurate genetic information is not simple. Studies have shown that up to 30% of the data publicly available in GenBank contains errors. If a BLAST search reveals numerous identical results from multiple sources, then chances are good the identification is correct; however if only a few results are produced, the identification is questionable. Another major point to consider

when using molecular sequencing for species identification is that the origin of all genetic data in GenBank is, in fact, morphological (which is actually ironic). Because GenBank is an open forum for submission, whoever submits the first sequence for an arthropod species designates it based on his or her own morphological identification. If the arthropod is misidentified, the data associated with it is incorrect. This can become a serious issue, as corrections are rarely made after an original submission to GenBank is misidentified. No submissions to GenBank ever require publication, so submissions are thereby immune from peer review. Therefore a misidentification could potentially persist unnoticed or uncorrected for decades.

Both morphological and molecular identification techniques have pros and cons, but ultimately a decision to use one or the other must be made based on the available equipment, knowledge, and resources. In addition, the purpose of the identification must be considered; if determination of the developmental stage is required, or if physical characteristics or general fitness are a point of concern, morphological analysis will provide the best information. However, if a robust identification or secondary verification of a species is needed, molecular analysis will work well. Ideal situations employ both methods together and provide the most data; in this way, one can consider both physical characteristics and genetic information to arrive at the most accurate determination. In either case, if identification seems to be an anomaly, further analysis is necessary.

II. Brief Review of Arthropod Morphology

A. Introduction

There is tremendous variety among the arthropods, but identification of the major groups is not difficult if one is familiar with their general morphology. In fact, it is rarely necessary to use a microscope to separate the classes of Arthropoda (e.g., insects, spiders, scorpions). Table 8.2 presents a summary of the key characteristics of these classes. On the other hand, more specific identification of specimens beyond the

Table 8.2 Key Characteristics of Adults of Selected Arthropod Groups

Arthropod Group	Key Characteristics
Insects	Six legs Three body regions—head, thorax, abdomen Some with wings
Spiders	Eight legs Two body regions—cephalothorax, abdomen
Mites and ticks	Eight legs (as adults) One (apparent) globose or disk-shaped body region No true head, mouthparts only
Scorpions	Eight legs Broad, flat body region and posterior tail with stinger
Centipedes	Numerous legs—hundred leggers One pair of legs per body segment Often dorsoventrally flattened
Millipedes	Numerous legs—thousand leggers Two pair of legs per body segment Often cylindrical

class can be difficult without the aid of proper literature and, perhaps, special training. Identification to the species level often must be done by specialists who conduct taxonomic research with that particular group. For outside help with identifications, try a local university entomology department (usually at the land-grant universities), a military base (the U.S. Army and Navy employ over 100 professional entomologists), or the entomology department of the Smithsonian Institution.

B. Characteristics of Insects (Class Insecta or Hexapoda)

Like all arthropods, insects possess a segmented body and jointed appendages. Beyond that, however, there is tremendous variation: long legs, short legs; four wings, two wings, no wings; biting mouthparts, sucking mouthparts; soft bodies, hard bodies; etc. Compounding identification problems, immature forms of most insects look nothing like the adults. Despite the diversity, adults can at once be recognized as insects by having three pairs of walking legs and three body regions (or tagmata): head, thorax (bearing legs, and wings if present), and abdomen (Figure 8.2). No other arthropods have wings. Some insect groups, however, might never have had wings or may have lost them through adaptation. In particular, several medically important species are wingless (e.g., lice, fleas).

Identification of immature insects presents entirely different problems. In those insects that have simple metamorphosis (grasshoppers, lice, true bugs), the immatures (called nymphs) look like the adults, get larger with each molt, and develop wings (if present) during later molts (Figure 8.3). Identifying nymphs as insects and placing them in their proper orders is generally not a problem; however, in those groups with complete metamorphosis (e.g., beetles, flies,

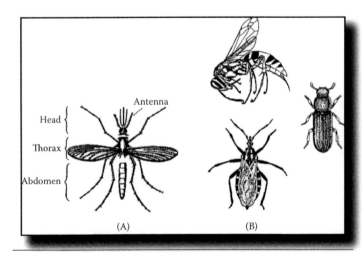

Figure 8.2 (A) Insect with body regions labeled, and (B) several representative insects. (From CDC, *Pictorial Keys to Arthropods, Reptiles, Birds, and Mammals of Public Health Significance*, U.S. Centers for Disease Control and Prevention, Atlanta, GA, 1963.)

bees and wasps, moths and butterflies, fleas), the immature stage, or larva, looks nothing like the adult (Figure 8.4). Often, the larva is wormlike. The three body regions are never as distinct as they are in adults, but generally the three pairs of walking legs are evident, although they are often extremely short. Fly larvae (maggots) lack walking legs. Although some (such as mosquitoes) have three body regions, others (such as house flies, blow flies, etc.) have no differentiated areas (Figure 8.5A). Caterpillars and similar larvae often appear to have legs on some abdominal segments (Figure 8.5D). Close examination of these abdominal legs (prolegs) reveals that they are unsegmented fleshy projections, with or without a series of small hooks (crochets) on the plantar surface and structurally quite unlike the three pairs of segmented walking legs on the first three body segments behind the head.

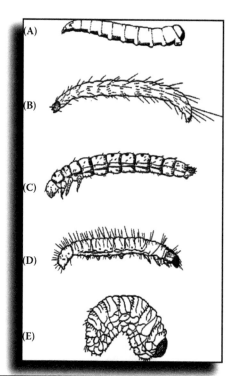

Figure 8.5 Various types of insect larvae: (A) fly, (B) flea, (C) beetle, (D) moth, and (E) another kind of beetle. (From CDC, *Pictorial Keys to Arthropods, Reptiles, Birds, and Mammals of Public Health Significance*, U.S. Centers for Disease Control and Prevention, Atlanta, GA, 1963.)

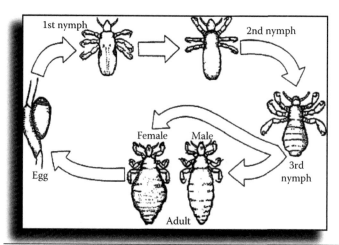

Figure 8.3 Head lice life cycle—example of simple metamorphosis. (Figure courtesy of U.S. Centers for Disease Control and Prevention, Atlanta, GA.)

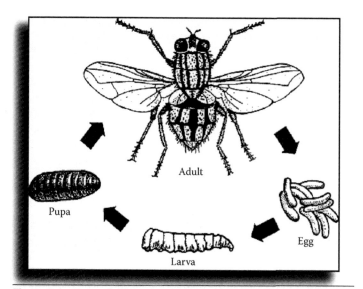

Figure 8.4 House fly life cycle—example of complete metamorphosis. (Figure courtesy of U.S. Centers for Disease Control and Prevention, Atlanta, GA.)

C. Characteristics of Spiders (Class Arachnida, Order Araneae)

A spider has two body regions: an anterior cephalothorax and posterior abdomen connected by a waistlike pedicel (Figure 8.6). The former bears the head and thorax structures, including the four pairs of walking legs, generally eight simple eyes on the anterior dorsal surface, and mouthparts. The mouthparts, called chelicerae, are generally fanglike and are used to inject poison into prey, all spiders being predaceous. Located between the chelicerae and the first pair of walking legs are a pair of short leglike structures called pedipalpi, which are used to hold and manipulate prey. Pedipalpi may be modified into copulatory organs in males. The abdomen is usually unsegmented and bears spinnerets for web production at the posterior end. Immatures look the same as adults, except that they are smaller. Harvestmen, or daddy longlegs (order Opiliones), have many characteristics in common with true spiders; however, they differ in that the abdomen is segmented and is broadly joined to the cephalothorax (not petiolate). Most species have extremely long, slender legs.

D. Characteristics of Mites and Ticks (Class Arachnida, Subclass Acari)

These small arachnids characteristically have only one apparent body region (cephalothorax and abdomen fused), the

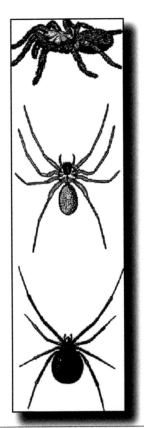

Figure 8.6 Various spiders: top, tarantula; middle, brown recluse; bottom, black widow dorsal view. (From CDC, *Pictorial Keys to Arthropods, Reptiles, Birds, and Mammals of Public Health Significance*, U.S. Centers for Disease Control and Prevention, Atlanta, GA, 1963.)

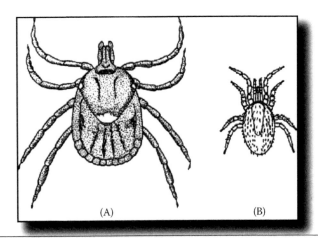

Figure 8.7 (A) Tick and (B) mite. (From CDC, *Pictorial Keys to Arthropods, Reptiles, Birds, and Mammals of Public Health Significance*, U.S. Centers for Disease Control and Prevention, Atlanta, GA, 1963.)

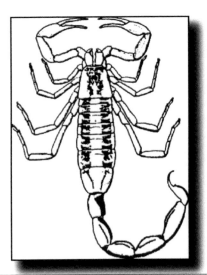

Figure 8.8 A typical scorpion. (From CDC, *Pictorial Keys to Arthropods, Reptiles, Birds, and Mammals of Public Health Significance*, U.S. Centers for Disease Control and Prevention, Atlanta, GA, 1963.)

overall appearance being globose or disk-shaped in most instances (Figure 8.7). This general appearance quickly separates the Acari from other arthropods. The body may be segmented or unsegmented with eight walking legs present in adults. Their larvae, which hatch out of eggs, have only six (or, in rare cases, fewer) legs, but their single body region readily separates them from insects. They attain their fourth pair of legs at the first molt and thereafter are called nymphs until they become adults. As in spiders, immature ticks and mites are generally similar in appearance to adults. In general, ticks are considerably larger than mites. Adult ticks are generally pea-sized; mites are about the size of a grain of sand (often even smaller).

E. Characteristics of Scorpions (Class Arachnida, Order Scorpiònes)

The body of a scorpion is easily separable into an anterior broad, flat area and a posterior tail with a terminal sting (Figure 8.8). Although these outward divisions do not correspond to actual lines of tagmatization, they do provide an appearance sufficient to distinguish these arthropods from most others. Like spiders, the mouthparts are

chelicerae and the first elongate appendages are pedipalpi. The pedipalpi are modified into pinchers and are used in prey capture and manipulation. Four pairs of walking legs are present. Immatures are similar to adults in general body form.

F. Characteristics of Centipedes and Millipedes (Classes Chilopoda and Diplopoda)

Centipedes and millipedes bear little resemblance to the other arthropods previously discussed. They have hardened, elongated wormlike bodies with distinct heads and many pairs of walking legs. Centipedes—often called hundred leggers—are swift-moving predatory organisms with one pair of

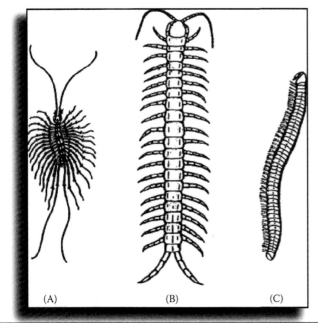

Figure 8.9 (A, B) Centipedes and (C) millipede. (From CDC, *Pictorial Keys to Arthropods, Reptiles, Birds, and Mammals of Public Health Significance*, U.S. Centers for Disease Control and Prevention, Atlanta, GA, 1963.)

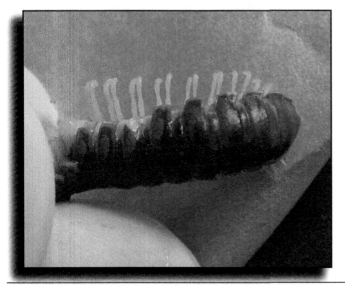

Figure 8.10 Millipede showing two legs per segment.

long legs on each body segment behind the head (Figure 8.8). Millipedes (thousand leggers), on the other hand, are slow-moving omnivores or scavengers that have two pairs of short legs on each body segment (after the first three segments, which only have one pair each; see Figures 8.9 and 8.10). Immatures are similar to the adults.

III. Identification Key to Common Arthropods

As mentioned in Part I, the identification of a particular arthropod to place it in its correct class, order, family, genus, or species is achieved by answering paired series of questions set out in the form of keys. Most keys are dichotomous, giving the reader two choices in each couplet, and the reader is then referred to another couplet depending on the answer. This flowchart continues until an identification is reached. The following chart is an updated and revised Centers for Disease Control and Prevention (CDC) Pictorial Key to the arthropods of medical importance. This chart is provided to enable the nonentomologist reader to identify an arthropod to (at least) its major group. In addition to this chart, morphological characteristics and helpful identification hints for each of the arthropod groups discussed in this book are provided in each of the respective chapters dealing with those groups.

References

1. Herbert PDN, Ratnasingham S, deWaard JR. Barcoding animal life: Cytochrome c oxidase subunit 1 divergences among closely related species. *R. Soc. London Ser. B Suppl.* 2003;270:S96–S99.
2. Will KW, Rubinoff D. Myth of the molecule: DNA barcodes for species cannot replace morphology for identication and classification. *Cladistics.* 2004;20:47–55.
3. Sperling F. Butterfly molecular systematics: From species definitions to higher-level phylogenies. In: Boggs CL, Watt WB, HEhrlich PR, eds. *Butterflies: Ecology and Evolution Taking Flight.* Chicago, IL: University of Chicago Press; 2003:431–458.

KEY TO SOME COMMON CLASSES AND ORDERS OF ARTHROPODS OF MEDICAL IMPORTANCE

1

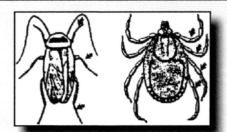

Three or four pairs of legs
see Set 2

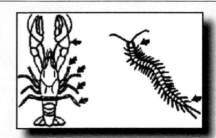

Five or more pairs of legs
see Set 22

2

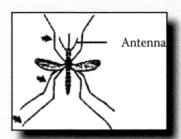

Three pairs of legs, with antennae
Class Insecta
(Insects)
see Set 3

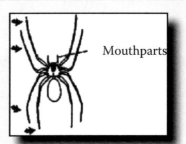

Four pairs of legs, without antennae
Class Arachnida
(Scorpions, Spiders, Ticks, etc.)
see Set 20

3

Wings present, well developed
see Set 4

Wings absent or rudimentary
see Set 12

4

One pair of wings
Order Diptera
see Set 5

Two pairs of wings
see Set 6

**KEY TO SOME COMMON CLASSES AND ORDERS OF
ARTHROPODS OF MEDICAL IMPORTANCE**

5

Wings with scales
Mosquitoes

Wings without scales
Other flies

6

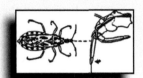

Sucking mouthparts, consisting of an elongated
proboscis
see Set 7

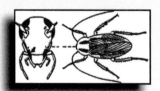

Biting/chewing mouthparts
see Set 8

7

Wings usually totally covered with scales, proboscis coiled up under head
Order Lepidoptera (Butterflies and Moths)

Wings not covered with scales, proboscis directed backward between
front legs when not in use
Order Hemiptera (True Bugs and
Kissing Bugs)

8

Both pairs of wings membranous and similar in
structure, although they may differ in size
see Set 9

Front pair of wings leathery or shell-
like, serving as covers for the membranous hind wings
see Set 10

9

Hind wings much smaller than front wings
Order Hymenoptera (Bees, Wasps, Hornets,
and Ants)

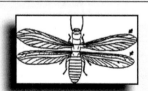

Both pairs of wings similar in size
Order Isoptera (Termites)

**KEY TO SOME COMMON CLASSES AND ORDERS OF
ARTHROPODS OF MEDICAL IMPORTANCE**

10

Front wings somewhat hardened without distinct veins, meeting in a straight line down the middle
see Set 11

Front wings leathery or paperlike with a network of veins, usually overlapping at the middle
Order Orthoptera (Cockroaches, Crickets, Grasshoppers)
(Note: cockroaches now placed in order Blatteria)

11

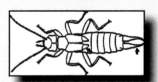

Abdomen with prominent forceps, wings shorter than abdomen
Order Dermaptera (Earwigs)

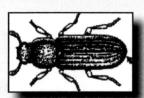

Abdomen without forceps, wings typically covering abdomen
Order Coleoptera (Beetles)

12

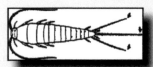

Abdomen with three elongate, tail-like appendages at tip, body usually covered with scales
Order Thysanura (Silverfish and Fire Brats)

Abdomen without three long tail-like appendages at tip, body not covered with scales
see Set 13

13

Abdomen with narrow waist
Order Hymenoptera (Ants)

Abdomen without narrow waist
see Set 14

KEY TO SOME COMMON CLASSES AND ORDERS OF
ARTHROPODS OF MEDICAL IMPORTANCE

14

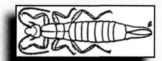

Abdomen with prominent pair of forceps
Order Dermaptera (Earwigs)

Abdomen without forceps
see Set 15

15

Body strongly flattened from side to side,
antennae small, fitting into grooves in side
of head
Order Siphonaptera (Fleas)

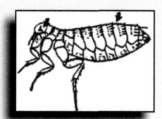

Body not strongly flattened from side
to side, antennae projecting from side
of head, not fitting into grooves
see Set 16

16

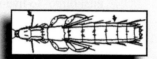

Antenna with nine or more segments
see Set 17

Antenna with three to five segments
see Set 18

17

Pronotum covering head
Order Blatteria (Cockroaches)

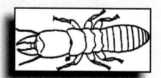

Pronotum not covering head
Order Isoptera (Termites)

KEY TO SOME COMMON CLASSES AND ORDERS OF
ARTHROPODS OF MEDICAL IMPORTANCE

18

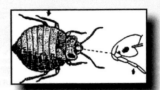

Mouthparts consisting of tubular jointed beak,
tarsi three- to five-segmented
Order Hemiptera (Bed bugs)

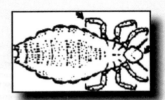

Mouthparts retracted into head or
of the chewing type, tarsi one- or
two-segmented
see Set 19

19

Mouthparts retracted into head, adapted for
sucking blood, external parasites of mammals
Order Phthiraptera (Sucking Lice)

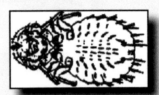

Mouthparts of the chewing type,
external parasites of birds and
mammals
Order Phthiraptera (Chewing Lice)

20

Body oval, consisting of a single saclike region
Subclass Acari (Ticks or Mites)

Body divided into two distinct
regions, a combined head-thorax and
an abdomen
see Set 21

KEY TO SOME COMMON CLASSES AND ORDERS OF
ARTHROPODS OF MEDICAL IMPORTANCE

21

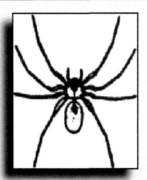

Abdomen broadly joined to cephalothorax, abdomen distinctly segmented, ending with a stinger
Order Scorpiones (Scorpions)

Abdomen joined to head-thorax (cephalothorax) by a slender waist, abdomen with segmen-
tation indistinct or absent, stinger absent
Order Araneae (Spiders)

22

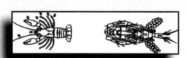

Five to nine pairs of legs in some species, swimmerets in others, one or two pairs of
antennae present, principally aquatic organisms
Subphylum Crustacea (Crabs, Crayfish, Shrimp and Copepods)

23

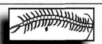

Ten or more pairs of legs, swimmerets absent,
one pair of antennae present, terrestrial organisms, body segments each with only one pairs
of legs
Class Chilopoda (Centipedes)

Ten or more pairs of legs, swimmerets absent, one pair of antennae
present, terrestrial organisms, body
segments each with two pairs of legs
Class Diplopoda (Millipedes)

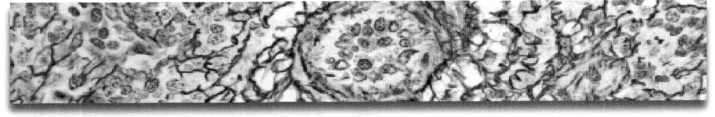

9

CLINICIAN'S GUIDE TO COMMON ARTHROPOD BITES AND STINGS*

I. Introduction

A. Background

The human body may react in various ways to foreign substances such as arthropod venom or saliva. Most often these reactions are manifested as cutaneous lesions. In fact, of the top dermatologic diagnoses in returning travelers, arthropod-related lesions are the third most common.[1] One might think, therefore, that a careful listing of signs and symptoms of arthropod-induced problems or a well-written algorithm concerning lesions would be useful to the practicing clinician in making diagnoses. Unfortunately, these reactions are rarely specific to one particular offending arthropod; however, the presence of certain signs and symptoms may alert physicians to the various diagnostic possibilities. Here we present a set of diagnostic tools and a list of some common bites and reactions. By no means should it be considered definitive or a complete differential diagnosis.

B. Pathophysiological Mechanisms

Arthropod bite or sting reactions can be loosely grouped into four basic categories: mechanical trauma, secondary infection, immunologic (hypersensitivity) reactions, and direct toxic effects.[2] For more detail, see Chapters 1–4 in which bites, stings, and allergic reactions and their underlying mechanisms are discussed. Mechanical trauma may be caused by large or damaging mouthparts, very often from phytophagous arthropods that do not usually bite people. Secondary infection results from people scratching streptococcal or staphylococcal bacteria into bite or sting lesions (Figure 9.1). Allergic

sensitization may develop from a wide variety of stinging arthropods and a few biting ones as well. Lastly, direct toxic effects may occur from excessive numbers of bites or stings, which result from salivary toxins or venom components.

II. Lesions from Bites, Stings, and Other Exposure

Bed bugs produce a mostly painless bite on exposed areas of the head, neck, face, and arms during the night, which is helpful information when performing a history and physical. Not all persons react to bed bug bites; perhaps as many as 50% do not.[3,4] Lesions sometimes occur in a line, but this is not due to a single bug biting and then moving a few millimeters to bite again. Instead, the linear nature of bed bug bites, often referred to as "breakfast, lunch, dinner bites", is due to numerous bugs feeding at the skin interface where they line up to feed. At the bite site, hemorrhagic puncta may be seen. Lesions may be macular, papular (Figure 9.2), nodular, urticarial, bullous, indurated plaques, or eczematous patches depending on the host's immune sensitivity to the bite and the length of exposure to the bug infestation. There is considerable evidence that the time-to-lesion development decreases upon subsequent biting events.[5,6]

Kissing Bugs, or the bloodsucking Triatominae, generally produce a painless bite. They bite uncovered skin so the distribution of lesions may follow the pattern of exposure. Hypersensitivity to kissing bug bites increases with exposure and lesions range from pruritic papules to hemorrhagic nodules or even bullae. Anaphylaxis is rare, but has been reported, especially from bites by *Triatoma protracta* in the

*Julie Porter Wyatt, M.D., Richard De Shazo, M.D., and Jerome Goddard, Ph.D.

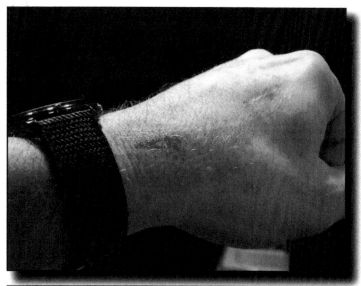

Figure 9.1 Scratching insect bites may lead to secondary infection. Here, a linear erosion with hemorrhagic crusting and surrounding erythema 5 days post insect assault. (Photo courtesy of Jerome Goddard, Ph.D.)

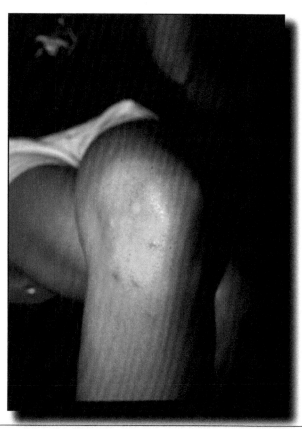

Figure 9.3 Flea bites of various stages: acutely induced dermal papules with surrounding erythema, mixed with darker older, erythematous papules on knee. Dimorphic lesions or crops of lesions exhibit repeated attacks. (Photograph copyright 2001 by Jerome Goddard, Ph.D.)

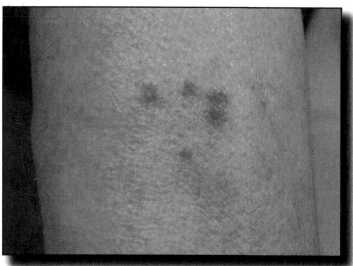

Figure 9.2 Bed bug bites, illustrated as grouped erythematous, dome-shaped papules 6 days post-bite. (Photograph courtesy Dr. Kristine T. Edwards, used with permission.)

western U.S.[7,8] In addition, kissing bugs are medically important arthropods as they can serve as vectors for *Trypanosoma cruzi,* the parasite that causes Chagas' Disease.

Centipedes can bite and cause intense pain with local redness and edema. There may be two puncta visible at the bite site (actually, centipedes bite with modified front legs). Symptoms of centipede bites are usually mild, including headache, malaise, dizziness, and anxiety, with the only real concern being secondary cutaneous infection.[9]

Fleas, specifically cat and dog fleas, can produce bite reactions ranging from papular urticaria to bullae and are often grouped in lines or clusters (Figure 9.3). The distribution of flea bites is helpful as cat/dog fleas usually affect the lower legs, although bites may also show up on the forearms or even be generalized. Physicians should be aware that flea infestations in homes have waned over the past two decades due to "spot-on" and other on-animal flea control products promoted through veterinary clinics. However, in the last few years flea problems (and thus, numbers of human bites) are increasing.

Black flies may bite exposed skin during the day causing intensely itchy and painful urticarial papules or nodules in sensitive individuals (Figure 9.4). The bite initially may leave a hemorrhage and a trickle of blood lasting a few minutes. One recent report followed black fly bites from the moment of attack through 48 hours.[10] Initially, there was erythema and a small hemorrhagic center. Four hours later, lesions developed into a central wheal with surrounding erythema, which was intensely itchy. The diagnosis was papular urticaria, presumably resulting from hypersensitivity to black fly salivary proteins. Over the course of 24 hours, the urticarial lesions resolved, leaving only an erythematous flattened lesion. By 48 hours, the lesions had completely faded.

Deer and horse flies produce painful bites that may lead to indurated papules or nodules (Figure 9.5). These may become secondarily infected with staphylococcal or streptococcal bacteria.[11]

Figure 9.4 Black fly bites at 48 hours post-bite show a central bright red dermal nodule with overlying punctum and surrounding erythema. (Photograph courtesy of Dr. Wendy C. Varnado, used with permission.)

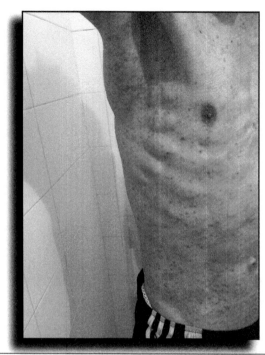

Figure 9.6 Body lice bites. Widespread eruption composed of erythematous, excoriated small papules and tiny urticarial papules with surrounding erythema. (Photograph courtesy of Faccini-Martinez et al. Emerg. Infect. Dis. 2017, 23: 1876–1879, used with permission.)

Figure 9.5 Horse fly bite on face, presenting as an acute edematous red nodule, 12 hours post-bite. (Photograph courtesy of Lindsey C. Goddard.)

Sand flies may cause a rash ranging from papular urticaria to bullae on exposed skin. In tropical countries, sand flies are important vectors of leishmaniasis, an intracellular parasite.

Human lice can bite and cause lesions. There are three types of lice associated with humans—head lice, pubic lice, and body lice. Head lice typically cause pruritus first at the post auricular scalp and then more diffusely on the scalp with a high risk of secondary impetiginization due to autoinoculation from scratching. Pubic lice will affect less densely placed hair in the eyelashes, eyebrows, axillary, and pubic regions. Along with pruritus, small blue macules may develop where the louse feeds on blood vessels—termed maculae ceruleae—and can be a helpful clinical clue. Body lice inhabit unwashed clothing and travel to the skin to feed. Affected persons will have intense diffuse itching with a rash composed of papules (Figure 9.6), eczematous patches, with more chronic exposure, and thickened or darker skin around the waist or groin if the exposure is chronic.[11]

Biting midges or "no-see-ums" are tiny gnats (flies) that usually produce small erythematous papules on exposed skin (Figure 9.7) but also can cause larger wheals, bullae, or nodules.

Biting mites refer to numerous mites that may bite people upon exposure to birds, rodents, or stored food products such as grains, meats, and cheeses. Some notable mites that bite people include the straw itch mite, *Pyemotes ventricosus*;

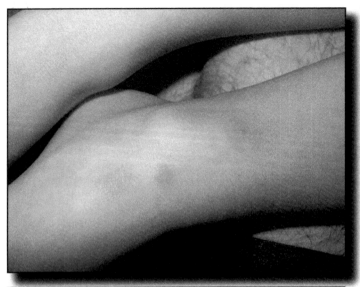

Figure 9.7 Bites by biting midges or no-see-ums. Here, pink dermal papules clustered in a group on exposed skin. (Photograph courtesy of Dr. Vidal Haddad, Jr., used with permission.)

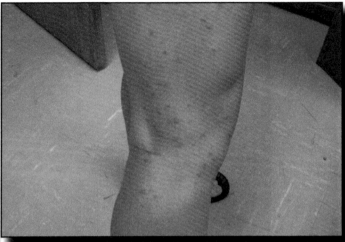

Figure 9.8 Pink edematous papules predominantly affecting the legs 48 hours after a chigger assault. (Photo courtesy Audrey Sheridan, Mississippi State University, used with permission.)

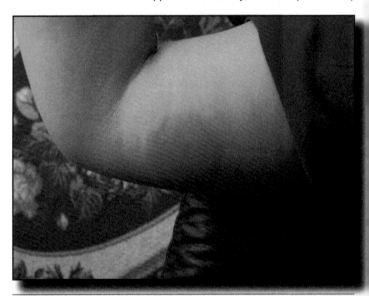

Figure 9.9 Erythematous and edematous round plaque resulting from mosquito bite 48 hours earlier. (Photograph copyright 2007 by Jerome Goddard, Ph.D.)

the stored product mite, *Acarus siro*; the chicken mite, *Dermanyssus gallinae*; the tropical rat mite, *Ornithonyssus bacoti*; and the tropical fowl mite, *Ornithonyssus bursa*. The variability of mite types and their varied exposures require clinicians to be on alert for cryptic mite induced eruptions. Despite the claims made on some websites, bird mites and rat mites will not make permanent residency on humans. The infestation is self-limiting after the source of exposure is eliminated. Another mite which sometimes bites humans upon exposure to pets is *Cheyletiella*, which may be found on dogs, cats, and rabbits. Humans are not a host but may develop a rash from bites when skin contact is made. Typically, the rash is seen on upper thighs, abdomen, and arms where a pet is held, and consists of papulovesicles with secondary excoriation due to intense pruritus.

Chigger mites or red bugs are a type of mite that often infests humans who are walking or working in grassy or brushy areas during the spring and summer months. It is the larval (1st) stage of chiggers that bite people; adults do not. They prefer thin skin like that on the feet, ankles, and genital/groin area as well as areas where there is tight-fitting clothing such as the waistline due to their weak attachment mechanism. These areas allow them to easily penetrate skin and push against articles of clothing to maintain their attachment. They do not burrow into the skin and therefore are easily washed or scratched off. However, even after being detached from the skin, there may be lesion development and intense itching for 2 weeks or more. The primary lesion from chiggers is an orange-red firm papule (Figure 9.8) or papulovesicle.

Mosquitoes can produce a variety of cutaneous reactions, depending on the host's immune response. Often, children have exaggerated responses, which lessen with age.[11] Anaphylaxis from mosquito bites is rare but serum

sickness has been reported. Severe late cutaneous reactions are not uncommon and include bullae, cellulitis-like plaques (Figures 9.9 and 9.10), and eczematization. The bite response may be exaggerated in the setting of other immunopathies such as chronic lymphocytic leukemia and infection with Human Immunodeficiency Virus.[11]

Spiders release their venom through a bite that initially may be painful or painless depending on the species. Puncta are not always evident at the bite site, with pain and well-demarcated erythema developing over subsequent minutes. Their venom is toxic to sodium channels leading to the over-release of neurotransmitters. Particularly dangerous spider species are the Australian funnel web spider, the Brazilian wandering spider, and the Australian red-backed spider.[9,12] Affected individuals may develop systemic symptoms consisting of

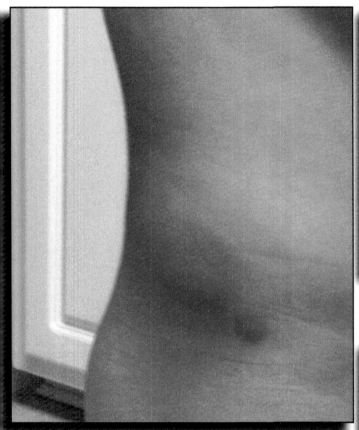

Figure 9.10   Two adjacent edematous red nodules with surrounding erythema on the hip in a young child following mosquito bites. (Photograph courtesy Dr. Wendy C. Varnado, used with permission.)

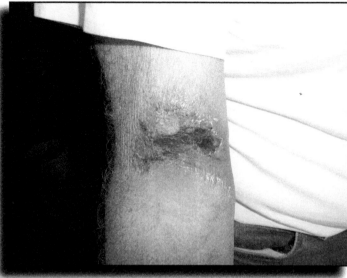

Figure 9.11   Large erythematous, exfoliative plaque with central eschar due to brown recluse spider bite, 17 days after bite. (Photo courtesy of Eugene Skiles, used with permission.)

Figure 9.12   Single indurated red nodule on the trunk, approximately 72 hours after a tick bite. (Photo by Jerome Goddard, Ph.D.)

hypertension, nausea and vomiting, priapism, cardiac arrhythmias, pulmonary edema, shock, and even death. Young children and elderly adults have the highest mortality. Wolf spiders and tarantulas are less toxic and therefore the pain and inflammation are localized. Tarantulas can cause a physical urticaria when they release (actually throw) abdominal hairs in self-defense. This condition appears as a papular dermatitis.[11]

Two other specific forms of arachnidism worth noting are lactrodectism and loxoscelism. Lactrodectism, caused by black widow spiders and related species, presents as increasing pain at the site of evenomation followed by crampy abdominal pain, sweating, incoordination, and paralysis.[11] There is usually no lesion at the bite site except perhaps two small puncta. As for loxoscelism, there are two main physical manifestations from fiddleback or brown recluse spider bites. One is the necrotic cutaneous type consisting of a violaceous plaque with an associated blister, surrounding pallor due to ischemia, and lastly, surrounding erythema, all forming what is often called the "red, white, and blue sign."[11] This progresses to full-thickness necrosis and ulceration of the skin with fatty areas such as the proximal thigh and buttock becoming more widely involved (Figure 9.11). Note: these lesions need to be differentiated from methicillin-resistant Staphylococcus aureus infections as localized tissue damage

is prominent as well. The second, more unusual, reaction from brown recluse spiders, called viscerocutaneous, is characterized by systemic symptoms of fever, headache, and restlessness followed by ecchymoses, jaundice, and hematuria indicating massive intravascular hemorrhage.[11]

Ticks use their mouthparts to cut through the epidermis, then penetrate further producing mixed, deep inflammation. Lesions can occur anywhere on the body and may vary from pruritic papules to more chronic nodules (Figure 9.12). Sometimes, a persistent nodule will develop at the bite site lasting 6–12 months and histologically may appear as a granuloma or lymphocytoma. Tick bites gain much attention because they may lead to a variety of diseases. For example, in boutonneuse fever, an eschar forms at the tick bite, called *tache noire,* where the causative bacterium, *Rickettsia conorii,* is injected into the human. This may be followed

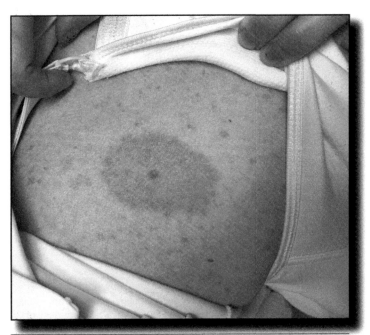

Figure 9.13 Annular erythematous plaque with central hemorrhagic macule after a tick bite in the southern U.S., usually occurring 1–4 days after the bite. Often classified by the Centers for Disease Control as southern tick-associated rash illness or STARI. (Photo courtesy Dr. Patrick Carrington, Shreveport, LA, used with permission.)

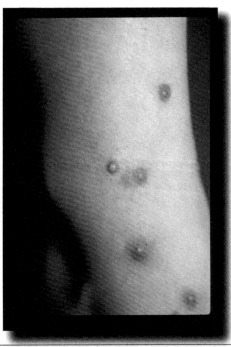

Figure 9.14 Numerous grouped erythematous papulopustules on the ankle due to fire ant stings, less than 24 hours post-assault. (Photo by Jerome Goddard, Ph.D.)

by a widespread non-pruritic maculopapular eruption and generalized malaise, fever, headache, abdominal pain, and myalgias.[11] There are many other systemic diseases caused by ticks, including Rocky Mountain spotted fever, tularemia, erlichiosis, Lyme disease, southern tick-associated rash illness or STARI (a Lyme-like eruption) (Figure 9.13), and babesiosis. Additionally, tick paralysis is an ascending flaccid paralysis caused by a neurotoxin injected by a feeding tick. When removing a tick (see Chapter 33), slow, gentle traction works best so that tick mouthparts are not left in the skin as this could lead to continued local inflammation. Retained ticks or tick parts can masquerade as a "melanoma" or cause prolonged localized inflammation.

Fire ants produce stings that are quite characteristic and whose clinical signs coincide well with the attack. In all individuals stung with adequate amounts of ant venom, the sting may be marked by a small puncta followed almost immediately by a local wheal and flare (local hive) reaction resulting from vasoactive amines in the venom. This is followed by development of a papule which becomes a sterile pustule within 24 hours. In some individuals, the pustule becomes surrounded by a large, erythematous, painful, and pruritic reaction called a late-phase allergic reaction. This lesion may persist for days. Fire ant venom contains a number of glycoprotein allergens that sensitize the majority of individuals who are stung, although only a small number of those individuals end up having systemic allergic reactions, some of which may be life-threatening. Adults who develop generalized urticaria, and all individuals who develop angioedema, laryngospasm, mental status changes, hypotension,

or other life-threatening symptoms of anaphylaxis, should be given intramuscular epinephrine immediately and referred to an allergist for immunotherapy which is very effective.[13] Stings and bites from other arthropods which are associated with these symptoms should be managed in a similar fashion (Figure 9.14).

Harvester ants sting in a similar method to fire ants with envenomation causing intense pain and a varying allergic reaction. Clinically, one may see grouped macules, papules, or vesiculobullous lesions depending on the victim's hypersensitivity to the venom.[11]

Caterpillars cause lesions through three mechanisms: mechanical irritation, pharmacologic evenomation (direct toxic effect), and hypersensitivity. Erucism is a term referring to the hairs or spines of the larvae or pupae affecting human skin. These hairs or spines may also be released into the air and dirt, as well as incorporated into cocoons, leading to respiratory problems.[14] Skin exposed to urticating caterpillars may develop a papular dermatitis that burns or stings and is usually found on an uncovered area of the body (Figure 9.15). The puss caterpillar and others are known to cause a grid-like linear hemorrhagic plaque when their hairs contact human skin.[14] For diagnostic and therapeutic purposes, one may use clear or cellophane tape to strip the skin and remove the hairs. Also of note, lonomism, caused by brightly colored *Lonomia* caterpillars found in Brazil and Venezuela, is characterized by local hypersensitivity that often advances to a severe systemic hemorrhagic syndrome.[14]

Scorpions sting with the tip of their tail, injecting venom mostly comprised of neurotoxins. Pain, swelling, and

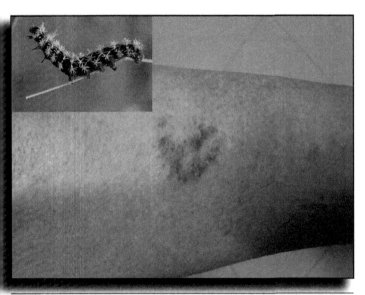

Figure 9.15 Urticating caterpillar sting presenting as linear hemorrhagic plaque. (Photograph courtesy of Dr. Vidal Haddad, Jr., used with permission.)

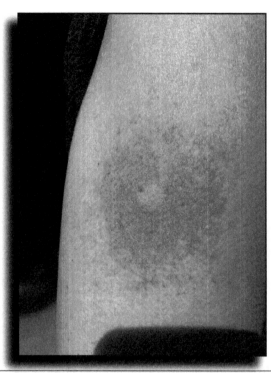

Figure 9.17 Honey bee sting lesion, 30 minutes post-sting, presenting as an edematous papule with surrounding deep red petechial plaque. (Photograph courtesy of Dr. Wendy C. Varnado, used with permission.)

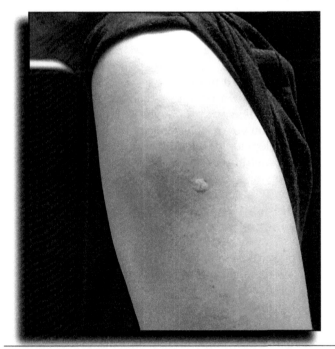

Figure 9.16 Initial honey bee sting lesion, 5 minutes post-sting, presenting as an edematous papule with surrounding erythematous halo. (Photograph courtesy of Dr. Wendy C. Varnado, used with permission.)

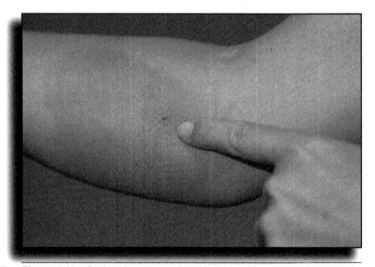

Figure 9.18 Dark red puncture site with large edematous red surrounding plaque due to a wasp sting 4 hours earlier. (Photograph copyright 2012 by Jerome Goddard, Ph.D., and courtesy of Audrey Sheridan.)

hyperesthesia occur immediately. Systemic poisoning presents as restlessness, sweating, muscle spasm, increased secretions, difficulty speaking, hypertension, cardiac arrhythmias, and pulmonary edema. If death occurs, it is usually seen in infants and young children due to respiratory or cardiac failure.[9]

Wasps (paper wasps, yellow jackets, hornets), honeybees, and bumblebees all inject venom through a sting apparatus, which may or may not be barbed. The honeybee has a barbed stinger; therefore, when it stings, the apparatus becomes lodged in the skin and avulses from the abdomen of the bee. The apparatus continues to release venom while embedded, so quick removal of the stinger is important. Honeybee stings often produce wheal and flare reactions (Figures 9.16 and 9.17). Wasps and bumblebees have non-barbed stingers and can therefore sting many times. These arthropods may induce hypersensitivity via an IgE-mediated immune response. Reactions may vary from a small puncture site with surrounding erythema and swelling (Figure 9.18), to

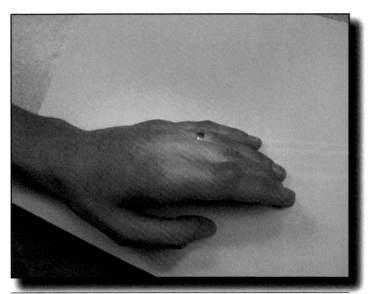

Figure 9.19 Deep broad edema with overlying erythema extending to periarticular tissue secondary to wasp sting 12 hours earlier. (Photograph courtesy of Dr. Wendy C. Varnado, used with permission.)

large local reactions (Figure 9.19) that can develop over hours, or anaphylaxis that occurs within minutes of the sting. There are several delayed reactions including generalized urticaria or a serum sickness-like reaction comprised of urticaria, joint swelling, and arthralgia.[11]

Blister beetles do not bite or sting, but contain vessicating fluid in their hemolymph which is released when handled or crushed. The toxins, pederin or cantharidin, depending on the species, cause exfoliation and a superficial epidermal blister in the skin often in a linear shape denoting external contact with the insect body.[9,15] The lesion may be associated with burning and pruritus, and usually evolves with crusting over 7 days, leaving a macular post-inflammatory erythema. Interestingly, health care providers utilize an extract from blister beetles called cantharidin. Application of cantharidin allows controlled damage of skin infected with viruses such as molluscum contagiousum or human papilloma virus.

III. Diagnostic Aids for Arthropod-Caused Problems

A. Questions to Ask the Patient

A thorough history may be very informative as to causes of mysterious bites/stings/lesions. Some helpful lines of questioning are as follows:

1. Any vacations/trips lately? **Consider arthropods from where patient traveled**
2. Where do you work? Nature of work? **Occupational exposure to arthropods or irritating fibers such as fiberglass**
3. Any new construction inside work or home? **Fibers from such activities may cause irritation**

4. What time of year did lesions occur? **Spring and summer—arthropods likely**
5. Do you know precisely when the sting/bite occurred?
   a. **Calculate the time since sting/bite—helpful to know if infectious agent involved**
6. Where were you and what were you doing at the time of the sting/bite?
   a. Outdoors working or recreating, etc.? **Mosquitoes, other biting flies, fire ants, other stinging ants, ticks, chiggers**
   b. Indoors? **Fleas, mites, bed bugs**
   c. Asleep? **Bed bugs, mosquitoes**
7. Did you see the offending insect/arthropod?
   a. Was it wormlike? **Centipedes, caterpillars**
   b. Did it fly? **Mosquitoes, other biting flies, bees, wasps, hornets**
   c. Was it beetle-like? **Blister beetles**
   d. Was it very small (pinhead size or smaller)? **Mites, immature ticks, immature bed bugs**
   e. Was it firmly attached? **Ticks, pubic lice**
   f. Did it jump greatly? **Fleas**
8. If you did not see the offending insect/arthropod, then...
   a. What time of day did the lesion occur? **If night, mosquitoes, especially if outdoors or unscreened houses**
   b. Do you own pets (especially ones that go outside)? **Fleas, ticks, cheyletid mites**
   c. Did you go hiking, fishing, hunting prior to lesion development? **Ticks, mosquitoes, other biting flies, chiggers**
   d. Did you work in the yard or garden or do other outdoor activities prior to lesion development? **Biting flies, mosquitoes, fire ants, other stinging ants**
   e. Were you cleaning out an attic, garage, or shed prior to lesion development? **Spiders, scorpions**
   f. Are there bats or bird nests in your attic or under the eaves of your windows? **Mites, soft ticks, bat bugs**
   g. Have you stayed in hotels or hostels lately? **Bed bugs**

B. Guidance for the Clinician

The number and location of skin lesions may provide clues for determining the cause of mysterious bites/stings/lesions. Tables 9.1 and 9.2 provide information on arthropod bite possibilities based upon these two factors. In addition to the number of lesions and their location, the color, distribution, or pattern of the bites may also be diagnostic (Table 9.3).[16] For example, expanding circular lesions resulting from arthropod bites are more common than one might think (Figure 9.20). They can be indicative of Lyme disease, but also of a number of other entities (Table 9.4). Sometimes

Table 9.1 Clues to Biting Arthropods Based on Number of Lesions on the Body

Number of Lesions	Possible Arthropods
Single	Tick
	Spider
	Centipede
	Wheel bug
	Kissing bug
Few	Fleas
	Mosquitoes
	Stable flies
	Horse flies and deer flies
	Kissing bugs
	Sand flies
Multiple	Bed bugs
	Mosquitoes
	Black flies
	Biting midges (*Culicoides*)
	Fleas
	Lice
	Chiggers
	Seed ticks
	Mites
	Scabies

Table 9.2 Clues to Biting Arthropods Based on Location of Lesions on the Body

Location of Lesions	Possible Arthropods
Predominantly on left side of body (if right-handed) or right side of body (if left-handed)	Imaginary bugs (see Chapter 7)
Legs or feet	Fleas
	Mosquitoes
	Spiders
	Chiggers
	Centipedes
Trunk	Chiggers
	Bed bugs
	Scabies
	Ticks
	Body lice
	Pubic lice
	Spiders
	Cheyletiella mites
Genitals[a]	Scabies
	Chiggers
Arms or hands	Mosquitoes
	Black flies
	Mites
	Biting midges
	Spiders
	Fleas
	Cheyletiella mites
	Centipedes
	Sand flies
	Wheel bugs
Head, neck, or face	Bed bugs[b]
	Mosquitoes
	Black flies
	Biting midges
	Sand flies
	Head lice
	Kissing bugs

[a] The presence of crusted, pruritic papules on the penis and buttocks is highly indicative of scabies.
[b] Mysterious bites on the face, neck, or upper trunk noticed after arising from sleep are highly indicative of bed bugs.

arthropod bites may be purple (purpura) (Figure 9.21). Residual effects of arthropod bites that occurred days or weeks earlier may also be classified and used as useful hints for diagnosis (Table 9.5).

IV. Treatment of Arthropod Bites and Stings

A. Local Reactions

The mainstay of therapy for local bites and stings is keeping the area clean and applying topical antibiotics to prevent secondary infection (bacitracin or mupirocin are often used). The patient should seek immediate medical attention if progressive tenderness, erythema, linear streaking, rigor, eschar, or lymphadenopathy develop, particularly at days 5–7 after the initial bite or sting. It is also important to diminish inflammation, which can be done in several ways. This should begin with the most conservative approach—RICE therapy, which stands for Rest, Ice, Compression, and Elevation.[11] Elevation allows gravity to drain cutaneous fluid, while the cold temperature and external compression induce vasoconstriction, keeping toxins and hypersensitivity fluids from arthropods local in addition to minimizing swelling. The next over-the-counter option is oral antihistamines to combat itching and swelling. Topical antihistamines have little efficacy and have increased sensitizing potential so they should be avoided. Potent to super-potent topical corticosteroids are quite effective anti-inflammatory agents and can be used immediately after the insult, as well as on an ongoing basis. Their potency can be increased by occlusion with a bandage. In envenomation where pain is quite impressive, such as scorpion stings or black widow bites, analgesics and topical or injectable anesthetics can be quite helpful.

B. Systemic Reactions

As discussed above, the first line of therapy for systemic reactions to insect bites or stings is appropriate doses of intramuscular epinephrine, most frequently given with either an adult or pediatric autoinjector as appropriate (see Chapters 1 and 2 and key medical literature[13] for more detail). A delay

Table 9.3 Diagnostic Patterns of Arthropod Bites

Pattern of Bite Lesions	Possible Arthropods
Scattered	Mosquitoes
	Horse flies and deer flies
	Black flies
	Biting midges
	Head and body lice
Grouped	Fleas
	Pubic lice
	Chiggers
	Scabies
Linear	Bed bugs
	Chiggers

Source: Adapted in part from Frazier, C.A., Insect Allergy: Allergic Reactions to Bites of Insects and Other Arthropods, Warren H. Green, St. Louis, MO, 1969, chap. 9. With permission.

in epinephrine administration to determine "if the reaction is severe enough to require epinephrine" would be imprudent as the progression of these reactions is unpredictable. Antihistamines may also be given, but are second line treatment as they do not dependably stop progression from mild to severe systemic reactions. Depending on the severity of the reaction, oral steroids may need to be utilized particularly when it is deemed an allergic reaction to envenomation rather than a direct toxicity effect. Allergen immunotherapy is available for most venoms and is highly protective. All individuals who present with systemic allergic reactions should be referred to an allergist–immunologist for patient education and consideration for immunotherapy.

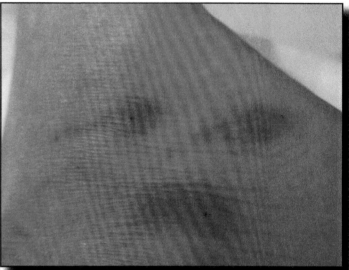

Figure 9.21 Grouped individual dark red puncta surrounded by purpuric plaques resulting from insect bites. (Photograph courtesy of Brook Burton, used with permission.)

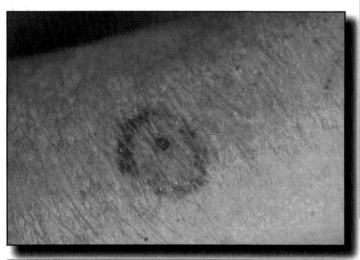

Figure 9.20 Targetoid hemorrhagic plaque resulting from an insect bite. (Photograph courtesy of Dr. Frank Davis, used with permission.)

Table 9.4 Differential Diagnosis of Erythema Migrans-like Lesions

Condition	Description
Tick-bite erythema migrans lesion from Lyme disease endemic areas	Expanding, often flat, annular lesion with or without central clearing 5–30 cm in diameter; persists 3–4 weeks; tick rarely still attached.
Multiple erythema migrans lesions from Lyme disease endemic areas	Disseminated Lyme disease may result in multiple EM lesions, even far removed from the site of tick bite.
Tick-bite erythema migrans-like lesion, from areas non-endemic for Lyme disease	Often southern tick-associated rash illness (STARI); expanding, often raised and vesicular annular lesion with or without central clearing; pruritic; tick often still attached; persists 1–2 weeks
Other hypersensitivity reaction to tick bite	Small lesion, non-expanding; present at time of tick bite or soon thereafter
Tinea (ringworm)	Pruritic rash with raised margins and scale on edges
Insect bite	Pruritic raised papule or nodule; non-expanding
Spider bite	Necrotic lesion; often asymmetrical
Erythema multiforme	Multiple small lesions, often on the mucosa, palms, and soles

Source: Adapted from Shapiro, *N. Engl. J. Med.* 370(18): 1724–1731, 2014.

Table 9.5 Cutaneous Sequelae Resulting from Arthropod Bites

Sequelae	Location on Body	Possible Arthropods
Single nodule	Scalp or trunk	Tick
Multiple nodules	Legs or ankles	Black fly
Bullae	Anywhere	Bed bug
Purpuric spots	Trunk	Bed bug
Bluish spots	Pubic and perianal area	Pubic lice
Hyperpigmentation	Waistline and genitals	Chigger
Hyperpigmentation	Trunk, arms, legs	Bed bug, body lice

Source: Adapted in part from Frazier, C.A., Insect Allergy: Allergic Reactions to Bites of Insects and Other Arthropods, Warren H. Green, St. Louis, MO, 1969, chap. 9. With permission

References

1. Manhorter SD, Longworth DL. Cutaneous lesions. In: Guerrant RL, Walker DH, Weller PF, eds. *Tropical Infectious Diseases.* 3rd ed. Philadelphia, PA: Saunders Elsevier; 2011.
2. Gordon RM. Reactions produced by arthropods directly injurious to the skin of man. *Br. J. Med.* 1950;2:316–318.
3. Goddard J. Cutaneous reactions to bed bug bites. *SkinMed.* 2014;12(3):141–143.
4. Goddard J, de Shazo RD. Bed bugs (*Cimex lectularius*) and clinical consequences of their bites. *J. Am. Med. Assoc.* 2009;301:1358–1366.
5. Reinhardt K, Kempke RA, Naylor RA, Siva-Jothy MT. Sensitivity to bites by the bedbug, *Cimex lectularius. Med. Vet. Entomol.* 2009;23:163–166.
6. Sheele JM. Antibody and cytokine levels in humans fed on by the common bed bug, *Cimex lectularius. Parasite Immunol.* 2017;doi:10.1111/pim.12411.
7. Marshall NA, Liebhaber M, Dyer Z, Saxon A. The prevalence of allergic sensitization to *Triatoma protracta* in a southern California community. *J Med Entomol.* 1986;23:117–124.
8. Moffitt JE, Venarske D, Goddard J, Yates AB, deShazo RD. Allergic reactions to *Triatoma* bites. *Ann. Allergy Asthma Immunol.* 2003;91(2):122–128; quiz 128–130, 194.
9. Haddad V, Jr., de Amorim PCH, Haddad WT, Jr., Cardoso JLC. Venomous and poisonous arthropods: Identification, clinical manifestations of envenomation, and treatments used in human injuries. *Rev. Soc. Brasil Med. Trop.* 2015;48:650–657.
10. Goddard J, Stewart PH, Deerman JH, Nations TM, Varnado WC. Development and resolution of cutaneous lesions caused by black flies. *Am. J. Med.* 2018;doi:10.1016/j.amjmed.2018.06.004.
11. Burns T, Breathnach S, Cox N, Griffiths C. *Rook's Textbook of Dermatology.* 8th ed. London: Wiley Blackwell; 2004.
12. Diaz JH. The global epidemiology, syndromic classification, management, and prevention of spider bites. *Am. J. Trop. Med. Hyg.* 2004;71(2):239–250.
13. Golden DB, Demain J, Freeman T, et al. Stinging insect hypersensitivity: A practice parameter update 2016. *Ann. Allergy Asthma Immunol.* 2017;118(1):28–54.
14. Hossler EW. Caterpillars and moths, Part II. dermatologic manifestations of encounters with Lepidoptera. *J. Am. Acad. Dermatol.* 2010;62:13–28.
15. Alexander JO. *Arthropods and Human Skin.* Berlin: Springer-Verlag; 1984.
16. Frazier CA. Diagnosis of bites and stings. *Cutis.* 1968;4:845–849.

Part III

Arthropods of Medical Importance

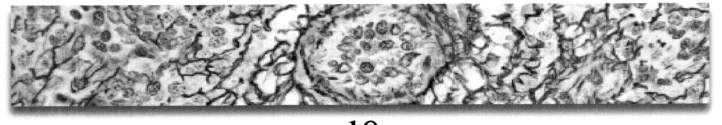

10

ANTS

I. Ants in General

Ants (family Formicidae) are very successful organisms occurring in tremendous numbers worldwide in terrestrial habitats. They are eusocial insects comprising three castes: queens, males, and workers. Their highly developed social structure makes them an interesting and much-studied group. Two excellent sources of information on the biology and taxonomy of ants are Hölldobler and Wilson[1] and Bolton.[2] All ant species may bite (if they are physically large enough), and some species sting. Ant stings can be significant. The sting of the giant tropical bullet ant, *Paraponera clavata* (Figure 10.1), is the most painful and debilitating of any known insect.[3,4] Indigenous people say it takes 2 weeks to recover from one sting. In addition, some ant species emit a foul-smelling substance when handled. Imported fire ants, harvester ants, and velvet ants (not actually ants) are discussed in this chapter. Many other species may be of public health importance; a few of them are listed in Table 10.1.

II. Fire Ants

A. General and Medical Importance

There are some native fire ants in the United States, but the imported ones, *Solenopsis invicta* and *S. richteri*, are among the worst pests.[5] (An invasive European fire ant is now causing problems in the northeastern United States; see box.) Imported fire ants (IFAs) sting aggressively and inject a necrotizing venom consisting primarily of alkaloidal compounds, which they use to paralyze or kill their prey. The ants characteristically boil out of their mounds in great numbers at the slightest disturbance (Figure 10.2). Worker IFAs attach to the skin of their victim with their mandibles and lower the tip of their abdomen to inject the stinger forcefully (Figure 10.3);

therefore, IFAs both bite and sting, but their stings cause the subsequent burning sensation and wheal. Alexander[6] reported that the characteristic symptom of an IFA sting is a burning itch. Within 24 hours, a pseudopustule develops that may persist for a week or longer.[7] Individuals have sustained up to 10,000 stings without systemic toxic or immunological reactions;[8] however, hypersensitivity to IFA venom may result in severe allergic reactions from just a few stings (see Chapter 2), and the elderly or infirm in nursing homes may die from fire ant attacks.[9] In addition, some evidence exists that fire ant venom may act as a neurotoxin. One 4-year-old boy in good health had two grand mal seizures 30 minutes after being stung on the foot by 20 IFAs.[10]

B. General Description

The first, and sometimes second, segment of the abdomen of all ants is nodelike. Adult worker fire ants (see box) have two nodes, are approximately 4–6 mm long, and are reddish-brown (red form, the most widely distributed form) or brownish black (black form). Outdoors, fire ants are best recognized by the appearance of their mounds, which are elevated earthen mounds 8–90 cm high surrounded by relatively undisturbed vegetation (Figure 10.4).

C. Geographic Distribution

Both IFA species were accidentally introduced into the United States (in the Mobile, Alabama, area) in the period from 1918 to 1940. The red form, *Solenopsis invicta*, was brought in from Brazil, and the black form, *S. richteri*, came from Uruguay. The ants may have originally been imported in infested nursery stock or dirt used for ship ballast. Through subsequent unintentional dissemination by people and through dispersal by flooding and mating flights, the red form has spread rapidly, and its range now extends from the southern Atlantic coast westward to California. At least 250 million acres in the United States are currently infested. The black form currently occurs only in northeastern Mississippi, northwestern Alabama, and a small portion of southern Tennessee. States currently reporting IFA infestations are Arizona, Alabama, Arkansas, California, Florida, Georgia, Louisiana, Mississippi, New Mexico, North Carolina, Oklahoma, South Carolina, Tennessee, Texas, and Virginia (Figure 10.5).

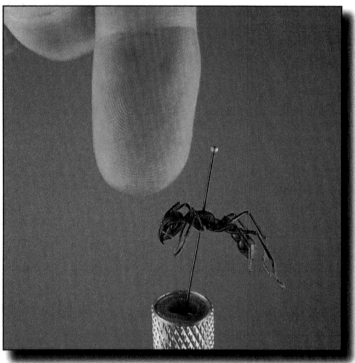

Figure 10.1   Bullet ant. (Photograph copyright 2011 by Jerome Goddard, Ph.D.)

D. Biology and Behavior

An IFA colony is usually started by a single fertilized queen with an initial burrow; the first eggs are laid from 24 to 48 hours after completion of the burrow. Normal development from egg to worker takes approximately 24–30 days, depending on temperature and available soil moisture. These first workers open the nest and begin to deepen the tunnel. After about 60 days, a few minor workers appear, and at about 5 months a few major workers can be found. The nest is further excavated and the characteristic mound begins to form during the 5-month period after initial colony formation. After 1 year, individuals in the nest may number 10,000 or more, and the colony typically increases from about 30,000 workers after 1 to 1½ years to around 60,000 workers after 2 to 2½ years. Major workers are 3–5 mm long and live 90–180 days; queens are a bit larger, live 2–5 years, and can lay 1,500 eggs a day. A colony is considered to be mature after 3 years and may contain as many as 200,000 workers or more in some areas. A single mature colony can produce 4,500 potential new queens during a year. Some heavily infested areas may contain 50–400 or more mature colonies per acre. Reproduction in the red IFA may occur throughout the year but peaks from late May through August. During nuptial flights, newly inseminated queens select a landing site suitable for colony formation, and burrow excavation is begun shortly thereafter.

Fire ants are omnivorous in their feeding habits but prefer insects, spiders, other small arthropods, and earthworms over plant feeding. In heavily infested areas, where food is in short supply, fire ants will eat the germplasm of newly germinated seeds, girdle the stems of small seedlings, and consume hatchling vertebrates. They will freely come indoors looking for food or water. Fire ant mounds occurring against nursing home building foundations are especially a risk factor

Table 10.1   Some Ants That Are Known to Sting People

Scientific Name	Common Name	Occurs Where
Solenopsis invicta	Red imported fire ant	South America, southern United States
S. richteri	Black imported fire ant	South America, southern United States
S. xyloni	Southern fire ant	California to South Carolina down into parts of Florida
S. geminata	Fire ant	Texas to South Carolina down to South America
Pogonomyrmex barbatus	Red harvester ant	Western United States
P. californicus	California harvester ant	Western United States
P. occidentalis	Western harvester ant	Western United States
P. badius	—	Eastern United States
Myrmica rubra	—	Europe (now North America)
M. ruginodes	—	United Kingdom
Pseudomyrmex spp.	Twig ants	North America
Paraponera clavata	Bullet ant; viente-cuatro hora hormiga	South America
Monomorium bicolor	—	Africa
Myrmecia gulosa	Red bulldog ant	Australia
M. pyriformis	Bulldog ant	Australia
M. forficata	—	Australia

IMPORTED FIRE ANTS

Imported fire ant. (Figure courtesy of Mississippi State Department of Health.)

Figure 10.3 Fire ants bite in order to gain leverage for the act of stinging. (From USDA, Insects, in The Yearbook of Agriculture, U.S. Department of Agriculture, Washington, DC, 1952.)

IMPORTANCE

Painful stings; allergic reactions

DISTRIBUTION

Imported to much of southern United States; native to South America

LESION

Immediate—Erythema and central wheal
Later—Pustules

DISEASE TRANSMISSION

None

KEY REFERENCES

Lofgren, C.S., Banks, W.A., and Glancey, B.M., Biology and control of imported fire ants, *Annu. Rev. Entomol.*, 20, 1–30, 1975

AntWeb, http://www.antweb.org/

Figure 10.4 Huge fire ant mound.

TREATMENT

Local—Pain relievers; antipruritic lotions
Systemic—May require antihistamines, epinephrine, and other supportive measures

Figure 10.2 Hand in fire ant mound for 10 seconds.

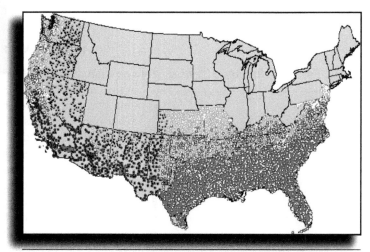

Figure 10.5 USDA distribution of the imported fire ant. Red is certain; green is possible; blue is uncertain. (http://www.ars.usda.gov/research/docs.htm?docid=9165).

in human stinging incidents (Figure 10.6). Foraging tunnels 15–25 m long originating from mounds are used by workers to collect food for the colony. These tunnels are excavated at or just below the soil surface and extend outward from the mound in all directions (Figure 10.7). Foraging workers travel through these tunnels, emerge from an opening, and search for a food source. Once a food source is located, the foraging worker returns to the tunnel, laying a trail of pheromone for other worker ants to follow. If a person steps on, or even stands near a fire ant mound or feeding trail, they are likely to receive multiple stings.

E. Treatment of Stings

Fire ant stings are usually treated in a manner similar to that for wasp or bee stings; however, the pustules characteristically produced by IFA stings may need to be covered with bandages to avoid excoriation. Ice packs, antibiotic ointments, or hydrocortisone creams are often used for local reactions, and epinephrine, antihistamines, and other supportive measures are indicated for more serious, systemic reactions (see Chapter 2).

Figure 10.6 Fire ant mounds next to healthcare facility foundations are especially troublesome due to the propensity of the ants to get inside.

Figure 10.7 Fire ants make feeding tunnels extending outward from their mound in all directions. Arrows indicate mounds (nests).

EUROPEAN FIRE ANT

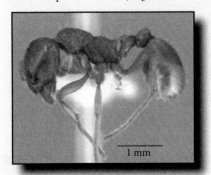

The European fire ant, *Myrmica rubra* (see figure), is a Palearctic species introduced into the northeastern United States in the early 1900s. It has been a nuisance pest along coastal Maine for quite some time but is now spreading west and south where recently it has been found in many northeastern states and as far south as Washington, DC. The ants typically make their nests under woody debris or leaf litter that can retain moisture needed for survival. Colonies may contain over 20,000 workers and 600 queens (polygynous colonies), and nest densities in an area can reach $4/m^2$. The European fire ant is a health concern due to its aggressiveness and painful sting. Numbers of complaints generated from human stinging events in Maine rose dramatically from 1993 through 2003.[1] Human health impacts from this invasive ant are likely to increase in number and spread geographically over time.

1 mm

European fire ant. (Photograph courtesy of AntWeb, http://www.antweb.org/, and used with permission.)

REFERENCE

1. Groden, E., Drummond, F.A., Garnas, J., and Franceour, A., Distribution of an invasive ant, *Myrmica rubra*, in Maine, *J. Econ. Entomol.*, 98, 1774–1784, 2005.

HARVESTER ANTS

Worker harvester ant.

IMPORTANCE

Painful stings; possible allergic reactions

DISTRIBUTION

In the United States, most species are located in western states (although there is one species east of the Mississippi River).

LESION

Variable

DISEASE TRANSMISSION

None

KEY REFERENCES

Creighton, W.S., The ants of North America, *Bull. Mus. Comp. Zool. Harvard*, 104, 1–585, 1950
AntWeb, http://www.antweb.org/

TREATMENT

Local—Wash wound; pain relievers
Systemic—May require antihistamines, epinephrine, and other supportive measures

F. Protecting Patients in Healthcare Facilities

Extremes in weather that cause movement of fire ants into inhabited dwellings can be especially problematic for health care facilities, such as hospitals and nursing homes. During the spring, when soils become saturated, IFA colonies may move inside seeking drier conditions. Similar movement of ants may occur during periods of drought, when they will travel inside toward moisture. Another important factor facilitating movement of IFAs to the inside of a building is proximity of ant mounds to the foundation of a building.

Most people are able to detect fire ant stinging and can move, jump, or run away to avoid further injury, but special care is required to ensure that patients in healthcare facilities

are not stung by fire ants.[9,11] Patients in these facilities may not be aware of their surroundings, may be immobilized by disease, or may be otherwise incapacitated, unable to respond to ant attack. Once foraging fire ants come in contact with a patient, a variety of external stimuli, including movement of the patient, might trigger a stinging event that leads to multiple stings in a very short period.

Some common sense suggestions for prevention of indoor fire ant infestations include:

1. Watching for IFA infestations indoors during weather extremes
2. Keeping patients' beds and linens away from walls and floors
3. Limiting food in beds
4. Requiring all food in the room to be kept in a well-sealed (airtight) container

Fire ant management also includes a systematic plan for keeping the pests out of healthcare facilities and, if they enter, ways to mitigate their effects. Close coordination with a licensed pest control firm is critical. Once fire ants are found on a patient, clinical evaluation is needed as well as possible transport (depending on findings) to the nearest emergency department.

III. Harvester Ants

A. General and Medical Importance

Harvester ants, *Pogonomyrmex* spp., are dangerous ant species that readily sting people and animals (see box). The reaction to their stings is not always localized; it may spread along the lymph vessels, producing intense pain in the lymph nodes in the axilla or groin long after the original sting pain subsides. Although relatively uncommon, systemic hypersensitive reactions may result from harvester ant stings.

B. General Description

Harvester ants are red to dark brown ants and are two to three times larger than fire ants; they have large, wide heads (Figure 10.8). Several species characteristically have many small, parallel ridges on the head. In addition, a good way to distinguish between harvester and fire ants is by the shape and size of their mounds (Figures 10.4 and 10.9). Harvester ant mounds are usually flat or slightly elevated and are surrounded by an area of no vegetation 1–3 m or more in diameter. Fire ant mounds are distinctly elevated and composed of excavated dirt.

C. Geographic Distribution

There is only one harvester ant species east of the Mississippi River, *Pogonomyrmex badius*, which is distributed throughout

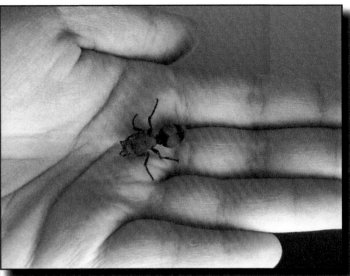

Figure 10.10 Female velvet ant.

Figure 10.8 Microscopic view of harvester ants. (Photograph courtesy of Joe MacGown, Mississippi State University, and used with permission.)

Figure 10.9 Harvester ant nests have a cleared-out area with centralized entrances. (Photograph courtesy of Tom Mann, Mississippi Museum of Natural Science, and used with permission.)

the southeastern states. In the western United States, approximately 20 species occur. Three of the notorious offenders are the red harvester ant, *P. barbatus*; the California harvester ant, *P. californicus*; and the western harvester ant, *P. occidentalis*.

D. Biology and Behavior

These ants are diurnal soil-inhabiting insects that build their nests in open areas. They feed on seeds and store them in honeycomb-like chambers within their nests. Harvester ants clear out a large, vegetation-free, craterlike area on the ground surface, and one or more holes lead down into the nest. Like

other ants, their colonies consist of at least one reproductive queen, several males, and many nonreproductive female workers. The workers are the ones that sting to defend the nest.

E. Treatment of Stings

Local first aid for harvester ant stings consists of washing the site with soap and water and applying ice packs. Unlike most ant species (and similar to the honey bee), the stinger can be torn off in the wound upon envenomation by *Pogonomyrmex californicus*[12] and may need to be removed. Allergic reactions may require administration of epinephrine, antihistamines, and other supportive treatment (see Chapter 2).

IV. Velvet Ants

A. General and Medical Importance

Velvet ants (also known as woolly ants, cow killers, mule killers, or mutillid wasps) are actually wingless female wasps in the family Mutillidae; they only resemble huge ants (see box and Figure 10.10). The wingless females have a long stinger and can inflict a painful sting. They are sometimes a problem to barefoot bathers on sandy beaches.

B. General Description

Velvet ants do not have a node on the petiole, which is characteristic of the true ant. Also, females (the ones confused with ants) are generally larger than true ants. Both male and female velvet ants are covered with a bright red, orange, black, or yellow pubescence (short, fine hairs). There are several species, but the more commonly encountered ones are from 0.75 to 2.5 cm in length.

VELVET ANTS

Female velvet ant. (From CDC, Pictorial Keys to Arthropods, Reptiles, Birds, and Mammals of Public Health Significance, U.S. Centers for Disease Control and Prevention, Atlanta, GA, 1963.)

IMPORTANCE

Painful stings

DISTRIBUTION

In the United States, most species occur in the south and west.

LESION

Variable

DISEASE TRANSMISSION

None

KEY REFERENCE

Mickel, C.E., Biological and taxonomic investigations on the mutillid wasps, *Bull. U.S. Natl. Mus.*, 143, 1–351, 1928

TREATMENT

Local—Ice packs; pain relievers
Systemic—May require antihistamines, epinephrine, and other supportive measures

C. Geographic Distribution

Members of this wasp family are widely distributed, but most U.S. species occur in the southern and western areas of the country. One northern species, *Dasymutilla occidentalis*, is common on the sandy beaches of Lake Erie in the summer, causing much distress to barefoot beachgoers.[13]

D. Biology and Behavior

These ants are solitary, diurnal wasps that are often associated with dry, sandy environments. Male mutillids are winged, whereas females are wingless. Their larvae are parasites of bees and other wasps. Females run about in the open searching for a suitable place to lay their eggs.

E. Treatment of Stings

Being solitary wasps (not actually ants), stings by numerous velvet ants are rare. Disinfection of the sting site, application of ice packs, and administration of analgesics are often indicated for normal or local sting reactions. In the event of an allergic reaction, epinephrine, antihistamines, and other supportive measures may be needed.

References

1. Holldobler B, Wilson EO. *The Ants.* Cambridge, MA: Harvard University Press; 1990.
2. Bolton B. *Identification Guide to the Ant Genera of the World.* Cambridge, MA: Harvard University Press; 1994.
3. Schmidt JO. Chemistry, pharmacology, and chemical ecology of ant venoms. In: Piek T, ed. *Venoms of the Hymenoptera.* New York: Academic Press; 1986:425–509.
4. Schmidt JO. Hymenopteran venoms: Striving toward the ultimate defense against vertebrates. In: Evans DL, Schmidt JO, eds. *Insect Defenses, Adaptive Mechanisms, and Strategies of Prey and Predators.* Albany, NY: State University of New York Press; 1990:387–419.
5. Taber SW. *Fire Ants.* College Station, TX: Texas A&M University Press; 2000.
6. Alexander JO. *Arthropods and Human Skin.* Berlin: Springer-Verlag; 1984.
7. Favorite F. Imported fire ant. *Public Health Rep.* 1958;73:445–449.
8. Diaz JD, Lockey RF, Stablein JJ, Mines HK. Multiple stings by imported fire ants without systemic effects. *South. Med. J.* 1989;82:775–777.
9. Goddard J, Jarratt J, deShazo RD. Recommendations for prevention and management of fire ant infestation of health care facilities. *South. Med. J.* 2002;95(6):627–633.
10. Fox RW, Lockey RF, Bukantz SC. Neurologic sequelae following the imported fire ant sting. *J. Allergy Clin. Microbiol.* 1982;70:120–124.
11. Goddard J. New record for Ixodes texanus Banks in Mississippi, with a new host record. *Entomol. News.* 1983;94:139–140.
12. Ebeling W. *Urban Entomology.* Berkeley, CA: University of California Press; 1978.
13. Frazier CA. *Insect Allergy: Allergic Reactions to Bites of Insects and Other Arthropods.* St. Louis: Warren H. Green; 1969.

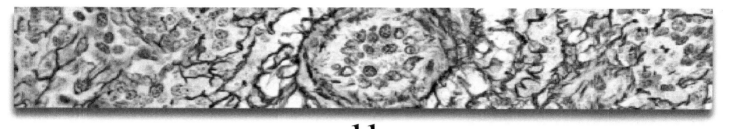

11
BEES

I. Honey Bees

A. General and Medical Importance

Honey bees, *Apis mellifera*, are commonly encountered insects and account for numerous stings (and even deaths due to allergy) in the United States each year.[1] The problem lies not with their aggressiveness but with their proximity to humans—backyard clovers, windowsill flowers, garden vegetable blooms, waste receptacle areas, etc. Unlike most other Hymenoptera, the honey bee worker has a barbed stinger and can sting only once. To escape, the bee must leave its entire stinging apparatus attached to the skin of its victim. Making matters worse, during stinging events honey bees release alarm pheromones associated with the sting gland, causing other bees in the vicinity of a pheromone-marked victim to attack and inflict multiple stings.

B. General Description

Honey bees are the familiar yellow-orange and black-striped bees with two membranous wings, commonly seen on flowers during spring and summer (see box). They have feathered hairs (when observed under magnification) and no spurs on their hind tibia. Worker honey bees are approximately 15–20 mm long, and drones are slightly larger and more robust. There are "races" of honey bees in the United States; the gold Italians and the black or gray Caucasian races make up the majority of bees found in this country. A close relative, the Africanized or "killer" bee, has recently entered the United States (Figure 11.1) (see Part II).

C. Geographic Distribution

Honey bees are not native to every continent (not to North America or the north of Mexico for example) but have

HONEY BEES

Worker honey bee. (Figure courtesy of Mississippi State Department of Health.)

IMPORTANCE
Painful stings; allergic reactions

DISTRIBUTION
Almost worldwide

LESION
Local swelling; central white spot with erythematous halo

DISEASE TRANSMISSION
None

KEY REFERENCE
Michener, C.D., Comparative social behavior of bees. *Annu. Rev. Entomol.*, 14, 299–342, 1969

TREATMENT
Local reactions—Ice packs, pain relievers
Systemic reactions—May require antihistamines, epinephrine, and other supportive measures

Figure 11.1 Africanized honey bee worker (left) and European honey bee worker (right). (Photograph courtesy of John Kucharski and the USDA Agricultural Research Service.)

been introduced virtually worldwide for pollination and honey production.

D. Biology and Behavior

Apis mellifera is a highly social insect. The colony consists of an egg-laying queen, drones to fertilize the queen, and workers to gather food and care for the young. Most honey bee colonies are artificial; however, wild colonies exist, mostly from escaped swarms, and they are usually found in hollow trees. Cells in a honey bee nest are in vertical combs and two cell layers thick. The colonies are perennial, with the queen and workers overwintering in the hive. Normally, there is only one queen per colony, and she may live for several years. When a new one is produced, it may be killed by the old queen, or one of the two may leave with a swarm of workers to build a new nest.

E. Treatment of Stings

Local treatment of honey bee stings consists of removing the stinger, disinfecting the sting site, and applying ice packs to slow the spread of venom. The stinger should be removed as quickly as possible. Conventional advice in scientific literature has emphasized that the stinger should be scraped off, never pinched, but one study[2] has shown that the method of removal is irrelevant, and even slight delays in removal caused by concerns about the correct procedure are likely to increase the dose of venom received. Other than stinger removal, Alexander[3] observed that nonallergic local reactions seldom require treatment. In the case of a severe or large local reaction, oral antihistamines and topical applications of corticosteroid creams may help, as well as rest and elevation of the affected arm or leg. Allergic (systemic) reactions to bee stings can be life-threatening events and may require administration of epinephrine, antihistamines, and other supportive treatment (see Chapter 2).

II. Africanized or "Killer" Bees

A. General and Medical Importance

The Africanized honey bee (AHB) is a subspecies of honey bee that was brought to South America from Africa to improve honey production. In 1956, 26 swarms of the AHB escaped from experimental colonies near Sao Paulo, Brazil, and they have subsequently multiplied and spread throughout much of South and Central America.[4] They are now in much of the southern and western United States. Winston[5] has presented a detailed review of the status of the AHB.

The AHB looks almost identical to our domesticated strains of European honey bees (see Figure 11.1), but it is more likely to attack with little provocation, stay angry for a longer time, and exhibit massive stinging behavior in colony defense. Although there are some differences in the venom of AHBs vs. European bees, AHB stings are not more toxic than those of domestic bees; however, death of humans and domestic or wild animals can result from toxic effects of multiple stings (400–1,000 stings) or from anaphylactic shock from an allergic reaction to very few stings (see Chapter 2).

As the AHB makes its way further into the United States, the numbers of sting-related deaths may increase. In

---

AFRICANIZED HONEY BEES

IMPORTANCE

Highly aggressive; painful stings; allergic reactions

DISTRIBUTION

South and Central America, Mexico, southern United States

LESION

Same as honey bee sting

DISEASE TRANSMISSION

None

KEY REFERENCE

Winston, M.L., The biology and management of Africanized honey bees, *Ann. Rev. Entomol.*, 37, 173–193, 1992.

TREATMENT

Local reactions—Ice packs, pain relievers
Numerous stings (nonallergic reactions)—Treat for histamine overdosage
Systemic reactions—May require antihistamine, epinephrine, and other measures

Venezuela, 12 deaths were attributed to honey bee stings in 1978 (before AHB); in contrast, in 1988 (after AHB) at least 100 bee-sting-related deaths were reported.[6] An increase in mortality of this magnitude may be avoided in the United States by better beekeeping practices and prompt medical management of stinging events.

B. General Description

Africanized honey bees look identical to common honey bees (see Figure 11.1). Only a specialist can tell them apart.

C. Geographic Distribution

Africanized bees are native to Africa. Since their accidental introduction in 1956, they have colonized most of South America, Central America, and Mexico. They are now reported in Texas, Louisiana, Arkansas, Florida, parts of New Mexico, Arizona, and California. Their spread into other southern U.S. states is only a matter of time.

D. Biology and Behavior

Most biological aspects of AHBs are similar to those of domestic honey bees (see preceding section); however, the mating habits of AHBs are such that Africanized queens tend to mate almost always with Africanized drones. They also reproduce at a high rate and swarm frequently. AHB colonies have more guard bees than European bee colonies; up to 50% of the bees in an AHB colony may be guard bees, which respond to a disturbance.

E. Treatment of Stings

Because the venoms are similar, AHB stings can be treated in the same manner as domestic honey bee stings. One note may be added: in the case of tens or hundreds of stings, extensive histamine release may result from venom action and not necessarily due to an allergic reaction. Accordingly, physicians may need to treat for histamine overdosage.

III. Bumble Bees

A. General and Medical Importance

There are several species of bumble bees in the genera *Bombus*, *Megabombus*, and *Pyrobombus*. Bumble bees, like other stinging Hymenoptera, will attack and sting when their nest is disturbed. A significant number of stinging incidents occur yearly, and some individuals experience allergic reactions; however, bumble bees are neither as aggressive nor as abundant as honey bees and therefore generally not dangerous.[7]

BUMBLE BEES

Bumble bee worker. (From Mitchell, T.B., Bees of the Eastern United States, Tech. Bull. No. 152, North Carolina Agricultural Experiment Station, Raleigh, 1962.)

IMPORTANCE
Painful stings; allergic reactions

DISTRIBUTION
Almost worldwide

LESION
Similar to other wasps and bees; central white spot and erythematous halo

DISEASE TRANSMISSION
None

KEY REFERENCE
Mitchell, T.B., *Bees of the Eastern United States*, Tech. Bull. No. 152, North Carolina Agricultural Experiment Station, Raleigh, 1962

TREATMENT
Local reactions—Ice packs, pain relievers
Systemic reactions—May require antihistamines, epinephrine, and other supportive measures

B. General Description

Bumble bees have feathered hairs on their body and have spurs on their hind tibia. They are robust bees, usually 20 mm or more in length, having black and yellow pubescence (covered with short, fine hairs) on their abdomen (see box and Figure 11.2). They are often confused with carpenter bees, which are similar in size and appearance, except

Figure 11.2 Female bumble bee. (Photograph courtesy of Dr. Blake Layton, Mississippi State University Extension Service.)

Figure 11.4 Tunnels in a piece of wood caused by carpenter bees.

only fertilized queens overwintering. These queens start new nests in the spring, and the colony builds to between 100 and 500 bees by late summer. The opening to their nests may appear mound-like from materials excavated by the bees, such that some nests have been noted to resemble fire ant mounds.

that carpenter bees have no pubescence on the abdomen (Figure 11.3). Carpenter bees rarely sting and are often seen forming galleries in wooden structures such as barns, sheds, stables, etc. (Figure 11.4).

C. Geographic Distribution

Bumble bees occur throughout the United States and over much of the world.

D. Biology and Behavior

Most bumble bees are diurnal plant feeders that nest in the ground, usually in loose fibrous habitats such as mouse nests, insulation, or grass clippings. Their colonies are annual, with

E. Treatment of Stings

Local treatment of bumble bee stings involves using ice packs and pain relievers, and washing to lessen the chances of secondary infection. In the case of a severe or large local reaction, oral antihistamines and topical applications of corticosteroid creams may help, as well as rest and elevation of the affected arm or leg. For allergic (systemic) reactions, administration of epinephrine, antihistamines, and other supportive treatment may also be required (see Chapter 2).

References

1. Harwood RF, James MT. *Entomology in Human and Animal Health.* 7th ed. New York: Macmillan; 1979.
2. Visscher PK, Vetter RS, Camazine S. Removing bee stings. *Lancet.* 1996;348(3 August):301–302.
3. Alexander JO. *Arthropods and Human Skin.* Berlin: Springer-Verlag; 1984.
4. Michener CD. The Brazilian honey bee – Possible problem for the future. *Clin. Toxicol.*1973;6:125–129.
5. Winston ML. The biology and management of Africanized honey bees. *Ann. Rev. Entomol.* 1992;37:173–193.
6. Gomez-Rodriguez R. Manejo de la Abeja Africanizada. Direccion General Desarrollo Ganadero, Caracas, Venezuela; 1986.
7. Biery TL. Venomous Arthropod Handbook. U.S. Air Force pamphlet No. 161–43, Brooks AFB, TX; 1977.

Figure 11.3 Carpenter bee. (Photograph courtesy of the Centers for Disease Control and Prevention, Atlanta, GA.)

12

BEETLES

I. Beetles in General

A. General and Medical Importance

Beetles belong to the insect order Coleoptera. At least 300,000 species have been described, representing 30–40% of all known insects. About 25,000 species of beetles occur in the United States and Canada.[1,2] Fortunately, beetles are of minor public health importance; however, some adults may inflict painful bites, several species may secrete irritating chemicals when touched or handled (discussed in the following section), and certain species may act as intermediate hosts for helminths. In addition, beetles such as grain beetles, weevils, and pantry beetles found in stored products may cause inhalational allergies. Some long-horned beetles (family Cerambycidae) have huge jaws and may inflict extremely painful bites if handled. Ladybugs, or lady beetles, have been reported to inflict unprovoked bites or cause stinging sensations by their defensive secretions.[3] Tapeworms, flukes, roundworms, and thorny-headed worms, which infect domestic and wild animals, use beetles as intermediate hosts. Humans may become accidentally infected owing to unhygienic practices or by ingesting the beetles. A case of acanthocephalan infection was reported to the Mississippi Department of Health in which an 8-cm worm was found in the diaper of an 11-month-old baby (Figure 12.1). Presumably, the infant ate infected beetles or consumed them in flour or bread products. The child suffered no ill effects from the parasite.

B. Prevention and Treatment of Infestations

Good sanitation is the best way to prevent ingestion of beetles or their body parts. Food preparation areas should be thoroughly cleaned and sanitized prior to cooking. Cereal, nuts, candy, flour, and cornmeal can be placed in tight-fitting plastic containers to prevent beetle infestation. Ladybugs can be kept out of houses by thorough sealing of exterior cracks and crevices. Once inside, they may be vacuumed.

II. Blister Beetles

A. General and Medical Importance

Blister beetles are plant-feeding insects that contain a blistering agent in their body fluids. Most are in the family Meloidae and Oedemeridae, although some are in Staphylinidae.[4,5] The agent in meloid and oedemerid beetles, cantharidin ($C_{10}H_{12}O_4$), penetrates the skin, producing blisters in a few hours[6] (also see Chapter 5). The blistering agent in staphylinids is somewhat different chemically from cantharidin, called pederin ($C_{25}H_{45}NO_9$), and is actually produced by endosymbiotic bacteria in the beetles.[7] In any event, handling live beetles or having contact with their pulverized bodies may cause blistering. Cantharidin is poisonous to humans and other animals when ingested and may lead to abdominal pain, kidney damage, and sometimes death.[8] Horses may die from ingesting blister beetles in their hay (especially alfalfa).[9,10] Lehman et al.[4] reported tingling and burning prior to blister formation on human skin, but other reports have indicated an almost symptomless course except for the blisters.[11] Dermatoses from blister beetle contact are seasonal, with most cases in the United States occurring in July, August, and September. Alexander[12] noted that workers harvesting potato crops, as well as children running around barefoot, are particularly vulnerable.

The Spanish fly is a blister beetle in southern Europe that has been used for millennia to make an extract called cantharides, which is used as a sexual stimulant.[13] The aphrodisiac properties of cantharides results from its irritant effects upon the body's genitourinary tract. Sexual enhancement is believed to be due to inhibition of phosphodiesterase and protein phosphatase activity and to stimulation of

BLISTER BEETLES

Blister beetles: (A) striped blister beetle, (B) ash-gray blister beetle, and (C) oil beetle. (Adapted from CDC, Pictorial Keys to Arthropods, Reptiles, Birds, and Mammals of Public Health Significance, U.S. Centers for Disease Control and Prevention, Atlanta, GA, 1963.)

IMPORTANCE

Production of a chemical compound called cantharidin; contact causes blisters on skin; ingestion may cause poisoning

DISTRIBUTION

Many areas of the world, including the United States

LESION

Generally painless, large blisters

DISEASE TRANSMISSION

None

KEY REFERENCES

Lehman, C.F., Pipkin, J.L., and Ressmann, A.C., Blister beetle dermatitis, *Arch. Dermatol.*, 71, 36–41, 1955

Frank, J.H. and Kanamitsu, K., *Paederus*, sensu lato (Coleoptera: Staphylinidae): natural history and medical importance, *J. Med. Entomol.*, 24, 155–191, 1987

TREATMENT

Skin reactions are generally not serious; topical antibiotics and bandaging may prevent secondary infection; poison control center consultation needed in cases of blister beetle ingestion.

beta-receptors, inducing vascular congestion and inflammation.[14] Alternatively, the idea that cantharides is an aphrodisiac may stem from priapism, an erection lasting hours, caused by damage to the urinary tract in the final stages of cantharidin poisoning shortly before death. There have been many historical accounts of human poisoning and deaths caused by cantharides when used as a sexual stimulant.[15] More recent

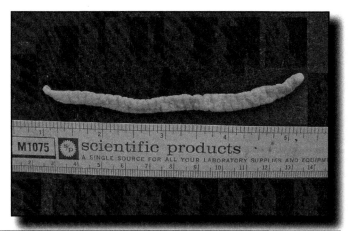

Figure 12.1 Acanthocephalan submitted to the Mississippi Department of Health that was found in the diaper of an 11-month-old baby.

reports include a case where 4 young adults were hospitalized after drinking Kool-Aid tainted with Spanish fly. One of the victims had added it to test the effects before using it with his girlfriend.[16]

B. General Description

Blister beetles are elongate, soft-bodied specimens in which the pronotum (section between the head and wings, viewed from above) is narrower than the head or wings (Figure 12.2). Two of the common blister beetle species are potato beetles: one with orange and black longitudinal stripes (see box, Figure A) and another black with gray wing margins (see box, Figure B). Members of the genus *Meloe* are called oil beetles because they exude an oily substance when disturbed. Oil beetles are approximately 20–25 mm long and black, with no hind wings, giving the appearance that their wings are very short (see box, Figure C).

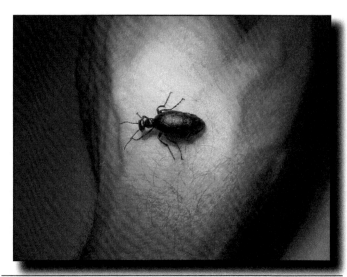

Figure 12.2 Blister beetle on hand. (Photograph copyright 2005 by Jerome Goddard, Ph.D.)

C. Geographic Distribution

The ash-gray beetle (*Epicauta fabricii*) and striped blister beetle (*E. vittata*) are common in the central and southeastern areas of the United States. Oil beetles and other large, rounded species are common in the southwestern United States from Texas to southern California. Many other blister beetle species are common in the western United States. In Europe, three important blister beetles are *Paederus limnophilus*, *P. gemellus*, and *Lytta vesicatoria* (the Spanish fly).

D. Biology and Behavior

Blister beetles have a very unusual biology. Larvae of the meloid beetles feed on the eggs of grasshoppers and in the nests of solitary bees. After several larval stages, they pupate and later emerge as adult beetles. Adults are plant feeders that may emerge in large numbers and appear in gardens and field crops quite suddenly.

E. Treatment of Exposed Areas

Blisters from exposure to the beetles are generally not serious and will be reabsorbed in a few days if unruptured. Even if the blisters are ruptured, complete clearing normally occurs within 7–10 days. Affected areas should be washed with soap and water and bandaged until the blisters reabsorb. Antibiotic ointments or creams may help prevent secondary infection. Large blisters and those occurring on the feet where they might be rubbed may need to be drained, treated with antiseptics, and bandaged.[17] When the beetles or beetle products are ingested, cantharidin may cause nausea, diarrhea, vomiting, and abdominal cramps. If large amounts are ingested, cantharidin poisoning may be gravely serious with kidney failure, cardiorespiratory collapse, coma, and death.[18] As for treatment for cantharidin ingestion, one paper reported frusemide to treat for renal edema and IV fluids.[8]

References

1. White RE. *A Field Guide to the Beetles of North America.* Boston: Houghton Mifflin; 1983.
2. Arnett RH, Jr. Present and future of systematics of the Coleoptera of North America. In: Kosztarab M, Schaefer CW, eds. *Systematics of the North American Insects and Arachnids: Status and Needs.* Blacksburg, VA: Virginia Polytechnic Institute and State University; 1990:165–173.
3. Southcott RV. Injuries from Coleoptera. *Med. J. Aust.* 1989;151:654–659.
4. Lehman CF, Pipkin JL, Ressmann AC. Blister beetle dermatitis. *Arch. Dermatol.* 1955;71:36–41.
5. Young DK. Cantharidin and insects: A historical review. *Great Lakes Entomol.* 1984;17:187–194.
6. Harwood RF, James MT. *Entomology in Human and Animal Health,* 7th ed. New York: Macmillan; 1979.
7. Piel J. A polyketide synthase-peptide synthetase gene cluster from an uncultured bacterial symbiont of *Paederus* beetles. *Proc. Nat. Acad. Sci.* 2002;99:213–220.
8. Tagwireyi D, Ball DE, Loga PJ, Moyo S. Cantharidin poisoning due to "blister beetle" ingestion. *Toxicon.* 2000;38(12):1865–1869.
9. Capinera JL, Gardner DR, Stermitz FR. Cantharidin levels in blister beetles associated with alfalfa in Colorado. *J. Econ. Entomol.* 1985;78:1052–1055.
10. Helman RG, Edwards WC. Clinical features of blister beetle poisoning in equids: 70 cases (1983–1996). *J. Am. Vet. Med. Assoc.* 1997;211(8):1018–1021.
11. Giglioli MEC. Some observations on blister beetles, family Meloidae, in Gambia, West Africa. *Trans. Royal. Soc. Trop. Med. Hyg.* 1965;59:657–661.
12. Alexander JO. *Arthropods and Human Skin.* Berlin: Springer-Verlag; 1984.
13. Falck B. Spanish fly-cantharidin's alter ego. *JAMA Dermatol.* 2018;154(1):51.
14. Sandroni P. Aphrodisiacs past and present: A historical review. *Clin. Auton. Res.* 2001;11(5):303–307.
15. Nickolls LC, Teare D. Poisoning by cantharidin. *Br. Med. J.* 1954;2:1384–1386.
16. Karras DJ, Farrell SE, Harringan RA, Henretig FM, Gealt L. Poisoning from "Spanish fly" (cantharidin). *Am. J. Emerg. Med.* 1996;14:478–483.
17. Frazier CA, Brown FK. *Insects and Allergy.* Norman, OK: University of Oklahoma Press; 1980.
18. Presto AJ, Muecke EC. A dose of Spanish fly. *J. Am. Med. Assoc.* 1970;214:591–592.

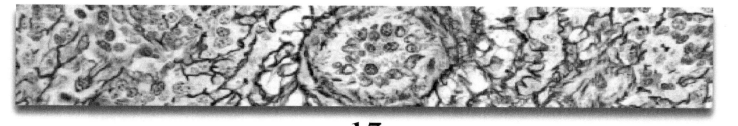

13

BUGS (THE TRUE BUGS)

I. Bed Bugs

A. General and Medical Importance

There are at least 91 species of bat, swallow, and bed bug species worldwide, but most occur only on bats and birds.[1,2] The common bed bug, *Cimex lectularius* (see box), however, has adapted to human feeding and has been a peridomestic associate of humans for thousands of years. Historically, the blood-sucking parasites have been common throughout the world, especially in areas of extreme poverty where people do not have the proper resources to control them. Scientific books and papers date back over 150 years concerning ways to prevent or avoid bed bugs, indicating their importance as household pests.[3–5] With the advent of DDT and other powerful pesticides, bed bugs nearly disappeared in developing countries until recently, where, in the last three decades or so, they have been making a progressively rapid comeback. In many areas, they are now the number one urban pest.[6–9] One report of an infested house in Great Britain contained a description of literally thousands of the bugs under and within the bed and in the mattress seams (Figure 13.1).[10] According to the authors, the area where the bed was against the wall was black with a layer of bed bug excrement, cast skins, and eggs several millimeters thick. Infestations like this can lead to anemia in affected individuals[11] and often to apartment managers throwing out expensive furniture (although this is actually not necessary) (Figure 13.2). In addition to being household pests, bed bug infestations may also occur in public transit systems (trains and subways), movie theaters, libraries, and healthcare facilities. Recent studies have described the frequency and costs associated with bed bugs in hospital emergency rooms.[12,13]

Bed bugs have been suspected in the transmission of more than 40 disease organisms such as those causing anthrax, plague, hepatitis, and typhus,[14] and have been found naturally infected with many of these agents (especially the hepatitis B virus).[15–19] However, there is little evidence that they are significant vectors of human diseases. The most likely disease agents which could be transmitted by bed bugs are *Trypanosoma cruzi* (Chagas' disease) and *Bartonella quintana* (trench fever).[20–22] Their principal medical importance is emotional trauma from biting incidents and the itching and inflammation associated with their bites. There is limited evidence that bed bug attacks may cause psychological effects similar to posttraumatic stress disorder.[23] The most common cutaneous reactions are small, pruritic maculopapular, erythematous lesions at bed bug feeding sites, one per insect. These usually itch and, if not scratched extensively, resolve within a week or so.[24–26] The intensity of bite reactions may increase in some individuals who experience repeated bites.[27–29] Some people experience more complex, even serious, cutaneous reactions. Reports of these have included pruritic wheals (local urticaria) around a central punctum, papular urticaria, and diffuse urticaria at bite sites usually noted on arising in the morning.[30–34] Bullous lesions may occur upon subsequent biting events days later.[35–39] The timing of cutaneous reactions to bed bugs may change with subsequent exposures and appears to reflect host immunological responses to bed bug salivary proteins.[25,35,40–45] Usinger[2] fed a colony of bed bugs on himself weekly for 7 years and noted that his reactions progressed from delayed to immediate, with no evidence of desensitization. Reinhardt et al.[46] reported that, with repeated exposure, the latency between bite and skin reaction decreased from 10 days to a few seconds. Interestingly, people who experience subsequent bed bug biting sometimes have old lesions that "re-inflame" upon new biting anywhere on the body. This "re-lighting-up" phenomenon at sites of previous lesions has been anecdotally reported previously[47,48] but is poorly understood. It presumably results from antigens residing at the bite site for an extended period of time and from these antigens responding to inflammatory mediators circulating to new bites.

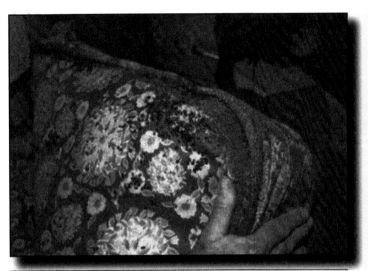

Figure 13.1 Numerous bed bugs congregating on the underside of a mattress. (Photograph courtesy of Ian Dick and the London Borough of Islington, Environmental Health Department.)

Figure 13.2 Mattresses and box springs are often discarded as a result of bed bug infestations.

There are a few reports of systemic reactions from bed bug bites, including asthma, generalized urticaria, and anaphylaxis.[37,49–52] One patient awakened during the night at a hotel with severe itching and urticaria on his arm and neck and bed bugs were found in his room.[53] He developed angioedema and hypotension, was hospitalized, and had transient anterolateral ischemia on electrocardiogram. Eight months later, after an experimental bed bug bite, he developed a wheal at the bite site and generalized itching that required epinephrine administration to resolve his symptoms. A home evaluation of another man who had asthma revealed bed bugs in his bedding, and an intradermal allergy skin test with an extract of bed bugs was positive.[51] Interestingly, after his bedding was changed, the asthma attacks ceased.

B. General Description

Adult bed bugs are approximately 5 mm long, oval shaped, and flattened. They may increase in size 150–200% while

BED BUGS

Adult bed bug, *Cimex lectularius*. (From CDC, Pictorial Keys to Arthropods, Reptiles, Birds, and Mammals of Public Health Significance, U.S. Centers for Disease Control and Prevention, Atlanta, GA, 1963.)

IMPORTANCE

Nuisance, irritation, and emotional effects from bites

DISTRIBUTION

Worldwide

LESION

Linear grouping of red blotches, urticarial wheals; sometimes, bullous lesions

DISEASE TRANSMISSION

Unlikely

KEY REFERENCES

Usinger, R.L., *Monograph of the Cimicidae (Hemiptera: Heteroptera)*, Vol. 7, Thomas Say Foundation, Entomological Society of America, College Park, MD, 1966

Goddard, J. and deShazo, R.D., Bed bugs (*Cimex lectularius*) and clinical consequences of their bites, *JAMA*, 301(13), 1358–1365, 2009

TREATMENT

Antiseptic or antibiotic creams or lotions; antihistamines and topical steroids for pruritic or urticarial reactions; elimination of infestations

feeding (Figure 13.3). They somewhat resemble unfed ticks or small cockroaches (Figure 13.4). Adults are reddish-brown (chestnut) in color; the immatures resemble adults but are yellowish white. First-stage nymphs are extremely tiny (Figure 13.5). Bed bugs have a pyramidal head with prominent compound eyes, slender antennae, and a long proboscis tucked backward underneath the head and thorax. The prothorax (dorsal side, first thoracic segment) bears rounded, winglike lateral horns on each side with numerous bristles. Bed bugs

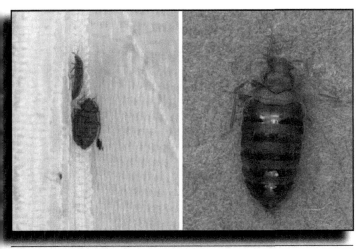

Figure 13.3 Bed bugs before and after feeding.

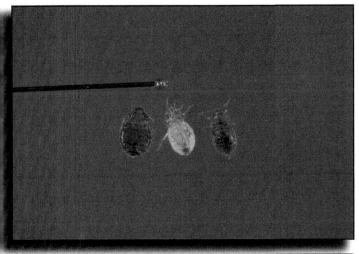

Figure 13.4 Bed bug nymphs and cast skins.

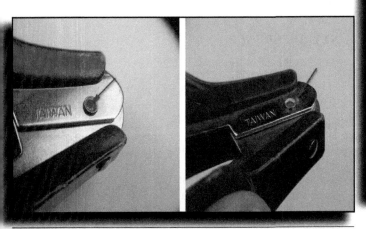

Figure 13.5 Bed bug nymph hiding in small hole in a staple puller. (Photograph copyright 2011 by Jerome Goddard, Ph.D.)

can be differentiated from bat bugs by the length of these pronotal bristles (Figure 13.6). Hind wings on bed bugs are absent entirely; the forewings are represented by two small pads.

C. Geographic Distribution

The common bed bug is cosmopolitan, occurring in temperate regions worldwide. Another human-biting bed bug species, *Cimex hemipterous*, is also widespread but is mostly found in the tropics. Many other bed bug species occur on bats and swallows, but they do not usually bite people.

D. Biology and Behavior

Bed bugs possess stink glands and emit an odor. Homes heavily infested with the bugs may have this distinct odor. Bed bugs mostly feed at night, hiding in crevices during the day. Hiding places include seams in mattresses, crevices in box springs, and spaces under baseboards or loose wallpaper. Although there is some evidence that bed bugs may actively disperse, moving from one place to another,[54] they primarily hitchhike from place to place in people's belongings, suitcases, or used furniture (Figure 13.7). There are five nymphal stages that must be passed before they develop into adults. Once an adult, the life span is 6–12 months. Each nymphal stage must take a blood meal in order to complete development and molt to the next stage. The bugs take about 5–10 minutes to obtain a full blood meal. Bed bugs can survive long periods of time without feeding (months), and when their preferred human hosts are absent they may take a blood meal from any warm-blooded animal, including pets.[55]

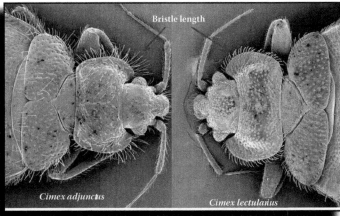

Figure 13.6 Bat bug vs. bed bug comparison by scanning electron micrograph. (Photograph copyright 2011 by Jerome Goddard, Ph.D., and courtesy of Dr. G.T. Baker.)

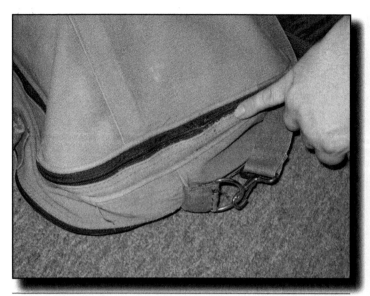

Figure 13.7 Bed bugs on a suitcase. (Photograph courtesy of Dr. Mike Potter, University of Kentucky, and used with permission.)

E. Treatment of Infestation and Bites

Bed bugs have piercing–sucking mouthparts typical of the insect order Hemiptera. Accordingly, bites from the bugs often produce welts and local inflammation, probably because of allergic reactions to salivary proteins injected via the mouthparts during feeding (see Chapter 4). On the other hand, for many people (estimated to be 30–50%) the bite is nearly undetectable, leaving no discernible lesion.[25,56,57] Treatment of common and complex cutaneous reactions is usually symptomatic and not necessarily evidence-based. If lesions are pruritic, topical application of over-the-counter or prescription antipruritic agents (paroxime, doxepin) or intermediate potency corticosteroids (triamcinolone) may be helpful. Secondary infections may benefit from topical mupirocin or systemic antibiotics as appropriate. Systemic reactions to bed bug bites are treated as insect-induced anaphylaxis.[40,53,58] That treatment includes intramuscular epinephrine first and antihistamines and corticosteroids where appropriate. Patients with previous generalized reactions should be instructed in the use of an epinephrine autoinjector device, which is to be kept available whenever traveling, and these patients should be referred to an allergist.

Bed bug infestations are best handled by pest management professionals who have at their disposal a wide array of treatment tools. Homeowner efforts to find and treat the pests with over-the-counter insecticides invariably fail. Insecticides are effective against bed bugs, except in certain cases where populations of the insects are resistant to the chemicals. Heat treatments, wherein the entire house or apartment is heated to approximately 130°F, are reportedly very successful.

KISSING BUGS

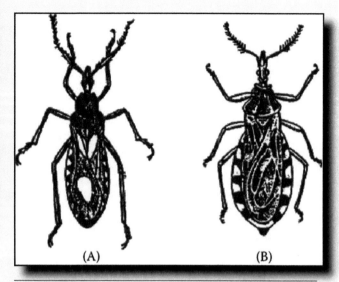

(A) Assassin bug, and (B) kissing bug. (Figures courtesy of U.S. Centers for Disease Control and Prevention, Atlanta, GA.)

IMPORTANCE

Blood-feeding on humans, often at night; anaphylaxis has been reported from bites

DISTRIBUTION

Most medically important species that occurs in the Americas

LESION

Variable (papular, nodular, or bullous)

DISEASE TRANSMISSION

Chagas' disease

KEY REFERENCE

Lent, H. and Wygodzinsky, P., Revision of the Triatominae and their significance as vectors of Chagas' disease, *Bull. Am. Mus. Nat. Hist.*, 163, 123–520, 1979

TREATMENT

Local reactions—Antihistamines, antipruritics, topical steroids
Systemic reactions—Antihistamines, epinephrine, and other supportive measures as needed

II. Conenose Bugs (Assassin and Kissing Bugs)

A. General and Medical Importance

Many members of the family Reduviidae have an elongate (cone-shaped) head, hence the name conenose bugs. Most reduviids "assassinate" or kill other insects. Some of the assassin bugs occasionally bite people, producing very painful lesions. A common offender is *Melanolestes picipes*. There is a report of two of these bugs biting an 8-year-old, causing intense pain for hours, swelling at the bite site for days, and, later, ulcers.[59] A relatively small but important group of reduviids in the subfamily Triatominae feeds exclusively on vertebrate blood. Notorious members of this group are frequently in the genus *Triatoma*, but not all (see Figure 13.8 for a classification scheme of this group). Triatomines are called kissing bugs because their blood meals are occasionally taken from the area around the human lips (Figure 13.9). Other sites of human attack, in order of frequency, are the hands, arms, feet, head, and trunk. Kissing bugs are not able to feed through clothing. Very often, their bites are painless; however, reactions to their bites range from a single papule, to giant urticarial lesions, to anaphylaxis, depending on the degree of allergic sensitivity.[60]

Chagas' Disease. Kissing bugs may transmit *Trypanosoma cruzi* (Figure 13.10), the agent of Chagas' disease, or American trypanosomiasis, one of the most important

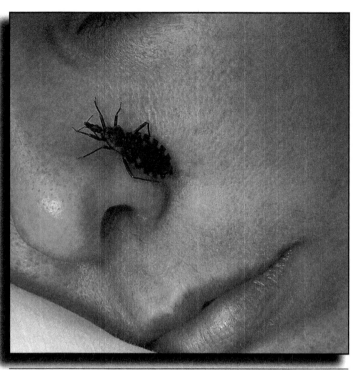

Figure 13.9 Kissing bugs often feed on the face. (Photograph copyright 2005 by Jerome Goddard, Ph.D.)

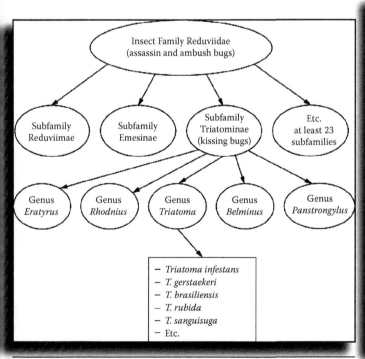

Figure 13.8 Partial classification scheme for insect family Reduviidae.

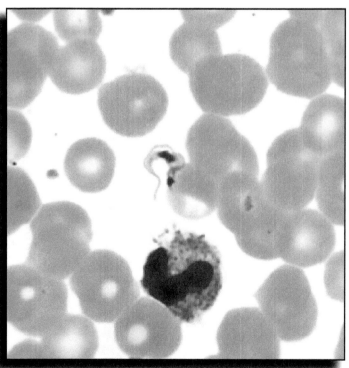

Figure 13.10 *Trypanosoma cruzi* organism in blood smear.

arthropod-borne diseases in tropical America. Chagas' disease is a zoonosis (originally a parasite of wild animals) mostly occurring in Mexico and Central and South America (Figure 13.11), but at least 25 indigenous (locally acquired) cases have been reported in the southern U.S. as far north as Oklahoma.[61-63] At present, the disease affects about 8 million people in Latin America,[64] and at least 300,000 U.S. residents who have emigrated from Latin American countries.[65,66] Chagas' disease has both acute and chronic forms, but it is perhaps best known for the myocardial damage it causes with cardiac dilation, arrhythmias, and major conduction abnormalities, as well as digestive tract involvement such as megaesophagus and megacolon. The digestive form of Chagas' disease is seen almost exclusively south of the Amazon basin and is rare in Central America and Mexico.[67]

B. General Description

Assassin bugs are often black or brown in color. They have elongate heads, and the portion behind the eyes is narrowed and necklike (see box). The beak is short and three-segmented, and its tip fits into a groove in the venter of the thorax (Figure 13.12). In nontriatomines, the beak is thick and curved. The abdomen is often widened in the middle, exposing the lateral margins of the segments beyond the wings. Kissing bugs are similar in appearance to many assassin bugs and may have orange and black markings where the abdomen extends laterally past the folded wings (Figure 13.13) as well as a thin, straight proboscis. In addition, the dorsal portion of the first segment of the thorax consists of a conspicuous triangular-shaped pronotum. Most adult kissing bugs are 1–3 cm long and are good fliers. An excellent monograph containing descriptions of the triatomines is provided by Lent and Wygodzinsky,[68] and a checklist and geographic distributions of U.S. kissing bugs has been published.[69]

C. Geographic Distribution

Numerous species of assassin bugs occur nearly worldwide. The wheel bug belongs to this group (see Part III). *Reduvius senilis*, the tan assassin bug, is found in the desert areas of the southwestern United States and Mexico. *Reduvius personatus* occurs throughout the United States and southern Europe. *Melanolestes picipes* is widely distributed in the United States.

Several species of kissing bugs will attack humans, and some are capable of transmitting *Trypanosoma cruzi*, the

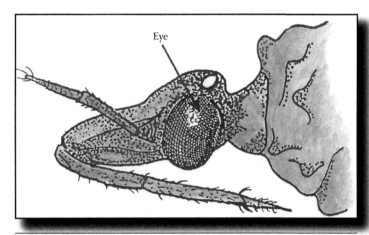

Figure 13.12 Kissing bugs have a three-segmented beak that fits into a groove under the thorax.

Figure 13.11 Approximate geographic distribution of Chagas' disease.

Figure 13.13 Kissing bugs showing variously colored abdomen extending laterally past folded wings.

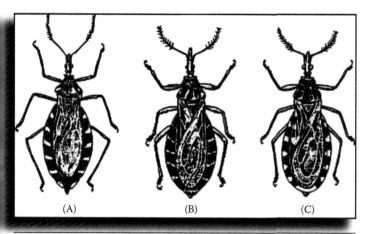

Figure 13.14 Adult (A) *Triatoma gerstaeckeri*, (B) *T. protracta*, and (C) *T. sanguisuga*. (From CDC, Pictorial Keys to Arthropods, Reptiles, Birds, and Mammals of Public Health Significance, U.S. Centers for Disease Control and Prevention, Atlanta, GA, 1963.)

causative agent of Chagas' disease; however, only a few species are efficient vectors. The four principal vectors of Chagas' disease in Central and South America are *Panstrongylus megistus*, *Rhodnius prolixus*, *Triatoma infestans*, and *T. dimidiata*. The most important vector in Mexico is *T. barberi*. In the southwestern United States, *T. gerstaeckeri* and *T. protracta* (Figure 13.14) are important kissing bug species, and *T. sanguisuga* occurs throughout much of the southern United States. Figure 13.15 shows the U.S. distribution of five common kissing bugs.

D. Biology and Behavior

Assassin bugs are predaceous on other insects and are often found lying in wait for their prey on various plants or flowers. They, along with kissing bugs, undergo simple metamorphosis; the developing nymphs look very much like adults, except that they are smaller.

Kissing bugs are nocturnal insects that are able to fly to their hosts with speed and agility. Both sexes bite, and they take their blood meals primarily at night, hiding in any available crack or crevice between feedings (about 36 hours). *Triatoma*, as a group, normally feed on a wide variety of small mammals but will readily feed on humans (Figure 13.16). Infection rates with *Trypanosoma cruzi*, the causative agent of Chagas' disease, may be as high as 80–100% in some adult triatomine bug populations.[70,71] Human kissing bug bites are especially common in poor, underdeveloped areas with dilapidated or poorly constructed huts or shacks. Infection is not by the salivary secretions associated with the bite but by fecal contamination of the bite site; however, other routes of transmission may be overlooked. In some communities in Mexico, for example, people believe that bug feces can cure warts or that the bugs have aphrodisiac powers.[71,72] In addition, Mexican children often play with triatomine bugs collected in their houses, and in Jalisco, reduviid bugs are eaten with hot sauce by the Huichol Indians.[71]

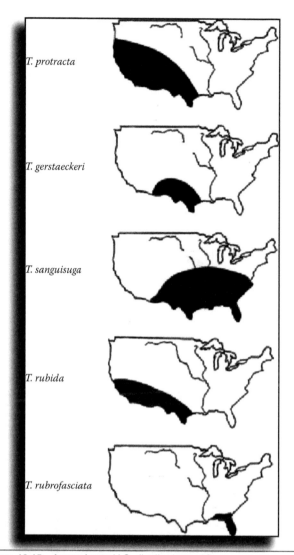

Figure 13.15 Approximate U.S. distributions of five commonly encountered kissing bugs.

E. Treatment of Infestation and Bites

Treatment for conenose bug bites involves washing the wound with soap and water. Itching lesions may be relieved with topical palliatives such as Caladryl® and oral antihistamines. Mild systemic reactions may require oral or intramuscular antihistamines. Anaphylactic shock has also been reported,[60] which should be treated with epinephrine, antihistamines, and other supportive measures (see Chapter 2). Follow-up for possible development of Chagas' disease may be needed for patients with kissing bug bites in endemic areas as well as for persons in the southern United States finding kissing bugs in their bedrooms. Benznidazole and nifurtimox are the only drugs with proven efficacy against Chagas' disease, although persons taking these drugs commonly experience adverse effects.[73] Benznidazole is currently considered the safer of the two and is widely considered the better first-line treatment;[67] however, it is ineffective against advanced cases of Chagas'.[74] Fortunately, two new drugs are being tested for Chagas'

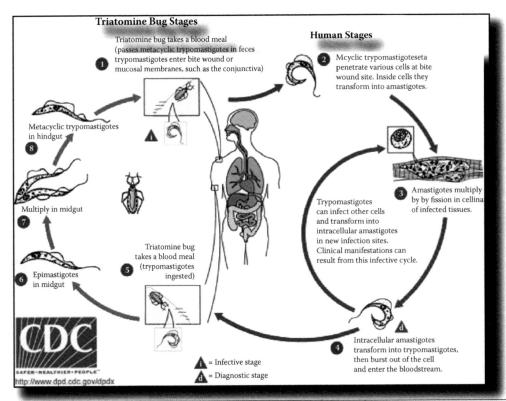

Figure 13.16 Life cycle of *Trypanosoma cruzi*. (From Image Library, Division of Parasitic Diseases, U.S. Centers for Disease Control and Prevention, Atlanta, GA.)

WHEEL BUGS

Adult wheel bug. (From CDC, Pictorial Keys to Arthropods, Reptiles, Birds, and Mammals of Public Health Significance, U.S. Centers for Disease Control and Prevention, Atlanta, GA, 1963.)

IMPORTANCE

Painful bite

DISTRIBUTION

United States

LESION

Swollen, inflamed, and indurated at bite site

DISEASE TRANSMISSION

None

KEY REFERENCE

Hall, M.C., Lesions due to the bite of the wheel bug *Arilus cristatus*, *Arch. Intern. Med.*, 33, 513, 1924

TREATMENT

None may be needed; oral analgesics, topical corticosteroids may help

treatment—posaconazole and ravuconazole.[73] In general, posaconazole does not seem more effective than benznidazole, but does produce fewer side effects; it has been shown to be efficacious against strains of *T. cruzi* that are resistant to benznidazole and nifurtimox. Physicians with questions about the most current guidelines for diagnosis and treatment of Chagas' disease may call the Centers for Disease Control and Prevention, Parasitic Diseases Division (404-718-4745).

III. Wheel Bugs

A. General and Medical Importance

The wheel bug, *Arilus cristatus*, is also in the family of true bugs called Reduviidae. They bite humans only in self-defense; however, the bite is characterized by immediate,

Figure 13.17 Wheel bug, showing cogwheel-like crest on upper side. (Photograph courtesy of Dr. Blake Layton, Mississippi State University Extension Service.)

intense pain.[75,76] Inflammation may become chronic, producing lesions at the bite site resembling papillomas.

B. General Description

Wheel bugs are gray and approximately 3 cm long (see box). The insects, as their name implies, have a cogwheel-like crest on the dorsal side of their prothorax (Figure 13.17). This gives the appearance of a half-wheel on their upper body. They also have a small, narrow head and long, piercing–sucking mouthparts tucked under their head and prothorax.

C. Geographic Distribution

The wheel bug is commonly found from New Mexico through the southern and eastern United States.

D. Biology and Behavior

These insects attack and eat soft-bodied insects. They inject a salivary fluid into their prey via their long beak. Being true bugs (Hemiptera), wheel bugs develop by simple metamorphosis, meaning that immatures are called nymphs and are almost identical to the adults.

E. Treatment of Bites

Bites of the wheel bug, although painful, are usually localized and self-limiting. Specific treatment measures are probably not necessary.

References

1. Reinhardt K, Siva-Jothy MT. Biology of the bed bugs. *Ann. Rev. Entomol.* 2007;52:351–374.

2. Usinger RL. *Monograph of Cimicidae.* Vol 7. College Park, MD: Entomological Society of America, Thomas Say Foundation; 1966.

3. Blackwall J. How to prevent attacks of the bed bug, *Cimex lectularius. J. Nat. Hist.* 1848;2(11):357–359.

4. Butler E. *Our Household Insects.* London: Longmans, Green, and Co.; 1893.

5. Jaeger B, Preston HC. *The Life of North American Insects.* New York: Harper and Brothers; 1859.

6. Potter MF. The perfect storm: An extension view on bed bugs. *Am. Entomol.* 2006;52:102–104.

7. Doggett S, Russell RC. The resurgence of bed bugs, *Cimex* spp., in Australia: Experiences from down under. *Proceed. 6th Inter. Conf. Urban pests*, Budapest, Hungary, 13–16 July, pp. 1–30; 2008.

8. Potter MF. The business of bed bugs. *Pest Management Professional Magazine.* 2008; January issue: 24–29.

9. Potter MF, Rosenberg B, Henriksen M. Bugs without borders: Defining the global bed bug resurgence. *Pest World.* 2010, Sept/Oct: 8–20.

10. King F, Dick I, Evans P. Bed bugs in Britain. *Parasitol. Today* 1989;5:100–102.

11. Pritchard MJ, Hwang SW. Severe anemia from bed bugs. *CMAJ.* 2009;181(5):287–288.

12. Sheele JM, Gaines S, Maurer N, et al. A survey of patients with bed bugs in the emergency department. *Am. J. Emerg. Med.* 2017;35(5):697–698.

13. Totten V, Charbonneau H, Hoch W, Sheele JM. The cost of decontaminating an ED after finding a bed bug: Results from a single academic medical center. *J. Emerg. Med.* 2016;34:649.

14. Burton GJ. Bed bugs in relation to transmission of human diseases. *Public Health Rep.* 1963;78:513–524.

15. Delaunay P, Blanc V, Del Giudice P, Levy-Bencheton A, Chosidow O, Marty P. Bedbugs and infectious diseases. *Clin. Infect. Dis.* 2011;52:200–210.

16. El-Masry SA, Kotkat AM. Hepatitis B surface antigen in *Cimex lectularius. J. Egypt. Public Health Assoc.* 1990;55:230–236.

17. Jupp PG, Lyons SF. Experimental assessment of bedbugs and mosquitoes as vectors of human immunodeficiency virus. *AIDS.* 1987;1:171–174.

18. Jupp PG, McElligott SE. Transmission experiments with hepatitis B surface antigen and the common bed bug. *S. Afr. Med. J.* 1979;56:54–57.

19. Jupp PG, Prozesky OW, McElligott SE, Van Wyk LA. Infection of the common bed bug with hepatitis B virus in South Africa. *S. Afr. Med. J.* 1978;53:598–600.

20. Goddard J, de Shazo RD. Bed bugs (*Cimex lectularius*) and clinical consequences of their bites. *J. Am. Med. Assoc.* 2009;301:1358–1366.

21. Leulmi H, Bitam I, Berenger JM, et al. Competence of *Cimex lectularius* bed bugs for the transmission of *Bartonella quintana*, the agent of trench fever. *PLOS Negl. Trop. Dis.* 2015;9(6).

22. Salazar R, Castillo-Neyra R, Tustin AW, Borrini-Mayori K, Naquira C, Levy MZ. Bed Bugs (*Cimex lectularius*) as vectors of *Trypanosoma cruzi. Am. J. Trop. Med. Hyg.* 2015;92(2):331–335.

23. Goddard J, de Shazo RD. Psychological effects of bed bug attacks (*Cimex lectularius* L.) *Am. J. Med.* 2012;125(1):101–103.

24. Kemper H. Die Bettwanze und ihre bekampfung. *Schriften uber Hygienische Zoologie, Z Kleintierk Pelztierk.* 1936;12:1–107.

25. Ryckman RE. Dermatological reactions to the bites of four species of triatominae (hemiptera: reduviidae) and *Cimex lectularius* L. (hemiptera: cimicidae). *Bull. Soc. Vector Ecol.* 1985;10:122–125.

26. Masetti M, Bruschi F. Bedbug infestations recorded in Central Italy. *Parasitol. Int.* 2007;56(1):81–83.

27. Liebold K, Schliemann-Willers S, Wollina U. Disseminated bullous eruption with systemic reaction caused by *Cimex lectularius. J. Eur. Acad. Dermatol. Venereol.* 2003;17:461–463.

28. Bartley JD, Harlan HJ. Bed bug infestation: Its control and management. *Mil. Med.* 1974;139:884–886.

29. Cestari TF, Martignago BF. Scabies, pediculosis, and stinkbugs: Uncommon presentations. *Clin. Dermatol.* 2005;23:545–554.

30. Alexander JO. *Arthropods and Human Skin.* Berlin: Springer-Verlag; 1984.

31. Brasch J, Schwarz T. 26-year-old male with urticarial papules. *J. Dtsch. Dermatol. Ges.* 2006;4(12):1077–1079.

32. Gbakima AA, Terry BC, Kanja F, Kortequee S, Dukuley I, Sahr F. High prevalence of bedbugs *Cimex hemipterus* and *Cimex lectularis* in camps for internally displaced persons in Freetown, Sierra Leone: A pilot humanitarian investigation. *West. Afr. J. Med.* 2002;21(4):268–271.

33. Honig PJ. Arthropod bites, stings, and infestations: Their prevention and treatment. *Pediatr. Dermatol.* 1986;3:189–197.

34. Rook AJ. Papular urticaria. *Ped. Clin. NA.* 1961;8:817–820.

35. Cooper DL. Can bedbug bites cause bullous erythema? *J. Am. Med. Assoc.* 1948;138:1206.

36. Hamburger F, Dietrich A. Lichen urticatus exogenes. *Acta. Paediat.* 1937;22:420.

37. Kemper H. Beobachtungen ueber den Stech-und Saugakt der Bettwanze und seine Wirkung auf die menschliche Haut. *Zeitschr f Desinfekt.* 1929;21:61–65.

38. Kinnear J. Epidemic of bullous erythema on legs due to bedbug. *Lancet.* 1948;255:55.

39. Patton WS, Evans A. *Insects, Ticks, and Venomous Animals of Medical and Veterinary Importance, Part I.* Croydon, U.K.: H.R. Grubb Ltd.; 1929.

40. Churchill TP. Urticaria due to bed bug bites. *J. Am. Med. Assoc.* 1930;95:1975–1976.

41. Elston DM, Stockwell S. What's eating you? Bed bugs. *Cutis.* 2000;65:262–264.

42. Hect O. Die hautreaktionen auf insektenstiche als allergische erscheinungen. *Zool. Anz.* 1930;87:94, 145, 231.

43. Ryckman RE. Host reactions to bug bites (hemiptera, homoptera): A literature review and annotated bibliography, Part I. *California Vect. Views.* 1979;26:1–24.

44. Ryckman RE, Bently DG. Host reactions to bug bites: A literature review and annotated bibliography, Part II. *California Vect. Views.* 1979;26:25–49.

45. Sheele JM. Antibody and cytokine levels in humans fed on by the common bed bug, *Cimex lectularius. Parasite Immunol.* 2017; doi:10.1111/pim.12411.

46. Reinhardt K, Kempke RA, Naylor RA, Siva-Jothy MT. Sensitivity to bites by the bedbug, *Cimex lectularius. Med. Vet. Entomol.* 2009;23:163–166.

47. Goddard J, Edwards KT, de Shazo RD. Observations on development of cutaneous lesions from bites by the common bed bug, *Cimex lectularius* L. *Midsouth Entomol.* 2011;4:49–52.

48. McKiel JA, West AS. Nature and causation of insect bite reactions. *Ped. Clin. NA.* 1961;8:795–814.

49. Bircher AJ. Systemic immediate allergic reactions to arthropod stings and bites. *Dermatology.* 2005;210(2):119–127.

50. Jimenez-Diaz C, Cuenca BS. Asthma produced by susceptibility to unusual allergens. *J. Allergy.* 1935;6:397–403.

51. Sternberg L. A case for asthma caused by the *Cimex lectularius. Med. J. Rec.* 1929;129:622.

52. Minocha R, Wang C, Dang K, Webb CE, Fernández-Peñas P, Doggett SL. Systemic and erythrodermic reactions following repeated exposure to bites from the common bed bug *Cimex lectularius* (Hemiptera: Cimicidae). *Austral. Entomol.* 2016; doi:10.1111/aen.12250.

53. Parsons DJ. Bed bug bite anaphylaxis misinterpreted as coronary occlusion. *Ohio State Med. J.* 1955;51:669.

54. Cooper R, Wang C, Singh N. Mark-release-recapture reveals extensive movement of bed bugs (*Cimex lectularius* L.) within and between apartments. *PLoS ONE.* 2015;10(9).

55. Little SE, West MD. Home infestation with *Cimex lectularius,* the common bed bug, affecting both dog and client (Abstract No. 61). *Amer. Assoc. Vet. Parasitol. Annual Meeting,* New Orleans, LA, July 19–22 2008.

56. Goddard J, de Shazo RD. Multiple feeding by the common bed bug, *Cimex lectularius* L., without sensitization. *Midsouth Entomol.* 2009;2:90–92.

57. Potter MF, Haynes KF, Deutsch M, Hardebeck E, Partin D, Harrison R. The sensitivity spectrum: Human reactions to bed bug bites. *Pest Control Technol.* 2010;February issue:70–75.

58. Oswalt ML, Kemp SF. Anaphylaxis: Office management and prevention. *Immunol. Allergy Clin. N. Am.* 2007;27:177–191.

59. Eads RB. An additional report of a reduviid bug attacking man. *J. Parasitol.* 1950;36:87.

60. Moffitt JE, Venarske D, Goddard J, Yates AB, deShazo RD. Allergic reactions to *Triatoma* bites. *Ann. Allergy Asthma Immunol.* 2003;91(2):122–128; quiz 128–130, 194.

61. Bradley KK, Bergman DK, Woods JP, Crutcher JM, Kirchoff LV. Prevalence of American trypanosomiasis among dogs in Oklahoma. *J. Am. Vet. Med. Assoc.* 2000;217:1853–1857.

62. Cantey PT, Stramer SL, Townsend RL, et al. The United States *Trypanosoma cruzi* infection study: Evidence for vector-borne transmission of the parasite that causes Chagas' disease among United States blood donors. *Transfusion.* 2012;52:1922–1930.

63. Gunter SM, Murray KO, Gorchakov R, et al. Likely autochthonous transmission of *Trypanosoma cruzi* to humans, south central Texas, USA. *Emerg. Infect. Dis.* 2017;23:500–503.

64. Albajar-Vinas P, Dias JCP. Advancing the treatment for Chagas' disease. *N. Engl. J. Med.* 2014;370:1942–1943.

65. Kuehn BM. Putting Chagas' disease on the U.S. radar screen. *JAMA.* 2015;313:1195–1197.

66. Bern C. Chagas' disease. *N. Engl. J. Med.* 2015;373:456–466.

67. Rossi AJ, Rossi A, Marin-Neto JA. Chagas' disease. *Lancet.* 2010;375:1388–1402.

68. Lent H, Wygodzinsky P. Revision of the Triatominae and their significance as vectors of Chagas' disease. *Bull. Am. Mus. Nat. Hist.* 1979;163:123–520.

69. Schmidt JO, Stevens L, Dorn PL, Mosbacher M, Klotz JH, Klotz SA. Kissing bugs in the United States. *The Kansas School Naturalist*, Volume 57(2): 1–15, Emporia State University, Emporia, KS; 2011.

70. de Shazo T. A survey of *Trypanosoma cruzi* infection in *Triatoma* spp collected in Texas. *J Bacteriol.* 1943;46:219–220.

71. Schettino PMS, Arteaga IDH, Berrueta TU. Chagas disease in Mexico. *Parasitol. Today.* 1988;4:348–352.

72. Salazar-Schettino PM. Customs which predispose to Chagas' disease and cysticercosis in Mexico. *Am. J. Trop. Med. Hyg.* 1983;32:1179–1180.

73. Urbina J. The long road towards a safe and effective treatment of chronic Chagas' disease. *Lancet Infect. Dis.* 2018;18:363–365.

74. Maguire JH. Treatment of Chagas' disease – Time is running out. *N. Engl. J. Med.* 2015;373:1369–1370.

75. Biery TL. *Venomous Arthropod Handbook.* U.S. Air Force pamphlet No. 161-43, Brooks AFB, TX; 1977.

76. Hall MC. Lesions due to the bite of the wheel bug *Arilus cristatus. Arch. Intern. Med.* 1924;33:513–515.

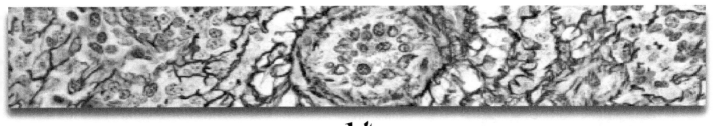

14
CATERPILLARS (URTICATING)

I. General and Medical Importance

At least 12 moth or butterfly families have species whose caterpillars possess urticating hairs or spines that secrete a poison when exposed to human skin[1,2] (see also Chapter 5). An excellent overview has been provided by Diaz,[3] and Alexander[4] presented detailed discussion of the nature of the various venoms and hair/spine structures. Caterpillar dermatitis from larval lepidopterans is often termed eurcism. Often, exposure to urticating caterpillars is accidental or incidental, but in some cases there is deliberate contact (e.g., children playing with caterpillars). Rash is the most common clinical manifestation,[5,6] and in many species there is a severe burning sensation immediately following the sting. This may be followed by swelling, numbness, urticaria, and intense stabbing pain that radiates to a nearby axillary or inguinal region; lymphadenitis may also be present. Usually the effects of these toxic hairs are limited to burning and inflammation of the skin, but they may progress to systemic reactions such as headache, nausea, vomiting, paralysis, acute renal failure, and shock and convulsions (rare).[1,7] Days later, the lesion may show a pattern similar to that of the spines of the offending specimen (see box and Figure 14.1). At least one group of moths in the genus *Lonomia* (giant silkworm moths) in South America have anticoagulants in their venom which can cause multiple intracerebral hemorrhages leading to death.[8] Other caterpillars, such as gypsy moth and tussock moth larvae, are not "stinging caterpillars" in the strictest meaning of the term, but their hairs may cause dermatitis, especially in sensitive persons[5] (see Chapter 26). Apparently, there has been an increase over the last few decades of reports of dermatologic, pulmonary, and systemic reactions following caterpillar encounters in the southern United States.[9]

II. General Description

Five common urticating caterpillars will be discussed in this section, although there are many others (Figure 14.2 and Table 14.1). *Automeris io* caterpillars are larvae of the io moth. Full-grown caterpillars are about 5–8 cm long, pale green, with lateral stripes of red or maroon over white running the length of the body (Figure 14.3A). The brown-tail moth larva, *Euproctis chrysorrhoea*, is a mostly black caterpillar with brown hairs and a row of white tufts on each of its sides. The puss caterpillar, *Megalopyge opercularis*, also sometimes called the opossum bug, asp, Italian asp, or el perrito, is about 3 cm long, tan to dark brown in color, and completely covered dorsolaterally with hair that causes it to resemble small tufts of cotton (Figure 14.3B). They really do not look like caterpillars. Intermingled among all those fine hairs on the back are clusters of venomous spines. *Sibine stimulea*, the saddleback caterpillar, is 2–3 cm long, has a brown sluglike body, and is covered middorsally with markings that resemble a brown or purplish saddle sitting on a green and white saddle blanket (Figure 14.3C). Gypsy moth larvae, *Lymantria dispar*, are approximately 3–5 cm long and gray to brown in color with yellow stripes (Figure 14.3D). They also may appear to have red or blue spots along the sides and top of the body. As in the other members of the family Lymantriidae, gypsy moth and tussock moth larvae have long tufts of hair along the body (Figure 14.4).

URTICATING CATERPILLARS

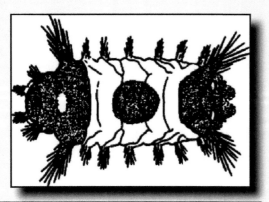

Saddleback caterpillar, *Sibine stimulea*.

IMPORTANCE

Venomous hairs or spines that cause stings or irritation, possibly allergic reactions

DISTRIBUTION

Many species involved worldwide

LESION

Variable (papular eruption, erythema, local swelling)

DISEASE TRANSMISSION

None

KEY REFERENCES

Henwood, B.P. and MacDonald, D.M., Caterpillar dermatitis, *Clin. Exp. Dermatol.*, 8(1), 77–93, 1983
Rosen, T., Caterpillar dermatitis, *Dermatol. Clin.*, 8(2), 245–252, 1990

TREATMENT

Local reactions—Topical anti-itch products, antihistamines, corticosteroids, and pain relievers
Systemic reactions—Antihistamines, epinephrine, and other supportive measures are required

Note: Eye lesions may be particularly serious and should be seen by a specialist

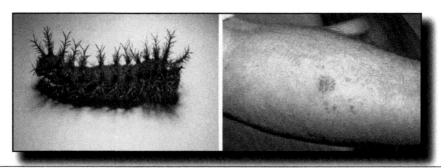

Figure 14.1 Buck moth caterpillar (left) and lesion it caused on a patient's arm (right).

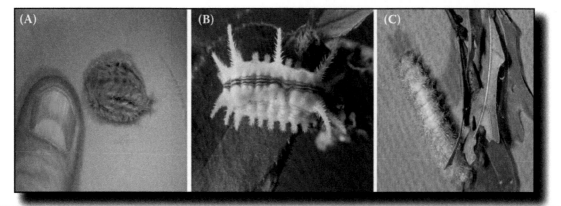

Figure 14.2 Some urticating caterpillars: (A) puss caterpillar, (B) stinging rose caterpillar, and (C) io moth caterpillar. (Io moth photograph courtesy of Centers for Disease Control and Prevention, Atlanta, GA.)

Table 14.1 Some Species of Lepidoptera Whose Larvae Are Known to Sting or Cause Dermatitis

Species	Condition	Occurs Where
Megalopyge opercularis	Sting	Southern United States, Central and South America
Several other *Megalopyge* spp.	Sting	South America
Sibine stimulea	Sting	Neotropical, Nearctic
Automeris io	Sting	Nearctic
Several other *Automeris* spp.	Sting	South America
Hemileuca spp.	Sting	Nearctic, South America
Euproctis chrysorrhoea	Sting	Nearctic, Palearctic
E. similis	Sting	Nearctic, Palearctic
E. edwardsii	Sting	Australia
E. flava	Sting	Oriental
Thaumetopoea wilkinsonii	Sting	Palearctic, Afrotropical
T. pityocampa	Sting	Palearctic, Afrotropical
Orchrogaster contraria	Sting	Australia
Several *Hylesia* spp.	Sting and dermatitis	Argentina
Lymantria dispar	Dermatitis	Europe, Eastern United States
Orgyia pseudotsugata	Dermatitis	Northwestern United States
O. leucostigma	Dermatitis	Nearctic

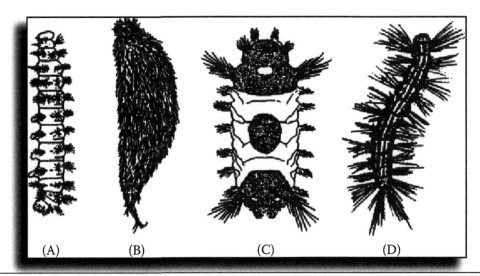

Figure 14.3 Line drawings of caterpillars that sting (A–C) or cause dermatitis (D): (A) io moth larva, (B) puss caterpillar, (C) saddleback caterpillar, and (D) gypsy moth larva. (Adapted from CDC, *Pictorial Keys to Arthropods, Reptiles, Birds, and Mammals of Public Health Significance*, U.S. Centers for Disease Control and Prevention, Atlanta, GA, 1963.)

III. Geographic Distribution

The io moth is found in the Nearctic region, including southern Canada and parts of Mexico; it is quite common in the eastern United States. Brown-tail moth caterpillars are bothersome pests in Europe and eastern portions of the United States, where they have been accidentally introduced. The saddleback caterpillar is found in many parts of the world. In the United States, it is generally distributed southeast of a diagonal line drawn from Massachusetts to the middle of Texas. Puss caterpillars are primarily found in the southern United States and southward into Central and South America.

They seem to be especially a problem in Texas. Gypsy moths occur in Europe and the eastern United States.

IV. Biology and Behavior

Io moth larvae feed on the leaves of a variety of plants, including corn and willow. In most areas they produce only one annual generation, but in the southernmost areas of their distribution there may be a second generation; therefore, depending on the area of the country, larval stages can be found anytime from spring to fall. The brown-tail moth is a

CASE HISTORY

PUSS CATERPILLAR STING

During the fall, the author was collecting insects at the Copiah County Game Management Area in central Mississippi when his son inadvertently brushed his right forearm against a puss caterpillar. Several of the caterpillars had been seen on a fallen log and some bushes just prior to the sting incident. The offending specimen was collected and identified at the site (see figure). Within 5 minutes the patient experienced intense, throbbing pain at the sting site. An erythematous spot approximately 4 cm in diameter developed, containing a few raised papule-like structures, presumably where the venomous hairs or setae contacted the skin. After 20 minutes, the patient was complaining of severe pain in the right axilla and kept clutching himself under that arm. No other symptoms developed, and the pain subsided within 45 minutes. The erythematous spot resolved in 24 hours.

Comment: Many moth families have species whose larvae possess stinging or urticating spines or hairs. One of the most troublesome of these is the puss caterpillar, *Megalopyge opercularis*, a member of the flannel moth family. The puss caterpillar is widely distributed in the southern United States, extending down into Mexico; members of the species feed on a variety of deciduous trees and shrubs. Interestingly, its appearance is not what one might expect for a caterpillar. Instead of being worm-like, or wormlike with prominent spines, the puss caterpillar is shaped like a teardrop and looks like a tuft of cotton or fur. It may vary in color from light yellow to gray or reddish-brown. Stings from the puss caterpillar may cause immediate, intense local burning pain (often referred to as "shooting" pain), headache, nausea, vomiting, lymphadenopathy, lymphadenitis, and sometimes shock and respiratory distress. Stings can be especially severe in children. From reports in entomological and medical literature, this case was fairly typical, except perhaps for the lack of local swelling. Certainly, a person with hypersensitivity to the venom could react in a much more profound manner, possibly leading to shock and respiratory distress. Even in the absence of an allergic reaction, however, puss caterpillar stings can lead to temporary severe pain and partial immobilization.

Puss caterpillar that stung a patient.

Source: Adapted from Goddard, J., Stings by the puss caterpillar, *Am. Fam. Physician*, 52, 86, 1995. Copyright American Academy of Family Physicians.

serious pest of forest and shade trees, as well as many varieties of fruit trees. Caterpillars of this species are most active between April and July.

The saddleback caterpillar may be found feeding on the leaves of a variety of trees, shrubs, and other plants from May to November. The puss caterpillar also feeds on the leaves of a wide range of trees and shrubs. In the southern area of its range, it may have two generations per year. The first generation develops in the spring and early summer, and the second generation develops in the fall. They seem to be especially abundant from September to November. Every few years there are outbreaks of puss caterpillars that lead to numerous stings, especially among children. They are commonly found

on the exterior walls of houses, sheds, gates, and fences; thus, risk of human contact is high.

The gypsy moth was introduced into the eastern United States in the 1860s and has since become widely distributed throughout New England (and is now spreading southward), where it causes widespread damage to forest trees. Eggs are laid on tree trunks in July and August in masses covered with froth and body hairs from the female. Eggs overwinter and the tiny first-stage larvae emerge the following spring, usually in late April or early May. The females can fly, but weakly. Dispersal of the gypsy moth infestation is primarily by young larvae on silken threads traveling tree to tree or limb to limb by "ballooning" in the wind.

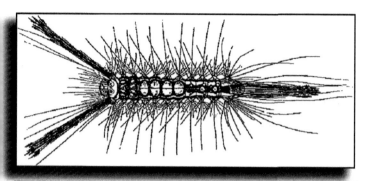

Figure 14.4 Tussock moth caterpillar. (From Steyskal, G.C. et al., Eds., *Insects and Mites: Techniques for Collection and Preservation*, USDA Misc. Publ. No. 1443, Agricultural Research Service, U.S. Department of Agriculture, Washington, DC, 1986.)

V. Treatment of Stings

Without additional or ongoing contact with the offending caterpillar, lesions resolve at varying times depending on the species involved. Generally, saddleback and io caterpillar dermatitis subsides within 2–8 hours, gypsy moth dermatitis within 48 hours, and brown-tail moth and puss caterpillar disease (various systemic manifestations) within 7–10 days.[10]

Local treatment of urticating caterpillar stings consists of careful repeated stripping of the sting site with adhesive or cellophane tape to remove the spines (if the offending species was stout-spined), application of ice packs, and oral administration of antihistamines to help relieve itching and burning sensations. Acute urticarial lesions may be further relieved by the application of topical corticosteroids, which help reduce the intensity of the inflammatory reaction. Allen et al.[11] reported good results with desoximetasone gel applied twice daily to the affected areas. Rosen[10] observed that systemic administration of corticosteroids in the form of intramuscular triamcinolone acetonide has been remarkably effective in relieving both severe itching due to gypsy moth dermatitis and pain due to puss caterpillar dermatitis.

For severe pain associated with the puss caterpillar or io moth larvae, physicians sometimes administer meperidine HCl, morphine, or codeine;[4] aspirin is reportedly not effective. Systemic hypersensitivity reactions such as hypotension or bronchospasm are usually treated with epinephrine, antihistamines, and other supportive measures (also see Chapter 2). Eye lesions (especially those resulting from forceful contact with caterpillars) may be very serious and should be seen by a specialist.

References

1. Keegan HL. Some medical problems from direct injury by arthropods. *Int. Pathol.* 1969;10:35–45.
2. Maschwitz UW, Kloft W. Morphology and function of the venom apparatus of insects – Bees, wasps, ants, and caterpillars. In: Bucherl W, Buckley E, eds. *Venomous Animals and their Venoms.* Vol. 3. New York: Academic Press; 1971:38–62.
3. Diaz JH. The evolving global epidemiology, syndromic classification, management, and prevention of caterpillar envenoming. *Am. J. Trop. Med. Hyg.* 2005;72(3):347–357.
4. Alexander JO. *Arthropods and Human Skin.* Berlin: Springer-Verlag; 1984.
5. CDC. Caterpillar-associated rashes in children – Hillsborough Coubty, Florida, 2011. *MMWR.* 2012;61:209–211.
6. French RNE, Brillhart D. Erucism due to Lepidoptera caterpillar envenomation. *N. Eng. J. Med.* 2015;373:e21.
7. Gamborgi GP, Metcalf EB, Barros EJ. Acute renal failure provoked by toxin from caterpillars of the species Lonomia obliqua. *Toxicon.* 2006;47(1):68–74.
8. Kowacs PA, Cardoso J, Entres M, Novak EM, Werneck LC. Fatal intracerebral hemorrhage secondary to *Lonomia obliqua* caterpillar envenoming: Case report. *Arq. Neuropsiquiatr.* 2006;64(4):1030–1032.
9. Diaz JH. The epidemiology, diagnosis, and management of caterpillar envenoming in the southern US. *J. Louisiana State Med. Soc.* 2005;157(3):153–157.
10. Rosen T. Caterpillar dermatitis. *Dermatol. Clin.* 1990;8:245–252.
11. Allen VT, Miller O, Tyler WB. Gypsy moth caterpillar dermatitis – Revisited. *J. Am. Acad. Dermatol.* 1991;24:979–981.

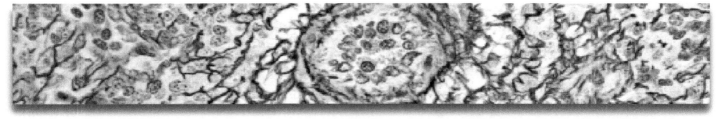

15
CENTIPEDES

I. General and Medical Importance

Centipedes are long, multisegmented arthropods that characteristically have one pair of legs per body segment (millipedes have two pairs per segment). They are agile, fast-moving creatures that can inflict a painful bite.[1] In fact, some of the larger species can produce extreme pain.[2-4] The venom is a cytolysin-based compound. Most human centipede bites result when a centipede is stepped on or picked up, or otherwise contacts the body. Centipede bites are rarely fatal to humans, but at least one death has been reported, in which a child from the Philippines was bitten on the head and the poison apparently entered the brain.[5] Members of the genus *Scolopendra* can be 20–25 cm long and produce bites with intense burning pain lasting 1–5 hours. The bite is characterized by two puncture wounds at the site of attack, often red and swollen. Other symptoms may include anxiety, vomiting, irregular pulse, dizziness, and headache.[6] Secondary infections can occur, and superficial necrosis at the bite site may persist for several days. There is at least one report of anaphylaxis resulting from a centipede bite (*Scolopendra subspinipes*).[7] In addition to their bites, large species have claws that can make tiny punctures if they crawl on human skin.

II. General Description

Currently, there are about 2,800 described species of centipedes worldwide, although the fauna is poorly known.[5] Centipedes are dorsoventrally flattened and have a distinct head, relatively long antennae, and one pair of legs per body segment (see box). They do not bite in the traditional sense in that no mouthparts are involved. Instead, the first body segment bears a pair of claws that contain ducts for the expulsion of a paralyzing venom contained in a gland at the base of the claw. The number of body segments, and thus number of legs, is variable depending on the species, but it usually ranges from 15 to over 100 pairs. Many species are 3–8 cm long, whereas some tropical species may reach 45 cm (Figure 15.1). The common house centipede, *Scutigera coleoptrata*, is approximately 4 cm long and has long antennae and fragile legs. From above, the relative lengths of the legs give it an oval appearance.

III. Geographic Distribution

Centipede species in the northern United States are small and generally harmless to humans, but larger species in the southern United States and tropics can inflict a painful bite. The common house centipede, *Scutigera coleoptrata*, occurs throughout southern Europe and the eastern parts of the

Figure 15.1 Large tropical centipede. (Photograph copyright 2010 by Jerome Goddard, Ph.D.)

CENTIPEDES

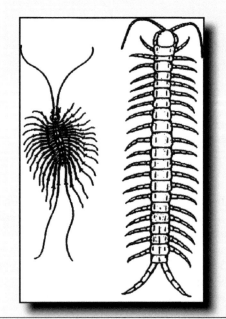

Centipedes: (A) *Scutigera coleoptrata*, and (B) *Scolopendra heros*. (From CDC, Pictorial Keys to Arthropods, Reptiles, Birds, and Mammals of Public Health Significance, U.S. Centers for Disease Control and Prevention, Atlanta, GA, 1963.)

IMPORTANCE

Painful bites

DISTRIBUTION

Numerous species worldwide

LESION

Often, two hemorrhagic punctures

DISEASE TRANSMISSION

None

KEY REFERENCE

Keegan, H.L., Some medical problems from direct injury by arthropods, *Intern. Pathol.*, 10, 35, 1969

TREATMENT

Analgesics; antibiotics; possibly tetanus prophylaxis

United States and Canada. *Scolopendra polymorpha* occurs in the southwestern United States, and *S. heros* occurs in southern California. *S. cingulata* is a common species around the Mediterranean and in the Near East. In Asia, *S. morsitans* is very large and can inflict severe bites, while *S. subspinipes*

CASE HISTORY

PAINFUL CENTIPEDE BITE

During July, a man in Jackson, Mississippi, was awakened during the night by a sharp, needlelike bite on his knee. Looking around in bed, he found the offending specimen: a centipede, which he collected for identification (see figure). A red and swollen lesion of a few centimeters in diameter appeared on his knee, followed by headache and dizziness which lasted approximately 20 minutes. The patient self-medicated with an antihistamine (Chlor-Trimeton® 4 mg), and all symptoms were completely resolved by morning. No further complications occurred.

Comment: Centipede bites are usually not life-threatening. The arthropods do not actually bite with their mouthparts, but instead with sharpened claws on their modified first pair of legs with which they hold their prey. They have a venom gland situated in the basal segment of the modified legs and inject venom through the claws. Clinical features of centipede bites vary with locality and species but generally are immediately painful, with some swelling and local tenderness for a few hours. More severe reactions rarely occur and may include anxiety, vomiting, irregular pulse, dizziness, and headache. This particular centipede was identified as *Hemiscolopendra punctiventris*, a member of the order Scolopendromorpha. This group is principally tropical; in the United States, they occur mainly in the southern states.

Centipede that bit a patient.

occurs on land around and within the Indian Ocean, Tropical and Subtropical Asia, Central and South America, and the Caribbean.

IV. Biology and Behavior

Centipedes lay their eggs in moist soil or vegetation. Development is slow, with about 10 instars. Adult centipedes may live 3–5 years. They usually hide during the day under rocks, boards, and bark, or in cracks, crevices, closets, etc. At night, they emerge to hunt for prey such as insects and other small arthropods. The larger species may feed on small vertebrates. This carnivorous habit brings centipedes into close contact with people as they frequently enter tents and buildings in search of prey. They inject venom through a pair of powerful claws on the first body segment. The claws are connected to poison glands located in the body trunk.

V. Treatment of Bites

Most centipede bites are uncomplicated and self-limiting.[8] Pain from bites usually subsides in 8–36 hours. Treatment recommendations include washing the bite site with soap and water, applying ice or cool wet dressings, and taking analgesics for pain. One reference reported pain relief after immersing the bite area in hot water.[3] Alexander[9] said that established inflammatory lesions require appropriate antibiotics.

References

1. Remington CL. The bite and habits of a giant centipede, *Scolopendra subspinipes*, in the Philippine islands. *Am. J. Trop. Med. Hyg.* 1950;30:453–454.

2. Acosta M, Cazorla D. Centipede envenomation in a rural village of semi-arid region from Falcon State, Venezuela. *Rev. Invest. Clin.* 2004;56:712–717.

3. Balit CR, Harvey MS, Waldock JM, Isbister GK. Prospective study of centipede bites in Australia. *J. Toxicol. Clin. Toxicol.* 2004;42:41–48.

4. Knysak I, Martins R, Bertim CR. Epidemiological aspects of centipede bites registered in greater S. Paulo, SP, Brazil. *Rev. Saude. Publica.* 1998;32:514–518.

5. Shelley RM. Centipedes and millipedes, with emphasis on the North American fauna. *The Kansas School Naturalist*, Vol. 45 (3), March issue, Emporia State University, Emporia, KS, pp. 3–15; 1999.

6. Harwood RF, James MT. *Entomology in Human and Animal Health.* 7th ed. New York: Macmillan; 1979.

7. Washio K, Masaki T, Fujii S, et al. Anaphylaxis caused by a centipede bite: A "true" type-I allergic reaction. *Allergol. Inter.* 2018;67(3):12–13.

8. Keegan HL. Some medical problems from direct injury by arthropods. *Int. Pathol.* 1969;10:35–45.

9. Alexander JO. *Arthropods and Human Skin.* Berlin: Springer-Verlag; 1984.

16
COCKROACHES

I. General and Medical Importance

Cockroaches are among the most important residential, commercial, institutional, and industrial pests today (see box). Several of the approximately 4,000 species in the world have become adapted to living in human habitations (Table 16.1). These are sometimes referred to as domestic or domiciliary cockroach species, and they breed in homes, institutions, or industrial areas, where they find food, water, shelter, and warmth. Being indoors, they can remain active throughout the year. They consume any human or animal food or beverage, as well as dead animal and plant materials, leather, glue, hair, wallpaper, fabrics, and the starch in bookbindings.

Cockroaches adversely affect human health in several ways: they sometimes bite feebly, especially gnawing the fingernails of sleeping children; they may enter human ear canals (Figure 16.1); they contaminate food, imparting an unpleasant odor and taste; and they may transmit disease organisms mechanically on their body parts.[1-3] Some health officials, however, see no association between cockroaches and disease and think that cockroaches are merely nuisance pests.[8] Many species of pathogenic or parasitic organisms such as bacteria, protozoans, fungi, and helminths have been found on cockroach body parts[4]; for example, Burgess[5] reported isolation of a strain of *Shigella dysenteriae* from German cockroaches that was causing a disease outbreak in Northern Ireland. Several researchers have also obtained data indicating that the insects may be most commonly implicated in the transmission of *Salmonella*.[6] In addition to disease transmission, cockroach excrement and cast skins contain a number of allergens to which sensitive people may exhibit allergic responses (also see Chapter 2). Symptoms exhibited by persons with a cockroach allergy are similar to those described by Wirtz[7] which include sneezing, runny nose, skin reactions, and eye irritation. Asthma-related health problems from cockroaches seem to be most severe among children in inner-city areas, but may also be significant in nonurban children.[8] In one study of 476 asthmatic inner-city children, 50.2% of the children's bedrooms had high levels of cockroach allergen in dust. That study also found that children who were both allergic to cockroach allergen and exposed to high levels of this allergen had 0.37 hospitalizations a year, as compared with 0.11 for other children.[9]

II. General Description

Cockroaches are dorsoventrally flattened, fast-running, nocturnal insects that seek warm, moist, secluded areas. They have prominent, multisegmented filiform antennae, cerci (small projections) on the abdomen, and two pairs of wings. The front wings are typically hardened and translucent, whereas the hind wings are membranous and larger. Cockroaches may be variously colored. Most domestic species are reddish-brown, brown, or black, although the "Cuban" cockroach is bright green (Figure 16.2). Many species can fly, but the domestic U.S. species rarely do so; however, the imported Asian cockroach in the Florida area both flies frequently and comes to lights.[10] Adult German and brown-banded cockroaches are approximately 15 mm long, whereas the American and Oriental cockroaches are 30–50 mm long (Figures 16.3 and 16.4). Some tropical species are even longer. Immature cockroaches look similar to adults (except they have no wings), and some of the first nymphal instars are so small as to be confused with ants.

III. Geographic Distribution

The German, Oriental, and American cockroaches are cosmopolitan in distribution. The German cockroach is probably the most important overall pest in human habitations worldwide. The American cockroach is an especially severe pest in

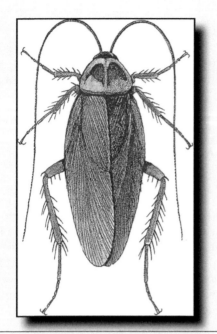

Adult American cockroach, a large, commonly encountered specimen.

COCKROACHES

IMPORTANCE

Contamination of food; cockroach allergies

DISTRIBUTION

Numerous species worldwide

DISEASE TRANSMISSION

Mechanical transmission of various bacteria and possibly viruses

KEY REFERENCES

Roth, L.M. and Willis, E.R., The medical and veterinary importance of cockroaches, *Smithsonian Misc. Coll.*, 134(10), 1–147, 1957

Schal, C., Cockroaches, in *The Mallis Handbook of Pest Control*, 10th ed., Hedges, S.A., Ed., GIE Media, Cleveland, OH, 2011

TREATMENT

Avoidance and control; possibly immunotherapy for allergies

Table 16.1   Some Cockroach Species Found in Close Association with Humans

Species	Common Name	Occurs Where
Blattella germanica	German cockroach	Cosmopolitan
B. asahinae	Asian cockroach	Far East and southern United States
Blatta orientalis	Oriental cockroach	Cosmopolitan
Periplaneta americana	American cockroach	Cosmopolitan
P. fuliginosa	Smoky brown cockroach	Central and southern United States
P. brunnea	Brown cockroach	Tropics, parts of United States
P. australasiae	Australian cockroach	Tropics, neotropics
Supella longipalpa	Brown-banded cockroach	Tropics, subtropics, parts of temperate zone

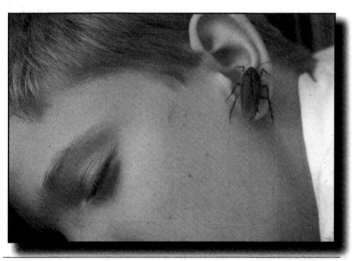

Figure 16.1   Cockroach by human ear. (Photograph courtesy of Joseph Goddard.)

the tropics and subtropics. Brown-banded cockroaches are a pest in the tropics and subtropics, and they are increasing as a pest in the temperate zone. They probably now infest the entire United States.

IV. Biology and Behavior

Cockroaches belong to the insect order Blattaria (formerly they were in the Orthoptera) and are closely related to crickets and grasshoppers. They develop by gradual metamorphosis in which the nymphs, when hatched, look similar to the adults, albeit smaller. Some cockroach species live outside and feed on vegetation and other organic matter; however, species that live in buildings are mostly scavengers, feeding on a wide variety of foods including starches, sweets, grease, meat products, glue, hair, and bookbindings. Cockroaches usually choose to

Figure 16.2 The Cuban cockroach, a green cockroach.

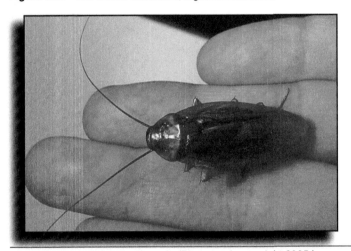

Figure 16.3 American cockroach. (Photograph copyright 2005 by Jerome Goddard, Ph.D.)

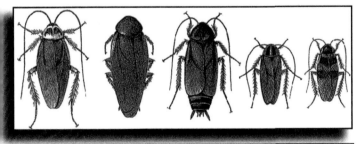

Figure 16.4 Five common adult cockroaches (L–R): American, smoky brown, Oriental, German, brown-banded.

live in protected areas that provide a warm and humid environment. American and Oriental cockroaches gather in large groups in protected areas such as wall voids or around steam pipes. The German cockroach spends most of its time hiding in cracks and crevices in dark, warm, and humid areas close to food and water. Brown-banded cockroaches are generally found on ceilings, high on walls, behind picture frames, or in electric appliances (if someone says cockroaches are living inside their telephone or radio, the brown-banded species is probably the culprit). These roaches do not require as close an association with moisture as German cockroaches do.

V. Treatment of Infestation

Disease transmission or allergic reactions due to cockroaches are best prevented by controlling cockroaches in the residential, institutional, and industrial environment. The best approach to cockroach control involves good sanitation, protecting against new entry, and a combination of the least toxic pest control options (baits, dusts, sticky traps, and residuals). In the last three decades, numerous cockroach bait products have been marketed that are extremely effective in reducing roach populations. One study showed that strategic placement of cockroach baits in homes of children with asthma eradicated cockroaches and improved asthma outcomes for children.[11]

References

1. Brenner RJ, Koehler PG, Patterson RS. Health implications of cockroach infestations. *Infect. Med.* 1987;4:349–360.
2. Lemos AA, Lemos JA, Prado MA, et al. Cockroaches as carriers of fungi of medical importance. *Mycoses.* 2006;49(1):23–25.
3. Tatfeng YM, Usuanlele MU, Orukpe A, et al. Mechanical transmission of pathogenic organisms: The role of cockroaches. *J. Vector. Borne. Dis.* 2005;42(4):129–134.
4. Roth LM, Willis ER. The medical and veterinary importance of cockroaches. *Smithson. Misc. Collect.* 1957;134 (10):1–147.
5. Burgess N. Biological features of cockroaches and their sanitary importance. Lectures Delivered at the International Symposium on Modern Defensive Approaches to Cockroach Control, Bajomi, D. and Erdos, G., eds., pp. 45–50. The Public Health Commission, Budapest, Hungary; 1982.
6. Rueger ME, Olson TA. Cockroaches as vectors of food poisoning and food infection organisms. *J. Med. Entomol.* 1969;6:185–192.
7. Wirtz RA. Occupational allergies to arthropods – Documentation and prevention. *Bull. Entomol. Soc. Am.* 1980;26:356–360.
8. Higgins PS, Wakefield D, Cloutier MM. Risk factors for asthma and asthma severity in nonurban children in Connecticut. *Chest.* 2005;128(6):3846–3853.
9. Rosenstreich DL, Eggleston P, Kattan M, et al. The role of cockroach allergy and exposure to cockroach allergen in causing morbidity among inner-city children with asthma. *N. Engl. J. Med.* 1997;336:1356–1360.
10. Brenner RJ, Koehler PG, Patterson RS. The Asian cockroach. *Pest Manag.* 1986;5:17–22.
11. Rabito FA, Carlson JC, He H, Werthmann D, Schal C. A single intervention for cockroach control reduces cockroach exposure and asthma morbidity in children. *J. Allergy Clin. Immunol.* 2017;140:565–570.

17

EARWIGS

I. General and Medical Importance

Earwigs are relatively harmless insects that are occasionally seen inside homes. They are included in this reference because of an old wives' tale that these insects enter human ears, causing much torment (hence the name earwig). Earwigs do not enter human ears, or even bite, but some of the larger species may pinch human skin with their abdominal cerci (abdominal pincers).[1] In addition, they have a frightful appearance, move rapidly around baseboards at ground level, and may emit a stinking, yellowish-brown liquid from their scent glands.

II. General Description

Earwigs are elongate, slender, flattened insects that are dark colored and have forcepslike cerci (see box). The forceps in females are straight sided, whereas male forceps are strongly curved (caliperlike) and larger. Earwigs are generally 4–20 mm long, and many species have short, stubby wings (Figure 17.1). There are actually two pairs of wings on winged specimens, the hind wings being fully developed and folded beneath the short front wings. Earwigs have chewing mouthparts and threadlike antennae. Nymphs resemble the adults with differences in abdominal segments and forceps structure. Earwigs may be mistaken for rove beetles (family Staphylinidae).

III. Geographic Distribution

There are more than 2000 species of earwigs worldwide. The European earwig, *Forficula auricularia*, and the ring-legged earwig, *Euborellia annulipes*, are cosmopolitan and commonly encountered. Other pest species include the striped earwig, *Labidura riparia*, and the seaside earwig, *Anisolabis*

EARWIGS

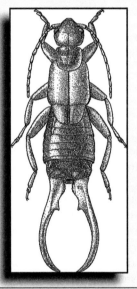

Adult earwig. (From Steyskal, G.C. et al., Eds., Insects and Mites: Techniques for Collection and Preservation, USDA Misc. Publ. No. 1443, Agricultural Research Service, U.S. Department of Agriculture, Washington, DC, 1986.)

IMPORTANCE

Harmless, but often mistakenly believed to invade human ears

DISTRIBUTION

Worldwide

LESION

None

DISEASE TRANSMISSION

None

KEY REFERENCE

Hoffman KM, Earwigs (Dermaptera) of South Carolina, with a key to the eastern North American species and a check-list of the North American fauna. *Proc. Entomol. Soc. Wash*. 1987;89:1–14

TREATMENT

None needed

Figure 17.1   Earwig. (Photograph by Jerome Goddard, Ph.D.)

maritima. In the United States, there are at least 20 species of earwigs, although only 2–3 are pests in human dwellings.[2]

IV. Biology and Behavior

Earwigs are mostly nocturnal feeders, feeding on dead and decaying vegetable matter. A few species feed on living plants, and some are predaceous. They may be destructive to garden vegetables, flowers, and greenhouse plants. Earwigs may enter homes and, once inside, may feed on sweet, oily, or greasy foods, as well as on house plants. In some areas, they may enter homes by the thousands and become significant pests. During the day, earwigs hide in cracks and crevices, under bark, or in piles of debris. It is not unusual to find them in private water well buildings and well casings. Accordingly, they can enter the system occasionally and even come out of a faucet. Earwigs lay their eggs in burrows in the ground or among debris, and the female tends them until hatching. This display of maternal care is rare among insects. Earwigs undergo simple metamorphosis and have as many as six nymphal stages. For further information, the reader is referred to Ebeling,[3] who discussed earwig species and their biology in detail.

References

1. Bishopp FC. Injury to man by earwigs. *Proc. Ent. Soc. Wash.* 1961;63:114–116.
2. Arnett RH, *American Insects.* 2nd ed. Boca Raton, FL: CRC Press; 2000.
3. Ebeling W. *Urban Entomology.* Berkeley, CA: University of California Press; 1978.

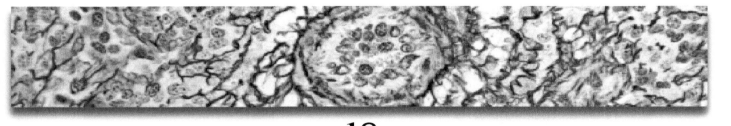

18

FLEAS

I. General and Medical Importance

Fleas are small, laterally flattened, wingless insects (see box) that are of great importance as vectors of disease in many parts of the world.[1,2] Because fleas are not very host specific (as compared to other insects such as lice), there is significant potential for zoonotic disease transmission. Public health workers are most concerned with fleas that carry the agents of bubonic plague and murine typhus from rats to people and fleas that transmit plague among wild rodents and secondarily to humans; however, there are other fleaborne diseases. The bacterium, *Rickettsia felis,* has been found worldwide in cat fleas. Its ability to cause disease in humans has been controversial,[3] but apparently it can infect humans, producing a murine typhuslike illness.[4–6] Certain rodent fleas are efficient vectors of *Bartonella* organisms.[7] Also, fleas may serve as intermediate hosts for helminths such as the dog tapeworm (Figure 18.1). Despite these disease threats, for many people (especially the lay public), the insidious attacks by fleas on people and domestic animals causing irritation, blood loss, and severe discomfort are equal in importance to disease transmission. Included here are discussions of a few of the more common species, comments as to their medical importance, and notes on their biologies.

The flea-bite lesion initially is a punctate hemorrhagic area representing the site of probing by the insect and may have a center elevated into a papule, vesicle, or even a bulla. Lesions may occur in clusters as the flea explores the skin surface, frequently stopping and probing. There is usually formation of a wheal around each probe site with the wheal reaching its peak in 5–30 minutes. Pruritus is almost always present. In most cases, there is a transition to an indurated papular lesion within 12–24 hours. Immunologically, flea bites may produce both immediate and delayed skin reactions.[8,9] In sensitized individuals, the delayed reaction appears in 12–24 hours, persisting for a week or more. The delayed papular

reaction with its intense itching is often the reason people with flea bites present to clinics.

A. Fleaborne Diseases

Plague. Plague, a zoonotic disease caused by the bacterium *Yersinia pestis,* has been associated with humans since the beginning of recorded history. Few diseases can compare to the devastating effects of plague on human civilization; for example, in the fourteenth century, approximately 25 million people died of plague in Europe. To this day, there are still hundreds of cases occurring annually over much of the world (Figure 18.2). In the United States, from 1970 to 1994, a total of 334 cases of indigenous plague were reported; the peak years were 1983 and 1984, in which there were 40 and 31 cases, respectively.[10] Cases are usually reported from locations west of the Mississippi River. In 2015, 16 cases of plague were reported to the CDC from Arizona, California, Colorado, Georgia, Michigan, New Mexico, Oregon, and Utah.[11] Sylvatic plague, sometimes also called campestral plague, is ever-present in endemic areas, circulating among rock and ground squirrels, deer mice, voles, chipmunks, and others. Transmission from wild rodents to humans can occur by direct contact with sick or dead animals, but this is rare; more commonly the agent is transmitted via fleas feeding. As an example of transmission by contact, in 2012 a seven-year-old girl contracted plague after playing with a dead ground squirrel in Colorado.[12]

Yersinia pestis inflicts damage on the host animal by an endotoxin present on its surface. Hematogenous dissemination of the bacteria to other organs and tissues may cause intravascular coagulation and endotoxic shock, producing dark discoloration in the extremities (thus, the name black death). Three clinical forms of plague are recognized: bubonic, septicemic, and pneumonic. The bubonic form is the most common in the Americas. In bubonic plague, lymph nodes

FLEAS

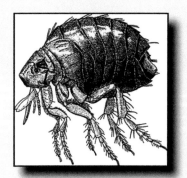

Sticktight flea. (Figure courtesy of the U.S. Department of Agriculture.)

IMPORTANCE

Biting; annoyance; papular urticaria; vectors of disease

DISTRIBUTION

Numerous species worldwide

LESION

Variable depending upon species; may be papular, vesicular, or bullous; in the United States, most lesions are irregular, very itchy red wheals

DISEASE TRANSMISSION

Plague; murine typhus; intermediate hosts of dog tapeworm

KEY REFERENCE

Lewis RE, Fleas (*Siphonaptera*). In: Lane RP and Crosskey RW, eds., *Medical Insects and Arachnids*. New York: Chapman & Hall; 1993:529–575

TREATMENT

Topical corticosteroids; antibiotics if secondary infection; tungiasis (tropics) may require excision of the embedded fleas

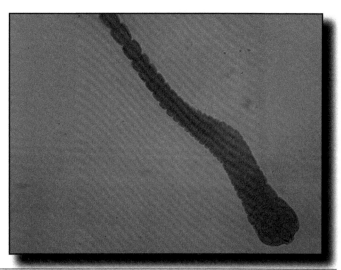

Figure 18.1 Fleas may serve as intermediate hosts for the dog tapeworm, *Dipylidium caninum*.

patients may develop secondary pneumonia and spread the infection by respiratory droplets, a condition known as pneumonic plague. This is the most serious form of plague (especially from the public health standpoint), leading to severe disease and frequently death.

There are several methods by which fleas transmit plague. Probably the most important method of infection occurs when fleas ingest plague bacilli along with host (rodent) blood. In the flea stomach, tremendous multiplication of the bacteria takes place (flea feces then also become infected). In some species—especially *Xenopsylla*—further multiplication of the bacteria occurs in the proventriculus, resulting in the flea becoming blocked. When blocked fleas try to feed, there is regurgitation of blood meal products from previous feedings. Blocked fleas become increasingly starved and repeatedly bite in order to get a blood meal. Also, as mentioned, flea feces can be infective, especially when rubbed into abrasions (such as bite wounds) in the skin. Plague bacilli can remain infective in flea feces for as long as 3 years.[13]

Murine Typhus. Murine typhus is a rickettsial disease transmitted to humans by fleas characterized by fever, malaise, headache, fatigue, myalgias, and rash, especially on the trunk. Although transmission of the agent can happen through flea bites, most infections are believed to occur when infectious flea feces are rubbed into the flea-bite wound or other breaks in the skin.[14] Murine typhus is generally mild with negligible mortality, except in the elderly. Complications, sometimes severe, can occur including bronchiolitis, pneumonia (even leading to acute respiratory distress syndrome), meningitis, septic shock, cholecystitis, pancreatitis, myositis, and rhabdomyolysis. The term murine indicates that the disease is related to rats. The classic cycle involves rat-to-rat transmission, with the Oriental rat flea, *Xenopsylla cheopis*, being the main vector. Murine typhus is one of the most widely distributed arthropod-borne infections endemic in many coastal areas and ports throughout the world.[6,15] Outbreaks have been reported

draining the flea-bite site enlarge, leading to local pain and swelling (the "buboes"). Other signs and symptoms common to many acute bacterial infections are present, including sudden onset of disease with a sharp rise in temperature, chills, headache, nausea, and increased pulse rate. The patient's condition may deteriorate rapidly, often leading to death within 3 days. Some patients are particularly susceptible to plague and develop a more widely disseminated and generalized infection called septicemic plague. Such cases are even more prone to death. Finally, a certain proportion of plague

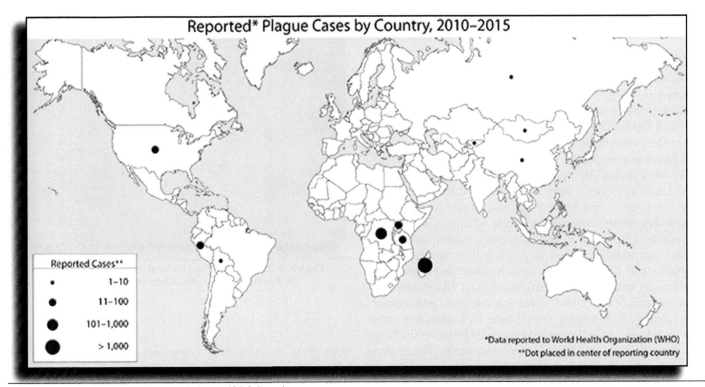

Figure 18.2 Worldwide distribution of plague (CDC figure).

from Australia, China, Greece, Israel, Kuwait, and Thailand. At one time, thousands of cases were reported annually in the United States; from 1931 to 1946, approximately 42,000 cases were reported.[2,16,17] After World War II, case numbers in the United States fell drastically to approximately 100 per year, but seem to be rising again.[18,19] A recent paper reported 90 cases of murine typhus from just 2 Texas hospitals over a 3-year period,[20] and there were 5 fatal cases reported from Texas during 2012–2015.[21] Almost all cases in the United States are now focused in central and southcentral Texas and Los Angeles and Orange counties in California;[22] however, physicians may encounter murine typhus in returning international travelers. Three cases in patients returning to Europe from Indonesia indicate that murine typhus should be considered as a possible cause of imported fever from Indonesia.[23] Interestingly, the ecology of this disease seems to be changing. The classic rat–flea–rat cycle seems to have been replaced in suburban areas by a peridomestic animal cycle involving free-ranging cats, dogs, opossums, and their fleas.[6]

B. Cat and Dog Fleas, *Ctenocephalides felis* and *C. canis*

Cat fleas are the fleas most often encountered by people in the United States (the dog flea is relatively rare in North America). Contrary to their name designation, dog fleas may feed on cats and cat fleas on dogs. In fact, in many areas the predominant flea species infesting dogs is the cat flea. Both species are mainly just pest species, although there is some evidence that cat fleas may transmit *Rickettsia felis* to humans.

In addition, cat fleas are intermediate hosts of the dog tapeworm, *Dipylidium caninum*, and their bites may produce papular urticaria. Children sometimes become infected with the tapeworm upon ingestion of an infected flea from a flea-infested dog or cat.

C. Oriental Rat Flea, *Xenopsylla cheopis*

This medically important flea is an ectoparasite of Norway rats and roof rats. Norway rats (also known as brown rats, sewer rats, or wharf rats) are found throughout the contiguous U.S., while the roof rat (also called the black rat) is generally found in the southeastern U.S. and along the west coast.[24] The Oriental rat flea is the primary vector of the agent of plague, *Yersinia pestis* (see earlier discussion) and is involved in the transmission of the agent of murine (endemic) typhus, *Rickettsia typhi*, from rat to rat and from rats to people.

D. Human Flea, *Pulex irritans*

Pulex irritans may also be a vector of the agent of plague. This flea occasionally becomes abundant on farms, especially in abandoned pigpens. Feingold and Benjamini[9] said *Pulex irritans* was one of two primary species involved in flea-bite allergic reactions in the San Francisco area; however, this species currently is only an infrequent parasite of humans in developed countries. In addition, records of *P. irritans* biting humans in the New World before 1958 may have been due to *P. simulans*, a closely related species.[25]

E. *Chigoe Fleas, Tunga penetrans,* and *Tunga trimamillata*

These fleas, also sometimes called jigger, nigua, chica, pico, pique, or suthi, burrow into the skin of people in tropical and subtropical regions.[8] Diagnosis is made by the presence of a blackish point (where the flea is embedded) surrounded by a whitish halo and a slight inflammatory reddening found at one of the preferential sites such as nail border of the toes or interdigital spaces and/or plantar zone or heel of the foot.[26] These fleas apparently originated in Latin America but have since been introduced into sub-Saharan Africa.[26,27] Ancient Peruvian pottery has been found depicting people examining holes on the soles of their feet (characteristic lesions for tungiasis).[27] Unlike most flea species, which spend only a small proportion of their lives on a host animal, *Tunga penetrans* and *T. trimamillata* remain embedded in the skin. The female remains embedded throughout blood engorgement and egg development, which leads to great enlargement of the flea body. Enlarging female fleas on human feet cause intense itching and inflammation and may produce swellings and ulceration. Secondary infection is common. Debilitating sequelae may develop, such as loss of nails and difficulty walking.[28]

F. Northern Rat Flea, *Nosopsyllus fasciatus*

The northern rat flea spends most of its adult life on Norway rats and roof rats. This flea is also involved in the transmission of murine typhus organisms among rats and to humans.[29]

G. Sticktight Flea, *Echidnophaga gallinacea*

This flea, sometimes called the hen flea, is primarily a pest of poultry, but humans are often attacked. As with the chigoe flea, *Echidnophaga gallinacea* attaches firmly to its host and engorges with blood. It may remain embedded in the integument of the host for some time.[30] Chickens frequently have dark flea-covered patches around the eyes, comb, or wattles (Figure 18.3).

H. Sand Fleas

The lay public uses the term sand fleas so much that they are included in this section. In the northern United States, what people call sand fleas are usually cat or dog fleas found in vacant lots and associated with stray cats or dogs. In the western United States, cat fleas or human fleas that are associated with deer, ground squirrels, stray cats and dogs, or prairie dogs are sometimes termed sand fleas. Cat and dog fleas, and occasionally sticktight fleas, are called sand fleas in the southern United States. Chigoe fleas are sometimes called sand fleas in South America. In addition, tiny crustaceans in the order Amphipoda occurring abundantly in seaweed along coastal beaches are also called sand fleas or beach fleas. These creatures are not fleas at all.

Figure 18.3 Sticktight fleas on head of a chicken. (Photo courtesy Dr. Kevin Maschek, Ceva Corporation, used with permission.)

II. General Description

Adult fleas have laterally compressed bodies, are between 2–6 mm long, and are usually brown or reddish-brown with stout spines on their head and thorax (Figures 18.4 and 18.5). Identification guides are provided by Holland[31] and Benton.[32] Fleas have a short, clublike antenna over each eye. Each segment of their three-segmented thorax bears a pair of powerful legs terminating in two curved claws. In most species, the hind legs are especially well developed for jumping (an exception is the sticktight flea). The chigoe flea is actually quite easily identified as it is so small (1 mm) and has a greatly shortened thorax. Figure 18.6 depicts the heads of the species discussed here. Most fleas move quickly on the skin or in hair and can jump 30 cm or more. They are readily recognized by their jumping behavior when disturbed.

III. Geographic Distribution

Most of the 2500 or so species of fleas are seldom seen or encountered, being found only on some small rodents, bats, or birds, often within a restricted range. The distributions of the seven species discussed here follow:

Cat and dog fleas—Worldwide in and around homes with pets. In certain regions, one species may occur to the exclusion of the other.

Oriental rat flea—Worldwide wherever *Rattus rattus* is found.

Human flea—Nearly cosmopolitan, although there may be large geographic areas apparently free of this species. This seems to be especially true in areas uninhabited by people.

Chigoe flea—Tropical and subtropical regions of North and South America, the West Indies, and parts of Africa. After a 40-year absence, this species has

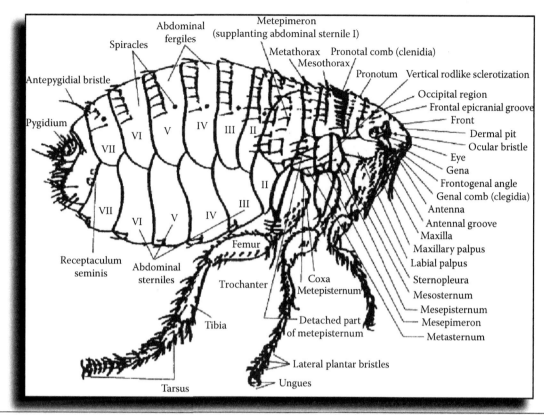

Figure 18.4 Adult cat flea, *Ctenocephalides felis*, with parts labeled. (From CDC, *Pictorial Keys to Arthropods, Reptiles, Birds, and Mammals of Public Health Significance*, U.S. Centers for Disease Control and Prevention, Atlanta, GA, 1963.)

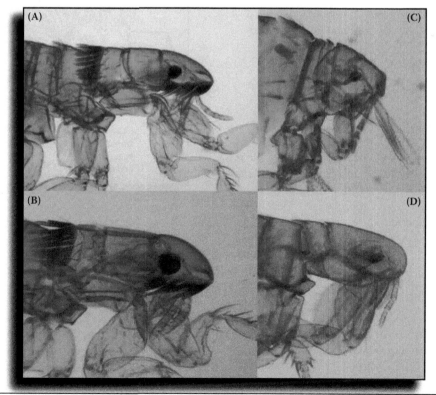

Figure 18.5 Color plate of selected fleas: (A) cat flea, (B) dog flea, (C) sticktight flea, and (D) Oriental rat flea. Note that the cat flea head slopes more than the dog flea head

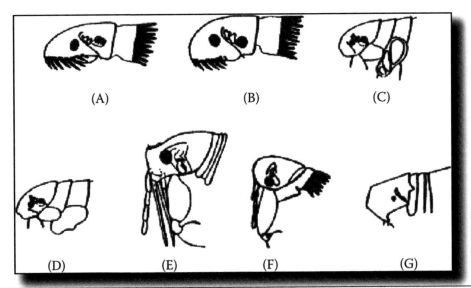

Figure 18.6 Line drawings of selected flea heads: (A) *Ctenocephalides felis*, (B) *Ctenocephalides canis*, (C) *Xenopsylla cheopis*, (D) *Pulex irritans*, (E) *Tunga penetrans*, (F) *Nosopsyllus fasciatus*, and (G) *Echidnophaga gallinacea*. (Adapted from CDC, *Pictorial Keys to Arthropods, Reptiles, Birds, and Mammals of Public Health Significance*, U.S. Centers for Disease Control and Prevention, Atlanta, GA, 1963; USDA, The Fleas of North America: Classification, Identification, and Geographic Distribution of These Injurious and Disease-Spreading Insects, Misc. Publ. No. 500, U.S. Department of Agriculture, Washington, DC, 1943.)

recently reappeared in Mexico.[33] There is one record of the chigoe flea in the United States (Texas).[34]

Northern rat flea—Widespread over Europe and North America and less commonly in other parts of the world.

Sticktight flea—Widely distributed in the world wherever chickens have been introduced as domestic animals.

IV. Biology and Behavior

Adult fleas have piercing–sucking mouthparts and feed exclusively on blood. The hosts of fleas are domesticated and wild animals, especially wild rodents. If hosts are available, fleas may feed several times daily, but in the absence of hosts adults may fast for months, especially at low to moderate temperatures. Some of the above-mentioned species have specialized life cycles, but in general, the life cycle of most fleas ranges from 30 to 75 days and involves complete metamorphosis.

As cat fleas are a notable pest and seemingly ubiquitous, their life cycle is presented here (Figure 18.7). Adult female fleas begin laying eggs 1–4 days after starting periodic blood feeding. Blood meals are commonly obtained from cats, dogs, and people, but other medium-sized mammals may be affected as well. Females lay 10–20 eggs daily and may produce several hundred eggs in their lifetime. Eggs are normally deposited in nest litter, bedding, carpets, etc. Warm, moist conditions are especially favorable for egg production. Eggs

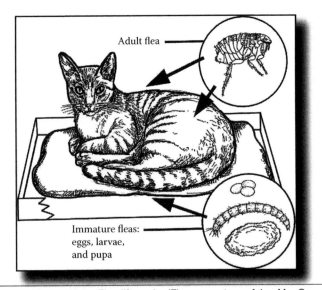

Figure 18.7 Flea life cycle. (Figure courtesy of Joe MacGown, Mississippi State University, and used with permission.)

quickly hatch into spiny, yellowish-white larvae. Flea larvae have chewing mouthparts and feed on host-associated debris including food particles, dead skin, and feathers. Blood defecated by adult fleas also serves as an important source of nutrition for the larvae. Larvae pass through three instars prior to pupating. Flea larvae are very sensitive to moisture and will quickly die if continuously exposed to less than 60–70% relative humidity. Pupating flea larvae spin a loose silken cocoon interwoven with debris. If environmental conditions are unfavorable, or if hosts are not available, developing adult fleas may remain inactive within the cocoon for extended periods. Adult emergence from the cocoon may be triggered by vibrations resulting from host movements.

V. Treatment of Bites and Infestation

Chigoe or burrowing fleas, *Tunga penetrans* and *T. trimamillata*, may need to be removed surgically. Within the first 48 hours of attachment, a sterile needle may be used to accomplish this. During maturation of the eggs, curettage of the contained flea and cautery of the hollow are sufficient. For a mature flea, Alexander[8] recommended total excision of the flea under local anesthesia. Most other fleas bite but do not remain attached or embedded. Dog or cat flea bites produce reddening, papules, and itching but generally require no specialized medical treatment. Oral antihistamines may help relieve the itching, and corticosteroids can aid in the resolution of the lesions. Scratching of flea bites may produce secondary infection and should be avoided. An antiseptic or antibiotic ointment may be indicated. Of primary concern, however, is to eliminate the source of flea infestation. This may involve sanitation, insecticidal treatment of pets, and the spraying of both indoor and outdoor premises.

References

1. Bibikova VA. Contemporary views on the relationships between fleas and the pathogens of human and animal diseases. *Ann. Entomol. Soc. Am.* 1977;22:1–34.
2. Harwood RF, James MT. *Entomology in Human and Animal Health.* 7th ed. New York: Macmillan; 1979.
3. Billeter SA, Metzger ME. Limited evidence for *Rickettsia felis* as a cause of zoonotic flea-borne rickettsiosis in southern California. *J. Med. Entomol.* 2017;54(1):4–7.
4. Reif KE, Macaluso KR. Ecology of *Rickettsia felis*: A review. *J. Med. Entomol.* 2009;46(4):723–736.
5. Perez-Osorio CE, Zavala-Velazquez JE, Arias-Leon JJ, Zavala-Castro JE. *Rickettsia felis* as emergent global threat for humans. *Emerg. Infect. Dis.* 2008;14:1019–1023.
6. Azad AF, Radulovic S, Higgins JA, Noden BH, Troyer JM. Flea-borne rickettsioses: Ecologic considerations. *Emerg. Infect. Dis.* 1997;3:319–327.
7. Brown KJ, Bennett M, Begon M. Flea-borne *Bartonella grahami* and *Bartonella taylori* in Bank voles. *Emerg. Infect. Dis.* 2004;10:684–688.
8. Alexander JO. *Arthropods and Human Skin.* Berlin: Springer-Verlag; 1984.
9. Feingold BF, Benjamini E. Allergy to flea bites. *Ann. Allergy.* 1961;19:1275–1279.
10. Craven RB, Maupin GO, Beard ML, Quan TJ, Barnes AM. Reported cases of human plague infections in the U.S. *J. Med. Entomol.* 1993;30:758–761.
11. CDC. Summary of notifiable infectious diseases and conditions – United States, 2015. *MMWR* 2017;64(53):1–144.
12. Golgowski N. Joy for girl, 7, who caught bubonic plague from dead squirrel. *The Daily Mail (UK).* 2012; 9 September 2012 issue. http://www.dailymail.co.uk/news/article-2200890/Sierra-Jane-Downing-Girl-7-caught-bubonic-plague-dead-squirrel-camping-trip-leave-hospital.html.
13. Service MW. *Medical Entomology for Students.* New York: Chapman and Hall; 1996.
14. Walker DH, Raoult D. Typhus group rickettsioses. In: Guerrant RL, Walker DH, Weller PF, eds. *Tropical Infectious Diseases: Principles, Pathogens, and Practice.* 3rd ed. New York: Saunders Elsevier; 2011:329–334.
15. Azad AF. Epidemiology of murine typhus. *Ann. Rev. Entomol.* 1990;35:553–569.
16. Pratt HD. The changing picture of murine typhus in the United States. *Ann. N. Y. Acad. Sci.* 1958;70:516–527.
17. Traub R, Wisseman CL, Azad AF. The ecology of murine typhus: A critical review. *Trop. Dis. Bull.* 1978;75:237–317.
18. Blanton LS, Vohra RF, Bouyer D, Walker DH. Reemergence of murine typhus in Galvaston, Texas, USA, 2013. *Emerg. Infect. Dis.* 2015;21:484–486.
19. Civen R, Ngo V. Murine typhus: An unrecognized suburban vectorborne disease. *Clin. Infect. Dis.* 2008;46:913–918.
20. Afzal Z, Kallumadanda S, Wang F, Hemmige V, Musher D. Acute febrile illness and complications due to murine typhus, Texas, USA. *Emerg. Infect. Dis.* 2017;23:1268–1273.
21. Pieracci EG, Evert N, Drexler NA, et al. Fatal flea-borne typhus in Texas: A retrospective case series, 1985–2015. *Am. J. Trop. Med. Hyg.* 2017;96:1088–1093.
22. Purcell K, Fergie J, Richman K, Rocha L. Murine typhus in children, South Texas. *Emerg. Infect. Dis.* 2007;13:926–927.
23. Parola P, Vogelaers D, Roure C, Janbon F, Raoult D. Murine typhus in travelers returning from Indonesia. *Emerg. Infect. Dis.* 1998;4:677–680.
24. Marsh RE. Roof rats. In: Hygnstrom SE, Timm RM, Larson GE, eds. *The Handbook of Prevention and Control of Wildlife Damage.* Lincoln, NE: University of Nebraska-Lincoln; 1994:367.
25. Durden LA, Hinkle NC. Fleas (Siphonaptera). In: Mullen GR, Durden LA, eds. *Medical and Veterinary Entomology.* 2nd ed New York: Elsevier; 2009:115–135.
26. Pampiglione S, Fioravanti ML, Gustinelli A, et al. Sand flea (*Tunga* spp.) infections in humans and domestic animals: State of the art. *Med. Vet. Entomol.* 2009;23(3):172–186.
27. Maco V, Tantalean M, Gotuzzo E. Evidence of tungiasis in pre-Hispanic America. *Emerg. Infect. Dis.* 2011;17:855–862.
28. Feldmeier H, Eisele M, Sabola-Moura RC, Heukelbach J. Severe tungiasis in underprivileged communities: Case series from Brazil. *Emerg. Infect. Dis.* 2003;9:949–955.

29. Pitalská E, Boldiš V, Mošanský L, Sparagano O, Stanko M. *Rickettsia* species in fleas collected from small mammals in Slovakia. *Parasitol. Res.* 2015;114:34–36.

30. Sanborn CE. The Chicken Sticktight Flea. Oklahoma State University, Agri. Exp. Sta. Bull. No. 123, 8 pp.; 1919.

31. Holland GP. The Fleas of Canada, Alaska, and Greenland. Memoirs of the Entomological Society of Canada, Publ. No. 130, Entomological Society of Canada, Ottawa; 1985.

32. Benton AH. *An Illustrated Key to the Fleas of the Eastern U.S.* Fredonia, NY: Marginal Media; 1983.

33. Ibanez-Bernal S, Velasco-Castrejon O. New records of human tungiasis in Mexico. *J. Med. Entomol.* 1996;33:988–998.

34. Ewing HE, Fox I. The Fleas of North America. U. S. Department of Agriculture, Misc. Publ. No. 500 1943:112.

19
FLIES (BITING)

I. Black Flies

A. General and Medical Importance

Black flies (also called buffalo gnats, turkey gnats, and Kolumbtz flies) are small, humpbacked flies that are important as vectors of disease and as nuisance pests.[1,2] In the tropics, black flies are vectors of the parasite *Onchocerca volvulus*, which causes a chronic nonfatal disease with fibrous nodules in subcutaneous tissues and sometimes visual disturbances and blindness (river blindness). The World Health Organization estimates that about 17 million people have onchocerciasis in Africa and Latin America (Figure 19.1).[3] Since 1987 (and still ongoing), onchocerciasis control has been greatly aided by mass drug administration of ivermectin, which has been a gold standard antiparasitic drug. More recently, a structurally similar new drug, moxidectin, has shown even more efficacy. There is a goal of total elimination of onchocerciasis from most of Africa by 2025.[4] Interestingly, an autoimmune response to infection with *Onchocerca* filarial worms has been reported, leading to a form of epilepsy and "nodding syndrome" in children in parts of Africa.[5]

In temperate regions, black flies are notorious pests, often occurring in tremendous swarms and biting humans and animals viciously (Figure 19.2). The older literature includes reports of human deaths from black fly biting, as well as one report of 400 mules dying within a few days after exposure to black fly swarms in Louisiana.[6] These historic outbreaks often led to drastic measures to prevent black fly biting such as "smokes," which were pots of burning wood or leaves (Figure 19.3). The bites may be painless at first, but they bleed profusely due to salivary components that prevent clotting.[7] Systemic reactions to black fly bites have been reported,

consisting of itching, burning, papular lesions accompanied by fever, leukocytosis, and lymphadenitis. Satellite bubos have been reported, and a syndrome called "black fly stiff neck" due to adenopathy in cervical lymph nodes.[6] Death can result from anaphylactic shock, suffocation, and toxemia.

B. General Description

Black flies vary in size from about 2–5 mm and are thus smaller than mosquitoes (see box). They are black, humpbacked flies with broad wings and stout bodies (Figure 19.4). Black flies have short legs and large compound eyes. Their antennae are short (although in 9–12 segments) and bare (lacking hairs or setae). Crosskey and Howard[8] provided a list of identification keys to black flies by zoogeographic region, and Adler et al.[9] provided an identification key to black flies of North America.

C. Geographic Distribution

There are numerous important species of black flies: *Prosimulium mixtum* is a serious pest of people and animals in much of the United States, as well as *Cnephia pecuarum* in the Mississippi Valley, and *Simulium meridionale* in the eastern and southcentral United States. In Mississippi, recent outbreaks of *S. meridionale* have killed chickens, purple martins, and other bird species. *Simulium vittatum* and *S. venustum* seriously annoy livestock, fishermen, and campers in the northern United States. In the Balkans region (former Yugoslavia), there have been severe outbreaks of *Simulium colombaschense* (the infamous golubatz fly) and *S. erythrocephalum*. Other notorious pests in Europe include *S. equinum*, *S. ornatum*, and *S. reptans*. In Africa, members of the *S. damnosum* and *S. neavei* complexes are important

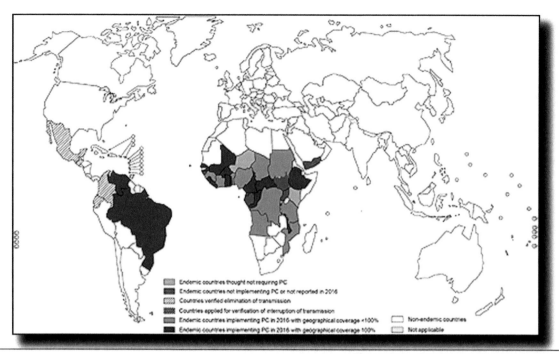

Figure 19.1 Distribution of onchocerciasis (WHO figure).

vectors of the agent of onchocerciasis. In Central and South America, vectors of onchocerciasis are *S. ochraceum* and *S. metallicum*.

D. Biology and Behavior

The larvae of black flies are aquatic and develop in shallow, fast-flowing streams, mainly in upland regions (Figure 19.5). The filter-feeding larvae have a circlet of numerous radiating rows of tiny hooks located on the tip of the abdomen (Figure 19.6). These hooks are used for attaching to silk, which the larvae secrete onto solid substrates, aiding them in attachment. Larvae also produce silk for construction of the pupal cocoon. There are usually six to nine larval instars. The densities of larvae can be tremendously high, and in some areas they can cover the entire substrate at the head of stream riffles. Life cycle length varies from 1 year to as little as 2 weeks depending on the species and geographic location. Males form small mating swarms during prenuptial behavior, and mating occurs in the air. Eggs are usually laid in groups of 150–600 in or near a water source. Adult black flies are most prevalent during late spring and early summer, when they may be present in large swarms (Figure 19.7). They are usually encountered near streams or lake outlets, but due to migration they may be dispersed and found tens of miles from the original water source. Black flies bite in the daytime, mostly in the early morning and near evening, and are hesitant to enter enclosures.

E. Treatment of Bites

Black fly bites may be itchy and slow healing (see Chapter 9). Cases in Mississippi seen by the authors have resulted in intensely pruritic papular urticaria, which resolved in 72 hours. Antiseptic and soothing lotions, as well as corticosteroids, may relieve pain and itching and help resolve lesions. Systemic reactions characterized by hives, wheezing, fever, leukocytosis, and widespread urticaria may require intensive evaluation and treatment. If the reaction is mild, oral antihistamine therapy may suffice, but severe reactions involving shock will probably require epinephrine (see Chapter 2). Persons bitten by black flies in Africa and Latin America may need follow-up for possible development of onchocerciasis.

II. Deer Flies

A. General and Medical Importance

Deer flies belong to the family Tabanidae (the same one as horse flies) but are usually much smaller. Deer flies are extremely annoying to people in the outdoors during summer months, often circling persistently around the head. Like horse flies, deer flies have scissorlike mouthparts and can inflict painful bites. Deer fly bites often become secondarily infected; in hypersensitive individuals, they have been known to produce systemic reactions characterized by generalized

BLACK FLIES

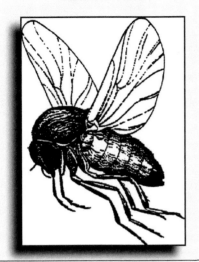

Typical adult black fly. (From Bowles, E., The Mosquito Book, Mississippi State Department of Health, Jackson, MS, 1989.)

IMPORTANCE

Fierce biters; possible toxic or systemic reactions

DISTRIBUTION

Numerous species worldwide

LESION

Variable—often small itching papules, sometimes erythematous wheals and swelling

DISEASE TRANSMISSION

Onchocerciasis (tropics)

KEY REFERENCES

Gudgel EF, Grauer FH. Acute and chronic reactions to black fly bites (*Simulium* fly). *Arch. Dermatol. Syphilol.* 1954;70:609–615

Adler PH, Currie DC, Wood DM. *The Black Flies of North America*. Ithaca, NY: ROM Publishers (Comstock); 2004

TREATMENT

Palliative creams or corticosteroids; oral antihistamines may relieve itching; systemic reactions may require antihistamines, epinephrine, and other supportive measures

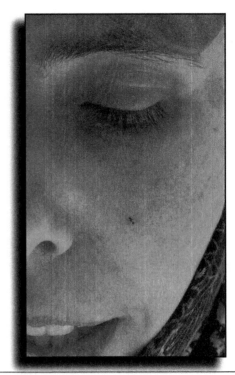

Figure 19.2 Black fly biting woman's face (Photo courtesy Wendy Varnado, Ph.D., with permission)

Figure 19.3 Farmer's using "smokes" to repel black flies in the Mississippi Delta, 1938. (Photo by George H. Bradley, USDA).

urticaria and wheezing. In the United States, the deer fly *Chrysops discalis* mechanically transmits tularemia organisms from rabbits to people by its bites, a condition sometimes referred to as deer fly fever.[10–12] During one outbreak, 64% of human cases of tularemia in Wyoming were attributed

Figure 19.4   Adult black fly.

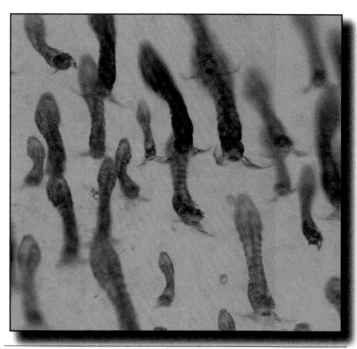

Figure 19.6   Black fly larvae. (Photograph courtesy of Elmer Gray, University of Georgia, Athens.)

Figure 19.5   Black flies breed in fast-moving water.

Figure 19.7   Pile of black flies collected in a mosquito trap.

to deer fly bites.[13] In the African equatorial rainforest, deer flies (particularly *C. silacea* and *C. dimidiata*) transmit the filarial nematode parasite *Loa loa* (Figure 19.8). Loiasis affects an estimated 10 million individuals and is characterized by Calabar swellings (localized nonpitting edema mainly on the wrists or ankles, 5–20 cm in diameter, lasting from a few hours to a few days), generalized pruritus, arthralgia, fatigue, hypereosinophilia, and sometimes serious central nervous system (CNS) involvement.[14,15] The pathognomonic symptom of loiasis, subconjunctival migration of the worm in the eye, is uncommon but still reported.[16]

B. General Description

Deer flies (about 8–15 mm long) are about half as large as horse flies (see box). They have wings with dark markings and have apical spurs on their hind tibiae (Figure 19.9). Many deer fly species are gray or yellow-gray in color with various arrangements of spots on the abdomen.

C. Geographic Distribution

Numerous deer fly species occur almost worldwide. In the United States, *C. discalis* occurs sparsely in the east but more so in the southwest. *Chrysops atlanticus* is a troublesome pest in the eastern United States, and *C. noctifer* is one of the most

DEER FLIES

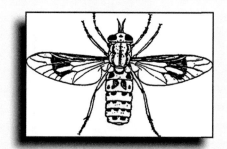

Adult deer fly, *Chrysops discalis*. (Figure courtesy of U.S. Centers for Disease Control and Prevention, Atlanta, GA.)

IMPORTANCE

Annoyance; painful bites

DISTRIBUTION

Numerous species worldwide

LESION

Deep and painful singular lesions, sometimes leading to cellulitis

DISEASE TRANSMISSION

Tularemia; loiasis (Africa)

KEY REFERENCES

Burger JF. Catalogue of Tabanidae (Diptera) of North America north of Mexico. *Contrib. Entomol. Int.* 1995;1:1–100

Foil LD. Tabanids as vectors of disease agents. *Parasitol. Today* 1989;5(3):88–96

TREATMENT

Systemic antibiotics of cellulitis, and allergic reactions may require antihistamines and epinephrine; otherwise, soothing ointments, lotions

common western species. *Chrysops silacea* and *C. dimidiata* occur in Africa.

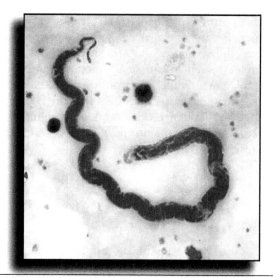

Figure 19.8 Deer flies may transmit the filarial worm *Loa loa* in tropical Africa. (Photo courtesy the CDC.)

Figure 19.9 Deer fly.

meal, whereas males feed on flower and vegetable juices. In the temperate zones, deer flies are active in the late spring and summer months, mostly in the early morning and late afternoon. Members of the genus *Chrysops* live in areas having a mixture of forest and open land and feed on various members of the deer family; however, they will aggressively pursue and bite people. Interestingly, in Africa, studies have shown that the presence of wood fires in local villages is attractive to *Chrysops silacea*.[17] Most bloodsucking insects are repelled by the smoke of a wood fire.

D. Biology and Behavior

Deer flies breed in moist or semiaquatic sites such as margins of ponds, damp earth, or other sites containing various amounts of mud and water. The wormlike carnivorous larvae spend their life in these wet, muddy habitats and then migrate to drier areas of soil to pupate. After the pupal stage, the adult flies emerge. The entire life cycle may take 2 years or more to complete in temperate regions. Females seek a blood

E. Treatment of Bites

Except for secondary infections, which require an appropriate antibiotic, deer fly bites are generally few and self-limiting. Antiseptic and soothing lotions may relieve pain and itching. Allergic reactions characterized by hives, wheezing, or widespread urticaria may require intensive evaluation and treatment. If the reaction is mild, oral antihistamine therapy may suffice, but severe reactions involving shock will probably require epinephrine (see Chapter 2). Follow-up may be prudent for persons bitten by deer flies in the southwestern United States (tularemia) or the African equatorial rainforest (*L. loa* filariasis). Gentamicin or streptomycin are drugs of choice for tularemia, while Hetrazan, diethylcarbamazine, is the first-line treatment for loiasis.[18] Ivermectin is sometimes used for loiasis, although it does not kill the adult worms.[15] Unfortunately, both treatments may produce fairly severe adverse effects.[19,20]

III. Horse Flies

A. General and Medical Importance

Horse flies (also family Tabanidae) are large, robust blood-sucking flies that are notorious pests of horses, cattle, deer, and other mammals. Several species of horse flies will also attack people. Horsefly bites have been known to produce systemic reactions in humans characterized by generalized urticaria and wheezing.

B. General Description

Horse flies look like giant robust house flies. They are often 20–25 mm long and have large prominent eyes (see box and Figure 19.10). Some species are called green heads because of their big green eyes. The antennae have only three sections. Their proboscis projects forward, and the female's mouthparts are bladelike for a slashing/lapping feeding method. Horse fly larvae are spindle shaped and generally white, tan, brown, or even greenish in color.

C. Geographic Distribution

Horse flies are distributed worldwide. Most people-biting species are in the genera *Tabanus*, *Hybomitra*, and *Haematopota*. Some of the notorious pest species include *Tabanus atratus* (the large black horse fly) (Figure 19.11), *T. lineola*, and *T. similis* (the familiar striped horse flies) in the eastern United States and *T. punctifer* and *Atylotus incisuralis* in the western United States.

D. Biology and Behavior

Horse fly biology is very similar to that of deer flies. Basically, they breed in moist or semiaquatic sites such as pond margins, damp earth, or rotten logs. The grublike larvae spend their life in wet mud, dirt, or shallow water and then pupate in drier patches of soil (Figure 19.12). Larval development may take a year or more. After a pupal stage, the adult flies emerge. Females seek a blood meal, whereas males feed on nectar from flowers and other vegetable juices. In the temperate region, horse flies are active only in the warmer months of the year.

HORSE FLIES

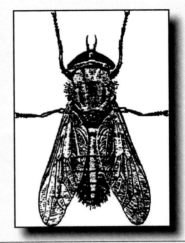

Adult horse fly. (From USDA, The Yearbook of Agriculture, U.S. Department of Agriculture, Washington, DC, 1952.)

IMPORTANCE

Annoyance; painful bites; possible systemic reactions

DISTRIBUTION

Numerous species worldwide

LESION

Deep and painful singular lesions sometimes leading to cellulitis

DISEASE TRANSMISSION

None

KEY REFERENCES

Burger JF. Catalogue of Tabanidae (Diptera) of North America north of Mexico. *Contrib. Entomol. Int.* 1995;1:1–100

Foil LD. Tabanids as vectors of disease agents. *Parasitol. Today* 1989;5(3):88–96

TREATMENT

Systemic antibiotics of cellulitis, and allergic reactions may require antihistamines and epinephrine; otherwise, soothing ointments, lotions

E. Treatment of Bites

Horse fly bites are generally few and self-limiting. If secondary infection/cellulitis develops, appropriate systemic antibiotics are indicated. Antiseptic and soothing lotions

Figure 19.10 Adult horse fly. (Photograph courtesy of Dr. Blake Layton, Mississippi State University.)

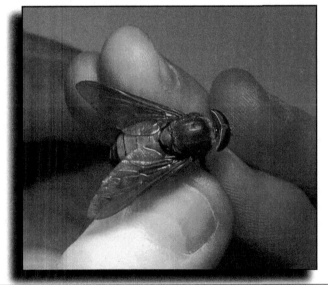

Figure 19.11 The large black horse fly, *Tabanus atratus*.

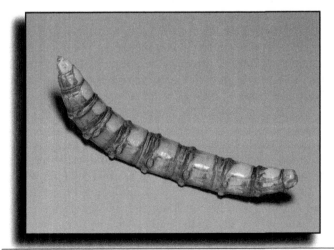

Figure 19.12 Typical horsefly larva. (Photograph courtesy of Dr. Blake Layton, Mississippi Cooperative Extension Service, Mississippi State University.)

may relieve pain and itching. Allergic reactions characterized by hives, wheezing, and widespread urticarial may require intensive evaluation and treatment. If the reaction is mild, oral antihistamine therapy may suffice, but severe reactions involving shock will probably require epinephrine (see Chapter 2).

IV. Midges (Biting Midges, Biting Gnats)

A. General and Medical Importance

The biting midges are very tiny slender gnats in the family Ceratopogonidae (see box); they are sometimes called punkies, no-see-ums, gnats, or flying teeth. In the Caribbean region and Australia, they are referred to as sand flies (not to be confused with psychodid sand flies; see Section V). Adult biting midges are vicious and persistent biters, and some persons have strong reactions to their bites. Their small size allows them to pass through ordinary screen wire used to cover windows and doors. These tiny insects are generally not involved in the transmission of disease agents to humans in the United States; however, in Africa and South America, certain species may be able to transmit filariae, protozoa, and viruses (particularly Shuni and Oropouche viruses in the Simbu serogroup of arboviruses). Oropouche fever, caused by a virus in the Simbu group of Bunyaviridae, is characterized by fever, headache, anorexia, dizziness, muscle and joint pain, and photophobia.[21] It is generally nonfatal but may be serious enough to lead to prostration. Since it was first isolated from a patient in Trinidad in 1955, this virus has been documented in numerous epidemics in the Amazon region of Brazil.[22] Over the past 60 years, there have been several outbreaks of Oropouche fever, with approximately 500,000 cases in the Americas.[23] Little is known about the animal reservoirs of Oropouche virus in nature, but some evidence implicates monkeys, wild birds, and sloths. The primary biting midge vector in urban outbreaks is *Culicoides paraensis*.[22]

B. General Description

Biting midges are typically gray in color (although some species may be yellowish), extremely small (0.6–1.5 mm), and delicate with narrow wings that have few veins and no scales. The wings may be clear or hairy, sometimes distinctly spotted (with pigment, not scales as in mosquitoes), and folded scissorlike over the abdomen at rest. The eyes on each side of the head are black, and the proboscis protrudes forward and downward. Biting midges somewhat resemble other small species of nonbiting gnats in the family Chironomidae, but they are not as large and mosquito-like as chironomids. Swarms of biting midges are small and inconspicuous. People attacked by this midge will often comment when outdoors, "Something is biting me but I can't see what it is."

CASE HISTORY

HORSE FLY BITE ON A SLEEPING VICTIM

In July, a woman in Montgomery, Alabama, was awakened during the night by a painful insect bite on her face. Looking around in the bedroom, she found a large horse fly on the wall. Earlier that evening, the horse fly had gotten inside the house and she and her husband had unsuccessfully tried to find it. Apparently, the horse fly later found the victim sleeping and bit her. Within 12 hours, the bite site was a 1.5-cm erythematous indurated plaque, hot to the touch (Figure 19.30). Other than a persistent itch for a few days, no further complications arose.

Comment. Horse flies are robust flies known for biting horses, cows, and other large mammals. Some species are almost 2 inches long. They do not have tubelike mouthparts for piercing–sucking but instead have a complex arrangement of bladelike mandibles and styliform maxillary galeae that cut or slash wounds in their host's skin (Figure 19.31). Horse flies then draw up the blood by means of pseudotracheae on labellar lobes. Blood uptake is quite inefficient, with blood often seen dripping from the wound. The lesion described

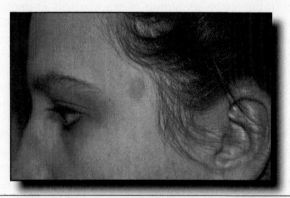

Figure 19.30   Lesion resulting from horsefly bite, 12 hours post-bite. (Photograph courtesy of Lindsey Carpenter Goddard.)

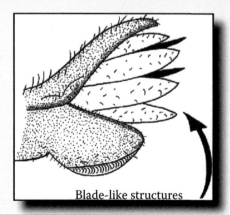

Blade-like structures

Figure 19.31   Diagrammatic representation of horse fly mouthparts.

previously was red and warm on palpation because of inflammation and vascular dilatation. Various immune cells attracted to the site (or formed) as part of the immune reaction release histamine, which causes blood vessels to dilate. Induration of a lesion like this is not only from the influx of cells; histamine also causes fluid to be leaked from the small skin blood' vessels, making the area feel hard and raised. Other than possible secondary infection, no human diseases are known to be caused by horse flies.

C. Geographic Distribution

There are more than 4000 species of Ceratopogonidae in temperate and tropical areas of the world. A worldwide list of the species, subspecies, and varieties in the genus *Culicoides* has been provided by Boorman and Hagen.[24] *Culicoides furens* is a vicious biter occurring in salt marshes along the Atlantic and Gulf coasts, from Massachusetts to Brazil, and along the Pacific Coast from Mexico to Ecuador. Further inland, *C. paraensis* is a troublesome species throughout much of the eastern United States. In California, *Leptoconops torrens* and *L. carteri* are very bothersome. *Culicoides pulicaris* is a serious pest of farmworkers in Europe, and *Forcipomyia* (*Lasiohelea*) *taiwana* is a severe pest in Taiwan and Japan. In the Near and Middle East, Central Asia, Africa, and southern Europe, *L. kerteszi* is an avid biter of people. Its bites may become vesicular, resulting in an open lesion that may exude moisture for a week or more. *Austroconops macmillani* is a vicious biter in western Australia, and *L. spinosifrons* is a beach pest in East Africa, Madagascar, India, Sri Lanka, and the Malay Archipelago.

D. Biology and Behavior

Only female ceratopogonids bite people; the males feed on the nectar of flowers. Females feed by the way of small cutting teeth on the elongated mandibles in their proboscis which they use to make a small cut in the skin. A chemical is present in the saliva to prevent blood clotting. Some of the cut capillaries bleed and form a tiny pool of blood that is then sucked up. Larval stages develop in highly organic detritus overlaying the bottom of shallow areas of water or in water-saturated soil high in organic material. They especially seem to be found in salt marshes. The larvae look more like worms than maggots (Figure 19.13). Female biting midges lay their eggs on the water or on a variety of objects overhanging larval habitats. Most species are active only in the warmer months of the year, but some may be active year-round. Lillie et al.[25] found that adult Culicoides host-seeking activity was greatest near sunrise and sunset, during the night when the moon was full, and during the afternoon hours in winter and early spring. They usually do not bite during the midday, except when the sky is heavily overcast and the winds are calm.

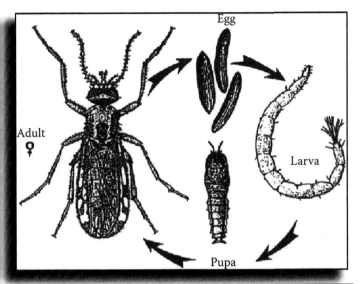

Figure 19.13 Life stages and life cycle of *Culicoides furens*. (Figure courtesy of Florida Medical Entomology Laboratory, Institute of Food and Agricultural Services, University of Florida, Gainesville.)

E. Treatment of Infestation

Alexander[26] suggested that simple antipruritics (oily calamine lotion or hydrocortisone with added anti-infective agents) are all that is needed for biting midge bites. Probably the best course of action for those affected by biting midges is a combination of avoidance and personal protection measures. Use of finer-mesh screen wire may prevent entry of these flies into dwellings. Repellents containing diethyltoluamide (DEET) or picaridin and long-sleeved shirts and long pants may provide some relief to those persons outside in infested areas. However, one study showed that traps surrounded by DEET-treated mesh caught 3–4 times more Culicoides midges than other treatments (so there wasn't any repellency).[27] Avon's Skin-So-Soft® bath oil is sometimes used as a repellent, and controlled studies indicate that the product provides some protection from Culicoides midges[28]; however, product effectiveness is not because it repels midges but because the oiliness traps the midges on the skin surface.

V. Sand Flies

A. General and Medical Importance

Sand flies (see box) are tiny bloodsucking flies in the family Psychodidae that transmit the causative agents of bartonellosis (Carrión's disease), sand fly fever, and leishmaniasis. Sandfly fever, a viral disease, occurs in those parts of southern Europe, the Mediterranean, the Near and Middle East, Asia, and Central and South America where the *Phlebotomus* vectors exist. Bartonellosis caused by the bacillus *Bartonella bacilliformis* occurs in the mountain valleys of Peru, Ecuador,

BITING MIDGES

Adult biting midge.

IMPORTANCE

Annoyance from biting

DISTRIBUTION

Numerous species occurring worldwide

LESION

Minute papular lesions with erythematous halo; wheals may occur in sensitized persons

DISEASE TRANSMISSION

Oropouche fever (tropics)

KEY REFERENCES

Kwan WH, Morrison FO. A summary of published information for field and laboratory studies of biting midges, *Culicoides* species. *Ann. Soc. Entomol. Quebec* 1974;19(13): 127–137

Downes JA. Feeding and mating in the insectivorous Ceratopogoninae (Diptera). *Memoirs Entomol. Soc. Canada* 1978;104:1–61

TREATMENT

Antipruritic lotions or creams

and southwest Colombia. Leishmaniasis occurs in tropical and subtropical areas over much of the world (Figures 19.14 and 19.15). There has been a recent resurgence of leishmaniasis; for example, cutaneous leishmaniasis (just one form of the illness) is endemic to more than 70 countries, with 90% of cases occurring in Afghanistan, Algeria, Brazil, Pakistan, Peru, Saudi Arabia, and Syria.[29] Global incidence is estimated to be 700,000–1,000,000 new cases reported annually with more than 20,000 deaths.[30] Contributing factors to the incidence of leishmaniasis include human settlement in zoonotic foci and

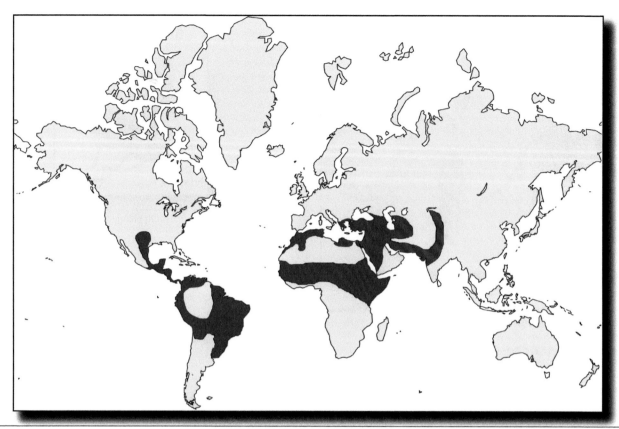

Figure 19.14 Cutaneous leishmaniasis distribution.

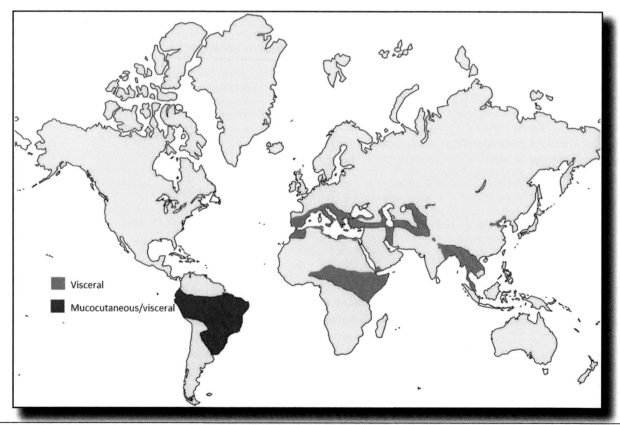

Figure 19.15 Visceral and mucocutaneous leishmaniasis distribution.

SAND FLIES

Adult sand fly. (From Bowles, E., The Mosquito Book, Mississippi State Department of Health, Jackson, MS, 1989.)

IMPORTANCE

Vectors of several important diseases

DISTRIBUTION

Numerous species almost worldwide

LESION

Sometimes red papules or urticarial wheals

DISEASE TRANSMISSION

Leishmaniasis, bartonellosis, sandfly fever

KEY REFERENCE

Young DG, Perkins PV. Phlebotomine sand flies of North America (Diptera: Psychodidae). *J. Am. Mosq. Control Assoc.* 1984;44(2):263–304

TREATMENT

Except for hypersensitivity reactions, palliative antipruritic lotions or creams

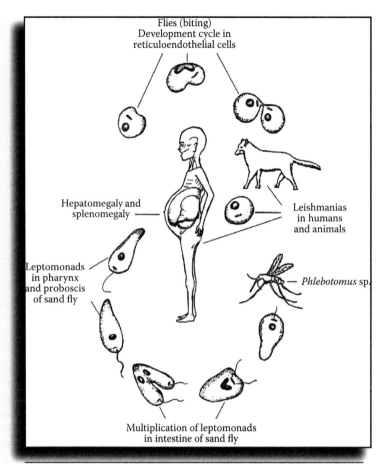

Figure 19.16 Life cycle of *Leishmania donovani*. (From Brooks, Jr., TJ. *Essentials of Medical Parasitology*. New York: Macmillan; 1963. With permission.)

urbanization up to the edge of forests. In many cases, the sylvatic cycles have now become peridomestic. The life cycle of *Leishmania donovani* is shown in Figure 19.16 to illustrate the involvement of sand flies in parasite life cycles.

Clinically, leishmaniasis manifests itself in four main forms: (1) cutaneous, (2) mucocutaneous, (3) diffuse cutaneous, and (4) visceral, and these forms have many different lay terms to describe them (Table 19.1). The cutaneous form may appear as small and self-limiting ulcers that are slow to heal. When there is destruction of nasal and oral mucosa, the disease is labeled mucocutaneous leishmaniasis (Figure 19.17). Sometimes there are widespread cutaneous papules or nodules all over the body, a condition termed diffuse cutaneous leishmaniasis. Finally, the condition in which the parasites invade cells of the spleen, bone marrow, and liver—causing widespread visceral involvement—is termed visceral leishmaniasis or kala-azar. Visceral leishmaniasis is fatal if left untreated in over 95% of cases.[30] Interestingly, there have been reports of visceral leishmaniasis in foxhounds in the United States,[31] but the only human sand-fly-transmitted disease in the United States is probably the few cases of cutaneous leishmaniasis diagnosed each year in south Texas.[32][32–34]

For a long time, these Old World and New World disease forms were correlated with various *Leishmania* species and geographic regions to make a well-defined classification. As is the case in many paradigms in science, this classification is turning out to be not so clear-cut. There is apparently a whole spectrum of disease—from cutaneous to visceral—depending on many factors such as species of Leishmania, numbers of parasites (parasite burden), and the predominant host immune response. The idea that a few Leishmania species each cause a distinct and separate clinical syndrome is no longer valid; for example, a visceral species, *L. chagasi*, has also been isolated from patients with cutaneous leishmaniasis in several Central American countries. However, the particular parasite species and geographic location may still serve as useful epidemiologic labels for the study of the

Table 19.1 Common Names for the Main Forms of Leishmaniasis

Cutaneous Leishmaniasis	
Aleppo boil	Delhi boil
Aleppo button	Jericho button
Aleppo evil	Kandahar sore
Bagdad boil	Lahore sore
Biskra button	Oriental button
Biskra nodule	Oriental sore
Calcutta ulcer	Pian bois
Chiclero ulcer	Uta

Visceral Leishmaniasis	
Black fever	Kala-azar
Dum-dum fever	

Figure 19.18 Adult sand fly. (Photograph courtesy of U.S. Armed Forces Pest Management Board.)

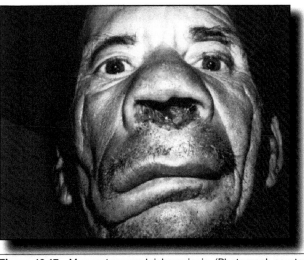

Figure 19.17 Mucocutaneous leishmaniasis. (Photograph courtesy of Armed Forces Institute of Pathology, Washington, DC, Negative No. 74-8873-1.)

disease complex, and generalized statements can be made. *Leishmania donovani* and *L. infantum*, for example, generally cause visceral leishmaniasis in the Old World, whereas *L. tropica* and *L. major* cause cutaneous lesions. In the New World, visceral disease is mainly caused by *L. chagasi*, cutaneous lesions by *L. mexicana* and related species, and mucocutaneous lesions by *L. braziliensis*. Further, by sorting out the species that do not cause human disease in a given area, researchers are more able to focus their studies on the ecology and behavior of those that do.

B. General Description

Sand flies are tiny (about 3 mm long), golden, brownish, or gray long-legged flies (Figure 19.18). The females have long, piercing mouthparts adapted for blood sucking. A distinctive feature of these flies is the way they hold their wings in a V-shape at rest. They have long, multisegmented antennae and hairs (not scales) covering much of the body and wing margins.

C. Geographic Distribution

Members of the sand fly genera *Phlebotomus* and *Sergentomyia* occur in the Old World, and *Lutzomyia*, *Brumptomyia*, and *Warileya* occur in the tropics and subtropics of the New World. *Phlebotomus argentipes* is the chief vector of visceral leishmaniasis in many areas of the Old World, but *P. langeroni* has recently been incriminated. *Lutzomyia longipalpis* is a major vector in the New World. Old World sand fly vectors of dermal leishmaniasis include *P. caucasicus*, *P. papatasi*, *P. longipes*, and *P. pedifer*. New World vectors of the malady include *L. olmeca*, *L. trapidoi*, *L. ylephiletrix*, *L. verrucarum*, *L. peruensis*, and others. Sandfly fever vectors are primarily *P. papatasi* and *P. sergenti*. Bartonellosis is transmitted by *L. verrucarum*. As *P. papatasi* is the most important and widespread vector of zoonotic cutaneous leishmaniasis caused by *Leishmania major* in the Old World, its distribution is given in Figure 19.19. *Lutzomyia anthophora* or *L. diabolica*, which generally occur in south and central Texas as well as in parts of Mexico, are thought to be the vectors of human cases of cutaneous leishmaniasis in Texas.

D. Biology and Behavior

Female sand flies are blood feeders; males are not, but they will suck moisture from available sources. Despite their medical importance, much of what is known about sand flies is speculative or based on limited field observations. Sand flies are usually found in microhabitats within larger biological communities. Examples of these microhabitats are caves, cavities, tree holes, burrows, pit latrines, animal enclosures, and other buildings. Breeding is believed to take place in these areas, and feeding on hosts occurs there or in close vicinity. Sand flies in the genus *Brumptomyia* feed on armadillos, whereas *Lutzomyia* spp. feed on both mammals and reptiles. Members of the genus *Sergentomyia* feed on reptiles and amphibians. *Phlebotomus* spp. are exclusively mammal feeders. Sand flies are generally active at night when there is little or no wind, but a few species may feed during the day if disturbed or under cloudy or

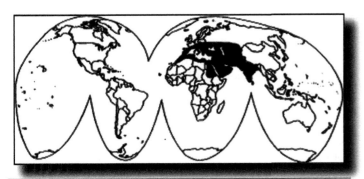

Figure 19.19 Approximate geographic distribution of *Phlebotomus papatasi*.

shaded conditions. They normally rest during the day in their microhabitats. Sand fly development is relatively slow, taking up to several months to complete the life cycle. *P. papatasi*, the primary vector of leishmaniasis in the Middle East region, is a nocturnal, desert species flying from sunset through dark. It enters dwellings from animal burrows up to 50 meters away. Its activity diminishes rapidly when relative humidity falls below 65%, temperature rises above 80°F, or wind velocity exceeds 5 miles per hour.

E. Treatment of Bites

Sand flies are relatively uncommon; thus, their bites are generally few and self-limiting. Soothing locations and antiseptic lotions may relieve itching and prevent secondary infection. Follow-up of patients with sand fly bites or persons with exposure to areas endemic with the several sand fly-transmitted diseases is needed to diagnose and treat these maladies. In the case of leishmaniasis, diagnosis is delayed by physicians unfamiliar with the clinical features and various forms of the disease. Cutaneous leishmaniasis is often initially described as a boil or ulcer that fails to heal. In addition, cutaneous leishmaniasis needs to be differentiated from other skin lesions such as those resulting from leprosy, tuberculosis cutis, sporotrichosis, histoplasmosis (*Histoplasma duboisii*), molluscum contagiosum, and cutaneous sarcoidosis. Treatment usually involves intralesional and parenteral pentavalent antimonals such as sodium stibogluconate and meglumine antimoniate, although these products may cause serious adverse side effects.[29] The most current treatment recommendations for leishmaniasis can be obtained by consultation with the U.S. Centers for Disease Control and Prevention, Parasitic Diseases Division (770-488-7775).

VI. Stable Flies

A. General and Medical Importance

The stable fly, *Stomoxys calcitrans*, is a significant medical and veterinary pest.[35] People who say "a house fly bit me" are usually mistakenly referring to the stable fly. The flies (sometimes also called dog flies) are fierce biters of people, pets, and

STABLE FLIES

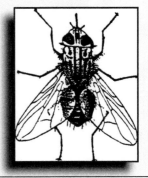

Adult stable fly.

IMPORTANCE
Nuisance biting

DISTRIBUTION
Most of the world

LESION
Small evanescent papules

DISEASE TRANSMISSION
None

KEY REFERENCE
Zumpt F. *The Stomoxyine Biting Flies of the World*. Jena, Germany: Gustav Fischer Verlag; 1973

TREATMENT
Except for hypersensitivity reactions (rare) or secondary infection, soothing and antiseptic lotions or creams sufficient

livestock and are a major pest in some seacoast areas, impeding development (Figure 19.20). Because of their bloodsucking habits, the flies have been suspected of transmitting a number of human diseases by mechanical action, but proof is lacking.

B. General Description

Stable flies are 5–6 mm long, have a dull gray thorax with four dark longitudinal stripes, and have a dull gray abdomen with dark spots (see box). They look very similar to house flies, but they are slightly larger and have a rigid proboscis projecting forward in a bayonetlike fashion (Figure 19.21). In contrast, house flies have sponging mouthparts that project downward.

Figure 19.20 Stable flies attacking woman's leg while walking on seacoast. (Photo courtesy Loretta Goddard, used with permission.)

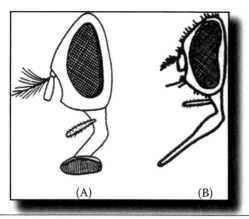

Figure 19.21 (A) House fly and (B) stable fly mouthpart comparison. (Figure courtesy of U.S. Centers for Disease Control and Prevention, Atlanta, GA.)

C. Geographic Distribution

The stable fly occurs in most parts of the world. There is a related species, *S. nigra*, occurring throughout Africa which attacks horses, donkeys, cattle, and people.

D. Biology and Behavior

Both male and female stable flies are vicious biters, and both take blood meals. It only takes 3–4 minutes for them to engorge to full capacity. The female lays her eggs in plant waste, cut grass, old haystacks, piles of fermenting seaweed, or manure (if there is sufficient straw or hay mixed in it). Larval development takes 8–30 days, depending on temperature. The larvae (maggots) look almost identical to those of house flies (Figure 19.22). Stable flies may overwinter as pupae, but normal pupal development takes about 1–3 weeks. Overall, under optimum conditions, the total time for development of stable flies from eggs to adults is about 35 days. Adults remain alive for a month or two. They bite people and animals during the day.

E. Treatment of Bites

Stable fly bites are generally few and self-limiting. Antiseptic and soothing lotions may relieve pain and itching. Allergic reactions characterized by hives, wheezing, and widespread urticaria are rare but may require intensive evaluation and treatment.

VII. Tsetse Flies

A. General and Medical Importance

There are over 20 species of flies in the genus *Glossina* that are called tsetse flies (see box). Most of the species are vectors of trypanosomes of people and animals (Figure 19.23); however, at least six species are of primary importance as vectors

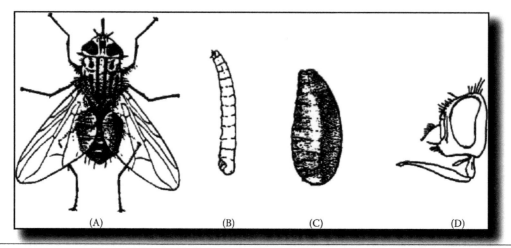

Figure 19.22 Stable fly life cycle: (A) adult, (B) larva; (C) pupa, and (D) close-up of head. (Figure courtesy of Alabama Cooperative Extension Service, Auburn.)

TSETSE FLIES

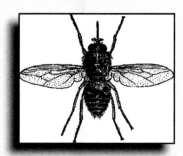

Adult tsetse fly. (Figure courtesy of U.S. Department of Agriculture.)

IMPORTANCE

Vectors of human African trypanosomiasis (HAT) or "sleeping sickness"

DISTRIBUTION

Tropical Africa

LESION

Small punctate hemorrhages unless sensitive to saliva (in which case, itchy wheals result)

DISEASE TRANSMISSION

African trypanosomiasis (HAT)

KEY REFERENCES

Buxton PA. *The Natural History of Tsetse Flies: An Account of the Biology of the Genus* Glossina *(Diptera)*. London: HK Lewis & Co.; 1955

Willet KC. Trypanosomiasis and the tsetse fly problem in Africa. *Annu. Rev. Entomol.* 1963;8:197–214

TREATMENT

Soothing lotions and antiseptic lotions or creams

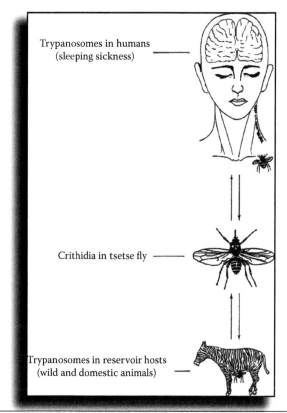

Figure 19.23 Lifecycle of *Trypanosoma rhodesiense* and *T. gambiense*. (From Brooks, Jr., T.J., *Essentials of Medical Parasitology*. New York: Macmillan; 1963. With permission.)

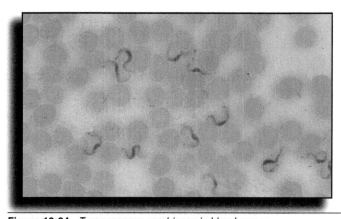

Figure 19.24 *Trypanosoma gambiense* in blood smear.

of human African trypanosomiasis (HAT), caused by subspecies of the protozoan *Trypanosoma brucei* (Figure 19.24). HAT is sometimes called sleeping sickness because meningoencephalitis associated with the disease causes apathy, fatigability, confusion, and somnolence. The patient may gradually become more and more difficult to arouse and finally becomes comatose. In 1998, the WHO estimated that at least 300,000 cases of African trypanosmiasis were undiagnosed and untreated; however, intensive tsetse fly trapping and control, as well as case surveillance/treatment, brought the number of cases in 2009 to below 10,000 for the first time in 50 years.[36] Fortunately, case numbers continue to decline, with only about 3000 reported in 2015.[37] Countries affected the most include the Democratic Republic of Congo, Angola, and Sudan (Figure 19.25). Unfortunately, there is rising resistance to melarsoprol, the only widely available drug for CNS involvement in African trypanosomiasis. Researchers are now exploring the use of fexinidazole as a viable alternative treatment. Nash[38] and Willet[39] provided very good reviews of the complicated African trypanosomiasis problem. Briefly, the chief vectors of *Trypanosoma brucei gambiense*, the cause of the Gambian form of sleeping sickness, are *Glossina palpalis*, *G. fuscipes*, and *G. tachinoides*. Cases of Gambian sleeping sickness occur in western and central Africa and are usually

Figure 19.25 Distribution of human African trypanosomiasis

more chronic. In eastern Africa, the Rhodesian (or eastern) form, which is virulent and rapidly progressive, is caused by *T. brucei rhodesiense*. The primary vectors of the Rhodesian form are *G. morsitans*, *G. swynnertoni*, and *G. pallidipes*. The eastern form is commonly contracted by travelers on game safaris or eco vacations. Other than the possibility for sleeping sickness transmission, the bites of tsetse flies are of minor consequence; however, some individuals become sensitized to the saliva, and subsequent bites produce welts.

B. General Description

Tsetse flies are 7–13 mm long and yellow, brown, or black. They fold their wings scissorlike over their back at rest, and this, along with other body features, gives them an appearance similar to wasps or honey bees. The arista arising from the short, three-jointed antennae has rays that are branched bilaterally (Figure 19.26). Tsetse flies have a long, slender proboscis that is held out in front of the fly at rest. The discal cell (first M) looks like a meat cleaver or hatchet (Figures 19.27 and 19.28).

C. Geographic Distribution

Tsetse flies are generally confined to tropical Africa between 15°N and 20°S latitude. *Glossina morsitans* is a bush species found in wooded areas and brush country in eastern Africa.

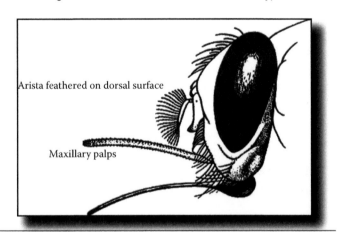

Figure 19.26 Side view of tsetse fly head. (From U.S. Naval Medical School, Laboratory Guide to Medical Entomology with Special Reference to Malaria Control. Bethesda, MD: National Naval Medical Center; 1945.)

In western and central Africa, where members of the *G. palpalis* group are the principal vectors, the flies are predominantly found near the specialized vegetation lining the banks of streams, rivers, and lakes.

D. Biology and Behavior

Tsetse flies feed on a wide variety of mammals and a few reptiles; people are not their preferred hosts. Both sexes feed on

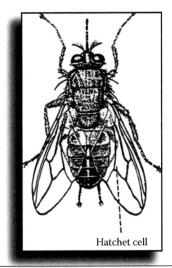

Figure 19.27 Adult tsetse fly. (From U.S.D.A. The Yearbook of Agriculture. Washington, DC: U.S. Department of Agriculture; 1952.)

Figure 19.29 Tsetse fly trap. (Photograph courtesy of Rachel Freeman Ford, RN, and used with permission.)

Figure 19.28 Adult tsetse fly. (Photo courtesy Whitney Cranshaw, Colorado State University, with permission.)

blood and bite during the day. Tsetse flies have a lifespan of about 3 months (less for males), and females give birth to full-grown larvae on dry and loose soil in places like thickets and sandy beaches. The larvae burrow a few centimeters into the substrate and pupate. The pupal stage lasts 2 weeks to a month, after which the adult fly emerges to continue the life cycle.

E. Treatment/Prevention of Bites

There have been intensive efforts to control tsetse flies with insecticides and traps (Figure 19.29), which have yielded measurable successes (see discussion in Section A about declining case numbers). Tsetse fly bites are generally self-limiting. Antiseptic and soothing lotions may relieve pain and itching. Persons sensitive to tsetse fly bites may require antihistamine and other treatment to control the immune response. People who have been bitten in endemic areas for sleeping

sickness should be followed up for development of the disease. Treatment for HAT differs according to the form and phase of the disease. Generally, it is treated with pentamidine, although suramin, eflornithine, nifurtimox, and melarsoprol may also be used. However, melarsoprol may be fatal in 3–10% of patients treated,[40,41] and there are other complicating factors with all these drugs. Physicians attempting to treat a patient with HAT should seek the most up-to-date treatment recommendations from the WHO or the Parasitic Drug Service, Centers for Disease Control, Atlanta, GA.

References

1. Crosskey RW. Blackflies. In: Lane RP, Crosskey RW, eds., *Medical Insects and Arachnids*. London: Chapman and Hall; 1996:241–287.
2. Nelson GS. Onchocerciasis. *Adv. Parasitol.* 1970;8:173–177.
3. WHO. Progress report on the elimination of human onchocerciasis, 2016–2017. *Weekly Epidemiol. Rec.* 2017;92:681–694.
4. Evans DS, Unnasch TR, Richards FO. Onchocerciasis and lymphatic filariasis elimination in Africa: Its about time. *Lancet.* 2015;385:2151–2152.
5. Vogel G. Parasitic worm may trigger mystery nodding syndrome. *Science (News Focus).* 2017;355:678.
6. Stokes JH. A clinical, pathological, and experimental study of the lesions produced by the bite of the black fly, *Simulium venustum. J. Cut. Dis. Incl. Syph.* 1914;32(11):1–46.
7. Cupp EW, Cupp MS. Black fly salivary secretions: Importance in vector competence and disease. *J. Med. Entomol.* 1997;34:87–94.
8. Crosskey RW, Howard TM. *A New Taxonomic and Geographical Inventory of World Blackflies*. Special Publ. London: The Natural History Museum; 1997:1–49.
9. Adler PH, Currie DC, Wood DM. *The Black Flies of North America*. Ithaca, NY: Comstock Publishing Associates (Cornell University Press); 2004.

10. Hopla CE. The ecology of tularemia. *Adv. Vet. Sci. Comp. Med.* 1974;18:25–53.

11. McDowell JW, Scott HG, Stojanovich CJ. *Tularemia*. Centers for Disease Control Training Guide, Washington, DC: U.S. Public Health Service; 1964: 65.

12. Francis E. Deer-fly fever or Pahvant Valley plague. *Public Health Rep.* 1919;34:2061–2062.

13. CDC. Tularemia transmitted by insect bites – Wyoming, 2001–2003. *MMWR*, 2005;54:170–173.

14. Pinder M. *Loa loa* – A neglected filaria. *Parasitol. Today.* 1988;4:279–284.

15. Metzger WG, Mordmuller B. *Loa loa* – Does it deserve to be neglected? *Lancet Infect. Dis.* 2014;14:353–357.

16. Shah AN, Saldana M. Ocular loiasis. *New Engl. J. Med.* 2010;363(11):e16.

17. Caubere P, Noireau F. Effect of attraction factors on sampling of *Chrysops silacea* and *C. dimidiata*, vectors of *Loa loa* filariasis. *J. Med. Entomol.* 1991;28:263–267.

18. Shookhoff HB, Dwork KG. Treatment of *Loa loa* infections with hetrazan. *Am. J. Trop. Med. Hyg.*. 1949;29(4):589–593.

19. Carme B, Boulesteix J, Boutes H, Puruehnce MF. Five cases of encephalitis during treatment of loiasis with diethylcarbamazine. *Am. J. Trop. Med. Hyg.* 1991;44(6):684–690.

20. Gardon J, Gardon-Wendel N, Demanga-Ngangue D, Kamgno J, Chippaux JP, Boussinesq M. Serious reactions after mass treatment of onchocerciasis with ivermectin in an area endemic for *Loa loa* infection. *Lancet.* 1997;350:18–22.

21. Mellor PS, Boorman J, Baylis M. *Culicoides* biting midges: Their role as arbovirus vectors. *Annu. Rev. Entomol.* 2000;45:307–340.

22. Mullen GR. Biting midges (Ceratopogonidae). In: Mullen GR, Durden LA, eds., *Medical and Veterinary Entomology. 2nd ed.* New York: Elsevier; 2009:169–188.

23. Azevedo R, Nunes M, Chiang J, et al. Reemergence of Oropouche fever, northern Brazil. *Emerg. Infect. Dis.* 2007;13:912–915.

24. Boorman J, Hagen DV. A name list of world *Culicoides. Inter. J. Dipterol.* 1996;7:161–192.

25. Lillie TH, Kline DL, Hall DW. Diel and seasonal activity of *Culicoides* spp. (Diptera: Ceratopogonidae) near Yankeetown Florida, monitored with a vehicle-mounted insect trap. *J. Med. Entomol.* 1987;24:503–507.

26. Alexander JO. *Arthropods and Human Skin.* Berlin: Springer-Verlag; 1984.

27. Murchie AK, Clawson S, Rea I, Forsythe WN, Gordon AW, Jess S. DEET (N,N-diethyl-meta-toluamide)/PMD (para-menthane-3,8-diol) repellent-treated mesh increases *Culicoides* catches in light traps. *Parasitol. Res.* 2016;115:3543–3549.

28. Schreck CE, Kline DL. Repellency determinations of four commercial products against six species of ceratopogonid biting midges. *Mosq. News* 1981;41:7–10.

29. Reithinger R, Dujardin JC, Louzir H, Pirmez C, Alexander B, Brooker S. Cutaneous leishmaniasis. *Lancet Infect. Dis.* 2007;7(9):581–596.

30. WHO. *Leishmaniasis.* World Health Organization, Global Health Observatory Data; 2017. http://www.who.int/gho/ne glected_diseases/leishmaniasis/en/.

31. Duprey ZH, Steurer FJ, Rooney JA, et al. Canine visceral leishmaniasis, United States and Canada, 2000–2003. *Emerg Infect Dis.* 2006;12:440–446.

32. Furner BB. Cutaneous leishmaniasis in Texas: Report of a case and review of the literature. *J. Am. Acad. Dermatol.* 1990;23:368–371.

33. Grimaldi G, Jr., Tesh RB, McMahon-Pratt G. A review of the geographic distribution and epidemiology of leishmaniasis in the New World. *Am. J. Trop. Med. Hyg.* 1989;41:687–693.

34. McHugh CP, Kerr SF. Isolation of *Leishmania mexicana* from *Neotoma micropus* collected in Texas. *J. Parasitol.* 1990;76:741–742.

35. Newson HD. Arthropod problems in recreational areas. *Ann. Rev. Entomol.* 1977;22:333–345.

36. WHO. *African trypanosomiasis.* World Health Organization, Media Center, Fact Sheet Number 25; 2010: 6.

37. Buscher P, Cecchi G, Jamonneau V, Priotto G. Human African trypanosomiasis. *Lancet.* 2017;390(10110):2397–2409.

38. Nash TAM. A review of the African trypanosomiasis problem. *Trop. Dis. Bull.* 1960;57:973–980.

39. Willet KC. African trypanosomiasis. *Ann. Rev. Entomol.* 1963;8:197–213.

40. Dumas M, Bouteille B. Current status of trypanosomiasis. *Med. Trop. (Mars).* 1997;57(3 Suppl.):65–69.

41. Enserink M. Welcome to Ethiopia's fly factory. *Science (News Focus).* 2007;317:310–313.

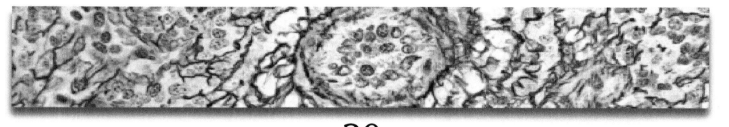

20
FLIES (NONBITING)

I. House Flies and Other Filth Flies

A. General and Medical Importance

The term filth flies refers to members of the families Muscidae, Sarcophagidae, and Calliphoridae, which are domestic non-biting flies commonly seen in and around human dwellings (see box). Although other species or groups could be considered filth flies, this topic will be restricted to those fly families (mentioned above) representing the house flies, flesh flies, and blow flies. These flies do not bite, but they are medically important in the mechanical transmission of disease agents from feces or dead animals directly to people or to food and food preparation areas. Throughout the world, these flies serve as carriers of organisms causing diseases such as typhoid, diarrhea, bacillary amoebic dysentery, cholera, giardiasis, pinworm, and tapeworm.[1] They may spread these agents via their sponging mouthparts (Figure 20.1), body hairs, or sticky pads of their feet, as well as through their vomitus or feces. Controlled studies have shown that fly management in communities leads to a reduction in the incidence of diarrhea.[2,3] The egg-laying habits of these insects may also lead to another human malady, myiasis, which is the infestation of people by the maggots of flies (see Chapter 6).

B. General Description

House flies are about 5–8 mm long, with a dull gray thorax and abdomen (not shiny). There are usually two pale spots laterally on the basal segments of the abdomen. The thorax has four longitudinal dark stripes (Figure 20.2). Mature house fly larvae are 10–13 mm long and usually creamy white in color (Figure 20.3). Overall, the larvae have a conical shape with two dark-colored mouth hooks at the narrow end and two oval spiracular plates at the broad posterior end. The three

FILTH FLIES

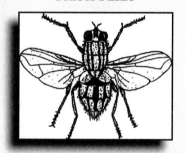

Adult house fly, *Musca domestica*, example of a filth fly.

IMPORTANCE

Nuisance; contamination; myiasis

DISTRIBUTION

Numerous species worldwide

LESION

Nonbiting, no lesion

DISEASE TRANSMISSION

Several agents mechanically

KEY REFERENCES

Greenberg B. *Flies and Disease*. Vol. I. *Ecology, Classification, and Biotic Associations*. Princeton, NJ: Princeton University Press; 1971

Greenberg B. *Flies and Disease*. Vol. II. *Biology and Disease Transmission*. Princeton, NJ: Princeton University Press; 1973

Figure 20.1 Blow fly sponging mouthparts (Photo copyright 2014 by Jerome Goddard, Ph.D.)

Figure 20.2 Adult house fly. (Photograph copyright 2011 by Jerome Goddard, Ph.D.)

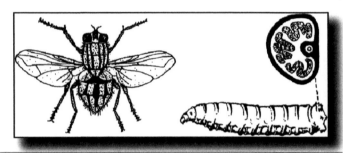

Figure 20.3 Adult (left) and larval (right) house fly. (Figure courtesy of U.S. Centers for Disease Control and Prevention, Atlanta, GA.)

"curly" sinuous slits in the spiracular plate are diagnostic features for this species (Figure 20.3).

Flesh flies look like house flies but are generally larger (11–13 mm long); they have three dark longitudinal stripes on their thorax, a checkerboard pattern of gray on the abdomen, and sometimes a reddish-brown tip on the abdomen (Figure 20.4). The larvae of flesh flies are similar to those of house flies, except that they have straight spiracular slits and often an incomplete ring around the spiracular plate (Figure 20.5). A diagnostic characteristic of larval flesh flies is that the spiracular slits do not point toward the opening of the plate (Figure 20.5).

Blow flies (also known as green flies or bluebottle flies) are about the same size as flesh flies, although some of the bluebottle flies (genus *Calliphora*) are larger and more robust (Figure 20.6). Blow flies, with the exception of the cluster fly, are metallic bronze, green, black, purplish, or blue colored (Figure 20.7). They are the commonly encountered green flies seen on flowers, dead animals, and feces or, occasionally, indoors. Blow fly maggots resemble both house fly and flesh fly maggots but have straight spiracular slits pointing toward the opening (or sometimes the "button") and often a complete sclerotized ring around the spiracular plate (see Figure 20.8 for comparisons).

C. Geographic Distribution

House flies, *Musca domestica*, occur worldwide in association with human dwellings. There are numerous blow fly species occurring over most regions of the world. *Lucilia* (formerly *Phaenicia*) *sericata* is a very common blow fly that is cosmopolitan in distribution. *Calliphora vicina* is one of the most common bluebottle species in Europe and North America. Various species of flesh flies also occur worldwide. One of the most common is *Sarcophaga haemorrhoidalis*, which is virtually worldwide in distribution and common in the United States.

D. Biology and Behavior

Like all Diptera, these flies exhibit a complete life cycle, having egg, larva, pupa, and adult stages (Figure 20.9). All three filth fly groups discussed in this section have similar biologies. Females lay their eggs, singly or in clusters, on or adjacent to an appropriate medium for larval development; however, flesh fly females deposit living larvae rather than eggs. One to a few days later the eggs hatch and the first-stage larvae emerge, entering the food source where they feed and develop. As the outer skin of a larva is nonliving and cannot grow, this skin (exoskeleton) must be periodically shed as the larva grows. Molting occurs three times in these filth fly families. Eventually, larval growth is completed and a pupa is formed inside the last larval skin. The larval growth period ranges from 2 days to 3 weeks, depending on temperature, and the pupal stage takes a similar length of time. The entire house fly life cycle can occur in 8–10 days under summer conditions. Emerging adults are active, moving from one attractant to another throughout most of the daylight hours. House flies are strongly attracted to feces,

Figure 20.4 Adult flesh fly.

Figure 20.5 Adult (left) and larval (right) flesh fly. (Adapted from Gorham JR. Insect and Mite Pests in Food: An Illustrated Key, Agriculture Handbook No. 655. Washington, DC: Agricultural Research Service, U.S. Department of Agriculture; 1991.)

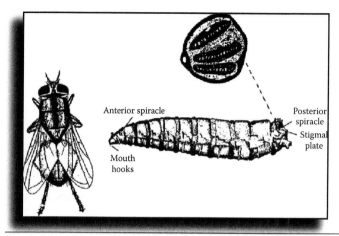

Figure 20.6 Adult (left) and larval (right) blow fly. (From U.S. Naval Medical School, Laboratory Guide to Medical Entomology with Special Reference to Malaria Control. Bethesda, MD: National Naval Medical Center; 1945.)

Figure 20.7 Adult blow flies are usually metallic green or blue.

decaying organic material, and foodstuffs. Garbage and pet feces are almost always the important breeding sources of house fly problems in urban settings. Flesh flies breed in decaying meat or animal excreta. Most blow flies lay their eggs on animal carcasses, and the developing maggots quickly dispose of the carrion (Figure 20.10); however, some blow fly species breed in dog manure or decaying organic matter. The entire blow fly life cycle requires 9–25 days or more.

E. Treatment of Infestation

Filth flies are best controlled by a combination of good sanitation, mechanical exclusion, ultraviolet light traps, and chemical control (when necessary). Good sanitation includes emptying and steam-cleaning dumpsters on a regular schedule and otherwise eliminating breeding sites. Mechanical exclusion methods include air curtains and properly fitted doors and screens. Ultraviolet light traps (some electrocute the flies, others stun them and trap them on glue boards) work well indoors, and the newer models are quite safe. Insecticidal treatments are best performed by competent pest management professionals.

II. Eye Gnats

A. General and Medical Importance

Members of the fly family Chloropidae are variously known as eye gnats, grass flies, eye flies, and fruit flies (although other flies are the true fruit flies). There are about 55 general and about 270 described species in the Nearctic region.[4,5] Members of the genus *Hippelates* are troublesome pests that are strongly attracted to the eyes of humans and other animals where they feed on the fluids of the eye surface and tear canal. Other species are attracted to serous discharges from sores, as well as excrement. They do not bite but they do have sponging mouthparts that are "spined"; these spines may cause eye irritation.[6] The lesions (minute scratches on the skin or eyeball) that result from these spines are thought to be entry points for disease agents such as that of pinkeye and yaws. Yaws, caused by *Treponema pallidum*, may cause weeping skin ulcers on the face, back, buttocks, and legs (Figure 20.11). The disease occurs in parts of Africa, Southeast Asia, and the Pacific (approximately 100,000 new cases each year), and is currently targeted for eradication by the World

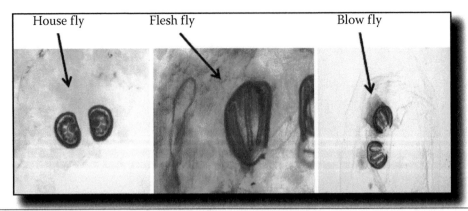

Figure 20.8   Filth fly spiracular plate comparisons.

EYE GNATS

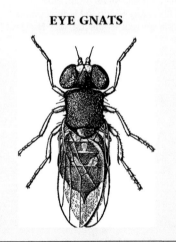

Typical eye gnat. (Redrawn after D.G. Hall.)

IMPORTANCE

Mechanical disease transmission

DISTRIBUTION

Numerous species almost worldwide

LESION

Nonbiting; sometimes cause minute scratch lesions on eyeball

DISEASE TRANSMISSION

Yaws and pinkeye

KEY REFERENCES

Greenberg B. *Flies and Disease*. Vol. I. *Ecology, Classification, and Biotic Associations*. Princeton, NJ: Princeton University Press; 1971

Greenberg B. *Flies and Disease*. Vol. II. *Biology and Disease Transmission*. Princeton, NJ: Princeton University Press; 1973

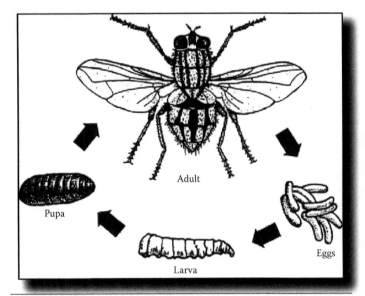

Figure 20.9   Lifecycle of the house fly. (Figure courtesy of U.S. Centers for Disease Control and Prevention, Atlanta, GA.)

Figure 20.10   Blow fly adults are attracted to decaying carcasses and corpses. (Photograph courtesy of Rosella M. Goddard.)

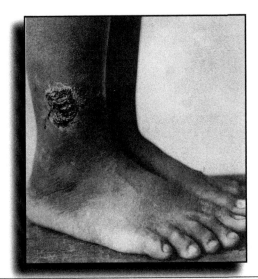

Figure 20.11 Yaws. (Photograph courtesy of Armed Forces Institute for Pathology, Washington, DC.)

Health Organization using a total community treatment strategy with azithromycin or intramuscular penicillin.[7]

B. General Description

These flies are small, bare flies about 2 mm long that are often shiny black (see box). Most *Hippelates* eye gnats have a large, black, curved spur on the hind tibia. Their antennae are short with bare aristae. The gnats are frequently seen on dogs around the eyes and anal–genital area.

C. Geographic Distribution

This is a large and widespread group composed of hundreds of species; numerous species occur almost worldwide.

D. Biology and Behavior

Eye gnats are common in agricultural areas, meadows, and other places where there is considerable grass. Larvae of most species feed in grass stems, decaying vegetation, and excrement. *Hippelates* larvae can be found feeding on decaying organic matter or roots of plants in the soil. Eggs hatch in about 3 days, the larval stage lasts about 2 weeks, and the pupal stage follows; adults emerge a few days later.

E. Treatment of Infestation

Because eye gnats do not actually bite, treatment involves only the incidental diseases transmitted (e.g., pinkeye and yaws). Avoidance of the gnats is important. Most insect repellents work reasonably well in minimizing their nuisance presence around humans (although it should be emphasized that repellents should not be sprayed on or near the eyes).

III. Nonbiting Midges

A. General and Medical Importance

Nonbiting midges (Chironomidae) (see box) are often confused with mosquitoes, but they are innocuous insects except during times when they are unusually abundant. They do not bite, but massive emergences may produce traffic hazards, bother livestock, and cover residences.[8,9] In addition, there may be an economic impact from damage to machinery, paint finishes, and airplanes.[10] Sometimes it is difficult to keep swarms out of the eyes or to avoid inhaling them. Persons with allergies may develop hypersensitivity to midge body parts or hemoglobin.[11]

NONBITING MIDGES

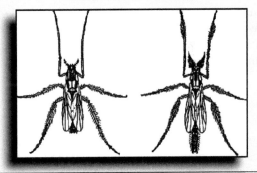

Adult nonbiting midges, Chironomidae.

IMPORTANCE

Massive emergences; airborne body parts induce hypersensitive reactions and asthma in some people

DISTRIBUTION

Numerous species almost worldwide

DISEASE TRANSMISSION

None

KEY REFERENCE

Ali A. Perspectives on management of pestiferous Chironomidae, an emerging global problem. *J. Am. Mosq. Control Assoc.* 1991;7:260–281

TREATMENT

Sensitive persons can avoid infested areas in spring and summer; stay indoors during emergences; wear dust masks when outdoors; see allergist for treatment

B. General Description

Midges are delicate insects, approximately the same size as mosquitoes, but they have a short proboscis that is not adapted for piercing (Figure 20.12). They have six-segmented antennae (possibly more, but no less), and in the males the antennae are plumose. Chironomids often have tiny hairs on the wing veins and on the membrane of the wing itself. Adult specimens are variously colored (usually brown or gray) and range in size from barely visible to mosquito-sized. Chironomids hold their wings rooflike over the abdomen at rest. They form huge swarms near water. Chironomid larvae are wormlike, having a hardened shell-like head capsule. Mature larvae have 12 body segments and prolegs on their first and last body segments. Some chironomid larvae are red and are frequently called bloodworms.

C. Geographic Distribution

Numerous species of chironomids occur worldwide. Three notorious pest species include *Tanytarsus lewisi* in northeastern Africa, *Glyptotendipes paripes* in the southeastern United States, and *Chironomus decorus* in the northeastern United States.

D. Biology and Behavior

Midges are extremely abundant around standing water (especially freshwater ponds, lakes, and reservoirs), as the larvae are aquatic. The life stages of chironomids include the egg, four larval instars, pupa, and adult. Depending on the species and time of year, development from egg to adult is 1 month to 2 years. The larva is the only feeding stage; larval food items are varied depending on species and include such items as small particulate organic matter, plants, or even other insects or tiny crustaceans. The larvae generally inhabit the bottom sediments in water sources. Most adult midges rest during daytime (Figure 20.13) and fly at night. They are attracted to light in great numbers, and sometimes huge swarms hover in the air, producing a humming sound.

E. Treatment

Persons sensitive to airborne chironomid body parts or emanations can sometimes move away from the source of the chironomid midges (e.g., a large freshwater reservoir). Efficient air-conditioning can help keep the indoor environment free of allergen. Sensitive persons may also gain protection from the offending particles by wearing a dust or particle mask when outdoors during chironomid emergences. In addition, sensitive persons should consult an allergist for advice and possible treatment.

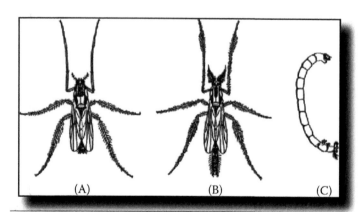

Figure 20.12 (A) Adult female, (B) male, and (C) larval nonbiting midge (Chironomidae).

Figure 20.13 Chironomid midges on blade of grass (Photo copyright 2013 by Jerome Goddard, Ph.D.)

References

1. Keiding J. The House Fly: Biology and Control. World Health Organization, Publ. No. WHO/VBC/76.650;1976:82.

2. Chavasse DC, Shler RP, Murphy OA, Huttly SRA, Cousens SN, Akhtar T. Impact of fly control on childhood diarrhea in Pakistan: Community-randomized trial. *Lancet*. 1999;353:22–25.

3. Watt J, Lindsay DR. Diarrheal disease control studies. I. Effect of fly control in a high morbidity area. *Public Health Rep*. 1948;63:1319–1334.

4. Sabrosky CW. The *Hippelates* flies or eye gnats: Preliminary notes. *Canadian Entomol*. 1941;73:23–27.

5. Sabrosky CW. Chloropidae. In: McAlpine JF, ed., *Manual of Nearctic Diptera*. Vol 2. Ottawa: Monograph No. 28, Research Branch, Agriculture Canada; 1987:1049–1067.

6. Harwood RF, James MT. *Entomology in Human and Animal Health*. 7th ed. New York: Macmillan;1979.

7. Kazura JW. Yaws eradication – A goal finally within reach. *N. Eng. J. Med*. 2015;372:693–695.

8. Bay EC, Anderson LD, Sugarman J. The batement of a chironomid nuisance on the highways at Lancaster, California. *Calif. Vector. Views*. 1965;12:29–32.

9. Beck EC, Beck WM Jr. The Chironomidae of Florida, II. The nuisance species. *Florida Entomol*. 1969;52:1–27.

10. Ali A. Perspectives on management of pestiferous Chironomidae, an emerging global problem. *J. Am. Mosq. Control. Assoc*. 1991;7:260–281.

11. Cranston PS. Allergens of non-biting midges: A systematic survey of chironomid haemoglobins. *Med. Vet. Entomol*. 1988;2:117–127.

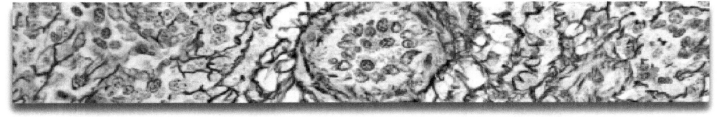

21

FLIES (THAT MIGHT CAUSE MYIASIS)

I. Human Bot Flies

A. General and Medical Importance

The condition of fly larvae occurring in human tissues is called myiasis. Some fly larvae develop in living flesh (also see Chapter 6); the human bot fly is one of them. This fly, *Dermatobia hominis*, is a parasite of humans, cattle, swine, cats, dogs, horses, sheep, other mammals, and a few birds in Mexico and Central and South America.[1,2] The larvae (Figures 21.1 and 21.2) burrow into the host's tissues, feeding and eventually emerging to drop to the ground and pupate. In people, the larvae have been recovered from the head, arms, back, abdomen, buttocks, thighs, and axilla. Human infestation is often characterized by painful discharging cutaneous swellings on the body. The condition is rarely fatal, except possibly in very young children (less than 5 years old); the larvae infesting the scalp may penetrate into the incompletely ossified skull and enter the brain.[3] Although the parasite does not occur in the United States, cases are occasionally seen in travelers to endemic areas, particularly Belize, Bolivia, and Brazil.[4] One such case was reported from Ohio in which a local physician submitted a second-stage larva to the Ohio Department of Health for identification.[5] The larva had been removed from a patient who had recently returned from Brazil. The Mississippi Department of Health has been involved in at least four cases of people returning from Belize.

B. General Description

Dermatobia hominis (see box) resembles a bluebottle fly (a large bluish member of the family Calliphoridae), is approximately 15 mm long, has a yellowish or brown head and legs, and has a plumose arista (a bristlelike branch off the antennae).

C. Geographic Distribution

Dermatobia hominis does not occur in the United States but is common in parts of Mexico and Central and South America. Sancho[6] reported its distribution from the northern provinces of Mexico (Taumalipas, bordering southern Texas) to the northern Argentine provinces of Misiones, Tres Rios, Corrientes, and Formosa—roughly between latitudes 25°N and 32°S. Vacationers may acquire this parasite in tropical America and return home before completion of larval development.

D. Biology and Behavior

Human bot flies have a unique and almost unbelievable life cycle (Figure 21.3). The adult flies catch various bloodsucking flies and attach eggs to their sides with a quick-drying glue. Common egg carriers are day-flying mosquitoes (*Psorophora* spp.) and muscid flies (*Sarcopromusca*, *Stomoxys*, and *Synthesiomyia* spp.).[7] These carrier flies later feed on humans (or other hosts) at which time the newly hatched bot fly larvae penetrate the host skin. The larvae feed inside a subdermal cavity of the host for 5–10 weeks. For respiration, they maintain a hole open to the external air. When mature, the larvae emerge, drop to the soil, and pupate. Interestingly, the fully fed larvae leave their hosts during the night or early morning, never during the afternoon, presumably to avoid desiccation. After a month or so, the adult flies emerge to mate and begin the life cycle over.

E. Treatment of Infestation

The recommended therapeutic procedure for all forms of myiasis is the direct removal of the maggots and treatment to prevent or control secondary infection (Figure 21.4). Alexander[8] said attempts to express *Dermatobia hominis* larvae by pressure on or around the lesion are fruitless and

HUMAN BOT FLY

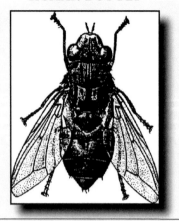

Human bot fly, *Dermatobia hominis.*

IMPORTANCE

Larvae cause myiasis in humans

DISTRIBUTION

Mexico, Central, and South America

LESION

Superficial, painful swelling with central opening

DISEASE TRANSMISSION

None

KEY REFERENCE

Sancho E. *Dermatobia*, the neotropical warble fly. *Parasitol. Today* 1988;4:242

TREATMENT

Excision of larvae; bacon therapy (see Brewer TF, et al. Bacon therapy and furuncular myiasis. *JAMA* 1993;270:2087–2088)

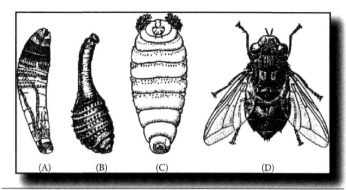

Figure 21.1 Human bot fly: (A) first-stage larva, (B) second-stage larva, (C) third-stage larva, and (D) adult. (From James, M.T., The Flies That Cause Myiasis in Man, USDA Misc. Publ. No. 631. Washington, DC: U.S. Department of Agriculture; 1947.)

Figure 21.2 Human bot fly larva, third stage. (Photograph copyright 2007 by Jerome Goddard, Ph.D.)

II. Screwworm Flies

A. General and Medical Importance

Two important species of screwworm flies that feed in living tissues are *Chrysomya bezziana*, the Old World screwworm, and *Cochliomyia hominivorax*, the New World screwworm (Figure 21.5). Larvae of these calliphorid flies, called screwworms, are obligate parasites of living flesh (humans and domestic and wild mammals), feeding during their entire larval period inside a host. Screwworms feed by the hundreds close together, making pockets in the live flesh, eating downward with their pointed ends (the head end), and leaving their rear ends exposed for breathing. Infested wounds give off a sickening odor and ooze blood continually.[11]

Female flies most often oviposit on or near a wound; however, human infestations have resulted from the flies ovipositing just inside the nostril while a person sleeps during the day, especially if there is a nasal discharge.[12] Upon hatching, the larvae begin feeding, causing extensive destruction of tissue and a bloody discharge. Tissues around the lesion become swollen, and pockets may be eaten out beneath the skin. The frontal and ethmoid sinuses may be entered and the cartilage and even the bone

painful; however, Smith[9] reported removing *D. hominis* larvae by suffocating them first with nail polish or petroleum jelly and then squeezing them out. Secondary infection or an allergic-type reaction may result if the larva is ruptured or killed within its cavity during the removal process. A traditional Central American remedy is occlusion of the punctum (breathing hole in the skin) with raw meat or pork, a treatment sometimes called bacon therapy. Apparently, the larvae will almost always migrate into the applied piece of meat. An American group modified this approach and was able to successfully extract 10 larvae after 3 hours of bacon therapy.[10]

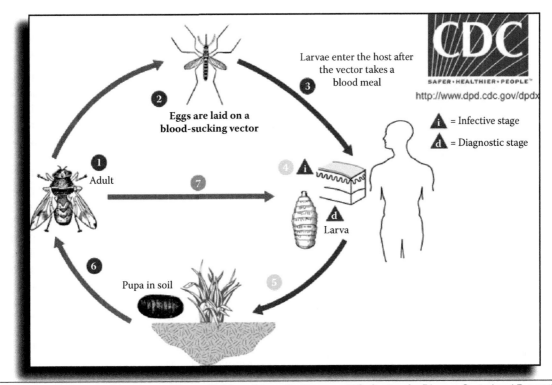

Figure 21.3   Human bot fly life cycle. (From Image Library, Division of Parasitic Diseases, U.S. Centers for Disease Control and Prevention, Atlanta, GA.)

attacked. Infested persons may die from tissue destruction. Human cases are uncommon but have occurred in areas where screwworm infestations occur in livestock. In 1935, 55 cases were reported during a large outbreak among livestock in Texas.[13] Because the New World screwworm fly has been eradicated from the United States, human cases are rare and due entirely to foreign travel. However, human cases are still reported from South America.[12] Old World screwworm myiasis cases are common in Asia but relatively rare in Africa.[14]

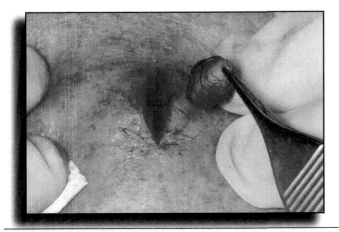

Figure 21.4   Human bot fly larva removed from a patient. (Photograph courtesy of Dr. Mark Mehrany, Department of Dermatology, Mayo Clinic, Rochester, MN.)

B. General Description

Adult *Chrysomya bezziana* are green to blue in color and have the base of the stem vein (radius) ciliate above, but they have at most only two narrow longitudinal thoracic stripes (see box and Figure 21.5A). They are approximately 8–12 mm

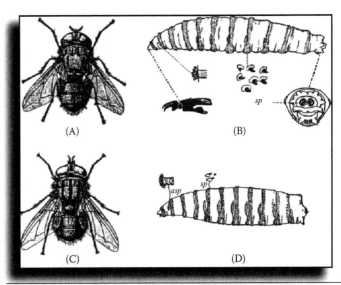

Figure 21.5   Screwworm flies: (A) adult female *Chrysomya bezziana*, (B) larval *C. bezziana*, (C) adult female *Cochliomyia hominivorax*, and (D) larval *C. hominivorax*. (From James MT. The Flies That Cause Myiasis in Man, USDA Misc. Publ. No. 631. Washington, DC: U.S. Department of Agriculture; 1947.)

SCREWWORM FLIES

Adult Old World screwworm fly, *Chrysomya bezziana*.

IMPORTANCE

Human myiasis—sometimes fatal

DISTRIBUTION

Central and South America (formerly in North America); Oriental and Afrotropical regions; parts of South Pacific region

LESION

Pocketlike sinuses with live maggots in tissues

DISEASE TRANSMISSION

Generally none

KEY REFERENCE

James MT. The Flies That Cause Myiasis in Man, USDA Misc. Publ. No. 631. Washington, DC: U.S. Department of Agriculture; 1947

TREATMENT

Removal of larvae by irrigation or surgical means

long. Full-grown *Chrysomya bezziana* larvae have the usual 12 segments with belts of dark spines encircling them (see Figure 21.5B). Adult *Cochliomyia hominivorax* are bluish-green, medium-sized flies with a yellowish-orange face and three dark stripes on the thorax (see Figure 21.5C). The base of the stem vein (radius) bears a row of bristlelike hairs on the dorsal side. Full-grown larvae are about 2 cm long, appear pinkish in color, and have prominent rings of spines around the body (see Figure 21.5D).

C. Geographic Distribution

The distribution of the New World screwworm once extended from the United States to southern Brazil, but it now has been

SCREWWORM OUTBREAK IN THE FLORIDA KEYS, 2016–2017

Beginning in August 2016, biologists at the National Key Deer Refuge on Big Pine Key, Florida noticed an unusual increase in nontravel-related deer mortality. In addition, a local veterinarian reported seeing an increase in myiasis in pets. Accordingly, refuge staff decided to submit fly larval samples to the USDA Animal and Plant Health Inspection Service (APHIS), National Veterinary Services Laboratory, where they were confirmed as New World screwworm fly larvae, *Cochliomyia hominivorax*. Cases of infestation in Key deer increased in the area during September, peaking in October (see Chapter 6 for more information). In fact, October 2016 had the highest mortalities of Key deer by any cause reported in any month over the past 10 years. In response, USDA APHIS personnel, in collaboration with the U.S. Fish and Wildlife Service, released 188 million sterile male screwworm flies into the area which successfully eliminated the New World screwworm by March 2017.

Comment: This was a classic example of introduced obligate myiasis which became established (more accurately, re-established). The screwworm fly was previously completely eradicated from the U.S. in the 1970s using the sterile male release technique. The pathway for introduction of this new outbreak in the Florida Keys remains unknown but may be due to a variety of factors, including movement of people or animals from screwworm-infested areas. This event demonstrates the need for increased diligence in surveillance and monitoring to prevent new infestations In the future.

limited to South and Central America due to an eradication program involving the sterile male release technique. The Old World screwworm occurs in the Afrotropical and Asian regions, extending south into Indonesia, the Philippines, and New Guinea (but not Australia).

D. Biology and Behavior

Females of both screwworm species are attracted to wounds in mammals and lay eggs at the edge of the wounds. Eggs are deposited in batches of 150–400 and hatch approximately 15 hours later. The larvae feed while embedded inside living tissue; sometimes, however, the peritremes (the plate surrounding the breathing tubes) are visible. The larvae emerge from the host as prepupae 4–7 days later and fall to the ground, where they pupate for a week or more. The entire life cycle from egg to egg takes about 24 days under optimum conditions. Adult screwworm flies are active all year-round but only fly during daylight.

E. Treatment

Treatment of screwworm fly myiasis usually involves removal of the larvae. This may be done in several ways, but often there is surgical exploration and removal of the larvae under local anesthesia.[15] A recent study reported good success (nonsurgically) by flushing the infested tissues with nitrofurazone.[16] Alexander[8] advocated irrigation with chloroform or ether under local or general anesthesia. Infestation of the nose, eyes, ears, and other areas may require surgery if larvae cannot be removed via natural orifices. Because screwworm flies lay eggs in batches, there could be tens or even hundreds of maggots in a wound.

III. Congo Floor Maggot

A. General and Medical Importance

The Congo floor maggot, *Auchmeromyia senegalensis* (= *luteola*), is a blow fly (see box). Its larvae suck human blood while their hosts are sleeping on the floor in huts or other dilapidated, infested dwellings. Although this blood-feeding behavior may not actually be myiasis in the strictest sense, this species is discussed in the following subsections. Congo floor maggots are not known to transmit disease agents. The bite of the larva is generally felt as a pinprick, but sensitive persons may experience pain, swelling, and irritation.

B. General Description

Adult *Auchmeromyia senegalensis* are yellow-brown flies that are 8–13 mm long. They are similar in appearance to the Tumbu fly (next section), but the second visible abdominal segment is longer than the third (Figures 21.6 and 21.7). Males of the Congo floor maggot fly have widely separated eyes. Full-grown larvae are up to 18 mm long.

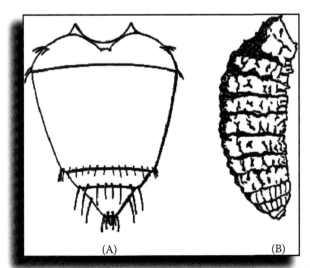

Figure 21.6 *Auchmeromyia senegalensis:* (A) abdomen of male showing exceptionally long second segment, and (B) larval stage. (From James MT. The Flies That Cause Myiasis in Man, USDA Misc. Publ. No. 631. Washington, DC: U.S. Department of Agriculture; 1947.)

C. Geographic Distribution

The Congo floor maggot is an African calliphorid occurring throughout Africa south of the Sahara Desert and in the Cape Verde islands.

D. Biology and Behavior

Congo floor maggot larvae occur in the earthen floor of huts or other similar dry-soil situations. They feed periodically at night by sucking blood from individuals sleeping on the ground. The larvae take about 20 minutes to feed on their host and then drop to the ground. They may feed daily if given the opportunity. The larvae go through three stages, which may include 6–20 blood feedings. They eventually pupate in the soil for about 2 weeks, after which the adults emerge. The entire life cycle is approximately 10 weeks under optimal conditions, and development continues year-round with no diapause. The adults feed on feces, fallen fruit, and

CONGO FLOOR MAGGOT

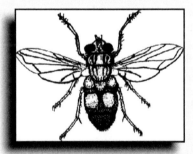

Congo floor maggot, *Auchmeromyia senegalensis*.

IMPORTANCE
Nocturnal blood-feeding maggots; nuisance

DISTRIBUTION
Sub-Sahara Africa; Cape Verde islands

LESION
Generally, trivial pinprick lesions

DISEASE TRANSMISSION
None

KEY REFERENCE
James MT. The Flies That Cause Myiasis in Man, USDA Misc. Publ. No. 631. Washington, DC: U.S. Department of Agriculture; 1947

TREATMENT
Generally, none needed

Figure 21.7 Congo floor maggot adult and larva. (Photograph copyright 2007 by Jerome Goddard, Ph.D.)

other fermenting materials. The adults mate and then the female lays her eggs in dry, dusty, or sandy soil, often in a hut. This fly has also been reported to be associated with warthogs, aardvarks, and hyenas.

E. Treatment of Bites

Blood feeding by the Congo floor maggot is generally not serious but may cause considerable discomfort. Because the larvae cannot climb vertically, avoidance is possible by sleeping in beds at least 10 cm off the ground.

IV. Tumbu or Mango Fly

A. General and Medical Importance

The Tumbu fly, *Cordylobia anthropophaga*, is a blow fly (family Calliphoridae) whose larvae can burrow into human subcutaneous tissues, producing a furuncular type of myiasis (see also Chapter 6). There are at least three other species of *Cordylobia* which can cause furuncular myiasis.[17] The lesion resembles a boil and often has a serous exudate. The first symptom is a slight itching, which occurs within the first day or two of the infestation. In a few days, a reddish papule develops, and then the typical boil-like lesion develops. With multiple infections, the lymph glands may become enlarged and fever may be present.[18] Children are most commonly affected, and lesions usually occur on areas of the body normally covered with clothing, as the adult flies tend to oviposit on soiled clothing. In an imported case seen in Maine, lesions appearing as boils approximately 25 mm in diameter contained second-stage Tumbu fly larvae.[18]

B. General Description

The adult Tumbu fly is similar in appearance to the adult Congo floor maggot as well as other flies in related genera; precise identification is therefore difficult (see box). The body of both sexes is yellow-brown and about 6–12 mm long (Figure 21.8). The second and third visible abdominal segments are of equal length. Males of this species have nearly contiguous eyes. (Congo floor maggot males have widely separated eyes.) Full-grown larvae are 13–15 mm long (Figure 21.8).

TUMBU FLY

Cordylobia anthropophaga adult female.

IMPORTANCE

Furuncular myiasis

DISTRIBUTION

Tropical Africa

LESION

Red or shiny nodules on skin; lesion very similar to those caused by *Dermatobia hominis*

DISEASE TRANSMISSION

None

KEY REFERENCE

James MT. The Flies That Cause Myiasis in Man, USDA Misc. Publ. No. 631. Washington, DC: U.S. Department of Agriculture; 1947

TREATMENT

Direct removal of larvae

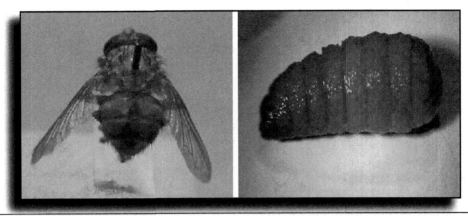

Figure 21.8 Adult and larval Tumbu fly. (Photograph copyright 2007 by Jerome Goddard, Ph.D.)

C. Geographic Distribution

Tumbu flies occur in the Afrotropical region.

D. Biology and Behavior

Tumbu flies cause cutaneous myiasis in several mammalian hosts, including humans. Dogs are the most common domestic hosts, and several species of wild rats are the preferred field hosts. The adult flies feed on plant juices, rotting plant or animal matter, and feces. They are most active in the early morning (7:00–9:00) and late afternoon (4:00–6:00). Adult flies lay their eggs in shady areas of sand contaminated with excrement or sometimes on soiled clothing. Improperly washed diapers placed in the shade to dry are a common oviposition site. The flies do not oviposit directly on the skin or hair of a host. The eggs hatch, and the larvae remain buried in the dirt or sand until a host approaches. In the presence of a host, the larvae actively search for and get on the host. Penetration of the skin is rapid and ordinarily unnoticeable. The larva develops in the subcutaneous tissues for about 10 days. As the larva develops, the lesion begins to look boil-like, and a clear fluid, occasionally stained with blood or larval feces, may ooze from the boil. Larvae (now prepupae) then exit the lesion and pupate in the ground. About 10–14 days later the adults emerge, mate, and begin laying new batches of eggs. The adult female lives only 2–3 weeks. People are most commonly parasitized during the rainy season, but fly development continues year-round.

E. Treatment of Infestation

As in other types of myiasis, the only known therapeutic procedure, other than applying local palliatives, is the direct removal of the maggots (surgically, if necessary) and treatment to prevent or control secondary infection. Service[19] observed that a standard method for Tumbu larvae removal is to cover the small hole in the swelling with medicinal liquid paraffin. This prevents the larva from breathing through its posterior spiracles and forces it to wriggle a little further out of the swelling to protrude the spiracles. In doing so, it lubricates the pocket (in the skin) and the larva can then be extracted by gently pressing around the swelling. Care must be taken not to break or damage the larva to prevent an intense immune response.

CASE HISTORY

MAGGOT IN URINARY TRACT

A large regional hospital sent in a fly maggot supposedly recovered from a urine sample. Laboratory personnel apparently did not think such an event was possible. "Is it really a maggot?" they asked. "What should we do? Are there more in the urinary tract?" Upon examination, the maggot was identified as a syrphid fly larva. Although not usually involved in urinary myiasis, the species was from a group of flies known to occasionally cause other forms of myiasis in people. Accordingly, the case was considered *urinary tract myiasis* and the physicians acted accordingly.

Comment: The maggots are generally expelled in cases of accidental urinary myiasis by noninvasive species; therefore, specific treatment may not be needed.

Source: Adapted from Goddard J. Direct injury from arthropods, *Lab. Med.* 1994;25:369. Copyright 1994, The American Society of Clinical Pathologists. With permission.

V. Sarcophagid Flies (*Wohlfahrtia* spp.)

A. General and Medical Importance

Two important sarcophagid flies that cause myiasis are *Wohlfahrtia magnifica* and *W. vigil*. *Wohlfahrtia magnifica* larvae produce traumatic myiasis throughout their host's

WOHLFAHRTIA FLIES

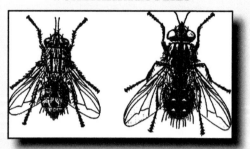

Adult female (A) *Wohlfahrtia magnifica* and (B) *W. vigil*. (From James MT. The Flies That Cause Myiasis in Man, USDA Misc. Publ. No. 631. Washington, DC: U.S. Department of Agriculture; 1947.)

IMPORTANCE

W. magnifica—myiasis similar to that from screwworm fly
W. vigil—furuncular myiasis

DISTRIBUTION

W. magnifica—southern Palearctic
W. vigil—northern United States and Canada

LESION

W. magnifica—deep lesions in the nose or in existing wounds
W. vigil—papule- or pustule-looking lesions containing larvae

DISEASE TRANSMISSION

None

KEY REFERENCE

James MT. *The Flies That Cause Myiasis in Man*, USDA Misc. Publ. No. 631. Washington, DC: U.S. Department of Agriculture; 1947

TREATMENT

Removal of the larvae by expression, direct picking, or irrigation

tissues, much like a screwworm fly. Tissues usually affected most commonly include the nose, ears, and eyes, and deafness, blindness, or facial disfiguration may result. Human fatalities due to this species have been reported.[2] *Wohlfahrtia vigil* larvae are not as invasive; they are usually limited to dermal tissues, producing a furuncular or boil-like lesion.

B. General Description

The *Wohlfahrtia* spp. look very similar to flesh fly adults (*Sarcophaga*), but instead of the checkerboard pattern on

their abdomen, they have clearly defined spots (see box and Figures 21.9 and 21.10). The abdomen of *W. vigil* is almost entirely black.

C. Geographic Distribution

Wohlfahrtia magnifica occurs in the southern Palearctic region. *Wohlfahrtia vigil* occurs in northern North America, primarily Canada and the northern United States.

D. Biology and Behavior

Wohlfahrtia magnifica is an obligate parasite in the wounds and natural orifices of warm-blooded animals, including humans. It never develops in carrion or rotting materials. Females of both *W. magnifica* and *W. vigil* are larviparous; that is, the adult fly deposits living larvae on potential hosts. Both tiny skin lesions (tick bite, scratch, etc.) and mucous

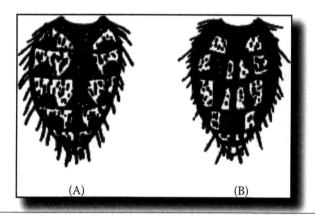

(A)   (B)

Figure 21.9   Comparison of (A) *Wohlfahrtia*, and (B) *Sarcophaga* abdomens.

Figure 21.10   *Wohlfahrtia magnifica*, showing abdomen. (Photograph copyright 2007 by Jerome Goddard, Ph.D.)

membranes of natural orifices are used as entry points by the larvae. The developing larvae feed for 5–7 days within a host, after which they emerge, fall to the ground and pupate. *Wohlfahrtia vigil* is also an obligate parasite, but it rarely penetrates deeper than dermal tissues. Dogs, cats, rodents, rabbits, mink, foxes, and humans are hosts of this pest. The larvae form a boil-like swelling under the skin with a circular opening through which the larva can be partially seen. After maturing, the larvae emerge, drop to the ground, and pupate. Multiple lesions are common on patients infested with *W. vigil*, with the average number being 12–14.[2]

E. Treatment

Wohlfahrtia vigil infestations are relatively easy to treat because the larvae are easily seen and removed. Unlike other forms of furuncular myiasis, expression seems easy and uncomplicated. *Wohlfahrtia magnifica* infestations are much more serious because they often involve numerous voraciously feeding larvae within the nose or in existing wounds. Irrigation or possible surgical exploration to remove the larvae will be necessary.[2]

VI. Other Flies That May Occasionally Cause Myiasis

A. General and Medical Importance

There are many examples of parasitic fly larvae that infest animals such as skin bot flies and nasal bot flies. Deer nasal bots are often encountered when hunters are skinning/processing deer. As part of their life cycle, these larvae reside in the retropharyngeal pouches in the throat of their hosts and may be seen emerging from the mouth or nose (Figure 21.11). Larvae of flies in the genus *Cuterebra* (Figure 21.12), often found in squirrels (called wolves) and rabbits, may rarely parasitize humans, forming a warblelike dermal tumor.[20,21] In a case Goddard consulted on (see Case History), a 3-year-old boy had two *Cuterebra* larvae—on his side and neck—forming boil-like lesions. Several healthcare providers examined the boy and diagnosed the lesions as either boils or larval migrans (from dog hookworm) because there were short migration trails visible in the skin. To everyone's surprise, one physician finally recognized the myiasis and expressed a larva from the neck lesion.

Other flies may occasionally cause myiasis in humans (Figures 21.13 through 21.17) (see also Chapter 6). This behavior is termed facultative myiasis. In some of these cases, the larvae enter living tissues after feeding in neglected, malodorous wounds. In other cases, ports of entry include natural orifices such as the ears, urinary opening, or anus. Goddard has even seen blow fly larvae removed from a (living) human eye socket (Figure 21.18). Table 21.1 provides a list of some species occasionally involved in facultative myiasis and notes

Figure 21.11 Nasal bots which emerged from a hunter-killed deer. (Photo courtesy Joseph Goddard and the Alabama Department of Conservation and Natural Resources.)

on their biologies. Sometimes these fly larvae can be identified by looking at the shape and pattern of their posterior spiracles (Figures 21.19 through 21.21).

One of the most commonly implicated fly groups is the Calliphoridae (blow flies). Several species of blow flies, and especially *Lucilia sericata* and *Phormia regina*, have been reported to cause facultative myiasis in humans. The discussion below will be limited to those two species.

B. General Description

Lucilia sericata is a typical-looking blow fly—shiny green or coppery green (Figure 21.22A). *Phormia regina*, also called the black blow fly, is more slender and is olive-colored or nearly

Figure 21.12 Adult *Cuterebra* fly. (From Image Library, Division of Parasitic Diseases, U.S. Centers for Disease Control and Prevention, Atlanta, GA.)

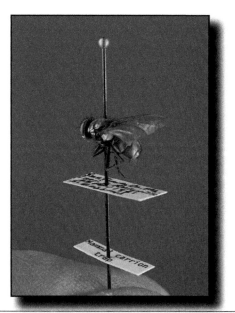

Figure 21.13 Typical blow fly, *Lucilia sericata*, usually shiny blue or green in color.

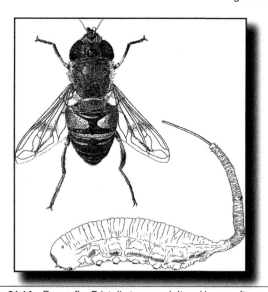

Figure 21.14 Drone fly, *Eristalis tenax*, adult and larva; often reported as a cause of accidental or facultative myiasis. (From Gorham JR. Insect and Mite Pests in Food: An Illustrated Key, Agriculture Handbook No. 655. Washington, DC: Agricultural Research Service, U.S. Department of Agriculture; 1991.)

Figure 21.15 Soldier fly larvae are often implicated in accidental or enteric myiasis cases. (Photograph copyright 2007 by Jerome Goddard, Ph.D.)

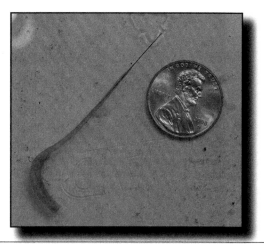

Figure 21.16 Rat-tailed maggot.

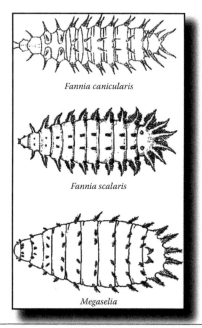

Fannia canicularis

Fannia scalaris

Megaselia

Figure 21.17 Fly larvae frequently involved in urinary or rectal myiasis. (From Gorham JR. Insect and Mite Pests in Food: An Illustrated Key, Agriculture Handbook No. 655. Washington, DC: Agricultural Research Service, U.S. Department of Agriculture; 1991.)

Figure 21.18 Blow fly larvae removed from an eye socket in a case of facultative myiasis. (Photograph copyright 2007 by Jerome Goddard, Ph.D.)

Table 21.1 Some Fly Species Not Ordinarily Causing Myiasis but Which May Opportunistically Be Found in Living Tissues or Hosts

Type of Myiasis	Fly Species	Fly Distribution	Comments	References
Urinary	*Fannia canicularis*	Cosmopolitan	Breeds in animal feces or decaying vegetable matter	James[2]
	F. scalaris	Cosmopolitan	Found outdoors more than *F. canicularis*; also known as the latrine fly	James[2]
	Musca domestica	Cosmopolitan	Breeds in animal feces, garbage, etc. Rarely found in corpses or carrion	—
	Muscina stabulans	Cosmopolitan	Breeds in animal feces and decaying vegetable matter	—
	Teichomyza fusca	Nearly cosmopolitan	Breeds in excrement, sewers, and drains	—
	Megaselia scalaris	Nearly cosmopolitan	Breeds in all kinds of decaying organic matter	Singh and Rana[23]
	Eristalis tenax	Cosmopolitan	Larvae called rat-tailed maggots	Mumcuoglu et al.[24]
Cutaneous or traumatic	*Lucilia (= Phaenicia) sericata*	—	Very common saprophagous flies attracted to fresh carrion	Anderson and Magnarelli[15] Greenberg[25] Merritt[26]
	L. cuprina[a]	—	Breeds in garbage and dog feces	—
	Phormia regina	—	One of the most common flies in the southern United States; common in cases of myiasis	Hall[27] Hall and Townsend[28]
	Cochliomya macellaria	Nearctic and neotropical regions	Very abundant carrion breeder	Alexander[8]
	Calliphora vicina	Holarctic	Widespread in U.S. and Europe during cool weather months	Alexander[8]
	Sarcophaga haemorrhoidalis	Nearly Cosmopolitan	Breeds In Carrion And Excrement	James[2]
	Chrysomyia ruffacies	Oriental and neotropical	Known as the hairy maggot blow fly	—
	Hermetia illucens	Nearly cosmopolitan	Adult known as the black soldier fly	Adler and Brancato[29]
	Cuterebra spp.	New World	Rodent and rabbit bots; infest living tissue, not decaying tissue	Rice and Douglas[21]
Rectal	*Eristalis tenax*	Cosmopolitan; most common in the holarctic	Larvae called rat-tailed maggots	—
	F. scalaris	Cosmopolitan	—	—
	F. canicularis	Cosmopolitan	—	—
	Musca domestica	Cosmopolitan	—	Goddard[30]
Aural	*Lucilia (= Phaenicia) sericata)*	Cosmopolitan	—	Davies[31] Keller and Keller[32]
	L. cuprina	Nearctic	—	Goddard[30]
	Sarcophaga citellivora		—	—
	Phormia regina	Holarctic	—	Damsky et al.[33]

[a] *Lucilia cuprina* occurring in North America and *L. cuprina* in Australia are morphologically identical but very different in biology and ecology

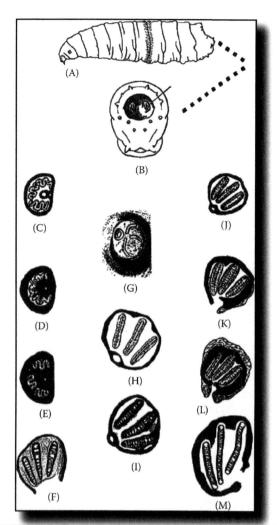

Figure 21.19 Posterior spiracles of various fly larvae occasionally involved in myiasis: (A) mature muscoid fly larva, (B) rear view of fly larva, (C) *Musca domestica*, (D) *Musca sorbens*, (E) *Musca crassirostris*, (F) *Wohlfahrtia vigil*, (G) *Muscina stabulans*, (H) *Cynomyopsis cadaverina*, (I) *Phaenicia caeruleiviridis*, (J) *Lucilia sericata*, (K) *Chrysomya chloropyga*, (L) *Chrysomya albiceps*, and (M) *Sarcophaga crassipalpis*. Note: D, E, K, and L occur in the Old World only. (Adapted in part from James MT. The Flies that Cause Myiasis in Man, USDA Misc. Publ. No. 631. Washington, DC: U.S. Department of Agriculture; 1947.)

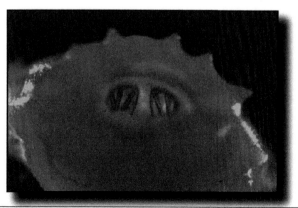

Figure 21.20 Posterior end of fly larva removed from the ear of a 4-month-old child. Note size and shape of spiracular slits. (Photograph courtesy of Dr. Alan Causey, University of Mississippi Medical Center, Jackson.)

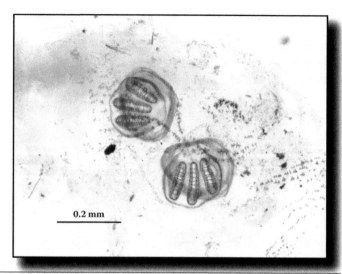

Figure 21.21 Posterior spiracles from fly larva found infesting a human nose. (Photograph copyright 2007 by Jerome Goddard, Ph.D.)

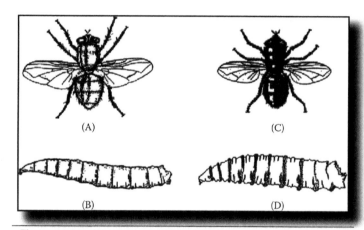

Figure 21.22 (A) Adult and (B) larval *Lucilia sericata*; (C) adult and (D) larval *Phormia regina*. (Figures courtesy of U.S. Centers for Disease Control and Prevention, Atlanta, GA; James MT. The Flies That Cause Myiasis in Man, USDA Misc. Publ. No. 631. Washington, DC: U.S. Department of Agriculture; 1947.)

black (Figure 21.22C). The pubescence (short, fine hairs) surrounding the anterior spiracle of *Phormia regina* is orange.

C. Geographic Distribution

Lucilia sericata is nearly cosmopolitan in distribution and is probably the most abundant species of *Lucilia* in North America. *Phormia regina* is holarctic in distribution. Goddard and Lago[22] found *Phormia regina* to be the most common blow fly in Mississippi, with specimens encountered throughout the year (peak numbers occurring in April and September).

D. Biology and Behavior

The general life history of blow flies is included in Chapter 20 on nonbiting filth flies. Other specific comments on these two species are presented here. *Lucilia sericata* is a

CASE HISTORY

ANIMAL BOT FLY LARVA IN CHILD

A 3-year-old boy complained of being "stung" on his side and neck (approximately 3 cm under the lower jaw bone) while watching television early one morning. His mother said that within 5 minutes, typical stinglike "welts" occurred at the places the child pointed out. Within a day or so, a line of vesicles extended away from the lesions—presumably where the larvae were migrating within the skin. The lesion on his side extended upward in a sinuous fashion about 10 cm, ending in a small papule. No further development occurred at the side lesion (apparently the developing larva died). The larva on his neck migrated about 4 cm laterally and began to enlarge (see Figure 21.23). After about 14 days, the dermal tumor was inflamed and contained an opening in the center about 3 mm in diameter, through which apparently the larva obtained air. The child cried often and complained of severe pain. Despite numerous trips to physicians, the myiasis was not diagnosed until almost 4 weeks after the initial stinging incident. An emergency room physician expressed the larva, which was ultimately forwarded to the health department for identification. Upon examination, the specimen (Figure 21.24) appeared to be a second-stage larva of a fly in the genus *Cuterebra* (the rabbit and rodent bot flies). Figure 21.25 is a drawing of the posterior spiracular slits of the specimen.

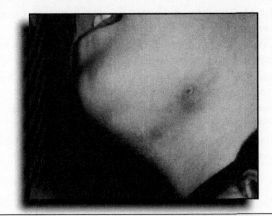

Figure 21.23   Lesion from *Cuterebra* fly larva removed from neck of 3-year-old child.

Figure 21.24   *Cuterebra* fly larva removed from neck of 3-year-old child.

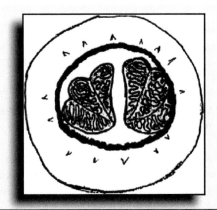

Figure 21.25   Drawing of posterior spiracles from *Cuterebra* larva infesting child.

species that is quick to appear at fresh carrion, although it is also attracted to a wide range of decaying substances. In many cases, facultative myiasis cases seem to be the result of *Lucilia sericata* being attracted to festering and malodorous wounds with subsequent invasion of healthy human tissue. *Phormia regina* is very common throughout the United States during the warmer months (April to September). It is also abundant near a wide variety of decaying substances, especially carrion.

E. Treatment of Infestation

Therapeutic measures for blow fly myiasis include removal of the maggots (surgically, if necessary) and treatment for secondary bacterial and fungal infections should they occur.

CASE HISTORY

TRAUMATIC MYIASIS

On February 17, 2000, a physician from a local hospital submitted a vial of insect larvae for identification. The tiny, wormlike specimens had been removed from a middle-aged woman found lying in the woods with a large open wound in her scalp. Upon microscopic examination, the larvae were found to be second-stage blow fly (insect family Calliphoridae) larvae.

Comment: This was a classic case of traumatic myiasis. It is not unusual for blow flies to lay eggs in wounds of people or animals, especially when the wound is neglected or malodorous. In nature, these flies oviposit on dead and rotting carcasses, so one can see how they might occasionally choose a festering wound on a living animal; however, this is still "opportunistic" myiasis and not purposeful, obligate myiasis.

References

1. Guimaraes JH, Papavero N. A tentative annotated bibliography of *Dermatobia hominis. Arq. Zool.* 1966;14:223–227.
2. James MT. The Flies that Cause Myiasis in Man. U.S. Dept. Agri. Misc. Publ. No. 631, Washington, DC; 1947: 175.
3. Rossi MA, Zucoloto S. Fatal cerebral myiasis caused by the tropical warble fly, *Dermatobia hominis. Am. J. Trop. Med. Hyg.* 1972;22:267–269.
4. Villalobos G, Vega-Memije ME, Maravilla P, Martinez-Hernandez F. Myiasis caused by *Dermatobia hominis*: countries with increased risk for travelers going to neotropic areas. *Inter. J. Dermatol.* 2016;55(10):1060–1068.
5. Anonymous. Surprise in a returning traveler. *Ohio Vector News.* 1991;10:2–3.
6. Sancho E. *Dermatobia* the neotropical warble fly. *Parasitol. Today.* 1988;4:242–246.
7. Catts EP, Mullen G. Myiasis (Muscoidae, Oestroidae). In: Mullen G, Durden L, eds. *Medical and Veterinary Entomology.* New York: Academic Press; 2002:317–348.
8. Alexander JO. *Arthropods and Human Skin.* Berlin: Springer-Verlag; 1984.
9. Smith SM. Treating infestations of the human botfly, *Dermatobia hominis. Lancet.* 2015;15(5):512.
10. Brewer TF, Wilson ME, Gonzalez E, Felsenstein D. Bacon therapy and furuncular myiasis. *J. Am. Med. Assoc.* 1993;270:2087–2088.
11. Boughton DC. The story of smear 62 (treatment for screw-worm infestation). *National County Agent and Vo-Ag Teacher Magazine* 1950; September issue: 14–29.
12. Baptista MAFB. Nasal myiasis. *N. Eng. J. Med.* 2015;372:e17.
13. Harwood RF, James MT. *Entomology in Human and Animal Health.* 7th ed. New York: Macmillan; 1979.
14. Zumpt F. *Myiasis in Man and Animals in the Old World.* London: Butterworths; 1965.
15. Anderson JF, Magnarelli LA. Hospital acquired myiasis. *Asepsis.* 1984;6:15.
16. Lima-Junior SM, Asprino L, Prado AP, Moreira RW, de Moraes M. Oral myiasis caused by *Cochliomyia hominivorax* treated non-surgically with nitrofurazone: Report of two cases. *Oral Surg. Oral Med. Oral Pathol. Oral Radiol. Endod.* 2010;109(3):70–73.
17. Pezzi M, Cultrera R, Chicca M, Leis M. Furuncular myiasis caused by *Cordylobia rodhaini*: A case report and literature review. *J. Med. Entomol.* 2015;52:151–155.
18. Rice PL, Gleason N. Two cases of myiasis in the U.S. by the African Tumbu fly, *Cordylobia anthropophaga. Am. J. Trop. Med. Hyg.* 1972;21:62–63.
19. Service MW. *Medical Entomology for Students.* New York: Chapman and Hall; 1996.
20. Goddard J. Human infestation with rodent botfly larvae: A new route of entry? *South. Med. J.* 1997;90:254–255.
21. Rice PL, Douglas GW. Myiasis in a man by *Cuterebra. Ann. Entomol. Soc. Am.* 1972;65:514–515.
22. Goddard J, Lago PK. An annotated list of the Calliphoridae of Mississippi. *J. Georgia Entomol. Soc.* 1983;18:481–484.
23. Singh TS, Rana D. Urogenital myiasis caused by *Megaselia scalaris*: A case report. *J. Med. Entomol.* 1989;26:228–229.
24. Mumcuoglu I, Akarsu GA, Balaban N, Keles I. *Eristalis tenax* as a cause of urinary myiasis. *Scand. J. Infect. Dis.* 2005;37(11–12):942–943.
25. Greenberg B. Two cases of human myiasis caused by *Phaenicia sericata* in Chicago area hospitals. *J. Med. Entomol.* 1984;21:615.
26. Merritt RW/. A severe case of human cutaneous myiasis caused by *Phaenicia sericata. California Vect. Views.* 1969;16:24–26.
27. Hall DG. *The Blowflies of North America.* Washington, DC: Thomas Say Foundation; 1948.
28. Hall RD, Townsend LHJ. *The Blowflies of Virginia.* Virginia Polytechnic Institute and State University, Research Bull. 1977;123:42.
29. Adler AI, Brancato FP. Human furuncular myiasis caused by *Hermetia illucens. J. Med. Entomol.* 1995;32:745–746.
30. Goddard J. Unpublished data. Mississippi Department of Health notes and records; 1989:1–5.
31. Davies DM. Human aural myiasis: A case in Ontario, Canada and a partial review. *J. Parasitol.* 1976;62:124.
32. Keller APJ, Keller API. Myiasis of the middle ear. *Laryngoscope.* 1970;80:646–648.
33. Damsky LJ, Baur H, Reeber E. Human myiasis by the black blowfly. *Minnesota Med.* 1976;59:303–304.

22

LICE

I. Body Lice

A. General and Medical Importance

The body louse, *Pediculus humanus humanus*, is a blood-feeding ectoparasite of humans. Body and head lice look almost identical (and are most likely one species[1]), but head lice remain more or less on the scalp and body lice on the body or in clothing (Figure 22.1). Body lice are relatively rare among affluent members of industrial nations, yet they can become severe in homeless populations[2] or under crowded and unsanitary conditions such as war or natural disasters. Body lice may transmit the agent of epidemic typhus, *Rickettsia prowazeki*, and there have been devastating epidemics of the disease in the past. Typhus is still endemic in poorly developed countries where people live in filthy, crowded conditions (Figure 22.2). Besides louse-borne typhus, body lice transmit the agents of trench fever and relapsing fever. Trench fever, caused by *Bartonella quintana*, is still widespread in parts of Europe, Asia, Africa, Mexico, and Central and South America, mainly in an asymptomatic form. However, it is now recognized as a reemerging pathogen among homeless populations (even in the U.S.),[3,4] where it is responsible for a wide spectrum of conditions such as chronic bacteremia, endocarditis, and bacillary angiomatosis.[5,6] Louse-borne relapsing fever (LBRF), caused by *Borrelia recurrentis*, occurs primarily in the Horn of Africa but may be seen in refugees and immigrants from that area.[7] Historically, LBRF had a much wider distribution. There were millions of cases during the two world wars of the twentieth century,[8] and 4972 cases with 29 deaths were reported worldwide in 1971.[9,10] Women who develop LBRF during pregnancy have a high incidence of spontaneous abortion. Genetic studies with *B. recurrentis* suggest that it is a degraded strain of tick-borne *B. duttoni*.[11]

Interestingly, none of the above agents is transmitted by the bites of body lice. Instead, they are transmitted by crushing lice onto human skin (LBRF) or scratching infected louse feces into human skin. Other modes of transmission, such as eating infected lice and thus mucous membrane exposure, cannot be ruled out. Walton and Horwitz[12] showed a picture of a Kikuyu native eating lice in Kenya. Aside from the possibility of disease transmission, body lice may cause severe skin irritation. The usual clinical presentation is pyoderma in covered areas. Characteristically, some swelling and red papules develop at each bite site (Figure 22.1B). There are intermittent episodes of mild to severe itch associated with the bites. Compounding this, some individuals become sensitized to antigens injected during louse biting, leading to generalized allergic reactions. Subsequent excoriation of the skin by the infested individual may lead to impetigo or eczema. Alexander[13] noted that long-standing infestations may lead to a brownish-bronze pigmentation of the skin, especially in the groin, axilla, and upper thigh regions.

B. General Description

An authoritative identification manual of all sucking lice in North America is available.[14] Human body lice are tiny (2–4 mm long), elongate, soft-bodied, light-colored, wingless insects (Figure 22.3). They are dorsoventrally flattened, with an angular ovoid head and a nine-segmented abdomen. The eggs are small (about 1 mm), oval, white or cream-colored objects with a distinct cap on one end. The head bears a pair of simple lateral eyes and a pair of short five-segmented antennae. Body lice are usually about 15–20% larger than head lice. They have modified claws enabling them to grasp tightly to hair shafts or clothing while they feed via piercing–sucking mouthparts. Body lice eggs are primarily laid in clothing.

BODY LICE

Adult body louse. (From CDC, Pictorial Keys to Arthropods, Reptiles, Birds, and Mammals of Public Health Significance. Atlanta, GA: U.S. Centers for Disease Control and Prevention; 1963.)

IMPORTANCE

Irritation; disease transmission

DISTRIBUTION

Essentially worldwide

LESION

Red papules, 3–4 mm in diameter; purpuric halo may be present

DISEASE TRANSMISSION

Epidemic typhus; trench fever; louse-borne relapsing fever

KEY REFERENCES

Kim KC, Pratt HD, Stojanovich CJ. *The Sucking Lice of North America*. University Park, PA: Pennsylvania State University Press; 1986

Bonilla DL, Durden L, Eremeeva ME, Dasch GA. The biology and taxonomy of head and body lice: implications for louse-borne disease prevention. *PLoS Pathogens.* 2013; 9:1–5

TREATMENT

Treatment of all clothing and bedding of infested individuals; pediculicidal shampoos and lotions

C. Geographic Distribution

Body lice may occur on humans worldwide.

D. Biology and Behavior

The body louse lives primarily on the clothing of infested individuals and moves to adjacent body areas to feed (Figure 22.1). The eggs are attached to fibers of clothing with a strong glue-like substance; they seem to be especially located along the

Figure 22.1 Body lice in clothing (Photo from Faccini-Martinez, et al. "Bartonella quintana and Typhus Group Rickettsiae Exposure among Homeless Persons, Bogotá, Colombia." *Emerging Infectious Diseases*, 2017, 23: 1876–1879. Used with permission)

seams inside of underwear or other places where clothing contacts the body. Eggs hatch in 5–11 days, depending on the temperature, and the young (called nymphs) begin feeding. Developing nymphs feed and molt several times before reaching the adult stage. The egg-to-adult cycle is about 3–5 weeks long. Body lice can survive in clothing even if it is habitually removed at night, but they usually will die if the clothing is not worn for several days. They seem to especially prefer woolen clothing.

E. Treatment of Infestation

Because body lice infest both the patient and his or her clothing, control strategies involve frequently changing clothing, washing infested garments in very hot water or having them dry-cleaned, and using pediculicidal lotions or shampoos.[15] The primary element of control is to ensure that all clothing and bedding of infested persons are sanitized or treated. Clothing and bedding can also be disinfected by spraying with pyrethrin preparations or other approved insecticides or dusts. When mass treatments are indicated (such as in a war or natural disaster), insecticide dusts may be applied directly to the body. Newer treatment strategies involve the use of oral ivermectin. One study demonstrated dramatic results using three doses (12 mg each), administered at 7-day intervals, in a cohort of homeless men living in a shelter in Marseilles, France.[16] Over a 14-day period, the number of lice found in the cohort fell from 1898 to 6.[16]

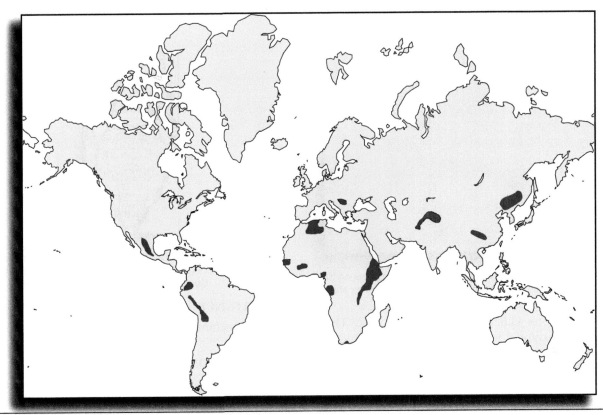

Figure 22.2 Approximate geographic distribution of epidemic typhus.

II. Head Lice

A. General and Medical Importance

The appearance and behavior of the head louse, *Pediculus humanus capitis*, are similar to the body louse (see box), but the head louse is confined to the scalp. Head lice have

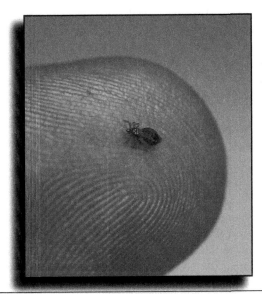

Figure 22.3 Body louse on a fingertip. (Photograph by Jerome Goddard, Ph.D.)

been pests of humans throughout the ages, as evidenced by the discovery of ancient hair combs dating back to the first century BC and containing the remains of lice and their eggs. The number of cases of head lice has increased worldwide since the 1960s, and hundreds of millions of cases are recognized annually.[17] Approximately 6–12 million people are treated for head lice each year in the United States.[17,18] Head lice generally pose no significant health threat, although heavily infested persons complain of severe itching and often have occipital and cervical adenopathy, sometimes accompanied by a generalized morbilliform eruption. The primary negative effects of lice infestation are embarrassment and social sanctioning.

B. General Description

Head lice are tiny (1–3 mm long), elongate, soft-bodied, light-colored, wingless insects. They are dorsoventrally flattened, with an angular ovoid head and a nine-segmented abdomen (Figure 22.4). The head bears a pair of simple lateral eyes and a pair of short five-segmented antennae. Head lice possess specially modified claws that enable them to grasp tightly to hair shafts while they feed through specially modified piercing–sucking mouthparts. Head lice look almost identical to body lice; however, they vary distinctly in behavior, as head lice occur chiefly on the head, whereas body lice occur on the body and clothing. Head lice eggs (nits) are about

Figure 22.4 Microscopic view of head louse. Note strong claws on two front legs especially suited for holding hair shafts

1-mm-long oval objects with a distinct cap on one end. They are white to cream-colored when viable and firmly attached to the hair (Figure 22.5). Under examination with the naked eye, they can be confused with dandruff, globules of hair oil or hairspray, and other substances, but they are easily identified under moderate magnification.

C. Geographic Distribution

Head lice occur on humans essentially worldwide.

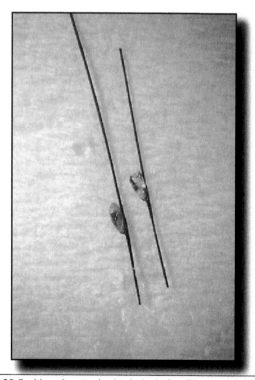

Figure 22.5 Lice nits attached to hair shafts. (Photograph by Jerome Goddard, Ph.D.)

HEAD LICE

Adult head louse. (From CDC, Pictorial Keys to Arthropods, Reptiles, Birds, and Mammals of Public Health Significance. Atlanta, GA: U.S. Centers for Disease Control and Prevention; 1963.)

IMPORTANCE

Irritation; itching; secondary infection

DISTRIBUTION

Essentially worldwide

LESION

Itchy scalp, often secondarily infected as lesions of impetigo contagiosa

DISEASE TRANSMISSION

None

KEY REFERENCES

Kim KC, Pratt HD, Stojanovich CJ. *The Sucking Lice of North America*. University Park, PA: Pennsylvania State University Press; 1986

Bonilla DL, Durden L, Eremeeva ME, Dasch GA. The biology and taxonomy of head and body lice: implications for louse-borne disease prevention. *PLoS Pathogens.* 2013;9:1–5

TREATMENT

Pediculicidal shampoos and heat devices

D. Biology and Behavior

The life cycle of head lice is fairly simple, with all stages being found on the human head (Figure 22.6). Lice live on the skin among the hairs on the head. Nits are laid on the shaft of the hair, near the base, and attached with a strong gluelike substance, most commonly behind the ears and at the nape of the neck. As long as adult lice remain on the scalp, they can live for about a month. Because they require the warmth

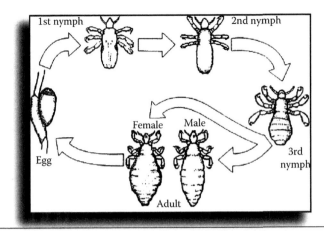

Figure 22.6 Head lice life cycle. (From CDC, Pictorial Keys to Arthropods, Reptiles, Birds, and Mammals of Public Health Significance. Atlanta, GA: U.S. Centers for Disease Control and Prevention; 1963.)

and blood meals afforded by the scalp, it is generally reported that lice can survive only about 24 hours if removed, but in one study involving hundreds of head lice removed from children none of the lice survived longer than 15 hours; most died between 6 and 15 hours.[19] Lice eggs hatch after 5–10 days, and the young (called nymphs) begin feeding immediately. Each feeding lasts several minutes. As the lice develop, they feed frequently, both day and night, particularly when the patient is quiescent. In as little as 18–20 days they mature through three molts and mate, and the females are ready to lay eggs to start the next generation. The egg-to-adult cycle averages 3 weeks.

E. Treatment of Infestation

Management of head lice infestations requires three general steps: (1) delousing infested individuals, with retreatment as necessary; (2) removing nits from the hair as thoroughly as possible; and (3) delousing personal items (clothes, hats, combs, pillows, etc.). An important principle of head lice management is to treat all infested members of a family concurrently. If an infested school-age child is the only family member treated, he or she may be quickly reinfested by a sibling or parent who is also unknowingly infested. Individuals with lice should be treated with one of the approved pediculicidal shampoos or cream rinses. Over-the-counter shampoos containing pyrethrins are generally effective; however, because these products are not sufficiently ovicidal, successful management may require 1 or 2 repeat applications at 7- to 10-day intervals to kill newly hatching lice. One over-the-counter product containing the synthetic pyrethroid permethrin (1%) is ovicidal (Nix® and generics) and is applied as a cream rinse after shampooing. Because Nix® can be ovicidal in addition to its pediculicidal effect, a single treatment may sometimes be sufficient for total eradication of lice and viable eggs. Many state health departments, however, still recommend a second treatment 7–10 days later—even

with Nix®. Another product, Kwell® (containing lindane, an organochlorine), is available by prescription as a shampoo; however, inappropriate use of lindane, such as frequent, repeated applications or ingestion, can result in neurotoxicity, seizures, and even death.[18,20] Kwell® is no longer considered first-line treatment for lice due to its potential side effects. Further, Brandenburg et al.[21] showed that single treatments with synthetic pyrethroids are more effective than treatments with lindane. Also, the U.S. Food and Drug Administration (FDA) reapproved malathion (Ovide® lotion) as a prescription drug for the treatment of head lice infestation, and there is now a 0.5% topical ivermectin (Sklice®) lotion on the market for head lice, which has shown good efficacy within 24 hours and most treated patients remaining louse-free through 2 weeks.[22] Any of these products, even lindane, are generally safe when used as directed. No matter which product they choose, patients should be instructed to follow the precautions and instructions on the product label.

Concerning the removal of nits, it is advisable to remove as many as possible for two reasons. First, reducing the number of viable ova in the hair reduces the chance of persistent infestation and treatment failure. Second, removal of nits lessens concern about continued infestation and risk of spread to others, thus allowing resumption of normal daily routine and school. Removal of all nits is probably impossible. For this reason, persons should be instructed to use the specially designed combs provided with the pediculicides and to follow the product instructions. Combing out nits is more successful if the hair is damp.

Finally, efforts should be made to delouse personal belongings of infested individuals. Washable clothing, hats, bedding, and other personal items should be washed properly and dried in a clothes dryer for at least 20–30 minutes. Nonwashable clothes should be dry-cleaned. Other personal items such as combs and brushes should be thoroughly washed in one of the pediculicidal products or soaked in hot water (130°F or more) for 5–10 minutes. Upholstery exposed to potential infestation should be vacuumed thoroughly. Use of certain appropriately labeled pyrethrin aerosols on furniture and carpeting and in other places where lice are suspected may kill a stray louse or two, but as the lice do not willingly leave their host and do not survive long off the host, these treatments would serve primarily as a psychological adjunct to other treatments and generally are not recommended. Furthermore, pyrethrin aerosols for environmental decontamination may cause severe bronchoconstriction in allergic individuals and should never be used on or near the head.[23]

Lice Resistant to Pediculicides   During recent years, there has been an increase in reports of head lice treatment failures. These failures often result from reinfestation or lack of adherence to already-established treatment protocols. However, there is resistance in certain lice populations to conventional agents such as permethrin, pyrethrins, and organochlorine insecticides.[24–27] In the United States, in particular, lice have become increasingly resistant to pyrethroids and lindane, but not to malathion.[28] Yoon et al.[29] found that 99% of lice sampled in the U.S. (2007–2009) had a gene coding for a phenotype resulting in permethrin resistance and 97% in Canada during 2008. In South America, a study of field-collected lice from Buenos Aires showed definite resistance to permethrin,[24] although it is unlikely that any particular population of lice is resistant to all pediculicides. The various lice treatment products have different active ingredients in various classes of insecticide; for example, Kwell® has lindane (chlorinated hydrocarbon), Ovide® has malathion (organophosphate), Nix® has permethrin (synthetic pyrethroid), and A200®, R&C®, and RID® have pyrethrins (botanical). Patients with suspected populations of resistant head lice should switch and be treated with a product containing a different class of active ingredient, again paying careful attention to label directions. Treatment failures with Nix®, for example, may be followed by treatment with Ovide®, as the products are from two totally different insecticide classes. Conversely, if over-the-counter products containing pyrethroids do not work, neither will prescription-strength (5%) permethrin. The patient must be treated with a product containing a different chemistry.

Recently, oral ivermectin has been shown to be extremely effective in treating resistant head lice.[30–32] Ivermectin is an antiparasitic drug discovered in the early 1970s that is a derivative of avermectin B1, a compound produced by *Streptomyces avermitilis*. It is a systemic parasiticide/acaricide that kills the pests as they feed on their host. Ivermectin already has approval in the United States for treatment of onchocerciasis and strongyloidiasis, but currently is not approved by the FDA for use as a pediculicide.[18] A well-designed study published in the New England Journal of Medicine, however, demonstrated good efficacy and safety of ivermectin for head lice control, even in children as young as 2 years old and weighing at least 15 kg.[31]

Nonpesticidal Lice Control Nonpesticidal treatments for head lice include shaving the head, coating the hair with a thick layer of mayonnaise or Vaseline® petroleum jelly, various combinations of lice combs/conditioners (wet hair combing), and hot-air machines. Anecdotal reports indicate that mayo or petroleum jelly indeed kills/smothers lice, but each must be left on a long time and is extremely difficult to subsequently remove from the hair. One person told the author that she had to wash her hair more than 200 times to remove all of the petroleum jelly. More recently, a product called Ulesfia® has become available which suffocates the lice in a manner similar to Vaseline but works faster.[33] Two treatments are still required about a week apart to kill any newly hatching lice. Wet-combing of the hair of lice-infested individuals along with conditioners has been advocated for several years by the U.K. Charity Community Hygiene Concern. Bug Buster kits containing fine-toothed lice combs and conditioner have been developed by the Community Hygiene Concern and are used in several locations in England, Wales, and Scotland. One evaluation of the Bug Buster kit showed that it delivers a 57% cure rate against head lice.[34] Hot air, delivered from a variety of hair dryers, has been proven effective against head lice,[35] although the LouseBuster™ was most effective, resulting in nearly 100% mortality of eggs and 80% mortality of hatched lice after treatment by experienced operators.[35,36]

III. Pubic or Crab Lice

A. General and Medical Importance

During the 1960s, widespread sexual freedom, particularly among young unmarried individuals, contributed significantly to an increased incidence of pubic lice infestation.[37] However, pubic lice infestation has subsequently declined due to a number of personal hygiene practices.[38] This condition is known as pediculosis pubis or pthiriasis and is caused by *Phthirus pubis*. The lice occur almost exclusively in the pubic, perianal, or coarse-haired areas of the body. Rarely they can be found on the eyelashes or eyebrows.[39] They feed on human blood through piercing–sucking mouthparts. Often, louse bites produce discrete, round, slaty-gray or bluish spots (maculae caeruleae). The bites may also cause intense itching due to the host's reaction to proteins in the louse saliva. Secondary infections may occur if the infestation is not treated. Excoriation of the skin from extensive scratching may lead to inflammation of the skin and lymph glands due to bacterial infection. If infestation occurs in the eyelashes, there may be blepharitis, an inflammation of the eyelids. Pubic lice are not known to transmit disease organisms; however, pediculosis pubis frequently coexists with other venereal diseases, particularly gonorrhea and trichomonas. One study indicated that one-third of patients with pubic lice may have other sexually transmitted diseases.[40]

B. General Description

Adult pubic lice are dark gray to brown in color. They are called crab lice because of their crablike shape. They are distinctly flattened, oval, and much wider than body or head lice (see box and Figure 22.7). As with head and body lice, the head bears a pair of simple lateral eyes and a pair of short five-segmented antennae. They are 1.5–2.0 mm long; their second and third legs are enlarged and contain a modified claw with a thumb-like projection, which aids them in grasping hair shafts (Figure 22.8). The individual egg, or nit, is dark brown in color, opalescent, and smaller than that of the body louse.

C. Geographic Distribution

Pubic lice occur on humans virtually worldwide.

PUBIC LICE

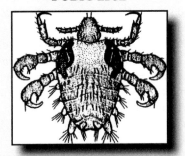

Adult pubic louse. (From CDC, Pictorial Keys to Arthropods, Reptiles, Birds, and Mammals of Public Health Significance. Atlanta, GA: U.S. Centers for Disease Control and Prevention; 1963.)

IMPORTANCE

Itching; irritation; secondary infection

DISTRIBUTION

Essentially worldwide

LESION

Variable—sometimes bluish spots

DISEASE TRANSMISSION

None

KEY REFERENCES

Kim KC, Pratt HD, Stojanovich CJ. *The Sucking Lice of North America*. University Park, PA: Pennsylvania State University Press; 1986

Witkowski JA, Parish LC. What's new in the management of lice? *Infect. Med.* 1997;14:287

TREATMENT

Pediculicidal shampoos

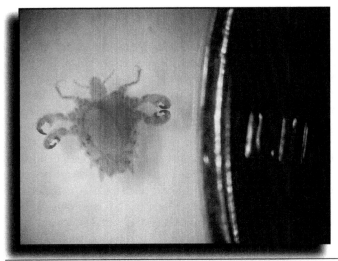

Figure 22.7 Adult pubic louse. (Photograph copyright 2012 by Jerome Goddard, Ph.D.)

Figure 22.8 Microscopic view of immature pubic louse holding hair shaft.

D. Biology and Behavior

Pubic lice require human blood to survive. They are only found on humans and do not infest rooms, carpets, beds, pets, etc. If lice happen to be forced off their host, they will die within 24–48 hours. In one study, 200 specimens removed from a man were kept in favorable conditions for 24 hours and only one survived.[41] Fisher and Morton[37] reported that off-host survival time is less than 20 hours. Female pubic lice deposit their eggs (nits) mainly on the coarse hairs of the pubic area and rarely on hairs of the chest, armpits, eyebrows, eyelashes, or mustache. In very rare cases, they have been found in the scalp. They lay approximately 30 eggs during their 3- to 4-week life span. There are three nymphal molts. Nymphs look almost identical to adults, only smaller. Pubic lice do not fly, jump, or even crawl very much. They often spend their entire life feeding in the same area where the eggs were deposited. Nuttall[41] reported that their maximum range is about 15 cm. Pubic lice are transmitted from person to person most often by sexual contact, although it is possible (though rare) for transmission to occur via toilet seats, clothing, or bedding.

E. Treatment of Infestation

As pubic lice infestations are usually transmitted through sexual contact, it is important to have the sexual contacts of the infested person examined and treated if needed. Likewise, as some family members all sleep in the same bed, if one member of a family has an infestation then all family members should be examined and the infested ones treated. As with head and body lice control products, some are sold over-the-counter and some are by prescription. Kwell® (lindane) shampoo, applied for 5 minutes in the shower, is quite effective for treatment of infestations in the pubic, perianal, chest, or underarm areas, but must be used cautiously due to potential side effects. Also effective are synergized pyrethrins (over-the-counter products), and permethrin (Nix® and generics). Treatment is usually repeated in 1 week. Treatments should closely follow directions on the box, bottle, or package insert. As is the case with head and body lice, oral ivermectin has been proposed as an effective treatment for pubic lice,[42] although to date the FDA has not approved the product for use as a pediculicide. For pubic louse infestations of the eyebrows or eyelashes, mechanical removal of nits may be used; twice daily applications of petrolatum for 7–10 days or possibly anticholinesterase eye ointments, yellow oxide of mercury, or fluorescein are also used.[39,43] At the same time of treatment (no matter location on the body), infested persons should wash all their underclothes and bedding in hot water for 20 minutes or more and dry them on the hottest setting. Because of the limited survivability of pubic lice off their hosts, insecticidal sprays, fogs, etc. in the patient's home, work, or school are not necessary.

References

1. Bonilla DL, Durden L, Eremeeva ME, Dasch GA. The biology and taxonomy of head and body lice: Implications for louse-borne disease prevention. *PLoS Pathog.* 2013;9:1–5.
2. Zlotnick C. Pediculosis corporis and the homeless. *J. Community Health Nurs.* 1987;4(1):43–48.
3. Voelker R. Lice-borne diseases in the homeless population. *JAMA.* 2014;312:1962.
4. Bonilla DL, Cole-Porse C, Kjemtrup A, Osikowicz L, Kosoy M. Risk factors for human lice and bartonellosis among the homeless, San Francisco, California, USA. *Emerg. Infect. Dis.* 2014;20(10):1645–1651.
5. Foucault C, Brouqui P, Raoult D. *Bartonella quintana* characteristics and clinical management. *Emerg. Infect. Dis.* 2006;12:217–223.
6. Faccini-Martinez AA, Marquez AC, Bravo-Estupinan DM, et al. *Bartonella quintana* and typhus group rickettsiae exposure among homeless persons, Bogata, Colombia. *Emerg. Infect. Dis.* 2017;23:1876–1879.
7. von Both U, Alberer M. *Borrelia recurrentis* infection. *N. Engl. J. Med.* 2016;375:e5.
8. Barbour A. Relapsing fever and other *Borrelia* diseases. In: Guerrant RL, Walker DH, Weller PF, eds., *Tropical Infectious Diseases: Principles, Pathogens, and Practice.* 3rd ed. New York: Saunders Elsevier; 2011:295–302.
9. Raoult D, Roux V. The body louse as a vector of re-emerging human diseases. *Clin. Infect. Dis.* 1999;29:888–911.
10. Harwood RF, James MT. *Entomology in Human and Animal Health.* 7th ed. New York: Macmillan; 1979.
11. Lescot M, Audic S, Robert C, et al. The genome of *Borrelia recurrentis*, the agent of deadly louse-borne relapsing fever, is a degraded subset of tick-borne *Borrelia duttonii. PLoS Genet.* 2008;4(9):e1000185.
12. Walton GA, Horwitz A. Possible extra human reservoirs of the relapsing fever spirochete *Borrelia recurrentis.* Proceedings of the International Symposium on the Control of Lice and Louse-borne Diseases, December 4–6. Washington, DC: Pan American Health Organization; 1972:117–129.
13. Alexander JO. *Arthropods and Human Skin.* Berlin: Springer-Verlag; 1984.
14. Kim KC, Pratt HD, Stojanovich CJ. *The Sucking Lice of North America.* University Park, PA: Pennsylvania State University Press; 1986.
15. Witkowski JA, Parish LC. What's new in the management of lice. *Infect. Med.* 1997;14:287–288,294–296.
16. Foucault C, Ranque S, Badiaga S, Rovery C, Raoult D, Brouqui P. Oral ivermectin in the treatment of body lice. *J. Infect. Dis.* 2006;193(3):474–476.
17. Mumcuoglu KY. The louse comb: Past and present. *Am. Entomol.* 2008;54:164–166.
18. Frankowski BL, Weiner LB. Head lice. *Pediatrics.* 2002;110(3):638–643.
19. Meinking TL, Taplin D, Kalter DC, Eberle MW. Comparative efficacy of treatments for pediculosis capitis infestations. *Arch. Dermatol.* 1986;122:267–271.
20. Sudakin DL. Fatality after a single dermal application of lindane lotion. *Arch. Environ. Occup. Health.* 2007;62(4):201–203.
21. Brandenburg K, Deinard AS, DiNapoli J, Englander SJ, Orthoefer J, Wagner D. One percent permethrin cream rinse vs 1% lindane shampoo in treating pediculosis capitis. *Am. J. Dis. Child.* 1986;140:894–898.
22. Pariser DM, Meinking TL, Bell M, Ryan WG. Topical 0.5% ivermectin lotion for treatment of head lice. *N. Engl. J. Med.* 2012;367:1687–1693.
23. Roberts RJ. Clinical practice: Head lice. *N. Engl. J. Med.* 2002;346(21):1645–1650.
24. Bailey AM, Prociv P. Persistent head lice following multiple treatment: Evidence for insecticide resistance in *Pediculus humanus capitis. Austr. J. Dermatol.* 2000;41:250–254.
25. Downs AMR, Stafford KA, Coles GC. Head lice: Prevalence in schoolchildren and insecticide resistance. *Parasitol. Today.* 1999;15:1–4.
26. Downs AMR, Stafford KA, Harvey I, Coles GC. Evidence for double resistance to permethrin and malathion in head lice. *Br. J. Dermatol.* 1999;141:508–511.
27. Picollo MI, Vassena CV, Casadio AA, Massimo J, Zerba EN. Laboratory studies of susceptibility and resistance to insecticides in *Pediculus capitis. J. Med. Entomol.* 1998;35:814–817.
28. Lebwohl M, Clark L, Levitt J. Therapy for head lice based on life cycle, resistance, and safety considerations. *Pediatrics.* 2007;119(5):965–974.
29. Yoon KS, Previte DJ, Hodgdon HE, et al. Knockdown resistance allele frequencies in North American head louse (Anoplura: Pediculidae) populations. *J. Med. Entomol.* 2014;51(2):450–457.

30. Burkhart KM, Burkhart CN, Burkhart CG. Update on therapy: Ivermectin is available for use against lice. *Infect. Med.* 1997;14:689.

31. Chosidow O, Giraudeau B, Cottrell J, et al. Oral ivermectin versus malathion lotion for difficult-to-treat head lice. *N. Engl. J. Med.* 2010;362(10):896–905.

32. Estrada B. Head lice: What about Ivermectin. *Infect. Med.* 1998;15:823.

33. Drugs.com. Ulesfia, a new treatment for head lice. Drugs.com consumer health information, http://www.drugs.com/ulesfia.html; 2010.

34. Hill N, Moor G, Cameron MM, et al. Single blind, randomised, comparative study of the Bug Buster kit and over the counter pediculicide treatments against head lice in the United Kingdom. *BMJ.* 2005;331(7513):384–387.

35. Goates BM, Atkin JS, Wilding KG, et al. An effective non-chemical treatment for head lice: A lot of hot air. *Pediatrics.* 2006;118(5):1962–1970.

36. Bush SE, Rock AN, Jones SL, Malenke JR, Clayton DH. Efficacy of the lousebuster, a new medical device for treating head lice. *J. Med. Entomol.* 2011;48:67–72.

37. Fisher I, Morton RS. *Phthirus pubis* infestation. *Brit. J. Vener. Dis.* 1970;46:326–329.

38. Dholakia S, Buckler J, Jeans JP, Pillai A, Eagles N, Dholakia S. Pubic lice: An endangered species? *Sex. Trans. Dis.* 2014;41(6):388–391.

39. Micali G, Lacarrubba F. Phthiriasis palpebrarum in a child *N. Engl. J. Med.* 2015;373:e35.

40. Chapel TA, Katta T, Kuszmar T. *Pthrius pubis* in clinic for treatment of sexually transmitted diseases. *Sex. Trans. Dis.* 1979;6:257–261.

41. Nuttall GHF. The biology of *Phthrius pubis. Parasitol.* 1918;10:383–405.

42. MayoClinic.com. Pubic lice: Treatment and control. Mayo Clinic health information, http://www.mayoclinic.com/health/pubic-lice-crabs/DS01072/DSECTION=treatments-and-drugs; 2011.

43. Buntin DM, Rosen T, Lesher JL, Jr., Plotnick H, Brademas E, Berger TG. Sexually transmitted diseases: Viruses and ecto-parasites. *J. Am. Acad. Dermatol.* 1991;25:527–534.

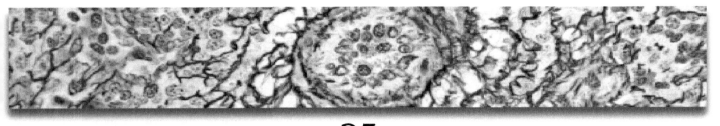

23

MILLIPEDES

I. General and Medical Importance

Millipedes, sometimes called thousand leggers, are elongate, wormlike arthropods (see box) that are commonly found in soft, decomposing plant matter.[1] They do not bite or sting, but some species secrete (and in some cases forcefully discharge) defensive body fluids containing quinones such as toluquinone and p-benzoquinones and/or hydrogen cyanide that may discolor and burn human skin (Figure 23.1). Alexander[2] observed that affected skin becomes yellowish brown in color, turning to a dark mahogany brown within 24 hours. The mahogany discoloration is attributed to oxidation of quinones on contact with the skin.[3] There may be blistering in a day or two, exfoliating to expose a raw surface.[4] Radford[5] provided an excellent review of millipede burns in people. We are aware of one documented case in the U.S. wherein a child in North Carolina picked up a millipede and it sprayed

Figure 23.1 Millipedes may stain human skin. (Photograph by Dr. Vidal Haddad and used with permission.)

MILLIPEDES

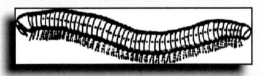

Millipede. (From CDC, Pictorial Keys to Arthropods, Reptiles, Birds, and Mammals of Public Health Significance, U.S. Centers for Disease Control and Prevention, Atlanta, GA, 1963.)

IMPORTANCE

Some species cause burns on skin

DISTRIBUTION

Numerous species worldwide

LESION

Yellow to brown discoloration; blisters

DISEASE TRANSMISSION

None

KEY REFERENCE

Radford AJ. Millipede burns in man. *Trop. Geogr. Med.* 1975; 27:279–280

TREATMENT

Skin—wash thoroughly to remove fluids, and apply antiseptics; eye—immediate, thorough washing, and consult an ophthalmologist for treatment guidelines him in the eye. A family member flushed the eyes with water which helped, but skin on the hands, as well as above and below the eye, was discolored and burned. After a couple of days, there was peeling, leaving new, pink skin (Elmer Gray, Department of Entomology, University of Georgia, personal communication).

II. General Description

Millipedes are somewhat similar to centipedes except that they have two pairs of legs on most body segments and are generally rounded instead of flattened (Figure 23.2). In addition, the mouthparts point downward, rather than forward as is the case with centipedes. Millipedes have one pair of antennae. Many species of millipedes are cylindrical (although some are flat) with a hardened, burnished metallic look. Some of the tropical species can attain a length of 30 cm.

III. Geographic Distribution

There are at least 7000 species of millipedes worldwide, although the fauna is poorly studied at this time.[6] Millipedes in tropical regions are much larger than those found in more northern areas (Figure 23.3). A genus commonly encountered in leaf litter in North America is *Narceus*, which contains several species. *Rhinochrichus latespargor* found in Haiti, *Spirostreptus* spp. in Indonesia, and *Orthoporus* spp. in Mexico are reported to cause human burns.

Figure 23.2 Millipedes have two pairs of legs on most body segments.

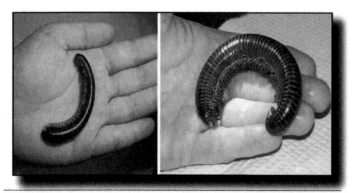

Figure 23.3 Millipede from the temperate region (left) vs. specimen from the tropics (right). (Photograph copyright 2011 by Jerome Goddard, Ph.D.)

IV. Biology and Behavior

Millipedes are commonly found under rocks, in soil, and in leaf litter in moist deciduous forests. They are mostly nocturnal in habit and may be active year-round; however, millipedes are more commonly encountered during the wet season. When uncovered, they coil up into a tight spiral. Some of the more slender, agile species, such as the common garden millipede, are attracted to light and may congregate by the thousands on front porches, patios, or sidewalks of homes. Interestingly, homes with these "millipede invasions" are most often 3–5 years old and have well-kept lawns.

Female millipedes lay their eggs in the soil. Upon hatching, immature millipedes have few body segments and three pairs of legs; additional legs are added with each molt. They go through 2–7 instars before reaching the adult stage. Depending on species, millipedes live from 1 to 7 years. Concerning the venomous secretions, for the majority of millipede species the secretions ooze out and form droplets around the foramina, but a few species (from genera *Spirobolida*, *Spirostreptus*, and *Rhinocrichus*) can squirt their secretions up to 0.5 meters.[5,6]

V. Treatment

Exposed skin should be washed with copious amounts of water as soon as possible. Alexander[2] and Radford[5] recommended using the solvents ether or alcohol to help remove the noxious fluids. Antiseptics may need to be applied. Antibiotics are indicated if secondary infection is suspected. Eye exposure is very painful and requires thorough irrigation with warm water as soon as possible. An ophthalmologist should be consulted for current treatment recommendations.

References

1. Borror DJ, Triplehorn CA, Johnson NF. *An Introduction to the Study of Insects.* 6th ed. Philadelphia: Saunders College Publishing; 1989.
2. Alexander JO. *Arthropods and Human Skin.* Berlin: Springer-Verlag; 1984.
3. Shpall S, Freiden I. Mahogany discoloration of the skin due to the defensive secretion of a millipede. *Pediatr. Dermatol.* 1991;8:25–26.
4. Harwood RF, James MT. *Entomology in Human and Animal Health.* 7th ed. New York: Macmillan; 1979.
5. Radford AJ. Millipede burns in man. *Trop. Geogr. Med.* 1975;27:279–287.
6. Shelley RM. Centipedes and millipedes, with emphasis on the North American fauna. *The Kansas School Naturalist,* Vol. 45 (3), March issue, Emporia State University, Emporia; 1999:3–15.

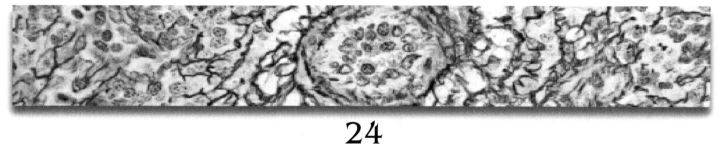

24

MITES

I. Mites in General

Mites are arachnids in the Subclass Acari which occur virtually everywhere on earth and display incredible diversity in form and habitat utilization.[1] Many are free-living, while some are parasitic on plants and animals causing a variety of harmful effects. In addition, a few species of mites may transmit disease agents (not counting ticks which are essentially big mites). Health effects resulting from mites include: 1) temporary skin irritation, 2) persistent dermatitis from mites in skin or hair follicles, 3) allergies, 4) disease transmission, 5) intermediate hosts for parasites, 6) invasion of lungs, ears, and other bodily openings, and 6) various fears and phobias in people who think mites are infesting them.[2] Although some mites bite people, only two species actually inhabit human skin (i.e., actually take up long-term residence on a person). These are covered here first, followed by other mites which may opportunistically attack people.

A. Generalized Mite Life Cycle

It is difficult to describe a "typical" mite life cycle and accompanying life stages as there is tremendous variation among the various orders; however, most mites generally display the following life cycle. Adult females lay eggs that hatch into larvae, pass through one to three nymphal stages, and finally become adults. The larvae have only three pairs of legs, whereas nymphs and adults have four pairs. The first nymphal stage is called a protonymph, the second is called a deutonymph, and the third nymphal stage, if present, is called a tritonymph. In some mite orders, one of the nymphal stages is a nonfeeding phoretic stage for passive transport (e.g., in the fur of animals, in bird feathers, on insects). Figure 24.1 is provided to familiarize the reader with some of the more prominent morphological characters of mites.

II. Scabies Mites (Human Itch or Mange Mites)

A. General and Medical Importance

Scabies, caused by *Sarcoptes scabei* (see box and Figure 24.2), is probably the most important disease caused by mites, with at least 100 million cases being reported annually.[3] It occurs worldwide, affecting all races and socioeconomic classes in all climates. The tiny mites burrow under the skin, leaving small open sores and linear burrows that contain the mites and their eggs (Figure 24.3). When a person is infested with scabies mites for the first time, there is little pathology for about a month, until sensitization develops. When that happens, there is severe itching, especially at night and frequently over much of the body. Large patches of erythema or rash may occur on the body. The patient's tissues apparently become sensitized to various proteins liberated by the mites. Interestingly, the generalized rash may not correspond to the sites where the mites are burrowing. The burrows are usually located on the hands, wrists, and elbows, especially in the webbing between the fingers and the folds of the wrists.[4,5] Alexander[6] observed that burrows are sometimes few in number and difficult to find. Genital lesions are common in scabies. The presence of crusted, excoriated, pruritic papules on the penis or buttocks is virtually pathognomonic.[7] In many cases, severe itching causes the patient to scratch himself or herself vigorously, leading to secondary infections such as impetigo, eczema, pustules, and boils. In children, scabies may affect the head, palms, soles, and genital area and lead to impetigo.[8,9]

The authors have frequently encountered scabies problems in nursing homes. In the elderly, reactions to the mite are often not inflammatory—as seen in younger people—and are muted. Accordingly, scabies is often missed by the attending healthcare provider. Bedridden patients with scabies

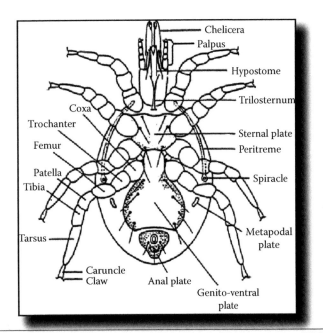

Figure 24.1 Female rat mite with parts labeled. (Figure courtesy of U.S. Centers for Disease Control and Prevention, Atlanta, GA.)

complain of intense itching. Mites may be found on the back of such patients (which is unusual).

Those who are immunocompromised (e.g., AIDS patients or those on immunosuppressant therapy) or who cannot scratch themselves may develop more serious scabies infestations in which millions of mites may inhabit thick crusts over the skin—a condition called Norwegian scabies or crusted scabies. Alexander[6] described two principal components of the eruption: localized horny plaques and a more diffuse erythematosquamous appearance. A recent report described a 60-year-old woman with cognitive impairment who had bilateral thick, adherent, white-gray scaling, and crusting on both hands, and web spaces with associated deep fissures.[10] Patients with Norwegian scabies are frequently inmates of

long-stay institutions and when presented to the dermatologist appear to be suffering from chronic exfoliative eczema.

It should be noted here that animal forms of scabies such as canine or equine are caused by "races" of *Sarcoptes scabei* (sometimes with devastating results; see Figure 24.4), but these mites cannot propagate in human skin. Canine scabies can be temporarily transferred to humans from dogs, causing itching and papular or vesicular lesions primarily on the waist, chest, or forearms; however, treatment or removal of the infested dog will result in a gradual resolution of this type of scabies.

SCABIES MITES

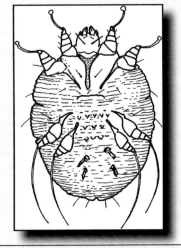

Human scabies mite.

IMPORTANCE

Intense itch; secondary infection

DISTRIBUTION

Worldwide

LESION

Burrows and papules where mites are located; generalized rash may occur in other areas

DISEASE TRANSMISSION

None

KEY REFERENCES

Alexander JO. *Arthropods and Human Skin.* Springer-Verlag, Berlin; 1984 (see section on scabies)

Mellenby K. *Scabies.* 2nd ed. London: EW Classey; 1972

TREATMENT

Scabicides (cream or lotion) per label instructions, sometimes Ivermectin

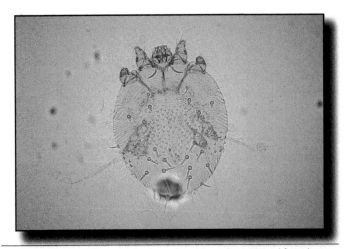

Figure 24.2 Microscopic view of scabies mite removed from human skin. (Photograph copyright 2012 by Jerome Goddard, Ph.D.)

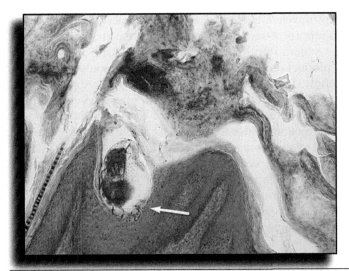

Figure 24.3 Microscopic view of scabies mite located in human skin sample (arrow points to mite leg).

Figure 24.4 Young canid with severe sarcoptic mange. (Photograph courtesy Joseph D. Goddard, Alabama Department of Conservation and Natural Resources, and used with permission.)

B. General Description

Sarcoptes scabei are very tiny (0.2–0.4 mm long), oval, saclike, eyeless mites (see box). Their legs are rudimentary; the anterior two pairs have bell-shaped suckers on their tips. The body is covered with striations and has several stout blunt spines and a few long setae. Scabies mite mouthparts are composed of toothed chelicerae and one-segmented palps fused to the central hypostome. Nymphs look almost identical to the adults, except that they are smaller.

C. Geographic Distribution

Human scabies mites occur worldwide. There are also several scabies mites that occur on domestic animals (dogs, cats, horses, and camels) worldwide. These "races" of *Sarcoptes*

scabei are virtually indistinguishable from the human form, but they do not produce sustained infestations on people.

D. Biology and Behavior

Scabies is transmitted by close, human-to-human contact with infested individuals. It is essentially a disease of overcrowding, where conditions are allow for prolonged skin-to-skin contact. There is some evidence that fomites can be important sources of infestation or reinfestation.[11] Touching or shaking the hands of infested persons is a major mode of transmission. The practice of several family members sleeping in one bed contributes to its spread, as does sexual activity. In addition, institutionalized children (daycare) and elderly (nursing homes) seem to be contributing to an increase in the incidence of scabies.

A female mite infests a new host by burrowing beneath the outer layer of skin and laying her eggs in the tunnels that she excavates. A six-legged larval stage emerges from each egg and molts to the first nymphal stage in 2 or 3 days. Nymphal and adult stages have eight legs. Larvae and nymphs are often found in short burrows or in hair follicles. After a few days, the nymphs molt to the next nymphal stage. After the second nymphal stage, adults are formed. The entire life cycle takes 10–17 days. The mites apparently eat human skin, although the immatures may feed on hair follicle secretions.

Survival of scabies mites off of a host is probably at most a few days, more likely hours. Studies have shown that the mites may survive 1–5 days at room conditions but may have a difficult time infesting a host after being off the host (presumably due to the mite's weakened condition).[11]

E. Treatment of Infestation

First of all, scabies should be confirmed by isolating the mites in a skin scraping, as other forms of dermatitis may resemble scabies infestation.[12] Scrapings should be made at the burrows, especially on the hands between the fingers and the folds of the wrists. Some dermatologists may choose to scrape burrows located on the feet.[6] To do the scraping, mineral oil is placed on a sterile scalpel blade and allowed to flow onto suspected lesions. By gentle scraping with the blade, the tops of burrows or papules are removed. The oil and scraped material are then transferred to a glass slide and a coverslip applied. Diagnosis can be made by finding mites, ova, or fecal pellets (Figure 24.3). Alternatively, mites can be extracted from a burrow by gently pricking open the burrow with a needle and working toward the end where the tiny mites usually are. A hand lens may be useful for this task.

Once a scabies infestation is confirmed, treatment can be initiated (see Chosidow[13] for a good review). As the mites cannot live very long off a human host, insecticide treatments of bedding, furniture, rooms, etc. are unnecessary. It is recommended, however, that upon initiation of treatment the patient's bedcovers, pillowcases, and undergarments be removed and washed on the hot wash cycle. If the patient is a

CASE HISTORY

SCABIES IN AN INSTITUTION FOR THE MENTALLY HANDICAPPED

During August, the medical director of a local institution for the mildly retarded called saying that several of his patients were obviously scratching themselves, even to the point of bleeding. He was wondering what different types of arthropod pests might be responsible. On August 13, Dr. Goddard visited the institution and found approximately 12 male patients complaining of intense itching. Most had obvious rashes on the arms and trunks. One man had numerous, self-inflicted, deep fingernail scratch marks on his body that were unbandaged and bleeding. Dr. Goddard told the medical director that scabies was likely the cause of the itching, but skin scrapings, performed by a dermatologist, would be needed for confirmation. Later he interviewed maintenance staff and asked questions about the layout of the building, history of pest (such as rat or bird) problems, and bird nesting (or lack thereof) in the attic. Nothing out of the ordinary was noted. Dr. Goddard then took dust samples—alcohol swipes of furniture, bedding, etc.—from the affected patients' rooms. Back at the lab, dust samples were examined microscopically for the presence of human-biting mites. A sample from one patient's bed contained dust, debris, pieces of excoriated skin, and a scabies mite, *Sarcoptes scabei*, thus confirming scabies infestation. Patients were subsequently treated with scabicidal lotions, and the infestation was eliminated.

Comment: Investigation of this event revealed several interesting features. The patients were itching so severely that they were scratching themselves to the point of bleeding. Perhaps this was due to their altered mental status or lowered immune status. Second, the infestation was so great that scabies mites could be found in the bed. Most scabies infestations involve only a few mites; finding them is a difficult task. Except in cases of Norwegian (also called *crusted*) scabies, it is extremely rare to find the mites on furniture, bedding, and the like.

child, then toys, stuffed animals, etc. should be removed from human contact for a week or so.

There have been several products used for scabies treatment in the past, such as sulfur ointment, benzyl benzoate, lindane, crotamiton, and thiabendazole. The most widely used in the United States today are lindane (Kwell®), permethrin (Elimite®), crotamiton (Eurax®), and (increasingly) ivermectin (Stromectol®).[14] There have been reports of lindane-resistant scabies, especially in cases of immigrants or recent travelers to Central and South America or Asia,[15] and also reports of serious, even fatal, adverse effects on the nervous system from misapplication of lindane.[16] Regardless of the product used, all package instructions should be followed carefully. For most scabicides, the product is applied to the entire body, except the head (see package insert; sometimes the head is treated), and left on for 14–48 hours, depending on instructions. After that, a cleansing bath may be taken. A second treatment may be called for in the instructions. Itching may persist for weeks or more after treatment and does not necessarily indicate treatment failure. The posttreatment itch/rash is often treated with cortisone cream. Some physicians prescribe oral antipruritic agents (antihistamines or hydroxyzine) simultaneously with the application of scabicides.

F. Management of Scabies in Nursing Homes

Scabies in long-term healthcare institutions can be a significant problem. One study found that 20% of 130 institutions in Ontario, Canada, had problems with scabies during a 1-year period.[17] Because scabies control in institutions can be difficult and frustrating, a brief outline of a control strategy is offered here. First of all, the diagnosis should be confirmed. Frequently, scabies control measures are implemented with only weak evidence of infestation (e.g., "Looks sorta like a scabies rash to me"); however, implementing a large-scale control effort on the suspicion that an outbreak is occurring in the nursing home is medically and administratively unsound. The active ingredients in scabicidal creams or lotions are pesticides. A dermatologist should be consulted to perform skin scrapings on affected patients. Once the infestation is confirmed, treatment with one of the scabicidal products can be initiated. Both patients and employees with direct, close contact with affected patients should be treated. If cases reappear, aggressive treatment strategies may have to be used such as simultaneously treating all patients and employees in the nursing home (and possibly even family members of employees).[18,19] Consultation with state or local health department epidemiology personnel is often helpful as well. Studies have demonstrated the effectiveness of oral ivermectin for treatment of human scabies both in the United States and abroad;[13,14,20,21] however, this antihelminthic drug is not approved by the U.S. Food and Drug Administration for scabies. Reports of successful off-label use of ivermectin for scabies (and lice) continue, even multiple doses separated by intervals of 1–2 weeks for cases of severe scabies or crusted scabies.[13,22,23]

III. Follicle Mites

A. Introduction

Although numerous species of *Demodex* (family Demodecidae) infest wild and domestic animals, only two species of the mites are specific human-associated mites: follicle mites.

FOLLICLE MITES

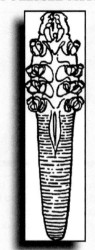

Hair follicle mite. (Figure courtesy of U.S. Centers for Disease Control and Prevention, Atlanta, GA.)

IMPORTANCE

Live in human skin but mostly harmless

DISTRIBUTION

Worldwide

LESION

Generally none

DISEASE TRANSMISSION

None

KEY REFERENCE

Desch C, Nutting WB. *Demodex folliculorum* (Simon) and
    D. brevis (Akbulatova) of man: redescription and reeval-
    uation. *J. Parasitol.* 1972;8:169–177

TREATMENT

Generally, none needed

These minute, wormlike mites live exclusively in hair follicles or sebaceous glands. They have no proven detrimental effect on humans, although some authors have attributed various pathological conditions of the skin to *Demodex* and a recent report described "demodectic frost" of the (human) ear from these mites[24] in which a patient displayed a frosted appearance of the skin on the ear caused by follicular-based scale. Alexander[6] provided a good review of this issue and concluded, "It should be emphasized that, in general, *Demodex* is a harmless saprophyte. It is only exceptionally that it appears to exercise a pathogenic influence, as, for example, when

excessive amounts of cosmetics prepare the ground for its proliferation or when it escapes into the dermis." Various estimates of the incidence of human *Demodex* infestation range from about 25–100%, and clinicians should be aware of mite appearance, as they may be seen during skin-scraping examination.

B. Discussion of species

Demodex folliculorum lives in the hair follicles and *D. brevis* in the sebaceous glands. Both species are similar in appearance (with the exception that *D. brevis* is a shortened form) and are elongated, wormlike mites with only rudimentary legs (see box and Figure 24.5). They are approximately 0.1–0.4 mm long and have transverse striations over much of the body. These mites most commonly occur on the forehead, malar areas of the cheeks, nose, and nasolabial fold, but they can occur anywhere on the face, around the ears, and occasionally elsewhere. Most people acquire *Demodex* mites early in life from household contacts—primarily maternal.

IV. Zoonotic Mites

A. Chiggers

1. General and Medical Importance   Larvae of mites in the family Trombiculidae, sometimes called chiggers, harvest mites, or red bugs (Figure 24.6), are medically important pests around the world (see box). Over 3000 species of chigger mites occur in the world, but only about 20 species cause dermatitis or transmit diseases such as the agent of scrub typhus. Scrub typhus, caused by *Orientia tsutsugamushi*, is a vector-borne zoonosis traditionally thought to only occur in the "tsutsugamushi triangle" from Pakistan to far eastern Russia to northern Australia. The disease threatens over a billion

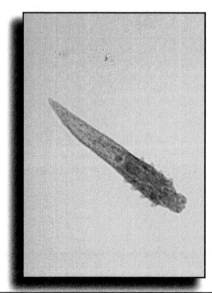

Figure 24.5 *Demodex* mite from skin scraping.

Figure 24.6 Larval chigger mite.

people and there are an estimated 1 million cases per year.[25] Recently, scrub typhus has been detected in the Middle East and South America, leading public health officials to worry about an ever-widening impact.[26,27] In addition, in northern India there is an encephalitis resulting from scrub typhus infection which affects hundreds of children every year.[28] Infestation with trombiculid larvae is called trombiculosis or sometimes trombidiosis. Larval chiggers crawl up on blades of grass or leaves and subsequently get on passing vertebrate hosts. Although humans are not ideal hosts,[29,30] chiggers may crawl to and attach where clothing fits snugly or where flesh is tender, such as ankles, groin, or waistline. Attachment seems to be especially common in the popliteal fossae[6] (Figure 24.7). Chiggers then attach to the skin with their mouthparts, inject saliva into the wound (which dissolves tissue), and then suck up this semidigested material. They do not burrow in human skin in the strictest sense; only the chelicerae penetrate the skin of the host. Following penetration, a straw-like feeding

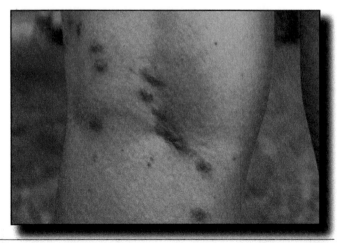

Figure 24.7 Chigger bites in the popliteal fossae appearing as palpable purpura. (Photograph courtesy Dr. Wendy C. Varnado, Mississippi Department of Health and used with permission.)

CHIGGERS

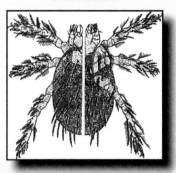

Chigger mite, *Leptotrombidium akamushi*, dorsal (left) and ventral (right) views. (From Gorham JR. Insect and Mite Pests in Food: An Illustrated Key, Agriculture Handbook No. 655. Washington, DC: Agricultural Research Service, U.S. Department of Agriculture; 1991.)

IMPORTANCE

Irritation; intense itching; disease transmission in the Far East

DISTRIBUTION

Numerous species worldwide

LESION

Variable—often a red, dome-shaped papule

DISEASE TRANSMISSION

None in United States; scrub typhus primarily in central, eastern, and southeastern Asia

KEY REFERENCE

Jenkins DW. Trombiculid mites affecting man. I. Bionomics with reference to epidemiology in the United States. *Am. J. Hyg.* 1948;48:22–35

TREATMENT

Antipruritics and/or topical corticosteroids

tube, called a stylostome, is formed through which the larva sucks digested host tissue and lymph. Chiggers do not take a bloodmeal. Feeding generally lasts 2–4 days, and salivary proteins may lead to intense itching in the host.[29,30] Chiggers attached to the skin may easily be removed by even the most casual scratching,[31] but itching may persist for a week or two after the offending specimen is gone, primarily as a result of material injected at the site of the bite. Most U.S. pest chiggers produce itching within 3–6 hours, and the rash consists of macules and wheals. At approximately 10–16 hours, red, dome-shaped papules appear, and itching increases in severity over the next 20–30 hours (Figure 24.8).[6] Not all medically

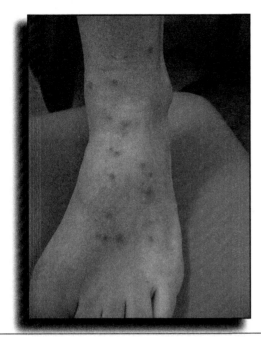

Figure 24.8 Chigger bites around foot. (Photograph courtesy Dr. Wendy C. Varnado, Mississippi Department of Health and used with permission.)

important chiggers produce the familiar itch reactions; those serving as vectors of scrub typhus (central, eastern, and southeastern Asia) are not associated with itching or skin reactions.[32]

2. Biology and Discussion of Species Adult chigger mites are oval shaped (approximately 1 mm long) with a bright red, velvety appearance, but it is the larval stage that attacks vertebrate hosts; adults do not bite. Accordingly, it is the larval stage or accompanying lesion that is ordinarily collected/seen in a clinical setting. Chigger larvae are very tiny (0.2 mm long), round mites with numerous setae (see box). The mites may be red, yellow, or orange in color and have a single dorsal plate (scutum) bearing two sensillae and four to six setae. Chigger species are mostly tropical and subtropical, although they occur from Alaska to New Zealand and from sea level to over 16,000 feet in altitude. Some particularly common pest chiggers are *Eutrombicula alfreddugesi* in the United States (particularly in the southern U.S.)[33] and parts of Central and South America, *Neotrombicula autumnalis* in Europe, and *E. sarcina* in Asia and Australia. Chiggers in the genus *Leptotrombidium* transmit the agent of scrub typhus in Japan, Southeast Asia, and parts of Australia. Chigger mites are found primarily outdoors in moist microenvironments within grassy, weedy, or wooded areas, especially forest edges and wild blackberry patches (at least in the southern United States) where their animal hosts occur. Hosts include lizards, snakes, toads, and a variety of birds and small mammals.[1,33] Adult chiggers are predaceous. The female lays eggs singly on soil or litter, and the eggs hatch in about 1 week. After hatching, the life cycle includes six stages: (1) the inactive prelarva (deutovum or maturing larva), (2) the parasitic larval stage, (3) quiescent protonymph/first nymphal stage (nymphochrysalis), (4) predaceous deutonymph/second nymphal stage, (5) quiescent tritonymph/third nymphal stage (imagochrysalis), and (6) the free-living adult stage. The entire life cycle from egg to adult may be completed in about 60 days. In the northern United States, the mites are active from about May through September, but in the southern United States they may be active essentially year-round. When feeding, larval chigger mites inject saliva that dissolves host cellular tissue. The mites then ingest this mixture of lymph, dissolved body tissues, and stray blood cells.

3. Treatment of Infestations After exposure to infested outdoor areas, hot soapy baths or showers will help remove any chiggers, attached or unattached (see Chapter 33 for prevention measures against chiggers). Various solutions, lotions, or ointments containing antihistamines, antiseptics, calamine, steroids, and/or anesthetics (benzocaine) are often used as treatments to minimize itching and reduce chances of secondary infection, which might require antibiotics.[6,34] However, it should be noted that in most cases, the chiggers themselves are no longer present at the attachment site.[30]

B. Other Biting Mites

1. General and Medical Importance As mentioned, there are only two human-associated parasitic mites, the scabies mite and the hair follicle mite (see the preceding sections on these two mites); however, many other species of mites are important to public health, such as chiggers, which were covered previously. This section includes a hodgepodge of the remaining mite species that may be involved in human bites or cases of dermatitis (clover mites are included—they do not bite, but often invade homes en masse). Most mites mentioned here are important only as causes of itch or dermatitis, but the house mouse mite, *Liponyssoides* (formerly *Allodermanyssus*) *sanguineus*, is a vector of rickettsialpox in Massachusetts, Connecticut, New York, Pennsylvania, and Ohio.[35]

2. Discussion of Species
Order Mesostigmata
Tropical Rat Mite *Ornithonyssus bacoti* is fairly easy to recognize. Females have scissorlike chelicerae, narrow, tapering dorsal and genitoventral plates, and an egg-shaped anal plate (see box). In addition, the protonymphal stage and adult females suck blood and become tremendously distended after feeding. They look like tiny engorged ticks. The tropical rat mite is an ectoparasite of rats. The adult mites produce a nonfeeding larval stage, a bloodsucking protonymphal stage, and a nonfeeding deutonymphal stage. A complete generation of this species usually takes about 2 weeks. Females can live unfed 10 days or more after rats have been removed from a building. Tropical rat mites

BITING MITES

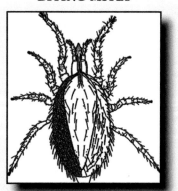

Adult tropical rat mite, *Ornithonyssus bacoti*. (From Gorham JR. Insect and Mite Pests in Food: An Illustrated Key, Agriculture Handbook No. 655. Washington, DC: Agricultural Research Service, U.S. Department of Agriculture; 1991.)

IMPORTANCE

Painful bites; itching; irritation

DISTRIBUTION

Numerous species worldwide

LESION

Variable—erythema, bright red papules of varying sizes

DISEASE TRANSMISSION

Generally none; some species transmit rickettsialpox

KEY REFERENCE

Alexander JO. *Arthropods and Human Skin*. Springer-Verlag, Berlin; 1984 (see section on biting mites)

TREATMENT

Generally palliative treatment only; eliminate source of exposure

bite people, readily producing a papulovesicular dermatitis with accompanying urticaria. They are widely distributed on all continents in association with rats.

Tropical Fowl Mite   Ornithonyssus bursa is similar in appearance to the tropical rat mite but has a wider dorsal plate (Figure 24.9). The biology of the tropical fowl mite is also similar to that of the tropical rat mite, except it is found on domestic and wild birds. It may be found on rodents, but rarely. This species infests poultry, but it is also a significant parasite of the English sparrow. Tropical fowl mites often infest wild birds roosting on roofs and around the eaves of homes and office buildings. Nestling birds are especially

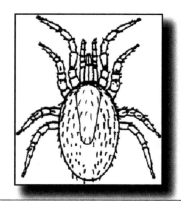

Figure 24.9   Adult tropical fowl mite. (From CDC, Pictorial Keys to Arthropods, Reptiles, Birds, and Mammals of Public Health Significance. Atlanta, GA: U.S. Centers for Disease Control and Prevention; 1963.)

infested with the mites.[36] These mites may subsequently enter human habitations and bite people when the birds abandon their nests. Tropical fowl mites are distributed over most of the world but are more commonly seen in tropical and subtropical areas.

Northern Fowl Mite   Ornithonyssus sylviarum also is similar in appearance to the tropical rat mite but has a much shorter sternal plate (Figure 24.10). This plate has only four setae; the setae on the dorsal plate are quite short. The northern fowl mite is a pest of domestic fowl, pigeons, sparrows, and starlings. The species overwinters in bird nests or cracks and crevices of buildings. Unlike the chicken mite, *Dermanyssus gallinae*, the northern fowl mite spends its entire life on the host. In poultry houses, the mites are usually only found on the birds, but they have been found on eggs and cage litter. Northern fowl mites cannot survive more than a month or so in the absence of their poultry hosts. The northern fowl mite occurs in temperate regions worldwide.

Spiny Rat Mite   Laelaps echidnina is easily recognized by its large genitoventral plate, with a concaved posterior margin into which the anal plate fits (Figure 24.11). Spiny rat mites are ectoparasites of the Norway rat and roof rat. This species is probably the most prevalent mite species occurring

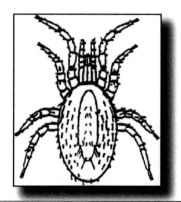

Figure 24.10   Adult northern fowl mite.

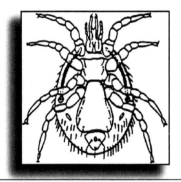

Figure 24.11 Adult spiny rat mite. (Figure courtesy of U.S. Centers for Disease Control and Prevention, Atlanta, GA.)

on rats in the United States, particularly in the central and northern regions of the country. The spiny rat mite is found worldwide.

Chicken Mite Dermanyssus gallinae has large dorsal and anal plates, a short sternal plate, and needlelike chelicerae (Figures 24.12 and 24.13). The chicken mite, also known as the red mite of poultry, is commonly found on domestic fowl, pigeons, English sparrows, starlings, and other birds. This mite is one of the most common species causing

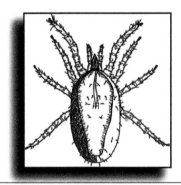

Figure 24.12 Adult chicken mite, *Dermanyssus gallinae*. (From Gorham JR. Insect and Mite Pests in Food: An Illustrated Key, Agriculture Handbook No. 655. Washington, DC: Agricultural Research Service, U.S. Department of Agriculture; 1991.)

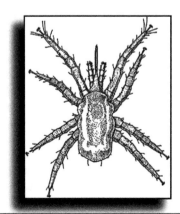

Figure 24.13 Nymphal stage of the chicken mite. (Figure courtesy of U.S. Department of Agriculture.)

human dermatitis, sometimes referred to as gamasoidosis, in poultry houses, farms, ranches, and markets where chickens are traded or sold. Apparently, the incidence of chicken mite infestations on humans is rising, particularly in Europe.[37] Poultry workers are often bitten on the backs of the hands and on the forearms. *D. gallinae* is nocturnal; during the day the mites hide in cracks and crevices in chicken houses or buildings where infested birds nest. Eggs are deposited in these hiding places. The chicken mite occurs worldwide.

House Mouse Mite Liponyssoides sanguineus looks similar to the chicken mite (*Dermanyssus gallinae*). Females of the house mouse mite can be distinguished from most other mites by the presence of two dorsal shields, a large anterior plate, and a small posterior plate bearing one pair of setae (Figure 24.14). The house mouse mite is primarily an ectoparasite of mice but has been collected from rats and other rodents. The protonymphs, deutonymphs, and adults all suck blood. The life cycle from egg to adult takes about 18–23 days. The house mouse mite occurs in northern Africa, Asia, Europe, and the United States (primarily the northeast).

Order Prostigmata

Straw Itch Mite Pyemotes tritici is an elongate species that has the first and second pairs of legs widely separated from the third and fourth pairs (Figure 24.15A). Also, this mite has club-shaped hair between the base of the first and second pairs of legs. The male is so small that it is almost invisible to the naked eye. Females with eggs may become enormously swollen (about 1 mm), with their abdomen resembling a tiny pearl (Figure 24.15B). Straw itch mites are parasites of several insect species but will bite people readily. Bites occur when humans come into contact with infested straw, hay, grasses, beans, peas, grains, and other materials. Accordingly, agricultural workers, persons who process or handle grain, or those who sleep on straw mattresses are most prone to infestation. Alexander[6] observed that severe

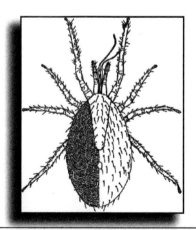

Figure 24.14 Adult house mouse mite, *Liponyssoides sanguineus*. (From Gorham JR. Insect and Mite Pests in Food: An Illustrated Key, Agriculture Handbook No. 655. Washington, DC: Agricultural Research Service, U.S. Department of Agriculture; 1991.)

CASE HISTORY

MYSTERIOUS MITES INFESTING A WORKPLACE

In early October in central Mississippi, a group of women working in a relatively new four-story building complained of seeing numerous tiny, almost invisible mites crawling around on their desks and office cubicles. As far as was known, the mites were not biting the workers. The affected office suite consisted of approximately 12 cubicles, 10 × 10 feet, each containing a desk, a filing cabinet, and a few personal items. The cubicle partitions were padded on each side with a carpet-like covering and stood 6 feet high. A hard plastic rim topped each partition. Careful inspection of the affected area revealed thousands of tiny tan- to yellow-colored mites, mostly seen walking along the upper edges of the partitions. Specimens were collected, examined microscopically, and identified as spider mites, family Tetranychidae. Upon further investigation at the office suite, a desk in one of the cubicles was found to have a potted plant on it that was heavily infested with spider mites and in contact with the partition. Mites were seen climbing the partition from that point.

Comment: Tetranychids (spider mites) are plant feeders that may reach up to 0.8 mm in length and vary in color from yellowish to greenish, orangish, reddish, or red (see figure). Some species are red only in winter. Spider mites receive their name from their ability to spin a fine web over the leaves of the plant upon which they feed. At times, the entire plant may be covered with this webbing. Potted plants are the source of a variety of insect and mite infestations in workplaces. In fact, a competent pest control technician will often ask to see indoor plants first when investigating a bug problem in a work setting.

Tetranychidae mites on plant leaf. (Photograph courtesy of Angus Catchot, Ph.D., Mississippi State University Extension Service.)

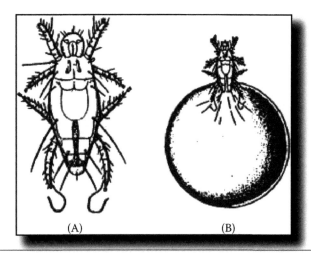

Figure 24.15 (A) Adult straw itch mite and (B) gravid adult female. (From U.S. Naval Medical School, Laboratory Guide to Medical Entomology with Special Reference to Malaria Control. Bethesda, MD: National Naval Medical Center, 1945.)

human infestations may occur, with upward of 10,000 lesions on an individual. Interestingly, several dermatologists have reported a diagnostic "comet sign" dermatitis associated with bites by this species.[38] Attacks are more frequent during the hot weather months. This mite has a curious life cycle. As the female feeds, the opisthosoma becomes enormously distended (Figure 24.15B). Also, all the nymphal stages occur within the egg so that, after hatching, active young occur in the body of the gravid female. Each adult female may produce 200–300 young. Males mate with emerging young females at or near the genital opening of the mother. Straw itch mites occur in most areas of the world.

In the last few years, there have been reports of another *Pyemotes* itch mite affecting thousands of people in Kansas, Missouri, Nebraska, Ohio, Oklahoma, Tennessee, Texas, and Pennsylvania.[39,40] These mites, identified by entomologists as *Pyemotes herfsi*, cause a rash in exposed individuals, primarily on the limbs, face, and neck. Approximately 77% of the rashes have been erythematous, well-demarcated, and papular; less than 22% have been pustular, macular, or confluent.[39] Rashes have been said to be extremely itchy and not responsive to a variety of treatments. Researchers from Kansas State University, Pittsburg State University, and the University of Nebraska determined that the mites were infesting gall-making midge larvae on oak trees. After itch mites feed on the midge larvae inside the galls, an estimated 16,000 mites can emerge from a single leaf! Such a high mite population is a ubiquitous source of exposure to persons living nearby.

Cheyletid Mite The mite family Cheyletidae includes species in the genera *Cheyleta* and *Cheyletiella*. *Cheyletiella yasguri*, *C. blakei*, and *C. parasitivorax* have fused cheliceral bases further fused with the subcapitulum, forming a capsular gnathosoma (Figures 24.16 and 24.17). This makes them look as though they are wearing a helmet. They have free and

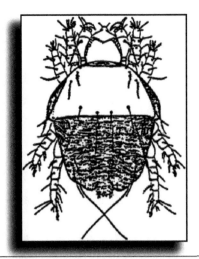

Figure 24.16 Adult cheyletid mite.

Figure 24.17 Scanning electron micrograph of cheyletid mite. (Photograph courtesy of Dr. Gerry Baker, Mississippi State University, and used with permission.)

highly developed palpi with strong curved claws that look like an extra pair of legs near the mouthparts. *Cheyletiella* mites are parasites of birds and various species of small mammals. The *Cheyletiella* spp. mentioned in this section are obligate parasites of small- or medium-sized mammals (including pet dogs, cats, and pet rabbits), living on the keratin layer of the epidermis; they do not burrow. These mites are not very host specific and may cause a mangelike condition on pets and a transient itching dermatitis on humans who handle these pets.[41,42] In most cases, the patient complains of itching and has papules or papulovesicles (2- to 6-mm diameter) on the flexor side of the arms, on the breasts, or on the abdomen. Mites are rarely found in scrapings of the lesions, as they have usually left the person by the time medical advice is sought. *Cheyletiella* eggs are attached to hairs on the host (pet dog, cat, etc.) about 2–3 mm above the host's skin. These mites cannot survive more than 48 hours off their hosts. In a review of the *Cheyletiella*,[43] specimens were reported from small

pets in the United States, South America, Western Europe, Australia, New Zealand, and few widely scattered spots in Africa, India, and Japan.

Order Astigmata

Grain and Flour Mites Although not truly zoonotic, there are several grain and flour mites that can cause grocer's itch, copra itch, and other lay-named itches as well as inhalational allergies in people who handle mite-infested grains or other stored products. They are tiny mites (0.5 mm or less) that are pale gray or yellowish white, and they have the first and second pairs of legs widely separated from the third and fourth pairs (Figures 24.18 through 24.20). Also, they have conspicuous long hairs. Mites causing grocer's itch, copra itch, vanillism, wheat pollard itch, and dried fruit dermatitis are basically scavengers on a wide variety of organic matter including flour, meal, grains, dried fruits, vanilla pods, meats, and other similar products. These tiny mites have brief

Figure 24.18 Adults, immatures, and eggs of the ham mite, *Tyrophagus putrescentiae*. (Photo courtesy of Salehe Abbar, Kansas State University, used with permission.)

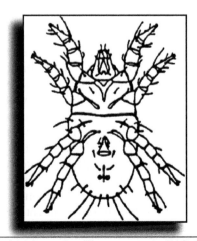

Figure 24.19 Adult ham mite. (Figure courtesy of U.S. Centers for Disease Control and Prevention, Atlanta, GA.)

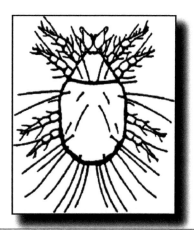

Figure 24.20 Adult cheese mite. (Figure courtesy of U.S. Centers for Disease Control and Prevention, Atlanta, GA.)

developmental times and can multiply into the billions in a stored product in a very short period of time. They do not suck blood, but they will penetrate the superficial epidermis, producing a temporary pruritus. Grain and flour mites occur worldwide wherever people have shipped food products.

3. Treatment of Infestation and Bites Human infestations with the previously mentioned mite species are mostly transitory, and the reaction is variable. Sensitive people may develop dermatitis. These mite species will not take up permanent residence on human skin and thus perpetuate the infestation. Accordingly, treatment primarily involves alleviation of the symptoms, and avoidance and/or eradication of the mites from the pet, home, or workplace. Acute urticarial lesions may respond satisfactorily to topical corticosteroid lotions and creams, which reduce the intensity of the inflammatory reaction. Oral antihistamines may relieve itching and burning sensations.

V. House Dust Mites

A. General and Medical Importance

Several species of mites have been reported from house dust, but only three are major sources of mite allergens and thus medically important: *Dermatophagoides farinae*, *D. pteronyssinus*, and *Euroglyphus maynei*. Since the mid-1960s, considerable research on house dust allergy has revealed that both *D. pteronyssinus* and *D. farinae* (the most familiar house dust mites) possess powerful allergens in the mites themselves, as well as in their secretions and excreta. The fecal pellets are especially allergenic. Although dust mites are harmless in that they do not sting or bite, a considerable amount of allergic rhinitis, asthma, and childhood eczema is attributable to their presence in the human environment.[44] Some studies have investigated the link between house dust mite allergens and atopic dermatitis.[45]

HOUSE DUST MITES

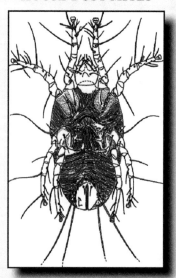

Adult female house dust mite, *Dermatophagoides farinae*. (From Gorham JR. Insect and Mite Pests in Food: An Illustrated Key, Agriculture Handbook No. 655. Washington, DC: Agricultural Research Service, U.S. Department of Agriculture; 1991.)

IMPORTANCE

Allergies

DISTRIBUTION

Almost worldwide

LESION

Generally none—nonbiting

DISEASE TRANSMISSION

None

KEY REFERENCE

Bronswijk JEMH, van Sinha, RN. Pyroglyphid mites (Acari) and house dust allergy—a review. *J. Allergy* 1971;47(1):31–52

TREATMENT

Dust and dust mite control; standard treatment for allergies

B. General Description

Adult house dust mites are white to light tan in color and about 0.5 mm long (see box and Figure 24.21). Their cuticle has numerous fine striations. The mites have plump bodies (not flattened), well-developed chelicerae, and suckers at the ends of their tarsi.

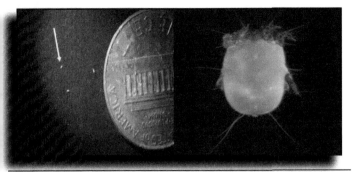

Figure 24.21 House dust mites, actual size (left) and magnified (right). (Photographs copyright 2011 by Jerome Goddard, Ph.D.)

C. Geographic Distribution

Contrary to its common name, the European house dust mite, *Dermatophagoides pteronyssinus*, actually is widespread throughout the world. It was the first species recognized as being significantly associated with house dust allergy.[46] The American house dust mite, *D. farinae*, is also a cosmopolitan inhabitant of houses. Very commonly, houses are coinhabited by both species.

D. Biology and Behavior

House dust mites are associated with furniture (especially mattresses, sofas, and recliner chairs) and debris in household carpets. They are generally more numerous in mattresses and bedrooms than other areas of a house. In a study in Hawaii,[47] dust samples taken from carpets always had more mites than those from noncarpeted floors. Old carpets not cared for properly contained more dust mites than new ones or those often cleaned. The preferred food source of the mites is believed to be shed human skin scales, although they will eat mold, fungal spores, pollen grains, feathers, and animal dander. The mites are most abundant in warm homes with high humidities. Laboratory populations exhibit maximum growth at 25°C and about 75% relative humidity (RH) and only survive 10 days or less at 40 or 50% RH.[48,49] High mite levels occur during periods of high RH, and these levels become low when RH drops below the critical level for extended periods. Both *Dermatophagoides pteronyssinus* and *D. farinae* have five developmental stages: egg, larva, protonymph, tritonymph, and adult. The life cycle is completed in about 1 month, and adults may live 2 months or so at optimum temperatures. Cultures of house dust mites are maintained in beds on a year-round basis, and seasonal fluctuations occur.

E. Treatment of Infestation

House dust mite allergy is managed by immunotherapy using mite extracts and by efforts to minimize the level of dust mites in the patient's home. As long-fibered carpets are difficult to clean, tile, wood, or other "hard" floor covering may be needed in those homes. One study demonstrated that carpet exclusion and mattress encasements lowered house dust mite allergen levels below the sensitization threshold.[50] Vacuuming should be done regularly, especially in the bedroom (although the allergic person probably should not be the one vacuuming). Double-thickness filters or HEPA filters are needed for maximum results. The mattresses should be vacuumed intensely. Studies have demonstrated that after mattresses were vacuumed, there was as much as an eightfold reduction in the number of mites that became airborne during bedmaking.[51,52] Synthetic pillows and mattress encasements should be used and perhaps replacement of mattresses should even occur yearly. Sheets and blankets should be cleaned with hot water on a regular basis. Efforts should also be made to reduce household humidity levels, as high humidity favors mite growth and reproduction. In fact, maintaining RH below 50% is one of the most common recommendations for reducing dust mites in a home. Although results have been equivocal, some studies have demonstrated the effectiveness of benzyl benzoate for carpet treatments to reduce mite allergen levels.[53,54] A newer miticide is currently under development for house dust mites.[55,56] It is important to note that reducing the number of mites does not always lower allergen levels right away.

VI. Imaginary Mites

A. Introduction

Patients claiming to be infested with mites (that subsequently cannot be seen, collected, or controlled) may be suffering from delusions of parasitosis (DOP) (see Chapter 7). This emotional disorder is covered more thoroughly in Part I of this book, but it basically involves the unwarranted belief that tiny, almost invisible insects or mites are present on the body. Characteristically, the patient presents to the clinic with pieces of tissue paper or small bags or cups containing the presumed mites or insects; however, these usually contain only dust, specks of dirt, dried blood, pieces of skin, and occasionally common (nonharmful) household insects or their body parts. Skin lesions may be present, but self-induced causes are possible. The majority of such patients are elderly females.

B. Treatment

Cases of imaginary insect or mite infestations must be carefully investigated to rule out actual arthropod causes. Skin scrapings of lesions should be examined for scabies mites. A competent entomologist should examine the patient's workplace and residence, looking for fleas or bed bugs. Careful attention should be given to rat- or bird-infested dwellings as they may harbor parasitic mites. If repeated collection attempts and insecticidal treatments are unsuccessful, and if the patient exhibits symptoms consistent with DOP, then

DOP should be strongly considered as the cause of the problem. After ruling out actual arthropod causes and any underlying medical conditions, physicians may refer the patient to a psychiatrist.

References

1. Krantz GW. *A Manual of Acarology.* 2nd ed. Corvallis, OR: Oregon State University Publication; 1978:509 pp.

2. Mullen GR, O'Connor BM. Mites (Acari). In: Mullen GR, Durden LA, eds., *Medical and Veterinary Entomology.* New York: Elsevier; 2009:433–492.

3. Vos T, Flaxman AD, Naghavi M, et al. Years lived with disability (YLDs) for 1160 sequelae of 289 diseases and injuries 1990–2010: A systematic analysis for the Global Burden of Disease Study 2010. *Lancet.* 2012;380(9859):2163–2196.

4. Bartley WC, Mellanby K. The parasitology of human scabies (women and children). *Parasitology.* 1944;35:207–220.

5. Johnson CG, Mellanby K. The parasitology of human scabies. *Parasitology* 1942;34:285–290.

6. Alexander JO. *Arthropods and Human Skin.* Berlin: Springer-Verlag; 1984.

7. Buntin DM, Rosen T, Lesher JL, Jr., Plotnick H, Brademas E, Berger TG. Sexually transmitted diseases: Viruses and ectoparasites. *J. Am. Acad. Dermatol.* 1991;25:527–534.

8. Bowers J. Attack of the arthropods. American Academy of Dermatology, Dermatology World, March issue, pp. 21–25; 2013.

9. Romani L, Steer AC, Whitfield MJ, Kaldor JM. Prevalence of scabies and impetigo worldwide: A systematic review. *Lancet.* 2015;15:960–967.

10. Parlette EC, Simonds RM, Barton M, Cummings LW, Leib AE. Diffuse, thick scale on both hands. *Am. Fam. Phys.* 2018;97:205–206.

11. Arlian LG. Biology, host relations, and epidemiology of *Sarcoptes scabei. Ann. Rev. Entomol.* 1989;34:139–161.

12. Mathison B, Pritt BS. Laboratory identification of arthropod ectoparasites. *Clin. Microbiol. Rev.* 2014;27:48–67.

13. Chosidow O. Scabies and pediculosis. *Lancet.* 2000;355(9206):819–826.

14. Bope ET, Kellerman R. *Conn's Current Therapy.* Philadelphia, PA: Elsevier Saunders; 2017.

15. Purvis RS, Tyring SK. An outbreak of lindane-resistant scabies treated successfully with permethrin 5% cream. *J. Am. Acad. Dermatol.* 1991;25:1015–1016.

16. Sudakin DL. Fatality after a single dermal application of lindane lotion. *Arch. Environ. Occup. Health.* 2007;62(4):201–203.

17. Holness DL, DeKoven JG, Nethercott JR. Scabies in chronic health care institutions. *Arch. Dermatol.* 1992;128:1257–1266.

18. Juranek DD, Currier R, Millikan LE. Scabies control in institutions. In: Orkin M, Maibach HI, eds. *Cutaneous Infestations and Insect Bites.* Vol. 4. New York: Marcel Dekker; 1985:139–156.

19. Paules SJ, Levisohn D, Heffron W. Persistent scabies in nursing home patients. *J. Fam. Prac.* 1993;37:82–86.

20. Fawcett RS. Ivermectin use in scabies. *Am. Fam. Phys.* 2003;68:1089–1092.

21. Meinking TL, Taplin D, Hermida JL, Pardo R, Kerdel FA. The treatment of scabies with ivermectin. *N. Engl. J. Med.* 1995;333:26–30.

22. Romani L, Whitfield MJ, Koroivueta J, et al. Mass drug administration for scabies control in a population with endemic disease. *N. Engl. J. Med.* 2015;373:2305–2313.

23. Currie BJ. Scabies and global control of neglected tropical diseases. *N. Engl. J. Med.* 2015;373:2371–2372.

24. Wallace MM, Guffey DJ, Wilson BB. Demodectic frost of the ear. *JAMA Dermatol.* 2017;153(3):356–357.

25. Kelly DJ, Fuerst PA, Ching WM, Richards AL. Scrub typhus: The geographic distribution of phenotypic and genotypic variants of *Orientia tsutsugamushi. Clin. Infect. Dis.* 2009;48(Suppl. 3):S203–S230.

26. Weitzel T, Dittrich S, Lopez J, et al. Endemic scrub typhus in South America. *N. Engl. J. Med.* 2016;375:954–961.

27. Walker DH. Scrub typhus – Scientific neglect, ever-widening impact. *N. Engl. J. Med.* 2016;375:913–915.

28. Pulla P. Disease sleuths unmask deadly encephalitis culprit. *Science (News Focus).* 2017;357:344.

29. Mullen G. Which chiggers attack humans in Alabama. Auburn University, Highlights of Agricultural Research 40(2): 16; 1993.

30. Mullen G. Fate of chiggers on human subjects. Alabama Vector Management Society Newsletter 4(1): 5; 1993.

31. Wharton GW, Fuller HS. A manual of the chiggers. *Mem. Entomol. Soc. Wash.* 1952;4:1–185.

32. Traub R, Wisseman CL. The ecology of chigger-borne rickettsiosis (scrub typhus). *J. Med. Entomol.* 1974;11:237–241.

33. Mullen G. Alabama chiggers. Alabama Vector Management Society Newsletter 4(1): 1–2.; 1993.

34. Juckett G. Arthropod bites. *Am. Fam. Physician.* 2013;88(12):841–847.

35. Huebner RJ, Jellison WL, Pomerantz C. Rickettsialpox – A newly recognized rickettsial disease. IV. Isolation of a rickettsia apparently identical with the causative agent of rickettsialpox, from *Allodermanyssus sanguineus,* a rodent mite. *Public Health Rep.* 1946;61:1677–1682.

36. Denmark HA, Cromroy HL. Tropical fowl mite, *Ornithonyssus bursa* (Berlese). Florida Depart. Agric. and Consumer Serv., Tallahassee, FL, Entomology Circular No. 299, pp. 1–4; 1987.

37. Flochlay AS, Thomas E, Sparagano O. Poultry red mite (*Dermanyssus gallinae*) infestation: A broad impact parasitological disease that still remains a significant challenge for the egg-laying industry in Europe. *Parasit. Vect.* 2017;doi:10.1186/s13071-017-2292-4.

38. Bellido-Blasco JB, Arnedo-Pena A, Valcuende F. Comet sign (and other) in *Pyemotes* dermatitis. *Emerg. Inf. Dis.* 2009;15:503.

39. CDC. Outbreak of pruritic rashes associated with mites – Kansas, 2004. *MMWR,* 2005;54:952–955;.

40. Sceppa JA, Young HL, Jacobs SB, Adams DR. What's eating you? Oak leaf itch mite (*Pyemotes herfsi*). *Cutis/* 2011;88:114–116.

41. Moxham JW, Goldfinch TT, Heath ACG. *Cheyletiella parasitivorax* infestation of cats associated with skin lesions of man. *N. Z. Vet. J.* 1968;16:50–52.

42. Kwochka KW. Mites and related disease. *Vet. Clin. NA Small Anim. Pract.* 1987;17:1263–1284.

43. Van Bronswijk JEMH, de Kreek EJ. *Cheyletiella* of dog, cat, and domesticated rabbit. *J. Med. Entomol.* 1976;13:315–319.

44. Wharton GW. House dust mites. *J. Med. Entomol.* 1976;12:577–621.

45. Cameron MM. Can house dust mite-triggered atopic dermatitis be alleviated using acaricides. *Br. J. Dermatol.* 1997;137:1–8.

46. Voorhorst R, Spieksma-Boezeman MIA, Spieksma FTM. Is a mite (*Dermatophagoides* sp.) the producer of the house dust allergen? *Allergy Asthma.* 1964;10:329–334.

47. Sharp JL, Haramoto FH. *Dermatophagoides pteronyssinus* and other acarines in house dust in Hawaii. *Proc. Hawaiian Entomol. Soc.* 1970;20:583–585.

48. Arlian LG. Dehydration and survival of the European house dust mite, *Dermatophagoides pteronyssinus. J. Med. Entomol.* 1975;12:437–442.

49. Murton JJ, Madden JL. Observations on the biology, behavior, and ecology of the house dust mite in Tasmania. *J. Aust. Entomol. Soc.* 1977;16:281–285.

50. Hill DJ, Thompson PJ, Stewart GA, et al. The Melbourne house dust mite study: Eliminating house dust mites in the domestic environment. *J. Allergy Clin. Immunol.* 1997;99:323–329.

51. Van Bronswijk JEMH, Schoonen JMCP, Berlie MAF. On the abundance of *Dermatophagoides pteronyssinus* in house dust. *Res. Popul. Ecol.* 1971;13:67–72.

52. Van Bronswijk JEMH, Sinha RN. Pyroglyphid mites and house dust allergy. *Res. Popul. Ecol.* 1971;13:67–73.

53. Chang JH, Becker A, Ferguson A, et al. Effect of application of benzyl benzoate on house dust mite allergen levels. *Ann. Allergy Asthma Immunol.* 1996;77:187–190.

54. Huss RW, Huss K, Squire EN, Jr., et al. Mite allergen control with acaricide fails. *J. Allergy Clin. Immunol.* 1994;94:27–32.

55. Mori T, Takada Y, Hatakoshi M, Matsuo N. New trifluoromethanesulfonanilide compounds having high miticidal activity against house dust mites. *Biosci. Biotechnol. Biochem.* 2004;68(2):425–427.

56. Yu SJ. *The Toxicology and Biochemistry of Insecticides.* Boca Raton, FL: CRC Press; 2008.

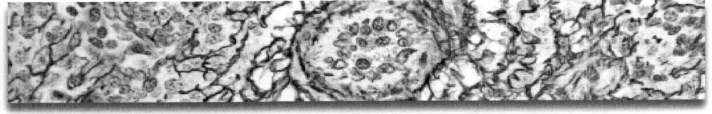

25

MOSQUITOES

I. General and Medical Significance

Mosquitoes are by far the most important of the bloodsucking arthropods worldwide, giving annoyance to and causing disease in humans, other mammals, and birds (see box and Figure 25.1). Swarms of snow-melt mosquitoes or salt marsh mosquitoes torment people in many areas throughout the world (Figure 25.2). In those areas, it is not unusual to experience a mosquito landing rate of over 200 mosquitoes per minute. Inland, rice fields and wetlands produce tremendous mosquito problems each year. Discarded automobile tires, ditches, bird baths, cemetery vases, and paint cans are excellent mosquito breeding grounds in urban areas (Figures 25.3 and 25.4). In many cases, mosquitoes may interfere with normal outdoor work and other activities due to the nuisance effects of their bites (Figures 25.5 and 25.6).

About 3500 species of mosquitoes have been described worldwide. Relatively few of them are significant vectors of human diseases; however, the mosquito-transmitted disease problem worldwide is quite severe. Table 25.1 details biological data on medically important mosquitoes in the United States. Below is a brief description of some of the major mosquito-transmitted diseases.

A. Malaria

Malaria has historically affected large areas of the world resulting in staggering case numbers and millions of deaths. The situation has greatly improved over the last decade due to use of indoor residual spraying (IRS), insecticide-treated bed nets (ITNs) and treated window screen and eave baffles for prevention[1] and artemisin-based combination therapies (ACTs) for treatment.[2] Although likely underreported, estimates of the annual number of clinical cases are now approximately 250 million, with about 500,000 deaths—mainly in children.[3-5] Malaria generally occurs in areas of the world

between 45°N and 40°S latitude. Although many countries are not entirely malarious, the WHO estimates that about 3.2 billion persons—well over 40% of the world's population—live in malarious areas.[6] The geographic distribution of malaria has shrunk significantly over the last 150 years, mainly from eradication efforts in temperate zones (Figures 25.7 and 25.8). Two-thirds of all malaria cases and approximately 90% of malaria deaths occur in just seven countries in sub-Saharan Africa.[7]

Climate change may be resulting in the spread of malaria to new areas; for example, malaria has recently spread into highland regions of East Africa where it previously did not exist. This spread presumably occurred because of warmer and wetter weather, resulting in high rates of illness and death because the disease was introduced into a largely non-immune population.[8,9] Complicating the malaria situation, mosquito vectors of malaria are becoming resistant to many of the pesticides being used to control them, as well as to the insecticides used in ITNs, and (lastly) the malaria parasites themselves are becoming resistant to the prophylactic drugs used to prevent the disease, even ACT combination drugs.

Previously, human malaria was thought to be caused by any one of four species of microscopic protozoan parasites in the genus Plasmodium—*Plasmodium vivax*, *Plasmodium ovale*, *Plasmodium malariae*, and *Plasmodium falciparum*, although we now know that a fifth species causes human malaria—*Plasmodium knowlesi*.[10-12] Not all species of *Plasmodium* occur in all places, nor do they produce identical disease syndromes. Generally, *P. vivax* is prevalent throughout all malaria-endemic areas, except sub-Saharan Africa, and is the form known for producing relapses. *Plasmodium ovale* is found chiefly in tropical areas of western Africa (occasionally western Pacific and Southeast Asia). *Plasmodium malariae* has a wide but spotty distribution around the world and is the most important cause of malaria resulting from blood transfusions. *Plasmodium falciparum* is the most virulent species

MOSQUITOES

Adult mosquito. (From Gjullin, C.M. et al., The Mosquitoes of Alaska, Agricultural Handbook No. 182, U.S. Department of Agriculture, Washington, DC, 1961.)

IMPORTANCE

Annoyance and disease transmission

DISTRIBUTION

Numerous species worldwide

LESION

Punctate hemorrhages, papular lesions, or large wheals with edema.

DISEASE TRANSMISSION

Number one arthropod vector of disease agents—malaria, dengue, encephalitis, etc.

KEY REFERENCES

Carpenter SJ, LaCasse WJ. *Mosquitoes of North America (North of Mexico)*. Berkeley: University of California Press; 1955

Darsie, Jr., RF, Ward RA. *Identification and Geographical Distribution of the Mosquitoes of North America, North of Mexico*. Gainesville, FL: University Press of Florida; 2005

TREATMENT

Generally, palliative creams or lotions; topical corticosteroids and antibiotic creams or ointments may be needed; oral antihistamines may be effective in reducing symptoms of mosquito bites

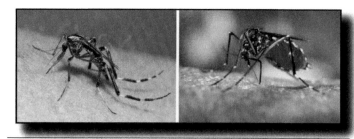

Figure 25.1 Adult *Psorophora* mosquito (left) and *Aedes* mosquito feeding (right). (Photograph on left is courtesy of Blake Layton, Ph.D., Mississippi State University Extension Service; photograph on right is courtesy of U.S. Centers for Disease Control and Prevention, Atlanta, GA.)

Figure 25.2 Snow melt can produce millions of mosquitoes.

Figure 25.3 Tire piles are excellent breeding sites for mosquitoes. (Photograph courtesy of Wendy C. Varnado, Mississippi Department of Health.)

and predominates in sub-Saharan Africa but is also common in Southeast Asia and South America. *Plasmodium knowlesi* occurs in Southeast Asia and is the species most often associated with long-tailed macaques (although humans can be infected and become quite ill). The first documented case of *P. knowlesi* in a U.S. traveler was documented in 2008.[13]

The malaria life cycle is quite complicated, and its description is fraught with technical terms (Figure 25.9). Malaria parasites are acquired and transmitted to humans by bites from *Anopheles* mosquitoes only, but not every species of *Anopheles* is a vector; less than half of the 470 or so known species are considered vectors. *Plasmodium falciparum* is

Figure 25.4 The author collecting mosquito larvae out of cemetery vases. (Photo courtesy Mississippi State University Extension Service.)

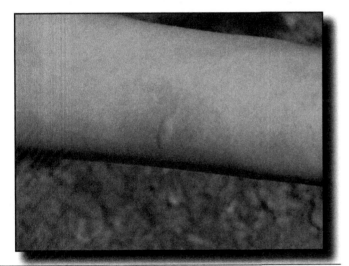

Figure 25.5 Mosquito bite on person's arm, 20 minutes after bite. (Photograph copyright 2005 by Jerome Goddard, Ph.D.)

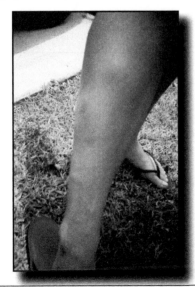

Figure 25.6 Huge indurations resulting from mosquito bites. (Photograph courtesy of Mallory Carter Pickering and used with permission.)

the worst of the five species and is often fatal in infants and young children. Most malaria occurring in the United States each year is a result of people having relapses from former cases or from cases recently acquired in foreign countries where malaria is endemic (introduced malaria). During 2015, 1390 malaria cases were reported in the United States, almost all of which were introduced cases.[14] Over the last couple of decades, there have been a few small foci of autochthonous malaria in the United States.[15–17]

There are two main ways drugs can be used to control malaria—*prevention* of clinical malaria (prophylaxis) and *treatment* of acute cases. Antimalarial drugs include chloroquine, primaquine, mefloquine, malarone, doxycycline, quinine (hardly used anymore), and artemisinin (in combination with other antimalarials). Because of increased parasite resistance to antimalarial drugs, treatment regimes have become quite complicated and vary tremendously by geographic region. In addition to the problem of resistance, serious side effects may occur with the use of some antimalarial products. The most effective current antimalarial treatment is artemisinin-based combination therapy (ACT) for countering the spread and intensity of *Plasmodium falciparum* resistance to chloroquine, sulfadoxine/pyrimethamine, and other malarial drugs,[18] however, there is resistance to ACT in parts of Southeast Asia.[3] Health care providers with questions about treatment should contact their local or state health department, the CDC, or the preventive medicine department at a local medical school for the most up-to-date malaria treatment recommendations.

Malaria Vaccine. Ever since the early 1900's, efforts have been directed toward producing a malaria vaccine. There are several points in the complex malaria life cycle where immunological interference with the multiplication of plasmodia could be attempted. Although many experimental vaccines have been developed and studied, no malaria vaccine is widely available for use. However, the most advanced of the current candidates against *P. falciparum* is the RTS,S vaccine which shows about 30% efficacy against malaria in infants and children.[19,20] There is another promising vaccine called PfSPZ.[5] Development of an effective malaria vaccine has been prevented through the years by several factors. For one thing, the persons at greatest risk of complications and death are young children, and most researchers think the initial immune responses elicited by a vaccine will be suboptimal. Second, even if a vaccine produces a strong humoral and cellular response, it does not necessarily provide sterile immunity. Even in naturally acquired infections, antibodies directed against the dominant antigen on the sporozoite surface do not prevent reinfection with sporozoites bearing the same dominant repetitive antigen. Third, for traditional vaccines, there is an inadequate number of adjuvants available for human use. For example, aluminum hydroxide is about the only adjuvant approved for human use. If other antigen—adjuvant combinations can be identified, which provide boosting with the re-exposures that occur repetitively with

Table 25.1 Biological Data on Medically Important Mosquitoes in the United States

Species	Larval Habitat(s)[a]	Biting Time[b]	Flight Range (miles)[c]	Disease Agent[d]
Aedes aegypti	AC	C, D	<0.5	DG, YF, CHIK, ZIKV
Ae. albopictus	AC, TH	D	<0.5	DG, YF, CHIK, ZIKV, (CE)
Ae. dorsalis	SM, LM	D	10–20	(WEE)
Ae. melanimon	IP, FW	D	1–2	WEE, CE
Ae. nigromaculis	IP, FW	D	1–2	(WEE), (CE)
Ae. sollicitans	SM	C	5–10	EEE
Ae. taeniorhynchus	SM	C, N	5–10	VEE, (CE), (WNV)
Ae. triseriatus	TH, AC	D	0.5–1	CE, (WNV)
Ae. trivittatus	GP, WP, FW	C, N	0.5–1	CE
Ae. vexans	FW, GP, IP	C, N	1–5	CE, (EEE), (WNV)
Anopheles crucians complex	SM, FS, LM	C	1–2	(VEE), (EEE)
An. freeborni	RF, DD	C	1–2	M, (WEE), (SLE)
An. quadrimaculatus complex	FW, GP, LM, RF	C	0.5–1	M, (WNV)
Coquillettidia perturbans	FS, GP, LM	C, N	1–2	EEE, (VEE)
Culex nigripalpus	GP, FW, DD	C	0.5–1	SLE, (WNV)
Cx. pipiens/quinquefasciatus	AC, SCB, GRP	C, N	<0.5	WNV, SLE, (WEE), (VEE)
Cx. restuans	WP, GRP, DD	C, N	1–2	WNV, (EEE), (WEE)
Cx. salinarius	GP, LM, FS	C, N	1–5	WNV, (EEE)
Cx. tarsalis	IP, RF, GRP	C, N	1–2	WEE, WNV, SLE
Culiseta inornata	GRP, DD	C,N	1–2	(WEE), (CE)
Mansonia titillans	FS, GP, LM	C, N	1–5	(VEE)
Psorophora columbiae	IP, RF, GRP	C, N	1–5	VEE, EEE, (WNV)

Source: Originally adapted from U.S. Air Force mosquito surveillance data.

[a] AC, artificial containers; DD, drainage ditches; FS, freshwater swamps; FW, floodwaters; GP, grassland pools; GRP, ground pools; IP, irrigated pastures; LM, lake margins; RF, rice fields; SCB, sewer catch basins; SM, salt marshes; TH, tree holes; WP, woodland pools.

[b] C, crepuscular (dusk and dawn); D, day; N, night.

[c] Values given are estimates of normal flight ranges. For some species, seasonal migratory flights may be ten times these values.

[d] Parentheses indicate secondary or suspected vectors, otherwise primary vectors. CE, California group encephalitis; CHIK, chikungunya; DG, dengue; EEE, Eastern equine encephalitis; M, malaria; SLE, St. Louis encephalitis; VEE, Venezuelan equine encephalitis; WEE, Western equine encephalitis; WNV, West Nile virus; YF, yellow fever; ZIKV, Zika virus.

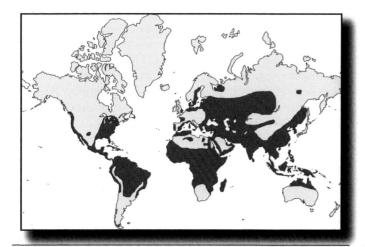

Figure 25.7 Distribution of malaria approximately 1850.

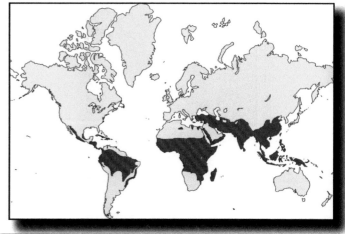

Figure 25.8 Distribution of malaria current.

OFTEN-ASKED QUESTION

WHY CAN'T MOSQUITOES TRANSMIT THE AIDS VIRUS?

Because human immunodeficiency virus (HIV) is a bloodborne pathogen, concerns have been raised about the possible transmission of HIV by blood-feeding arthropods. Laboratory studies and epidemiologic surveys indicate that this possibility is extremely remote. For biological transmission, the virus must avoid digestion in the gut of the insect, recognize receptors on and penetrate the gut, replicate in insect tissue, recognize and penetrate the insect salivary glands, and escape into the lumen of the salivary duct. In one study by Webb and colleagues, the virus persisted for 8 days in bed bugs.[1] Another study by Humphrey-Smith and colleagues[2] showed the virus to persist for 10 days in ticks artificially fed meals with high levels of virus ($\geq 10^5$ tissue culture infective doses per milliliter [TCID/mL]), but there was no evidence of viral replication. Intraabdominal inoculation of bed bugs and intrathoracic inoculation of mosquitoes was used to bypass any gut barriers, but again the virus failed to multiply.[1] Likewise, *in vitro* culture of HIV with a number of arthropod cell lines indicated that HIV was incapable of replicating in these systems. Thus, biological transmission of HIV seems extremely improbable.

Mechanical transmission would most likely occur if the arthropod were interrupted while feeding and then quickly resumed feeding on a susceptible host. Transmission of HIV would be a function of the viremia in the infected host and the virus remaining on the mouthparts or regurgitated into the feeding wound. The blood meal residue on bed bug mouthparts was estimated to be 7×10^{-5} mL, but 50 bed bugs, interrupted while feeding on blood containing 1.3×10^5 TCID/mL HIV, failed to contaminate the uninfected blood on which they finished feeding or the mouse skin membrane through which they refed.[1]

Within minutes of being fed blood with 5×10^4 TCID of HIV, stable flies regurgitated 0.2 µL of fluid containing an estimated 10 TCID.[3] The minimum infective dose for humans contaminated in this manner is unknown, but under conditions such as those in some tropical countries where there are large populations of biting insects and a high prevalence of HIV infection, transfer might be theoretically possible, if highly unlikely. In these countries, however, other modes of transmission are overwhelmingly important, and, although of fatal importance to the extremely rare individual who might contract HIV through an arthropod bite, arthropods are of no significance to the ecology of the virus.

An epidemiologic survey of Belle Glade, a South Florida community believed to have a number of HIV infections in individuals with no risk factors, provided no evidence of HIV transmission by insects.[4] Interviews with surviving patients with the infections revealed that all but a few had engaged in the traditional risk behavior (e.g., drug use and unprotected sex). A serosurvey for exposure to mosquito-borne viruses demonstrated no significant association between mosquito contact and HIV status, nor were repellent use, time outdoors, or other factors associated with exposure to mosquitoes related to the risk of HIV infection. A serosurvey for HIV antibodies detected no positive individuals between 2 and 10 years of age or 60 and older. No clusters of cases occurred in houses without other risk factors. There was thus no evidence of insect-borne HIV transmission.

REFERENCES

1. Webb PA, Happ CM, Maupin GO, Johnson BJB, Ou C-H, Monath TP. Potential for insect transmission of HIV: experimental exposure of *Cimex hemipterous* and *Toxorhynchites amboinensis* to human immunodeficiency virus. *J. Infect. Dis.* 1989;60: 970
2. Humphrey-Smith I, Donker G, Turzo A, Chastel C, Schmidt-Mayerova H. Evaluation of mechanical transmission of HIV by the African soft tick, *Ornithodoros moubata. AIDS* 1993;7:341
3. Brandner G, Kloft WJ, Schlager-Vollmer C, Platten E, Neumann-Opitz P. Preservation of HIV infectivity during uptake and regurgitation by the stable fly, *Stomoxys calcitrans* L. *AIDS-Forschung.* 1992;5:253
4. Castro KG, Lieb S, Jaffe HW, et al., Transmission of HIV in Belle Glade, Florida: lessons for other communities in the United States. *Science* 1998;239:193

Source: Adapted from McHugh CP. Arthropods: vectors of disease agents. *Lab. Med.* 1994;25:436. With permission.

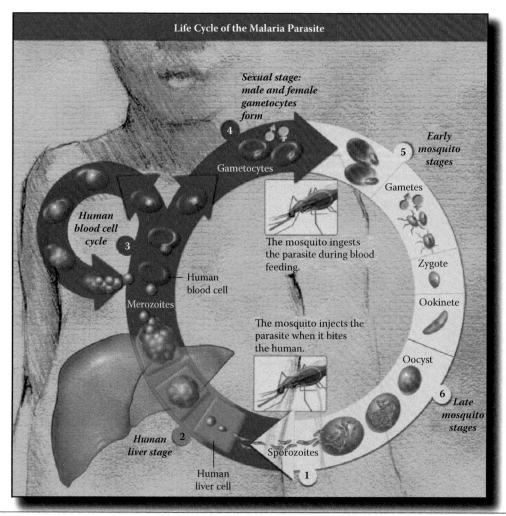

Figure 25.9 Life cycle of malaria. (Figure courtesy of U.S. National Institutes of Health, Bethesda, MD.)

natural reinfection under field conditions, then there is better hope for malaria control through vaccines.

B. Yellow Fever

Yellow fever (YF) is probably the most lethal of all the arboviruses and historically has had a widespread and devastating effect on human social development. The causative agent, a flavivirus, may still be found in tropical regions of Africa and South America (Figure 25.10) and is transmitted to primates by various *Aedes* mosquitoes in Africa and *Haemagogus* and *Sabethes* species in South America.[21] An urban form of the disease occurs when humans become infected and transmission occurs from person to person by *Aedes aegypti*. This urban form makes YF worrisome because *Ae. aegypti* is widespread in many urban areas and the disease could be easily imported by travelers. Mild cases of YF may be characterized by fever, headache, generalized aches and pains, and nausea. Persons with severe disease may exhibit high fever, headache, dizziness, muscular pain, jaundice, hemorrhagic symptoms, and profuse vomiting of brown or black material. There is often collapse and death. Even though there is an effective vaccine,

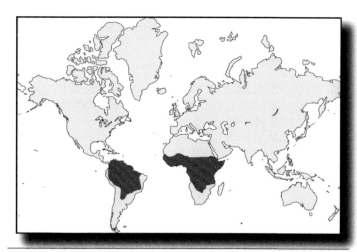

Figure 25.10 Approximate geographic distribution of yellow fever.

the WHO estimates that 200,000 people are infected with YF every year, resulting in at least 30,000 deaths.[22–24] Many large epidemics have occurred in the past, especially during the time of exploration of the New World; for example, 20,000 of 27,000 British troops were killed by YF during an expedition to conquer Mexico in 1741. Currently, there is no specific therapy for YF. Apparently, no antiviral medications block or decrease replication of YF virus.[25] Cases can only be treated symptomatically, and hematologic support is vital. In recent years, there have been large YF outbreaks in Angola in Africa and Brazil in South America,[26,27] and also a shortage of the U.S. Food and Drug Administration-approved YF vaccine (YF-VAX). Fortunately, there is an alternative vaccine called Stamaril® which is available through a limited number of U.S. vaccination clinics.[28]

C. Dengue Fever

Dengue fever, caused by a virus in the family Togaviridae, is responsible for widespread morbidity (break-bone fever) and some mortality (dengue hemorrhagic fever, or DHF) in much of the tropics and subtropics each year (Figure 25.11). There are four closely related, but antigenically distinct, virus sero-types: DEN-1, DEN-2, DEN-3, and DEN-4. DHF (sometimes called dengue shock syndrome in its most severe form) is a hemorrhagic complication occurring mostly in children and is thought to be a result of sequential infection by more than one dengue serotype or variations in viral virulence. Most dengue infections result in relatively mild illness character-ized by fever, headache, myalgia, rash, nausea, and vomiting. However, DHF is often severe and is characterized by pete-chiae, purpura, mild gum bleeding, nosebleeds, gastrointesti-nal bleeding, and dengue shock syndrome. The case fatality rate of DHF in most countries is about 5%, with most deaths occurring among children. In one study of DHF in the Cook Islands, deaths occurred in patients 14–22 years old and were due to acute upper gastrointestinal bleeding.[29] The virus is transmitted among humans by *Aedes aegypti* and *Ae. albop-ictus*. Dengue fever seems to be increasing in prevalence and geographic distribution, especially in Southeast Asia, India, the Caribbean, and Central and South America,[30] with an esti-mated 2.5 billion people currently at risk (Figure 25.12).[31] After an almost 60-year absence, there were more than 100 cases

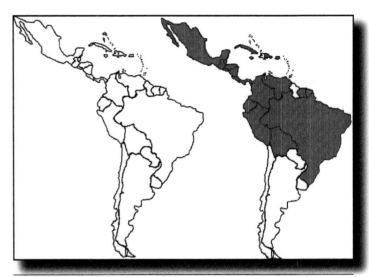

Figure 25.12 Increase in distribution of dengue hemorrhagic fever. (Figure courtesy of U.S. Centers for Disease Control and Prevention, Atlanta, GA.)

of locally acquired dengue in Hawaii in 2001.[32] There was also a relatively recent outbreak in the Florida Keys (see box) where *Ae. aegypti* is fairly common.[33] Worldwide, as many as 400 million cases of dengue occur annually and hundreds of thousands of cases of DHF.[34,35] Treatment of dengue and DHF is mostly symptomatic, although there is some progress toward developing a vaccine.[36]

Dengue Vaccine. There are several candidate dengue vaccines currently making their way through clinical trials.[37] The only one so far which has been registered (or licensed), named Dengvaxia®, has shown modest efficacy in prevent-ing dengue infection. In trials, Dengvaxia® reduced dengue disease by 56.5% but provided only limited protection against the Den 2 serotype.[38,39] Further, it did not have the hoped-for high and balanced efficacy over all age groups.[39] Making mat-ters worse, there have been safety concerns with Dengvaxia® and the company now has dramatically curtailed its use.[40]

D. Lymphatic Filariasis

Lymphatic filariasis (LF) (Figures 25.13 and 25.14) is an impor-tant human disease occurring in much of the world. Malayan filariasis, caused by *Brugia malayi*, is mostly confined to Southeast Asia (Figure 25.15), and the Bancroftian form, *Wucheria bancrofti*, is prevalent over much of the tropical world (Figure 25.16). In 2000, WHO estimated that 120 million people were infected with Bancroftian or Brugian filariasis with an additional 1.34 billion persons at risk.[41] In the Western Hemisphere, 80% of LF occurs in Haiti, likely imported from Africa with the slave trade.[42] There was at one time a small endemic center of LF near Charleston, SC, that has apparently disappeared; however, we still have an efficient vector in the United States, *Culex quinquefasciatus*, that could transmit the filarial worms should a reintroduction occur. Bancroftian

Figure 25.11 Approximate geographic distribution of dengue fever.

OUTBREAK OF DENGUE FEVER IN FLORIDA

Dengue fever, or "break-bone fever," is a mosquito-carried viral disease common to tropical areas worldwide. Symptoms include high fever, severe headache, pain behind the eyes, a rash, and pain in the bones and joints. A more severe form of the disease that sometimes occurs in people, *dengue hemorrhagic fever*, leads to persistent vomiting, severe abdominal pain, failure of the circulatory system, bleeding from natural orifices, shock, and possibly death. Dengue has been a major plague of tropical areas for a long time, with 100–300 million cases each year. Interestingly, the mosquito vectors of dengue, *Aedes aegypti* and *Aedes albopictus*, commonly occur throughout much of the southern United States; therefore, there has always been a threat for dengue to occur in the United States (locally acquired, not just ill travelers returning from tropical areas).

In 2009–2010, dengue fever occurred in Florida after a 75-year absence (although there have been a few cases along the U.S.–Mexican border through the years). Key West, Florida, had 27 cases in 2009 in an outbreak that ended with the mosquito season in mid-October, and between April and December of 2010, there were again at least 60 locally acquired cases in Key West.[1] Complicating matters, cases have now been reported further up into Florida, one in Broward County and one in Miami–Dade County.

Because there is no specific treatment for dengue and no vaccine, it is important to educate the public about dengue prevention. Several of the following suggestions might help:

1. Eliminate mosquito breeding around the home. Look for artificial containers such as buckets, old tires, pots, bird baths, etc., and empty or replace the water once a week.
2. Avoid outdoor activity when mosquitoes are most active. The dengue mosquitoes are daytime biters, but primarily early morning and late evening.
3. Wear long-sleeved shirts and long pants when possible, and apply insect repellents on exposed skin. Use products containing the active ingredient DEET, picaridin, oil of lemon eucalyptus, or IR3535. Note that some herbal concoctions and most mosquito control necklaces, bracelets, sonic repellers, and other such devices do not work.
4. If a resident of Florida and someone in your house becomes ill with dengue, take extra precautions to prevent mosquitoes from biting the patient and going on to bite others in the household. Perhaps you could sleep under a mosquito net, utilize appropriately labeled insecticidal space sprays in the home for mosquito control, or wear repellents.

Even though this recent occurrence of dengue in Florida was controlled, citizens should not drop their guard. Dengue is here to stay, and there will surely be future outbreaks of the disease in the southern United States.

REFERENCE

1. Radke EG, Gregory CJ, Kintzinger KW, Sauber-Schaz EK, Hunsperger EA, et al. Dengue outbreak in Key West, Florida, USA, 2009. *Emerg. Infect. Dis.* 2012;18:135–137

filariasis is an interesting disease in that there is no other known vertebrate host of the worms. It is transmitted solely by mosquitoes, and there is no multiplication of the parasite, only development, in the mosquito vector. In addition, the adult worms may live up to 10 years in humans.[43] In cases of filariasis it is important to distinguish between infection and disease. Infection refers to the establishment of worms in the human host, whereas disease indicates that the infection has caused some sort of pathologic response. The majority of infected individuals living in endemic countries do not exhibit overt signs of lymphatic filariasis. Generally, only about 10% of infected persons display swollen limbs, breasts, damage to the genitals, or extremely swollen limbs with thickened, hardened skin, a condition called elephantiasis.

E. Other Human-Infesting Filarial Worms

Numerous filarial worms are transmitted to humans and other mammals by mosquitoes, deer flies, and black flies.

Examples include the causative agents of Bancroftian and Malayan filariasis (discussed earlier), loiasis, onchocerciasis, and dirofilariasis (dog heartworm). Other filarial worms may or may not cause symptomatic disease and are less well known (and thus have no common name), such as *Mansonella ozzardi, M. streptocerca, M. perstans, Dirofilaria tenuis, D. ursi, D. repens*, and others. Beaver and Orihel[44] reported 39 such cases that were caused by *Dirofilaria immitis* (dog heartworm), other *Dirofilaria* spp., *Dipetalonema* spp., and *Brugia* spp. Recently, there seems to be an increase in cases of human subcutaneous dirofilariasis (HSD), and some researchers think this is an emerging disease. One paper reported a cluster of 14 cases of HSD in humans in Russia from February 2003 through July 2004, the highest number of cases of HSD diagnosed worldwide in such a short period.[45]

The dog heartworm, *Dirofilaria immitis*, occurs mainly in the tropics and subtropics but also extends into southern Europe and North America. This worm infects several canid

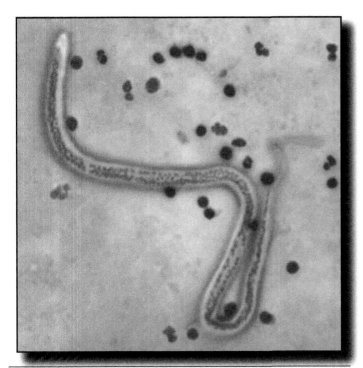

Figure 25.13 Filarial worm, *Wuchereria bancrofti*. (Photograph courtesy of U.S. Centers for Disease Control and Prevention, Atlanta, GA.)

Figure 25.15 Approximate geographic distribution of Malayan (Brugian) filariasis.

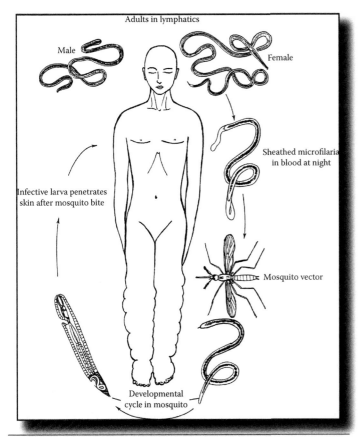

Figure 25.14 Life cycle of *Wuchereria bancrofti*. (From Brooks, Jr., TJ. *Essentials of Medical Parasitology*. New York: Macmillan; 1963. With permission.)

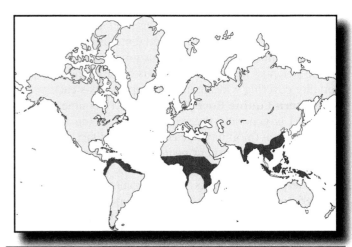

Figure 25.16 Approximate geographic distribution of Bancroftian filariasis.

species, sometimes cats, and, rarely, humans (Figure 25.17). Numerous mosquito species are capable of transmitting dog heartworm, especially those in the genera *Aedes*, *Anopheles*, and *Culex*. Mosquitoes pick up the microfilariae with their blood meal when feeding on infected dogs. In endemic areas, a fairly high infection rate may occur in local mosquitoes.

Undoubtedly, thousands of people in the United States are bitten each year by mosquitoes infected with *Dirofilaria immitis*. Fortunately, humans are accidental hosts, and the larvae usually die; however, they may occasionally be found as a subadult worm in the lung (seen as a coin lesion on

Figure 25.17 Dog heartworm in dog's heart. (Photograph copyright 2012 by Jerome Goddard, Ph.D.)

x-ray exam)[46] or (extremely rarely) as an adult worm in the cardiovascular system.[47] The incidence of dog heartworm in humans may be decreasing in the United States because of widespread—and fairly consistent—treatment of domestic dogs for heartworm. However, there is evidence of the development of resistance to the treatments used for heartworms in dogs.[48] The closely related *D. tenuis* is commonly found in the subcutaneous tissues of raccoons (again, mosquito-transmitted) and may accidentally infest humans as nodules in subcutaneous tissues.

F. Encephalitides

In temperate North America, the worst mosquito-borne diseases are probably the encephalitides. Certainly, not all cases of encephalitis are mosquito caused (enteroviruses and other agents are often involved), but mosquito-borne encephalitis has the potential to become a serious cause of morbidity and mortality covering widespread geographic areas each year.

Eastern Equine Encephalitis. Eastern equine encephalitis (EEE) is generally the most virulent, being severe and frequently fatal (mortality rate of 35–75%). In fact, half of EEE survivors suffer permanent neurologic sequelae and require long-term care costing millions of dollars.[49] Fortunately, large and widespread outbreaks are not common; between 1961 and 1985 only 99 human cases were reported.[50] There were six human cases reported in the U.S. in 2015.[14] EEE occurs in late summer and early fall in the central and northcentral United States, parts of Canada, southward along the coastal margins of the eastern United States and the Gulf of Mexico, and sparsely throughout Central and South America (Figure 25.18). Horses are especially susceptible to EEE infection and may serve as sentinel animals to indicate virus activity in an area; however, widespread use of the eastern–western tetanus vaccine (Figure 25.19) may bias surveys of horse cases of EEE. Recently, human cases of EEE seem to be occurring more northeasterly into Maine and Vermont. The ecology of EEE is complex. The virus circulates in wild bird populations, and the exact mosquito vectors responsible for spread to humans are not well known. Some species likely involved include *Aedes sollicitans*, *Coquillettidia perturbans*, *Culex salinarius*, and *Ae. vexans*,[51–53] although certain *Anopheles* species may serve as bridge vectors during epizootics.[54]

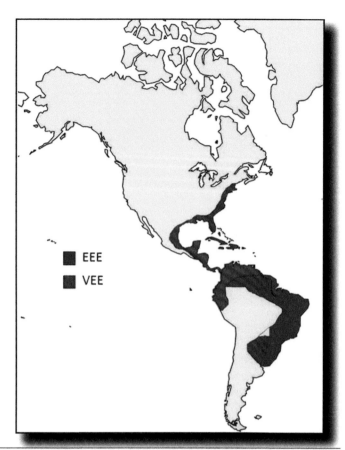

Figure 25.18 Approximate geographic distribution of eastern equine encephalitis and Venezuelan equine encephalitis.

St. Louis Encephalitis. St. Louis encephalitis (SLE) produces lower mortality rates than EEE (3–20%), but it occurs occasionally in large epidemics in much of the United States (Figure 25.20). As with EEE, most cases occur in late summer (Figure 25.21). In 1933, there were 1095 cases in St. Louis with more than 200 deaths,[55] and again in 1975–76, there were over 2000 cases reported from 30 states, primarily in the Mississippi Valley.[55] Most years, however, case numbers are lower and the distribution spotty. For example, during 2015, there were 23 cases reported nationwide and the presence of SLE was documented again in California after a 12-year absence.[14,56] This trend of re-emergence continues in the Americas. SLE is a bird virus that is transmitted by *Culex tarsalis* (western and southwestern United States), *Cx. quinquefasciatus* (central and southern United States), and *Cx. nigripalpus* (southeastern United States) (Figure 25.22).

West Nile Encephalitis. The West Nile virus (WNV) was identified for the first time in the Western Hemisphere in New York in 1999. By the end of the year, it had caused encephalitis in 62 people and numerous horses in and around New York City, resulting in seven human and ten equine deaths.[57] The virus continued to spread in subsequent years to all the continental United States, at least seven Canadian provinces, Mexico, the Caribbean, and portions of South America (Figure 25.23).[58] As far as severity of the disease, WNV is similar to SLE (one of our "native" encephalitis viruses). Approximately 80% of all

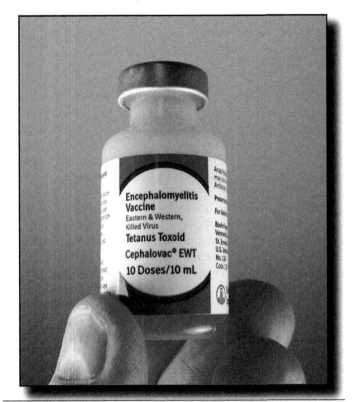

Figure 25.19 EWT vaccine used for horses protects against eastern equine encephalitis.

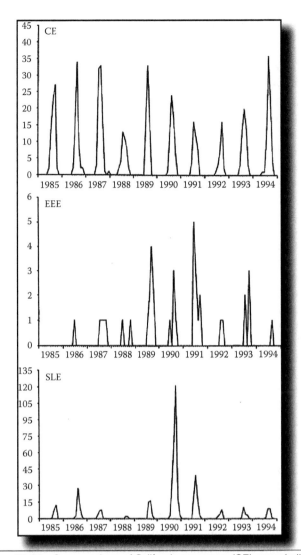

Figure 25.21 Case numbers of California serogroup (CE) encephalitis, eastern equine encephalitis (EEE), and St. Louis encephalitis (SLE), showing appearance of cases in mid-to-late summer. (From CDC, TB Prevention in Drug-Treatment Centers and Correction Facilities, DHHS Publ. No. 93-8017. Atlanta, GA: U.S. Department of Health and Human Services; 1993.)

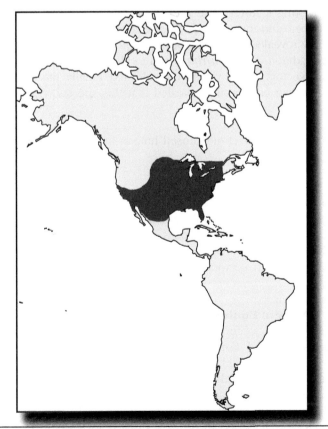

Figure 25.20 Approximate geographic distribution of St. Louis encephalitis.

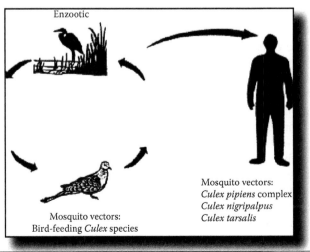

Figure 25.22 Life cycle of St. Louis encephalitis.

CASE HISTORY

FATAL CASE OF MOSQUITO-TRANSMITTED ENCEPHALITIS

An 11-year-old Native American from an Indian community developed a fever of 103°F and diarrhea on July 31. Gastroenteritis was reportedly "going around" in the community at the time. He was taken to a local emergency room and given symptomatic treatment. He was somewhat better until the night prior to admission when he developed a headache, stomachache, and decreased appetite. He went to bed early, which was unusual for him. The next morning he went with his family to a scheduled ophthalmologic exam and slept most of the 1-hour drive. He was drowsy and nauseated on arrival at the clinic, then "turned pale" and began grand mal seizure activity. He was taken to the emergency room, loaded with Dilantin®, and transferred to the admitting hospital. On admission, he was responsive but lethargic. His admission temperature was 102.2°F. His admission laboratory showed a white blood cell count of 19,500 with 59% neutrophils, 19% band forms, and 16% lymphocytes. The hematocrit was 34.4% and the serum glucose was 184 mg per 100 mL. Spinal fluid showed a white blood cell count of 980, with 91% neutrophils, no organisms seen on Gram stain, negative latex agglutination, a protein of 68, and a glucose of 105 mg per 100 mL. Additional blood, spinal fluid, and stool cultures were obtained. The patient was placed on Claforan® and Dilantin®. He remained febrile up to 105°F, but became more responsive and was ambulatory by the second day after admission. At approximately 2 p.m. on August 5, the patient experienced another seizure with eye deviation to the right and head turning to the right. CT showed enhancement of the cisterna, but only mild increased intracranial pressure. Respirations became irregular and the patient was electively intubated and hyperventilated. He was started on streptomycin, PZA, and INH for possible TB, and acyclovir for possible CNS herpes. His condition deteriorated over the next 24 hours until he showed no evidence of brain stem function. A lumbar puncture was performed for viral studies as none of his previous cultures was growing. He was taken off the ventilator on the evening of August 6. At autopsy, the patient's meninges were relatively clear, but cerebral edema was present. Confirmation of infection with eastern equine encephalitis (EEE) virus was made by the Centers for Disease Control in Ft. Collins; two separate serum samples indicated a fourfold rise in HI antibody to EEE virus, and ELISA tests indicated the presence of specific IgM.

Comment: Investigation of this EEE case revealed several interesting features. The patient lived in a house without window screens. This likely led to increased exposure to mosquitoes (and thus, biting)—a risk factor for any mosquito-borne disease. An environmental survey of the Indian community revealed numerous prime *Coquillettidia perturbans* (the suspected mosquito vector in this case) breeding sites. In addition, *Cq. perturbans* were collected by CDC light traps in the community at the time of the survey. No other known vectors of EEE virus were collected. Finally, mosquito trapping a year later (same month as patient's infection) at the Indian community revealed that this species was the predominant species in the area.

Source: Adapted from Goddard J, Currier MM. Case histories of insect or arachnid-caused human illness. *J. Agromed.* 1995;2:53–61. With permission.

WNV infections are asymptomatic, approximately 20% cause West Nile fever, and less than 1% cause West Nile neuroinvasive disease.[59] From 1999 to 2005, more than 8000 cases of neuroinvasive WNV disease were reported in the United States, resulting in over 780 deaths.[58] There were 2175 cases reported to the CDC in 2015.[14] In recent years, WNV has become cyclical with peaks occurring about every 5 years. As with SLE, WNV is more dangerous to older patients; people in the age range of 60–89 years old are particularly at risk. Interestingly, of the first five patients in New York City admitted to hospitals, four had severe muscle weakness and respiratory difficulty, a finding atypical for encephalitis.[60] Also, gastrointestinal complaints such as nausea, vomiting, or diarrhea occurred in four of five patients. In fatal cases, cardiac and pulmonary complications are the most common primary causes of death.[61] Ecologically, WNV is a bird disease transmitted from bird to bird by various species of mosquitoes (Figure 25.24). The house sparrow has been found to be one of the best amplifying hosts in nature, producing the highest viremias for the longest period of time. Although the virus has been isolated from many mosquito species, the main vectors to humans are believed to be *Culex pipiens*, *Cx. quinquefasciatus*, *Cx. salinarius*, *Cx. restuans*, and *Cx. tarsalis*.[62–64]

Western Equine Encephalitis. Western equine encephalitis (WEE), occurring in the western and central United States, parts of Canada, and parts of South America, has occurred in several large outbreaks (Figure 25.25). There were large epidemics in the northcentral United States in 1941 and in the central valley of California in 1952. The 1941 outbreak involved 3000 cases. From 1964 to 1997, 639 human WEE cases were reported to the U.S. Centers for Disease Control and Prevention (CDC), for a national average of 19 cases per

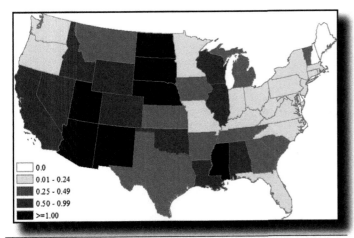

Figure 25.23 Incidence of West Nile virus in the United States, 2017. (Figure courtesy of U.S. Centers for Disease Control and Prevention, Atlanta, GA.)

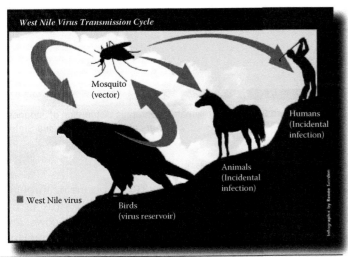

Figure 25.24 West Nile virus life cycle. (Figure courtesy of U.S. Food and Drug Administration, Washington, DC.)

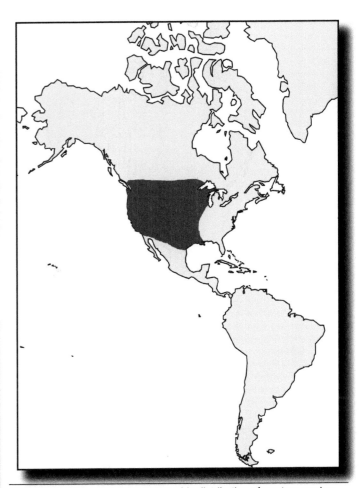

Figure 25.25 Approximate geographic distribution of western equine encephalitis in the United States.

year.[65] WEE is generally less severe than EEE and SLE, with a mortality rate of only 2–5%. Cases appear in early to midsummer and are primarily due to bites by infected *Culex tarsalis* mosquitoes. The incidence of WEE has declined significantly over the past few decades.

LaCrosse Encephalitis. LaCrosse encephalitis (LAC) has historically affected children in the midwestern states of Ohio, Indiana, Minnesota, and Wisconsin; however, it is increasingly being diagnosed in the southern states (Figure 25.26). The mortality rate of LAC is generally less than 1%, but seizures (even status epilepticus) and cerebral herniation may result from LAC infection. In fact, most LAC patients present with seizures. The national average for LAC is about 70 per year.[65] Cases occur in July, August, and September. LAC is a bunyavirus that is transmitted to humans primarily by *Aedes triseriatus*, but possibly also *Aedes albopictus* and *Ae. canadensis*. Interestingly, the virus may be transferred vertically from adult female *Ae. triseriatus* to her offspring through ovarial

contamination. Some amplification of the virus takes place in nature through an *Ae. triseriatus* small-mammal cycle.

Other California Group Encephalitis. Although LAC (described earlier) encephalitis is probably the most notorious, several other California group encephalitis viruses exist. North American forms include California encephalitis (CE), Jamestown Canyon (JC), Jerry Slough (JS), Keystone (KEY), San Angelo (SA), and Trivittatus (TVT), among others. Viruses in the California serogroup are primarily pathogens of rodents and lagomorphs. They are transmitted to people by several species of mosquitoes, but especially the tree hole, floodwater, and snow pool mosquitoes in the genus *Aedes*. California group encephalitis viruses generally produce only mild illness in humans (mortality rates of 1% or so).

Venezuelan Equine Encephalitis. Venezuelan equine encephalitis (VEE) is relatively mild in humans and rarely affects the central nervous system, but it is included here as an encephalitid. VEE is endemic in Mexico and Central and South America (Figure 25.18); epidemics occasionally reach the southern United States. Cases generally appear during the rainy season. Although the mortality rate is generally less than 1%, significant morbidity is produced by this virus. In an outbreak in Venezuela from 1962 to 1964, there were more

Figure 25.26 Approximate geographic distribution of LaCrosse encephalitis in the United States.

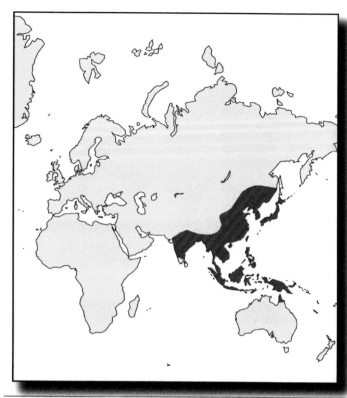

Figure 25.27 Approximate geographic distribution of Japanese encephalitis in the United States.

G. Other Arboviral Diseases

Rift Valley Fever. Rift Valley fever (RVF), occurring throughout sub-Saharan Africa and Egypt, is a Phlebovirus in the family Bunyaviridae. It causes abortions in sheep, cows, and goats and heavy mortality of lambs and calves. In humans, it may result in fever, myalgias, encephalitis, hemorrhage, or retinitis. Permanent visual impairment has been reported. Huge epidemics occasionally occur; from 1950 to 1951, there were an estimated 100,000 sheep and cattle cases and at least 20,000 human cases. Outbreaks continue; in 1998, there were 27,500 human cases and 170 deaths from RVF in Kenya.[75] Although RVF may be acquired by handling infective material of animal origin during necropsy or butchering, it is primarily a vector-borne zoonotic pathogen transmitted by *Aedes* mosquitoes.[76] The distribution of RVF seems to be expanding.

Ross River Virus Disease. Ross River (RR) disease, also called epidemic polyarthritis, occurs throughout most of Australia and occasionally New Guinea and some Pacific Islands such as Fiji, Tonga, and the Cook Islands. It causes fever, headache, fatigue, rash, and—most notably—arthritis in the wrist, knees, ankles, and small joints of the extremities. The disease is not fatal but may be debilitating, with symptoms occurring for weeks or months. RR disease is the most common arboviral disease in Australia with more than 5000 cases annually.[77] During 1997, there were 6683 cases reported.[78] Peak incidence of RR occurs from January through March, when the mosquito vectors are most abundant.

than 23,000 reported human cases with 156 deaths.[66] In 1971, an outbreak of VEE in Mexico extended into Texas, resulting in 84 human cases.[67] A more recent outbreak occurred in Colombia and Venezuela during the summer of 1995, with 75,000 to 100,000 human cases being reported.[68]

Japanese Encephalitis. Japanese encephalitis (JE) does not occur in the United States, but it is the principal cause of epidemic viral encephalitis in the world, with approximately 70,000 clinical cases each year and 10,000 deaths.[69–71] Returning U.S. travelers from Asia are sometimes infected with the virus.[72] JE epidemics have, at times, been widespread and severe. In 1924, there were 6125 cases with 3797 deaths.[73] JE occurs in Asia, roughly in a triangle from Pakistan to Indonesia, north to about Siberia (Figure 25.27). The JE virus is highly virulent. Approximately 25% of the cases are rapidly fatal, 50% lead to neuropsychiatric sequelae, and only 25% fully recover. In temperate zones, JE has a summer–fall distribution, but in the tropics no seasonal peak is apparent. There are several mosquito vectors of JE, but probably the most important is *Culex tritaeniorhynchus*, a rice-field-breeding species. Hogs may serve as amplifying reservoirs for JE.[74]

CASE HISTORY

SEIZURES ASSOCIATED WITH LACROSSE ENCEPHALITIS

In late June, a 6-month-old, previously healthy infant was brought to an emergency department (ED) with a several-hour history of fever up to 101.6°F. He was experiencing a focal seizure characterized by uncontrollable blinking of the left eye, twitching of the left side of the mouth, and random tongue movement. In the ED, seizures continued intermittently in spite of the administration of diazepam and lorazepam. The infant was started on phenytoin and admitted to the hospital. Examination of CSF upon admission showed 294 white blood cells (47% polymorphonuclear leukocytes, 41% histiocytes, 12% lymphocytes), and 3 red blood cells. Protein and glucose were within normal limits. Admission CT scan was read as normal. The infant was started on acyclovir because of the possibility of herpes encephalitis, and cefotaxime and vancomycin were also initiated to cover possible bacterial infection. The patient then became seizure free. He was taken off intravenous phenytoin and started on oral phenobarbital. Focal seizures, which progressed to generalized tonic clonic seizures, recurred on the fourth hospital day, at which time he had a therapeutic level of phenobarbital. After a repeat CT scan, which was read as within normal limits, the child was transferred to another hospital. On admission, the infant was noted to have continuous seizure-like movements of the chin and face. A repeat lumbar puncture revealed 307 white blood cells (33% polymorphonuclear leukocytes, 29% lymphocytes, and 38% histiocytes), 812 red blood cells, a protein of 104 mg/dL, and a glucose of 74 mg/dL. He was continued on acyclovir, cefotaxime, and vancomycin. The patient was intubated secondary to excessive secretions and to avoid respiratory compromise. He was again treated with lorazepam and restarted on phenytoin. Seizure activity ended. The patient continued to be intermittently febrile. He was extubated on the sixth continuous hospital day. Herpes PCR results from admission samples were negative. In addition, admission blood, urine, and CSF cultures were negative, and on the ninth hospital day, cefotaxime and vancomycin were discontinued. Patient's maximum temperature on that day was 100.1°F, and he was becoming more alert and playful. The patient subsequently became and remained afebrile and without seizure activity, so he was transferred back to the original hospital on the 12th hospital day for completion of 21 days of intravenous acyclovir. On day 21, he was discharged home on oral phenytoin. He had some residual left-sided weakness requiring several weeks of physical therapy. A sample of CSF was sent to a reference laboratory for testing for antibodies to various encephalitis agents including herpes and arboviruses. Results showed an indirect fluorescent antibody titer of 1:8 to LaCrosse (LAC) virus. Serum was sent to the CDC for confirmation, which showed the presence of IgM antibody to LaCrosse virus.

Comment: Differentiation of LAC infection must be made from other encephalitides (postvaccinal or postinfection), tick-borne encephalitis (not common in the United States), rabies, nonparalytic polio, mumps meningoencephalitis, aseptic meningitis from enteroviruses, herpes encephalitis, various bacterial, mycoplasmal, protozoal, leptospiral, and mycotic meningitides or encephalitides, and others. Any cases of encephalitis in late summer should be suspect. Specific identification is usually made (with the help of the Centers for Disease Control in Ft. Collins) by finding specific IgM antibody in acute serum or CSF, or antibody rises (usually HI test) between early and late serum samples. Serological identification of the particular virus is complicated because of cross-reactivity with heterologous viruses of the same group. For example, IgM antibody from patients with LAC virus infection has the highest titers to LAC virus itself, but it also reacts to a lesser extent with Snowshoe Hare virus and to a still lesser extent with Jamestown Canyon virus (two other viruses within the California serogroup).

Serological studies and laboratory research have indicated that most likely kangaroos and wallabies are natural hosts for RR virus. There are several mosquito vectors of RR virus in Australia, but particularly *Culex annulirostris* inland and *Aedes vigilax* and *Ae. camptorhynchus* in northern and southern coastal areas, respectively.

Chikungunya Fever. Chikungunya (CHIK) is a mosquito-transmitted Alphavirus which is not usually fatal but can cause severe fevers, headaches, fatigue, nausea, and muscle and joint pains.[79,80] There may be excruciatingly painful swelling of the joints in fingers, wrists, back, and ankles. The virus was first isolated during an epidemic in Tanzania in 1952, and the word chikungunya comes from Swahili, meaning "that which bends up," referring to the position patients assume while suffering severe joint pains.[80] The virus may be transmitted by *Aedes aegypti*, *Ae. albopictus*, and *Ae. polynesiensis* (Polynesian islands). The geographic distribution of CHIK has historically included most of sub-Saharan Africa, India, Southeast Asia, Indonesia, and the Philippines, although the disease is increasing both in incidence and geographic range. There were at least 300,000 cases on Reunion Island in the Indian Ocean during 2005–2006. India suffered an explosive outbreak in 2006 with more than 1.25 million cases. CHIK was found in Italy in 2007.[79,81] During 2013–14, there were thousands of cases reported in the Caribbean and Central, and South America.[82] In 2015, there were 896 cases of travel-related CHIK in the United States and one locally acquired case.[14] One of the mosquito vectors of CHIK, the Asian tiger mosquito, *Ae. albopictus*, is extremely abundant in the southern United States, raising fears of widespread outbreaks should local mosquitoes become infected.[81]

Zika Virus. Zika virus (ZIKV), a mosquito-borne human disease with only limited clinical severity, is transmitted by *Aedes aegypti, Ae. albopictus, Ae. polynesiensis*, and *Ae. hensilli*. The virus is a RNA virus in the family *Flaviviridae*, which includes several other viruses of clinical importance. ZIKV is closely related to the Spondweni virus, the only other member of its group. Genetic analysis shows that ZIKV can be classified into distinct African and Asian lineages; both emerged from East Africa during the late 1800s or early 1900s.[83] In humans, the incubation period is approximately 3–12 days. Infection is extremely mild and asymptomatic in 80% of cases, making reporting and (necessary) public health interventions difficult. People of all ages are susceptible to ZIKV (4 days–76 years), with a slight preponderance of cases in females.[83,84] If symptoms occur, they are typically temporary, self-limiting, and nonspecific. Commonly reported symptoms include rash, fever, arthralgia, myalgia, fatigue, headache, and conjunctivitis.[85] Rash, a prominent feature, is maculopapular and pruritic, and usually begins proximally then spreading to the extremities with spontaneous resolution within 1–4 days of onset.[83] There may be a low-grade fever (37–38°C). Symptoms generally resolve within 2 weeks and reports of longer persistence are rare. The most important medical issues associated with Zika are the severe clinical sequelae Gullian-Barré syndrome and microcephaly (Figure 25.28).[84] During the 2015 outbreak in Brazil, reports of infants born with microcephaly greatly increased (>3,800 cases; 20 cases/10,000 live births vs. 0.5/10,000 live births in previous years).[83] During 2015–2017, there were millions of cases of ZIKV throughout the Americas and approximately 275 locally acquired cases occurred in the United States.[85,86] All cases in the United States during 2016–2017 were associated with *Ae. aegypti* mosquitoes and not *Ae. albopictus* (which is fortunate because *Ae. albopictus* is much more widespread). Cases declined rapidly after 2017 but may resurface again after a few years since arboviruses are often cyclical.

II. Basic Biology and Ecology

Mosquitoes undergo complete metamorphosis, having egg, larval, pupal, and adult stages (Figure 25.29). Larvae are commonly referred to as wigglers and pupae as tumblers. Larvae and pupae of mosquitoes are always found in water. The water source may be anything from water in discarded automobile tires to water in the axils of plants, to pools, puddles, swamps, and lakes. Mosquito species differ in their breeding habits, biting behavior, flight range, etc.; however, a generalized description of their life cycle is presented here and will serve as a useful basis for understanding mosquito

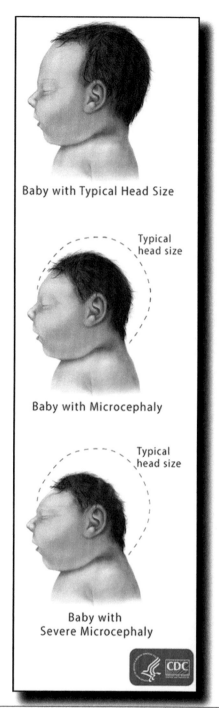

Figure 25.28 Microcephaly caused by Zika virus (CDC figure).

biology and ecology. Most larvae in the subfamily Culicinae hang down just under the water surface by a breathing tube (siphon), whereas anopheline larvae lie horizontally just beneath the water surface supported by small notched organs of the thorax and clusters of float hairs along the abdomen. They have no prominent siphon. Mosquito larvae feed on suspended particles in the water as well as microorganisms. They undergo four molts (each successively larger), the last of which results in the pupal stage. With optimum food and

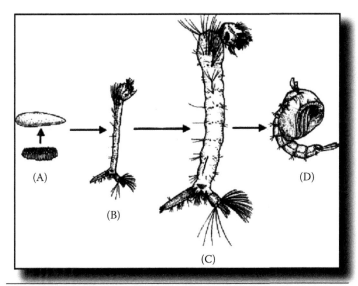

Figure 25.29 Immature stages of a mosquito. (Photograph courtesy of Joe MacGown, Mississippi Cooperative Extension Service, Mississippi State University.)

temperature conditions, the time required for larval development can be as short as 7 days.

Mosquito pupae are quite active and will "tumble" toward the bottom of their water source upon disturbance. Pupae do not feed. They give rise to adult mosquitoes in 2–4 days. This process begins with the splitting of the pupal skin along the back. The emerging adult must dry its wings and separate and groom its head appendages before it can fly away (Figure 25.30). This is a critical stage in the survival of mosquitoes. If there is too much wind or wave action, the emerging adult will fall over while leaving its skin, becoming trapped on the water surface to soon die.

Adult mosquitoes of both sexes obtain nourishment for basic metabolism and flight by feeding on nectar. In addition, females of most species obtain a blood meal from birds, mammals, reptiles, or amphibians for egg development. An excellent review of host-finding behavior has been provided by Bowen.[87] Breeding sites selected for egg laying differ by species, but generally, mosquitoes can be divided into three

major breeding groups: permanent water breeders, floodwater breeders, and artificial container/tree hole breeders. *Anopheles* and many *Culex* mosquitoes select permanent water bodies, such as swamps, ponds, lakes, and ditches that do not usually dry up. Floodwater mosquitoes lay eggs on the ground in low areas subject to flooding. During heavy rains, water collecting in these low areas covers the eggs, which hatch from within minutes up to a few hours. Salt marsh mosquitoes (*Aedes sollicitans*), inland floodwater mosquitoes (*Ae. vexans*), and dark rice field mosquitoes (*Psorophora columbiae*) are included in this group. Artificial container/tree hole breeders are represented by yellow fever mosquitoes (*Ae. aegypti*), Asian tiger mosquitoes (*Ae. albopictus*), tree hole mosquitoes (*Aedes triseriatus*), and others. However, several species of *Anopheles* and *Culex* may also occasionally oviposit in these areas. Some of these container-breeding species lay eggs on the walls of a container just above the water line. The eggs are flooded when rains raise the water level. Other species oviposit directly on the water surface.

Female *Anopheles* mosquitoes generally lay eggs on the surface of the water at night. Each batch usually contains 100–150 eggs. The *Anopheles* egg is cigar-shaped and about 1 mm long, and it bears a pair of air-filled floats on the sides (Figure 25.31). Under favorable conditions, hatching occurs within 1 or 2 days.

Aedes mosquitoes lay their eggs on moist ground around the edge of the water or, as previously mentioned, on the inside walls of artificial containers just above the waterline. *Aedes* eggs will die if they become too dry when first laid; however, after the embryo in the egg develops, the eggs can withstand dry conditions for long periods of time. This trait has allowed *Aedes* mosquitoes to use temporary water bodies for breeding, such as artificial containers, periodically flooded salt marshes or fields, tree holes, and stormwater pools. Also, *Aedes* mosquitoes have inadvertently been carried to many parts of the world as dry eggs in tires, water cans, or other suitable containers. The Asian tiger mosquito (*Ae. albopictus*) was introduced into the United States in the 1980s in shipments of used truck tire casings imported from Taiwan and Japan. After these tires were stacked outside and began to collect rainwater, the eggs hatched. A more recent introduction into the United States has been *Ae. japonicus*.[88]

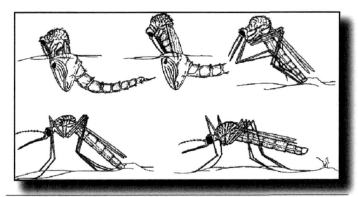

Figure 25.30 Adult mosquito emerging from pupal stage. (From Bowles E. *The Mosquito Book*. Jackson, MI: Mississippi State Department of Health; 2007.)

Figure 25.31 Scanning electron micrograph of an *Anopheles* egg. (Photograph courtesy of Dr. Gerry Baker, Mississippi State University, and used with permission.)

Psorophora mosquitoes also lay dry-resistant eggs. These mosquitoes are often a major problem species in rice fields. Eggs are laid on the soil and hatch once the field is irrigated.

Culex mosquitoes lay batches of eggs that are attached together to form little floating rafts (Figure 25.32). Upon close inspection of a suitable breeding site, these egg rafts can often be seen floating on the surface of the water.

In tropical areas, mosquito breeding may continue year-round, but in temperate climates, many species undergo a diapause in which the adults enter a dormant state similar to hibernation. In preparation for this, females become reluctant to feed, cease ovarian development, and develop a fat body. They may seek a protected place to pass the approaching winter. Some species, instead of passing the winter as hibernating adults, produce dormant eggs that can survive the harsh effects of winter.

Mosquitoes vary in their biting patterns as well. Most species are diurnal in activity, biting mainly in the early evening; however, some species, especially *Aedes aegypti* and *Ae. albopictus*, bite in broad daylight (although there may be a peak of biting very early and late in the day). Others, such as the salt marsh species and many members of the genus *Psorophora*, do not ordinarily bite during the day but will attack if disturbed (such as walking through high grass harboring resting adults).

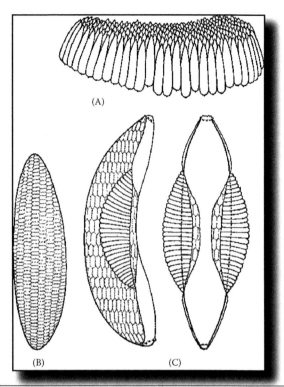

Figure 25.32 Various mosquito eggs: (A) *Culex*, (B) *Aedes*, and (C) *Anopheles*. (From Carpenter SJ, and LaCasse WJ. *Mosquitoes of North America (North of Mexico)*. Berkeley: University of California Press; 1955.)

III. General Description of Mosquitoes

Adult mosquitoes can be distinguished from other flies by several characteristics (Figure 25.33 is provided to familiarize the reader with these characteristics). They have long, 15-segmented antennae, a long proboscis for bloodsucking, and scales on the wing fringes and wing veins. Scale patterns on the "back" (scutum) of mosquitoes can sometimes be used to identify the species (Figure 25.34). The mosquito head is rounded, bearing large compound eyes that almost meet. Males can usually be distinguished from females by their

---

OFTEN-ASKED QUESTION

ARE SOME PEOPLE MORE ATTRACTIVE TO MOSQUITOES THAN OTHERS?

Everywhere we go, people seem eager to offer anecdotal evidence that mosquitoes show preferences in their biting behavior. In theory, certain people's skin components are more attractive to the bloodsucking pests than others. There is some evidence for this. Children infected with malaria parasites at just the precise time when the parasites are transmissible to mosquitoes are more attractive to *Anopheles* mosquitoes (the vector) than uninfected children.[1] Recent scientific evidence, however, suggests that instead of being more attractive to mosquitoes, some people produce natural insect repellent compounds in their skin and bodily odors. One scientific study revealed that three aldehydes (octenal, nonanal, and decanal) and two ketones (geranylacetone and 6-methyl-5-hepten-2-one) were present in higher quantities in volunteers that were "unattractive" to mosquitoes compared to "attractive" individuals.[2] Therefore, regardless of whether attractants or repellents are produced, differential attraction of mosquitoes to humans and other hosts is more than anecdotal.

REFERENCES

1. Lacroix R, Mukabana WR, Gouagna LC, Koella JC. Malaria infection increases attractiveness of humans to mosquitoes. *PLoS Biol.* 2005;3:1590–1593
2. Logan JG. Why do mosquitoes "choose" to bite some people more than others? *Outlooks Pest Manage.* 2008;1:280–283

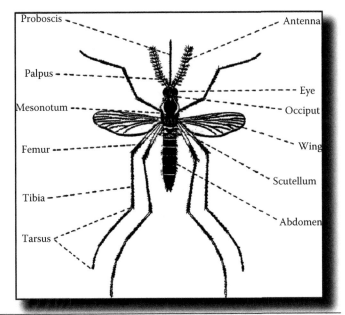

Figure 25.33 Adult mosquito with parts labeled. (From CDC, Pictorial Keys to Arthropods, Reptiles, Birds, and Mammals of Public Health Significance. Atlanta, GA: U.S. Centers for Disease Control and Prevention; 1963.)

bushy antennae (Figure 25.35C, D). Female *Anopheles* mosquitoes have palpi as long as the proboscis (Figure 25.35B); other groups do not have this characteristic. Larval mosquitoes, commonly called wiggle tails or wigglers, are white to dark gray in color and possess no legs or wings. They bear simple or branched tufted setae (hairs) along the body, anal gills, antennae, and chewing mouthparts. Culicine larvae have a prominent siphon tube for respiration at the water surface. Anopheline species do not and thus must lie horizontally just beneath the water surface. Pupal mosquitoes, often called tumblers, also occur in water. They are comma-shaped, with two prominent respiratory trumpets on the thorax and a set of paddles on the last abdominal segment.

IV. Discussion of Some Common U.S. Species*

A. Yellow Fever Mosquito, *Aedes aegypti* (Linnaeus)

Medical Importance. Known vector of yellow fever, dengue, chikungunya, and Zika viruses.

Distribution. Cosmotropical in distribution, occurring worldwide within the 20°C isotherms. The U.S. distribution is spotty across the southern states (Figure 25.36).

---

* The descriptions for each mosquito species are only general comments about their macroscopic appearance. Mosquito identification to the species level is quite difficult, utilizing a number of microscopic characteristics. Specific identifications should be performed by specialists at institutions that routinely handle such requests (e.g., universities, extension services, state health departments, the military).

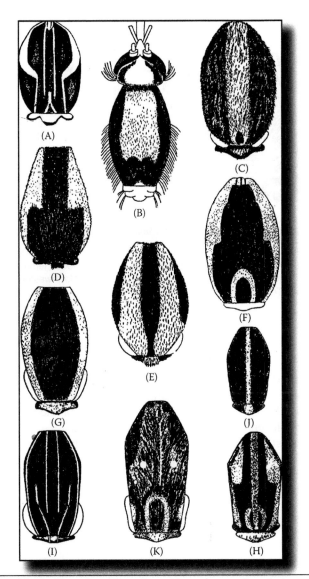

Figure 25.34 Examples of thoracic markings of mosquitoes (not necessarily drawn to scale): (A) *Aedes aegypti*, (B) *Aedes infirmatus*, (C) *A. atlanticus*, (D) *A. thibaulti*, (E) *A. trivittatus*, (F) *A. triseriatus*, (G) *Psorophora varipes*, (H) *P. ciliata*, (I) *Orthopodomyia signifera*, (J) *Uranotaenia sapphirina*, and (K) *Culex restuans*. (From King WV, et al. A Handbook of the Mosquitoes of the Southeastern United States, Agriculture Handbook. No. 173. Washington, DC: U.S. Department of Agriculture; 1960.)

Description. Typical *Aedes* mosquito with a pointed abdomen (at tip) and black and white rings on the legs. Small black species with a silver–white lyre-shaped figure on the upper sides of its thorax (Figure 25.37); also silver–white bands on the hind tarsi and abdomen.

Remarks. Breeds in shaded artificial containers around buildings such as tires, cans, jars, flowerpots, and gutters; usually bites during the morning or late afternoon; readily enters houses and prefers human blood meals, biting principally around the ankles or back of the neck; flight range from 100 feet to 100 yards; in many places in the United States where *Ae. albopictus* was introduced, this species diminished, apparently being displaced.

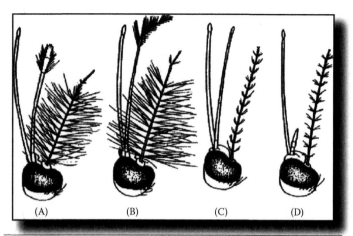

Figure 25.35 Mosquito head and appendages: (A) *Anopheles* male, (B) *Aedes* male, (C) *Anopheles* female, and (D) *Aedes* female. (From Gjullin, CM, Eddy GW. The Mosquitoes of the Northwestern United States, Technical Bulletin No. 1447. Washington, DC: U.S. Department of Agriculture; 1972.)

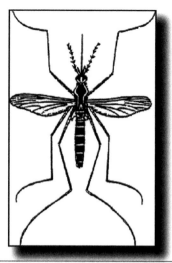

Figure 25.37 Adult *Aedes aegypti*.

B. Asian Tiger Mosquito, *Aedes albopictus* (Skuse)

Medical Importance. Aggressive, daytime biting mosquito which breeds in artificial containers; often found associated with piles of used automobile tires; vector of the agents of yellow fever, dengue, chikungunya, and Zika.

Distribution. Widely distributed in the Asian region, the Hawaiian islands, and parts of the southern and eastern United States, where it was accidentally introduced in 1986 (Figure 25.38); recently spreading throughout central Africa.

Description. Very similar in appearance to *Ae. aegypti* with a black body and silver–white markings (Figure 25.39); major difference between the two is that *Ae. albopictus* has a single, silver–white stripe down the center of the dorsum of the thorax instead of the lyre-shaped marking.

Remarks. Breeds in artificial containers such as cans, gutters, jars, tires, flowerpots, etc., and seems especially fond

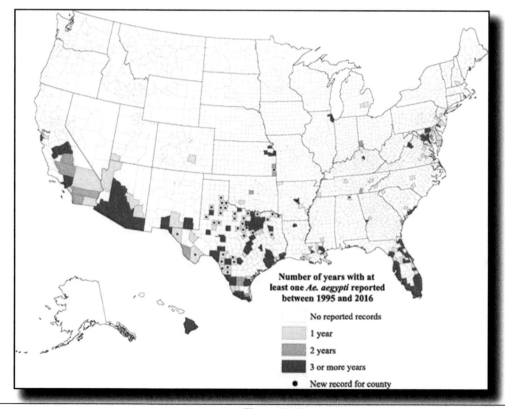

Number of years with at least one *Ae. aegypti* reported between 1995 and 2016

No reported records
1 year
2 years
3 or more years
● New record for county

Figure 25.36 Approximate U.S. distribution of *Aedes aegypti* (CDC figure).

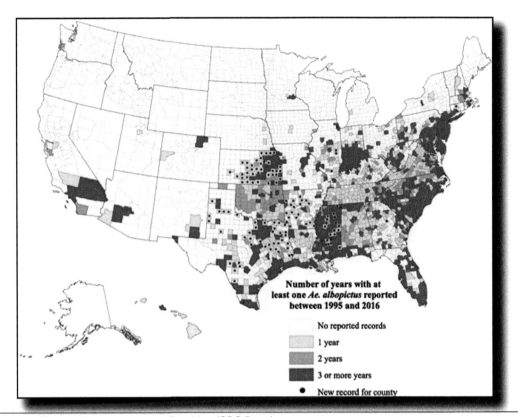

Figure 25.38 Approximate U.S. distribution of *Aedes albopictus* (CDC figure).

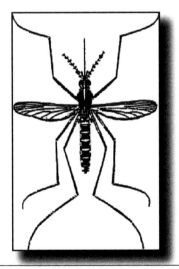

Figure 25.39 Adult *Aedes albopictus*.

of discarded tires. It is an aggressive mosquito, often landing and biting immediately, although not as anthropophilic as *Ae. aegypti*. Flight range is generally less than a quarter mile.

C. Salt Marsh Mosquito, *Aedes sollicitans* (Walker)

Note that Reinert[89] reclassified many *Aedes* species as genus *Ochlerotatus*, though most entomologists have reverted back to the traditional classification (*Aedes*).

Medical Importance. Fierce biter that can often discourage coastal development (in Louisiana, it is one of the most important pest species in coastal parishes[90]); known vector of the agent of EEE.

Distribution. Nearctic region along the eastern coastal and inland saline areas (Figure 25.40).

Description. Bronze–brown species with golden yellow markings (Figure 25.41); whitish–yellow bands on the upper side of the abdomen with a yellowish stripe down the center; short palps with small white tips, banded legs, and dark proboscis with a single band in the middle.

Remarks. Breeds primarily in salt marshes flooded by tides and rain; breeds throughout the year, although developmental time is longer in winter. It is an aggressive biter that feeds at night (however, adults rest on vegetation during the day and will readily bite when disturbed). Flight range is between 5 and 10 miles.

D. Inland Floodwater Mosquito, *Aedes vexans* Meigen

Medical Importance. Bothersome pest and possible vector of EEE and WN virus.

Distribution. Holarctic and Oriental regions, Pacific Islands, South Africa, Mexico, and parts of Central America (Figure 25.42).

Description. Typical *Aedes* mosquito with a pointed abdomen (at tip) and black and white rings on the legs. Medium-sized brown to golden-brown mosquito with light gray or white markings shaped like wide "B's" on the dorsum of the middle segments of the abdomen (Figure 25.43); hind

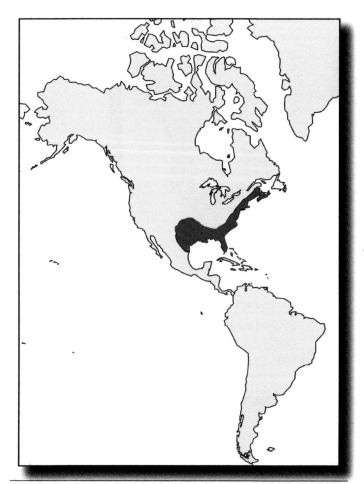

Figure 25.40 Approximate geographic distribution of *Aedes sollicitans*.

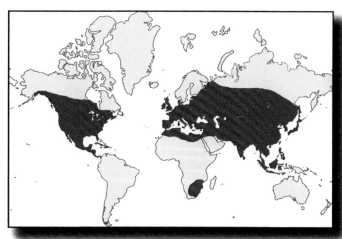

Figure 25.42 Approximate geographic distribution of *Aedes vexans*.

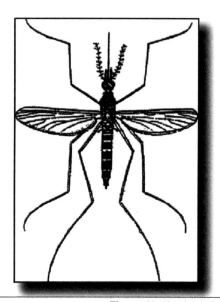

Figure 25.43 Adult *Aedes vexans*.

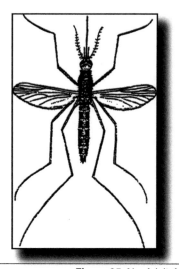

Figure 25.41 Adult *Aedes sollicitans*.

tarsi with narrow white rings on all segments and short and dark palps with a few white scales at the tip.

Remarks. Breeds in any temporary body of freshwater in both wooded and open areas; many broods are produced each year from May through September when breeding areas are flooded by rains. It is a vicious biter that is active mainly at dusk and just after dark. Flight range is 5–10 miles.

E. Malaria Mosquito, *Anopheles quadrimaculatus* Say

Medical Importance. One of many *Anopheles* mosquitoes that can transmit the agent of malaria to humans; principal vector in the eastern half of the United States until malaria was eradicated.

Distribution. Central and eastern United States, southern Canada, and Mexico (Figure 25.44).

Description. Actually, a complex of five sibling species (named *An. quadrimaculatus*, *An. smaragdinus*, *An. diluvialis*, *An. maverlius*, and *An. inundatus*) that all look alike but differ in behavior and ecology. They are medium to large dark brown mosquitoes with palps as long as the proboscis, dark unbanded legs, and wings marked by four dark

Figure 25.44 Approximate geographic distribution of *Anopheles quadrimaculatus* complex.

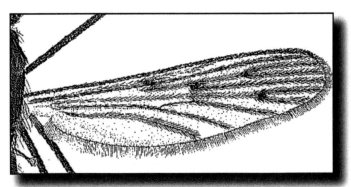

Figure 25.46 Wing of *Anopheles quadrimaculatus*, showing four dark spots. (From Carpenter SJ, LaCasse WJ. *Mosquitoes of North America (North of Mexico)*. Berkeley: University of California Press; 1955.)

spots (Figures 25.45 and 25.46); long maxillary palps in both sexes (typical characteristics of all *Anopheles* mosquitoes). They feed with their proboscis, head, and body in an almost straight line (many species feed almost perpendicular to the skin surface of the host); it looks as though they are standing on their heads when they bite (Figure 25.47) (also characteristic of all *Anopheles* mosquitoes).

Remarks. Breed in permanent freshwater sites such as ponds, pools, swamps, and rice fields containing emergent or floating vegetation (as for most *Anopheles* mosquitoes). Breeding may be continuous throughout the year, especially if winters are mild; however, peak populations usually occur during July or August (decline rapidly during September and October). Adults rest during the day in cool, dark, damp areas and come out to feed at night. Adults usually fly no more than 4 miles from their breeding area.

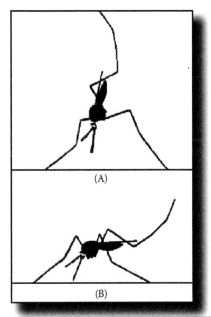

Figure 25.47 Feeding positions of mosquitoes: (A) *Anopheles*, and (B) *Culex*. (From King, WV, et al. A Handbook of the Mosquitoes of the Southeastern United States, Agriculture Handbook. No. 173. Washington, DC: U.S. Department of Agriculture; 1960.)

F. *Culex nigripalpus* Theobald

Medical Importance. Major vector of SLE and WN viruses in the southeastern United States; in 1990, central and southern counties of Florida experienced an outbreak due to this species causing 212 SLE cases and 10 deaths.[91]

Distribution. Southern United States, Antilles, Mexico, Central America, Trinidad, Ecuador, Colombia, Venezuela, the Guianas, Brazil, and Paraguay (Figure 25.48).

Description. Typical *Culex* mosquito with a blunt abdomen (at tip) and hardly any striking features; adults are medium-sized dark mosquitoes with a rounded abdomen, unbanded dorsally (Figure 25.49). White patches may be present laterally. They are difficult to distinguish from other small *Culex* mosquitoes.

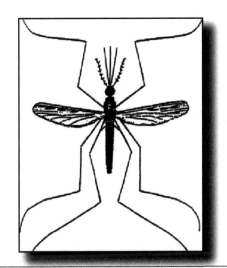

Figure 25.45 Adult *Anopheles quadrimaculatus*.

Figure 25.48 Approximate geographic distribution of *Culex nigripalpus*.

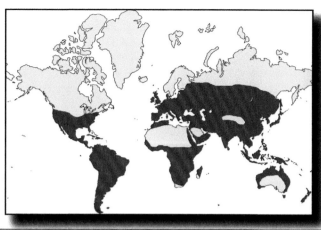

Figure 25.50 Approximate geographic distribution of *Culex quinquefasciatus*.

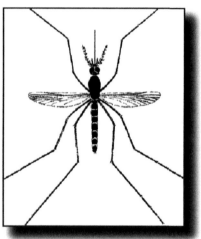

Figure 25.49 Adult *Culex nigripalpus*.

Remarks. Active in spring, summer, and autumn, but generally most abundant in autumn; breeds in shallow rainwater pools and semipermanent ponds.

G. Southern House Mosquito, *Culex quinquefasciatus* Say

Medical Importance. One of the major vectors of the SLE virus (numerous outbreaks of SLE are associated with this species); the primary vector of WNV in southern states. This is a highly anthropophilic species in many parts of the tropical and subtropical world, where it is also a major vector of lymphatic filariasis.

Distribution. *Culex pipiens* complex, which includes *Cx. quinquefasciatus*, is cosmotropical in distribution (Figure 25.50); in North America, this group is represented by three main members. The northern house mosquito, *Cx. pipiens*, occurs in areas above 39°N latitude; the southern house mosquito, *Cx. quinquefasciatus*, occurs at latitudes less than

36°N latitude; between 36° and 39°N latitudes, *Cx. pipiens* and *Cx. quinquefasciatus*, as well as hybrids between the two, are encountered.

Description. Typical *Culex* mosquito with a blunt abdomen (at tip) and hardly any striking features. It is a medium-sized brown mosquito with a few white bands on the upper side of the abdominal segments (Figure 25.51); bands are widest in the middle of the dorsum and narrow laterally. Legs and proboscis are dark and unbanded. This mosquito is difficult to distinguish from other *Culex* mosquitoes.

Remarks. Breeds in ditches, storm sewer catch basins, cesspools, and polluted water (also breeds in artificial containers around homes such as cans, jars, and tires); readily enters homes, producing the familiar "singing" sound around persons at night. Flight range is variable but may be as far as 1100 meters in a single night.

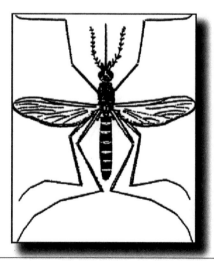

Figure 25.51 Adult *Culex quinquefasciatus*.

H. *Culex salinarius* Coquillett

Medical Importance. Often occurs in extremely high numbers and thus a nuisance pest; possibly a major bridge vector of WNV, and a secondary vector of SLE and EEE viruses.

Distribution. United States and Canada east of Rocky Mountains, as well as Bermuda and parts of Mexico (Figure 25.52); especially common along the Atlantic and Gulf coasts.

Description. Typical *Culex* mosquito with a blunt abdomen (at tip) and hardly any striking features; brown species has both proboscis and hind tarsi entirely dark (no white bands) (Figure 25.53). Dorsal abdominal segments have drab and often inconspicuous basal bands of yellowish or brownish scales; bands are usually irregular and narrow. Last two abdominal segments are often coppery colored.

Remarks. Breeds in ditches, pools, artificial containers, and often salty water, such as that in salt marshes; found breeding year-round in Gulf coastal states. It feeds readily on people, mostly outdoors, but occasionally inside buildings; feeding is heaviest at dusk and first hours of darkness. Flight range is 1 mile or more.

I. Encephalitis Mosquito, *Culex tarsalis* Coquillett

Medical Importance. Abundant and widespread species; main vector of WEE, WNV, and SLE viruses in central, western, and southwestern United States; probably the most ubiquitous mosquito in California.

Distribution. Most of United States, parts of southwestern Canada, and Mexico (Figure 25.54).

Description. Medium-sized mosquito, dark brown to black in color, having white bands on the legs and abdomen (Figure 25.55) and a broad white band in the middle of the proboscis. The vector of the abdomen is white with a black chevron on each segment.

Remarks. Primarily a rural species that breeds in both polluted and clear water in ground pools, grassy ditches, and artificial containers; in arid and semiarid regions, it is frequently found in canals and ditches associated with irrigation. Domestic and wild birds are the preferred food source, but people are also readily attacked. It is most active soon after dusk and active primarily in summer but may remain active all winter in southern parts of the range.

Figure 25.52 Approximate geographic distribution of *Culex salinarius*.

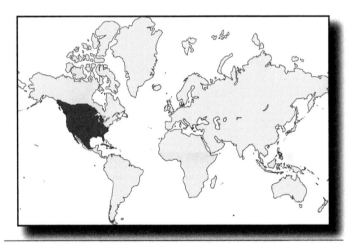

Figure 25.54 Approximate geographic distribution of *Culex tarsalis*.

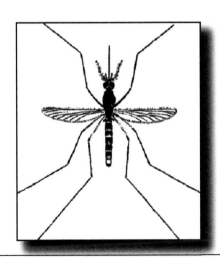

Figure 25.53 Adult *Culex salinarius*.

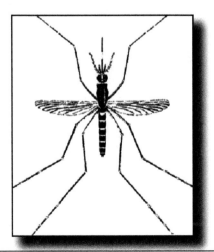

Figure 25.55 Adult *Culex tarsalis*.

J. Dark Rice Field Mosquito, *Psorophora columbiae* Dyar and Knab

Medical Importance. Although occasionally involved as a secondary vector of several encephalitides, the primary adverse effect on people is nuisance biting; it is a vicious biter and, thus, a severe nuisance pest.

Distribution. Much of the southeastern United States as far north as New York state and west to southern South Dakota, south to Texas and northern Mexico (Figure 25.56); reports from Arizona and New Mexico may be closely related species in the *Ps. confinnis* complex.

Description. Large mosquito, dark brown to black (Figure 25.57), with white bands on the legs and a yellow band on otherwise dark proboscis; the first segment of hind tarsus is brown with a white ring in the middle. Light and dark scales give much of the body and wings a "salt-and-pepper" effect.

Remarks. Breeds in open, temporary pools of freshwater such as ditches, rice fields, and low, flooded areas; eggs are deposited on moist soil subject to flooding either by rain or irrigation. Several broods are produced in the active season (April to October); it bites aggressively during the day or night and can fly 6–8 miles.

V. Discussion of Some Major Pest Species in Other Areas of the World

A. *Anopheles albimanus* Weidemann

Medical Importance. One of the most important vectors of malaria in Mexico, Central and South America, and Caribbean.

Distribution. Extreme southern United States (Texas and Florida), Mexico, Central America, South America (probably only Colombia, Venezuela, and Ecuador), Antilles, and the Caribbean (Figure 25.58).

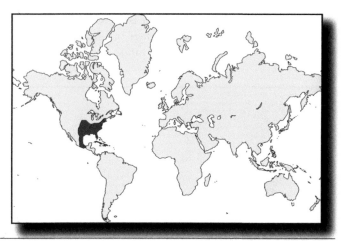

Figure 25.56 Approximate geographic distribution of *Psorophora columbiae.*

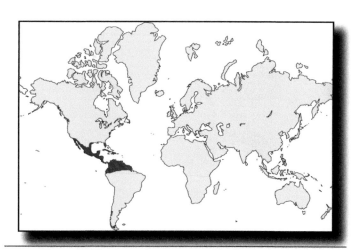

Figure 25.58 Approximate geographic distribution of *Anopheles albimanus.*

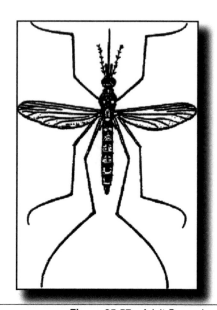

Figure 25.57 Adult *Psorophora columbiae.*

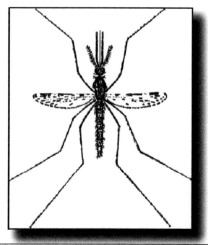

Figure 25.59 Adult *Anopheles albimanus.*

Description. All *Anopheles* females have long maxillary palps, and wings are often spotted; *An. albimanus* is a spotted species (Figure 25.59) with pale hind legs similar to many other related *Anopheles* species. It has a black proboscis and white terminal segment of palpi.

Remarks. Prefer hot, humid climates; most abundant during the wet season, when they breed in sunny pools, pits, puddles, ponds, marshes, lagoons, and artificial containers, especially those water sources containing floating or grassy vegetation. Females feed on people and domestic animals, both indoors and outdoors; they usually rest outdoors after feeding.

B. *Anopheles darlingi* Root

Medical Importance. A major contributor to endemic malaria in extreme southern Mexico and Central and South America; adults feed readily on people and often collect in large numbers inside houses.

Distribution. Mexico through Central America southward into Argentina and Chile (Figure 25.60).

Description. Last three segments of the hind tarsi are entirely white (Figure 25.61); it has a long, slender, black proboscis, white terminal segment of palpi, and a light spot at the cross vein on the front wing margin smaller than the preceding dark spot.

Figure 25.60 Approximate geographic distribution of *Anopheles darlingi*.

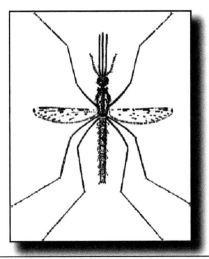

Figure 25.61 Adult *Anopheles darlingi*.

Remarks. Breeds in shaded areas of freshwater marshes, swamps, lagoons, rice fields, lakes and ponds, and the edges of streams, especially those with vegetation; feeds on humans indoors and rests indoors after feeding.

C. *Anopheles gambiae* Giles (Consists of Several Species in a Complex)

Medical Importance. Most efficient malaria transmitter in Africa (including all members of the complex). It is highly anthropophilic; blood meal analyses have indicated that more than 50% of fed females contain human blood.

Distribution. At least six members of the complex and their distributions in Africa are presented below (Figure 25.62):

1. *An. arabiensis*—Tropical Africa, southwest Arabia, the Cape Verde Islands, Zanzibar, Pemba, Madagascar, and Mauritius
2. *An. gambiae*—Tropical Africa, Bioko, Zanzibar, Pemba, and Madagascar
3. *An. melas*—Coastal West Africa
4. *An. merus*—Brackish water habitats in East and South Africa
5. *An. quadriannulatus*—Ethiopia, South Africa, Swaziland, Zambia, Zanzibar, and Zimbabwe
6. *An. bwambae*—Uganda

Description. Members of complex generally look like the one depicted in Figure 25.63; precise identification of members of this species complex is done by genetic methods such as PCR.

Remarks. Each member of the complex may exhibit different behaviors, but larvae of *An. gambiae* and *An. arabiensis* (formerly species A and B of the complex) occur in all types of water-containing depressions close to human habitations including pools, puddles, hoofprints, borrow pits, and even rice fields. They are especially abundant during the rainy season. Females have an average flight range of about 1 mile.

D. *Anopheles leucosphyrus* Group

Medical Importance. Group of *Anopheles* mosquitoes (containing at least 20 closely related species); contains several main vectors of malaria in Southeast Asia.

Distribution. Distributions and descriptions are based on two papers.[92,93] Distributions of the main malaria vectors in the group are given below (Figure 25.64):

1. *An. balabacensis*—Philippines (Balabac, Culion, and Palawan Islands), Malaysia (Sabah and Sarawak), Indonesia (Kalimantan, Java, Lombok), and Brunei
2. *An. dirus*—Thailand, Kampuchea, Vietnam, China (Hainan Island)

3. *An. leucosphyrus*—Sumatra, Indonesia
4. *An. lateens*—Malaysia (Sabah and Sarawak, and states in Peninsular Malaysia), Indonesia (Kalimantan), Thailand (southern provinces)
5. *An. baimaii*—Thailand (western and southern provinces), Myanmar, India (eastern states), Bangladesh

Description. Members of the group have numerous spots within the wings and on the legs (Figure 25.65); they have conspicuously white joints where the tibia and tarsus meet on each hind leg.

Remarks. Breeds in shaded freshwater pools in and among rocks, in hoofprints, vehicle ruts, and the like; many are forest species that bite humans and animals outdoors. Fed females generally rest outdoors; in the case of *Anopheles balabacensis*, the females rest on forest vegetation during the day, congregate around human dwellings at sundown, remain

Figure 25.62 Approximate geographic distribution of the *Anopheles gambiae* complex.

Figure 25.64 Approximate geographic distribution of the *Anopheles leucosphyrus* group.

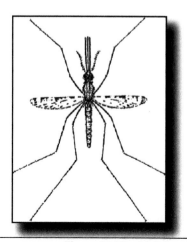

Figure 25.63 Adult *Anopheles gambiae*.

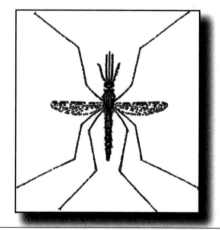

Figure 25.65 Anopheles balabacensis, a member of the *A. leucosphyrus* group.

quiescent on vegetation for some time after dark, and then enter the dwellings to feed on sleeping persons.

E. *Anopheles funestus* and *Anopheles minimus* Groups

Medical Importance. Important vectors of malaria in tropical Africa, the Indian subcontinent, and Southeast Asia.[94,95] Note: further studies are needed to settle the relationships between the African *An. funestus* group and the Asian *An. minimus* group.

Distribution. The distributions of three members of the *A. minimus* complex are given below (Figure 25.66):

1. *An. flavirostris*—Philippines
2. *An. funestus*—Most of tropical Africa
3. *An. minimus*—Burma, Bangladesh, India, Cambodia, Laos, Malaysia, Thailand, Vietnam, and Taiwan

Description. Small, dark specimens having narrow white bands on the female palps (Figure 25.67); *An. minimus* has a totally dark proboscis.

Remarks. *An. flavirostris* bites indoors but rests outdoors afterward; both *An. flavirostris* and *An. minimus* breed in grassy edges of small foothill streams, springs, irrigation ditches, or seepages. They both prefer shaded areas. *An. funestus* is an extremely important vector of malaria in tropical Africa (second only to *An. gambiae*). The larvae of *An. funestus* prefer marshes, swamps, and edges of streams; they feed indoors and outdoors but rest outdoors after feeding.

F. *Anopheles pseudopunctipennis* Theobald

Medical Importance. Important vector of malaria in Mexico, Central America, and parts of South America; primary vector of a large malaria outbreak in Oaxaca, Mexico, in 1998.

Distribution. Southern United States southward to Argentina, and Antilles (Figure 25.68).

Description. Females are large gray mosquitoes with black legs (Figure 25.69), two narrow white bands and a white

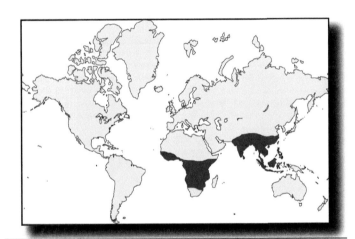

Figure 25.66 Approximate geographic distribution of the *Anopheles funestus* and *Anopheles minimus* groups.

Figure 25.68 Approximate geographic distribution of *Anopheles pseudopunctipennis*.

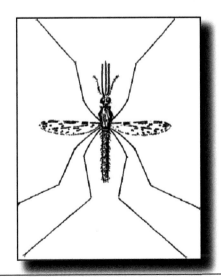

Figure 25.67 Adult *Anopheles minimus*.

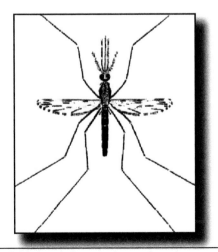

Figure 25.69 Adult *Anopheles pseudopunctipennis*.

tip on the palps, totally dark proboscis, entirely black hind tarsus, and two light spots on the front wing margin.

Remarks. Highland species occurring most abundantly during the dry season. Heavy rains during the wet season flush the larvae out of their breeding sites, which include shallow, quiet, or slow-moving water such as drying stream-beds. Larvae are often associated with green algae, such as Spirogyra, in stream pools. Adults enter houses and bite people readily.

G. *Anopheles stephensi* Liston

Medical Importance. Widely distributed urban vector of malaria.

Distribution. Afghanistan, Bangladesh, Burma, China, India, Indochina, Iraq, Nepal, Oman, Pakistan, Saudi Arabia, and United Arab Emirates (Figure 25.70); one of the most abundant species in the Persian Gulf region.

Description. Females have numerous white spots on both the legs and wings (Figure 25.71); hind tarsal segments have narrow white band apically (the last segment is dark). Females have white-tipped palps and one dark spot on the front margin of the wing proximal to the outermost spot; the abdomen is almost totally covered with scales.

Remarks. Breeds in artificial habitats such as wells, cisterns, gutters, pools, and many other peridomestic water sources; accordingly, malaria transmitted is usually localized and limited to urban centers. Adults bite people both indoors and outdoors and rest mainly indoors after feeding. Flight range usually does not exceed 1/2 mile.

H. *Culex tritaeniorhynchus* Giles

Medical Importance. Vector of JE virus in Far East; also most important nuisance mosquito during July and August in Japan.

Distribution. Saudi Arabia, Borneo, Celebes, China, India, Iran, Iraq, Japan, Okinawa, Israel, Lebanon, Syria, Turkey, Java, Sumatra, Philippines, Sri Lanka, Madagascar, Egypt, Gold Coast, Nigeria, Cameroon, Sudan, Zaire, Kenya, Tanzania, Angola, Namibia, Uganda, Zanzibar, Mozambique,

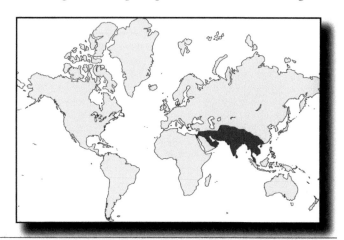

Figure 25.70 Approximate geographic distribution of *Anopheles stephensi*.

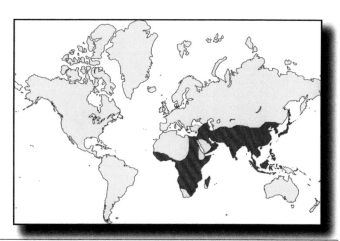

Figure 25.72 Approximate geographic distribution of *Culex tritaeniorhynchus*.

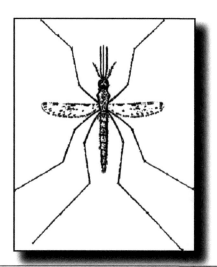

Figure 25.71 Adult *Anopheles stephensi*.

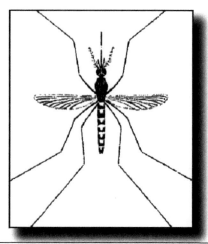

Figure 25.73 Adult *Culex tritaeniorhynchus*.

and Ivory Coast (Figure 25.72); especially abundant in Korea, Japan, and eastward through central China.

Description. Very small mosquito (average size is 3 mm, the Japanese forms may be larger); dark reddish-brown in color with a banded proboscis (Figure 25.73).

Remarks. Breeds in freshwater collections such as natural and artificial water impoundments, ground pools, and drainage and irrigation ditches; readily bites people and also feeds on horses, cows, pigs, and other animals.

I. *Culex annulirostris* Skuse

Medical Importance. Most medically important mosquito in Australia; main vector of the flaviviruses Murray Valley encephalitis and Kunjin (a subtype of West Nile virus) and the main inland vector of the alphaviruses Ross River and Barmah Forest.

Figure 25.76 Approximate geographic distribution of *Mansonia annulifera*.

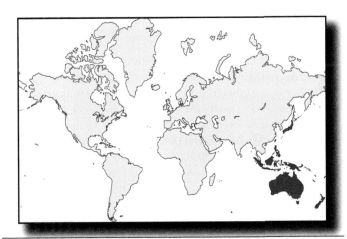

Figure 25.74 Approximate geographic distribution of *Culex annulirostris*.

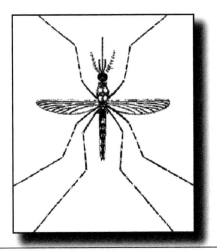

Figure 25.77 Adult *Mansonia annulifera*.

Distribution. Southern and western Australasian region (including the entire Australian mainland), Indonesia, Philippines (Figure 25.74).

Description. Medium-sized, brown mosquito; similar to the U.S. species, *Culex tarsalis*, tarsi banded, scutum dark, proboscis with a complete white ring in middle (Figure 25.75); may be difficult to distinguish from other, closely related Australian species.

Remarks. Active mid-spring to late fall; breeds in freshwater areas such as natural and artificial water impoundments, ground pools, and drainage and irrigation ditches. It

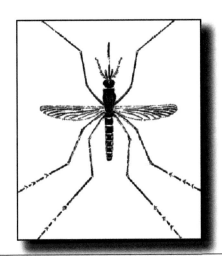

Figure 25.75 Adult *Culex annulirostris*.

does not breed in artificial containers. It readily bites people and also feeds on other mammals and birds.

J. *Mansonia annulifera* Theobald

Medical Importance. An important vector of *Wuchereria malayi* (lymphatic filariasis) in Asia, Philippines, and New Guinea; significant nuisance, being strongly anthropophilic.

Distribution. Borneo, Myanmar, Celebes, India, Indochina, Java, New Guinea, Philippines, Sumatra, and Thailand (Figure 25.76).

Description. Similar to other *Mansonia* mosquitoes (except lighter) with legs, palps, wings, and body covered with a mixture of brown and pale (white or creamy) scales, giving them a dusty appearance. This is a pale species, ranging in color from yellow to light brown; mottled wings have very broad and asymmetrical scales and banded legs (Figure 25.77). Very prominent silver–white scales are on mid-lobe of scutellum.

Remarks. Most commonly found in ponds, pools, backwaters, swamps, and marshes that contain the aquatic plants *Pistia* spp. and *Eichhornia* spp.; larvae obtain oxygen by puncturing the underwater stems of aquatic plants (similar to other members of genus).

References

1. Killeen GF, Masalu JP, Chinula D, et al. Control of malaria vector mosquitoes by insecticide-treated combinations of window screens and eave baffles. *Emerg. Infect. Dis.* 2017;23:782–789.
2. Anonymous. Artesunate for severe malaria. *Lancet.* 2011;377:1466.
3. Anonymous. Is malaria elimination within reach? *Lancet Infect. Dis.* 2017;17:461.
4. Baird JK. Telling the human story of Asia's invisible malaria burden. *Lancet.* 2017;389:781–782.
5. Greenwood B. Progress with the PfSPZ vaccine for malaria. *Lancet Infect. Dis.* 2017;17:463–464.
6. WHO. World malaria report. World Health Organization, http://www.who.int/malaria/publications/world_malaria_report_2014/en/; 2014.
7. Perry A. Quote from U.N. special envoy for malaria, in Epidemic on the Run. *Time Mag.* 2011;September 26 issue:47–49.
8. Lafferty KD. The ecology of climate change and infectious diseases. *Ecology.* 2009;90:888–900.
9. Shuman EK. Global climate change and infectious diseases. *N. Engl. J. Med.* 2010;362:1061–1063.
10. Collins WE, Barnwell JW. *Plasmodium knowlesi*: Finally being recognized. *J. Infect. Dis.* 2009;199(8):1107–1108.
11. Indra V. *Plasmodium knowlesi* in humans: A review of the role of its vectors in Malaysia. *Trop. Biomed.* 2010;27(1):1–12.
12. Kantele A, Jokiranta S. *Plasmodium knowlesi* – The fifth species causing human malaria. *Duodecim.* 2010;126(4):427–434.
13. CDC. Malaria surveillance – United States, 2008. *MMWR.* 2010;59(SS-7):1–16.
14. CDC. Summary of notifiable infectious diseases and conditions – United States, 2015. *MMWR.* 2017;64(53):1–144.
15. CDC. Multifocal autochthonous transmission of malaria – Florida, 2003. *MMWR.* 2004;53:412–413.
16. Robert LL, Santos-Ciminera PD, Andre RG, et al. Plasmodium-infected *Anopheles* mosquitoes collected in Virginia and Maryland following local transmission of *Plasmodium vivax* malaria in Loudoun County, Virginia. *J. Am. Mosq. Control. Assoc.* 2005;21(2):187–193.
17. Strickman D, Gaffigan T, Wirtz RA, et al. Mosquito collections following local transmission of *Plasmodium falciparum* malaria in Westmoreland County, Virginia. *J. Am. Mosquito. Contr. Assoc.* 2000;16:219–222.
18. Breman JG, Alilio MS, Mills A. Conquering the intolerable burden of malaria: what's new, what's needed: a summary. *Am. J. Trop. Med. Hyg.* 2004;71(2 Suppl):1–15.
19. Anonymous. What's next for the malaria RTS,S vaccine candidate? *Lancet.* 2015;386:1708.
20. Clemens J, Moorthy V. Implementation of RTS,S/AS01 malaria vaccine – the need for further evidence. *N Engl J Med.* 2016;374:2596–2597.
21. Barrett AD, Higgs S. Yellow fever: A disease that has yet to be conquered. *Ann. Rev. Entomol.* 2007;52:209–229.
22. Roberts L. Resurgence of yellow fever in Africa prompts a counterattack. *Science (News Focus).* 2007;316:1109.
23. WHO. Yellow fever. WHO/EPI/GEN/98.11, Geneva; 1998:6.
24. WHO. Yellow fever technical consensus meeting. WHO/EPI/GEN/98.08, Geneva; 1998:3.
25. Spira AM. Yellow fever: An update on risks, presentation, and prevention. *Infect. Med.* 2006;23:385–389.
26. Monath TP, Vasconcelos PFC. Yellow fever. *J. Clin. Virol.* 2015;64:160–173.
27. Barrett AD. The reemergence of yellow fever. *Science.* 2018;361:847–848.
28. CDC. Fatal yellow fever in travelers to Brazil, 2018. *MMWR.* 2018;67:1–3.
29. CDC. Dengue surveillance summary. *San Juan Laboratories Rep. (CDC).* 1991;62:1–3.
30. Calisher CH. Persistent emergence of dengue. *Emerg. Infect. Dis.* 2005;11:738–739.
31. Luz PM, Vanni T, Medlock J, Paltiel AD, Galvani AP. Dengue vector control strategies in an urban setting: An economic modelling assessment. *Lancet.* 2011;377(9778):1673–1680.
32. Effler PV, Pang L, Kitsutani P, et al. Dengue fever, Hawaii, 2001–2002. *Emerg. Infect. Dis.* 2005;11(5):742–749.
33. CDC. Locally acquired dengue – Key West, Florida, 2009–2010. *MMWR.* 2010;59:577–581.
34. Gubler DJ, Clark GG. Dengue/dengue hemorrhagic fever: The emergence of a global health problem. *Emerg. Infect. Dis.* 1995;1:55–57.
35. Bhatt S, Gething PW, Brady OJ, et al. The global distribution and burden of dengue. *Nature.* 2013;496(7446):504–507.

36. Guy B, Almond J, Lang J. Dengue vaccine prospects: A step forward. *Lancet.* 2011;377(9763):381–382.

37. Normile D. Hunt for dengue vaccine heats up as the disease burden grows. *Science (News Focus).* 2007;317:1494–1495.

38. Normile D. Dengue vaccine trial poses public health quandary. *Science (News Focus).* 2014;345:367.

39. Wilder-Smith A, Gubler DJ. Dengue vaccines at a crossroad. *Science (News Focus).* 2015;350:626–627.

40. Normile D. Safety concerns derail dengue vaccination program. *Science (News Focus).* 2017;358:1514–1515.

41. WHO. Global programme to eliminate lymphatic filariasis. *Wkly Epidemiol Rec.* 2010;85:365–372.

42. Roberts L. Relief among the rubble. *Science.* 2010;327.

43. Laurence BR. The global distribution of Bancroftian filariasis. *Parasitol Today.* 1989;5:260–264.

44. Beaver PC, Orihel TC. Human infection with filariae of animals in the United States. *Am J Trop. Med. Hyg.* 1965;14:1010–1014.

45. Kramer LH, Kartashev VV, Grandi G, et al. Human subcutaneous dirofilariasis. *Emer. Infect. Dis.* 2007;13:150–152.

46. Thomas JG, Sundman D, Greene JN, et al. A lung nodule: Malignancy or the dog heartworm? *Infect. Med.* 1998;15:105–106.

47. Harrison EG, Thompson JH, Schlotthauer JC, Zollman PE. Human and canine dirofilariasis in the United States. *Mayo. Clin. Proc.* 1965;40:906–916.

48. Bourguinat C, Keller K, Bhanb A, Peregrinec A, Gearya T, Prichard R. Macrocyclic lactone resistance in *Dirofilaria immitis. Vet Parasitol.* 2011;181:388–392.

49. Armstrong PM, Andreadis TG. Eastern equine encephalitis virus – Old enemy, new threat. *N. Eng. J. Med.* 2013;368:1670–1673.

50. Morris CD. Eastern equine encephalitis. In: Monath TP, ed., *The Arboviruses: Epidemiology and Ecology.* Boca Raton, FL: CRC Press; 1988:1–20.

51. Cupp EW, Klingler K, Hassan HK, Viguers LM, Unnasch TR. Transmission of eastern equine encephalomyelitis virus in central Alabama. *Am. J. Trop. Med. Hyg.* 2003;68(4):495–500.

52. Cupp EW, Tennessen KJ, Oldland WK, et al. Mosquito and arbovirus activity during 1997–2002 in a wetland in northeastern Mississippi. *J. Med. Entomol.* 2004;41:495–501.

53. Wozniak A, Dowda HE, Tolson MW, et al. Arbovirus surveillance in South Carolina, 1996–1998. *J. Am. Mosquito. Contr. Assoc.* 2001;17:73–78.

54. Molaei G, Farajollahi A, Armstrong PM, Oliver J, Howard JJ, Andreadis TG. Identification of bloodmeals in *Anopheles quadrimaculatus* and *Anopheles punctipennis* from eastern equine encephalitis virus foci in northeastern U.S.A. *MedVet Entomol.* 2009;23(4):350–356.

55. Chamberlain RW. History of St. Louis encephalitis. In: Monath TP, ed., *St. Louis Encephalitis.* Washington, DC: American Public Health Association; 1980: 3–61.

56. White GS, Symmes K, Sun P, et al. Reemergence of St. Louis encephalitis virus, California, 2015. *Emerg. Infect. Dis.* 2016;22(12):2185–2188.

57. CDC. Outbreak of West Nile-like viral encephalitis in New York. *MMWR.* 1999;48:845–848.

58. DeBiasi RL, Tyler KL. West Nile virus meningoencephalitis. *Nature Clin. Prac. Neurol.* 2006;2:264–275.

59. Mostashari F, Bunning ML, Kitsutani P. Epidemic West Nile encephalitis, New York, 1999: Results of a household-based seroepidemiological survey. *Lancet.* 2001;358:261–264.

60. Asnis DW, Conetta R, Teixeira A. The West Nile virus outbreak of 1999 in New York: The Flushing Hospital experience. *Clin. Infect. Dis.* 2000;30:413–417.

61. Sejvar JJ, Lindsey NP, Campbell GL. Primary causes of death in reported cases of fatal West nile fever, United States, 2002–2006. *Vector-borne Zoon Dis.* 2011;11:161–164.

62. Hayes EB, Komar N, Nasci RS, Montgomery SP, O'Leary DR, Campbell GL. Epidemiology and transmission dynamics of West Nile virus disease. *Emerg. Infect. Dis.* 2005;11:1167–1173.

63. Kilpatrick AM, Kramer LD, Campbell SR, Alleyne EO, Dobson AP, Daszak P. West Nile virus risk assessment and the bridge vector paradigm. *Emerg. Infect. Dis.* 2005;11(3):425–429.

64. Molaei G, Andreadis TG, Armstrong PM, Anderson JF, Vossbrinck CR. Host feeding patterns of *Culex* mosquitoes and West Nile virus transmission, northeastern United States. *Emerg. Infect. Dis.* 2006;12:468–474.

65. CDC. Summary of notifiable diseases – United States, 2006. *MMWR.* 2008;55:1–94.

66. Harwood RF, James MT. *Entomology in Human and Animal Health.* 7th ed. New York: Macmillan; 1979.

67. Anonymous. *Venezuelan Equine Encephalitis, A National Emergency.* Washington, DC: USDA, APHIS-81-1; 1972.

68. Weaver SC, Salas R, Rico-Hesse R, et al. Re-emergence of epidemic Venezuelan equine encephalomyelitis in South America. *Lancet.* 1996;348:436–439.

69. Burke DS, Leake CJ. Japanese encephalitis. In: Monath TP, ed., *The Arboviruses: Epidemiology and Ecology.* Boca Raton, FL: CRC Press; 1988:63–92.

70. CHPPM. *Japanese Encephalitis.* Aberdeen Proving Ground, Maryland: U.S. Army Center for Health Promotion and Preventive Medicine (USACHPPM), Facts Sheet Number 18-021-0305; 2005.

71. CDC. Japanese encephalitis surveillance and immunization – Asia and the Western Pacific. *MMWR.* 2013;62:658–662.

72. CDC. Japanese encephalitis among three U.S. travelers returning from Asia, 2003–2008. *MMWR.* 2009;58:737–740.

73. Rappleye WC. *Epidemic Encephalitis, Etiology, Epidemiology, Treatment.* New York: Third Report of the Matheson Commission, Columbia University Press; 1939.

74. Spira AM. Japanese encephalitis: A review of trends and preventive measures. *Infect. Med.* 2007;24:72–76.

75. Woods CW, Karpati AM, Grein T, et al. An outbreak of Rift Valley fever in northwestern Kenya. *Emerg. Infect. Dis.* 2002;8:138–142.

76. CDC. Rift Valley Fever outbreak – Kenya, November 2006–January 2007. *MMWR.* 2007;56:73–76.

77. Musso D, Rodriguez-Morales AJ, Levi JE, Cao-Lormeau V, Gubler DJ. Unexpected outbreaks of arboviral infections: Lessons learned from the Pacific and tropical America. *Lancet.* 2018;doi.org/10.1016/S1473-3099(18)30269-X.

78. Anonymous. *Ross River Fever*. Sydney, Australia: New South Wales Ministry of Health; 2002:1.

79. Enserink M. Tropical disease follows mosquitoes to Europe. *Science (News Focus)*. 2007;317:1485.

80. Weaver SC, Smith DW. Alphavirus infections. In: Guerrant RL, Walker DH, Weller PF, eds., *Tropical Infectious Diseases*. 3rd ed. London: Saunders (Elsevier); 2011:519–524.

81. Enserink M. Chikungunya: No longer a Third World disease. *Science (News Focus)*. 2007;318:1860–1861.

82. Leparc-Goffart I, Nougairede A, Cassadou S, Prat C, de Lamballerie X. Chikungunya in the Americas. *Lancet*. 2014;383:514.

83. Plourde AR, Bloch EM. A literature review of Zika virus. *Emerg. Infect. Dis.* 2016;22(7):1185–1192.

84. Petersen LR, Jamieson DJ, Honein MA. Zika virus. *N. Engl. J. Med.* 2016;375(3):294–295.

85. Gatherer D, Kohl A. Zika virus: A previously slow pandemic spreads rapidly through the Americas. *J. Gen. Virol.* 2016;97:269–273.

86. CDC. Cumulative Zika virus disease case counts in the United States, 2015–2017. CDC website, https://www.cdc.gov/zika/reporting/case-counts.html; 2017.

87. Bowen GS, Francy DB. Surveillance. In: Monath TP, ed., *St. Louis Encephalitis*. Washington, DC: American Public Health Association; 1980:473–499.

88. Peyton EL, Campbell SR, Candeletti TM, Romanowski M, Crans WJ. *Aedes (Finlaya) japonicus japonicus* (Theobald), a new introduction into the United States. *J. Am. Mosq. Contr. Assoc.* 1999;15:238–241.

89. Reinert JF. New classification for the composite genus *Aedes* (Diptera: Culcidae: Aedini), elevation of subgenus *Ochlerotatus* to generic rank, reclassification of the other subgenera, and notes on certain subgenera and species. *J. Am. Mosq. Contr. Assoc.* 2000;16:175–188.

90. Chapman HC, Johnson EB. The mosquitoes of Louisiana. Louisiana Mosquito Control Assoc. Tech. Bull. 1986;1.

91. Anonymous. St. Louis encephalitis outbreak in central Florida. *Florida Epigram*. 1991;1:1.

92. Sallum MAM, Peyton EL, Harrison BA, Wilkerson RC. Revision of the *Leucosphyrus* Group of *Anopheles* (*Celia*). *Rev Brasil Entomol*. 2005;32:23–178.

93. Sallum MAM, Peyton EL, Wilkerson RC. Six new species of the *Anopheles leucosphyrus* group, reinterpretation of *An. elegans*, and vector implications. *Med. Vet. Entomol.* 2005;19:158–199.

94. Garros C, Harbach RE, Manguin S. Morphological assessment and molecular phylogenetics of the *Funestus* and *Minimus* groups of *Anopheles* (*Cellia*). *J. Med. Entomol.* 2005;42(4):522–536.

95. Garros C, Harbach RE, Manguin S. Systematics and biogeographical implications of the phylogenetic relationships between members of the *funestus* and *minimus* groups of *Anopheles* (Diptera: Culicidae). *J. Med. Entomol.* 2005;42(1):7–18.

26

MOTHS (SPECIES WHOSE SCALES OR HAIRS CAUSE IRRITATION)

I. Introduction and Medical Importance

As discussed in Chapter 14, several species of Lepidoptera have larvae, commonly called urticating caterpillars, that possess stinging hairs or spines and can inflict a painful sting upon exposure to human skin (erucism).[1] In addition, some moth species, as adults, bear scales or hairs, which may detach, become airborne and cause urticaria and irritation in humans (see Chapter 5). Irritation from adult moths is called lepidopterism.[2] Several moth species in the families Notodontidae, Saturniidae, and Lymantriidae have been reported as causes of lepidopterism. Dermatitis may present as localized, widespread, or generalized erythematous macules that rapidly evolve into urticarial wheals.[3] Wheals are often, but not always, replaced by small infiltrated papules or papulovesicles. There was a recent outbreak of 40 cases of dermatitis, conjunctivitis, and pulmonary irritation in Europe due to the oak processionary caterpillar, Thaumetopoea processionea.[4] Lepidopterism is especially severe in South America during certain times of the year owing to moths in the genus Hylesia. Severe problems have occurred elsewhere. DeLong[5] reported an epidemic of 500,000 cases of caterpillar dermatitis in and around Shanghai, China, that was associated with an unexpected population explosion of yellow-tail moths in surrounding rural areas and appropriate climatic conditions conducive to the windblown spread of the disease.

In the United States, the tussock moths and their relatives in the family Lymantriidae may cause irritation. The female Douglas fir tussock moth, Orgyia pseudotsugata, covers her egg masses with froth and body hairs (see box). These hairs, along with other airborne hairs from the tips of the female abdomens, cause rashes, upper respiratory irritation, and eye irritation to forest workers in the Pacific Northwest.[6] A related species, the white-marked tussock moth, O. leucostigma, occurs throughout most of North America and may also be involved in cases of lepidopterism, especially in the East. In addition, airborne gypsy moth hairs, silken threads, and shed skins have been reported to cause a pruritic cutaneous reaction in sensitive individuals.

II. General Description of Some Species Involved

Moths in the genus Hylesia (Figure 26.1) are about 20 mm long and have a wingspan of about 40 mm. Alexander[7] stated that the northern species (Mexico and Central America) are a golden color, whereas those from Peru and Argentina have dark brown or black stripes on the body. The wings and body are covered with numerous scales and long hairs.

Tussock moths and their relatives are medium-sized moths (Figure 26.2) whose larvae are quite hairy. Female Douglas fir tussock moths are wingless. Males are a brownish-gray color with a wingspan of about 25 mm. The mature caterpillars are about 30 mm long and are gray to brown in color with a black head (Figure 26.3). They bear two long, dark tufts or pencils of hair, similar to horns, right behind the head; a similar but longer pencil is on the posterior. Four dense, buff-colored tussocks are located forward along the middle of the back.

The white-marked tussock moth is larger than the Douglas fir tussock moth. Females are wingless; males are gray and have lighter hind wings with a wingspan of about 30 mm. The mature caterpillars are about 35 mm long and are light brown with yellow and black stripes (Figure 26.4). In contrast to the Douglas fir tussock moth, Orgyia leucostigma caterpillars have a bright red head. Again, there are tufts of white hair on the dorsal side of the first four abdominal segments.

LEPIDOPTERISM

Douglas fir tussock moth, *Orgyia pseudotsugata*.

IMPORTANCE

Urticaria; irritation; respiratory problems from airborne moth hairs or scales.

DISTRIBUTION

Several species worldwide, especially those in genus *Hylesia* in South America.

LESION

Variable—often itchy papules or maculopapules.

DISEASE TRANSMISSION

None.

KEY REFERENCES

Delgado A. Venoms of Lepidoptera. In Bettini S, ed., *Arthropod Venoms*. Berlin: Springer-Verlag; 1978: 555

Rosen T. Caterpillar dermatitis. *Dermatol. Clin.* 1990;8:245

TREATMENT

Antihistamines, topical corticosteroids, or calamine products; occasional oral steroids may be indicated.

Figure 26.2   Tussock moths. (Photo courtesy Whitney Cranshaw, Colorado State University, Bugwood.org, used with permission)

Figure 26.3   Douglas fir tussock moth larva. (Photo courtesy Donald Owen and the California Department of Forestry and Fire Protection, Bugwood.org, used with permission)

The female gypsy moth is winged (in contrast to the tussock moths mentioned above) and has a wingspan of approximately 50 mm, with white and black markings (Figure 26.2C). The males are slightly smaller than females and are gray. Mature gypsy moth caterpillars, like other members of the Lymantriidae, have long tufts of hair along the body (Figure 26.2D). They are approximately 30–50 mm long and gray in color, and they have yellow stripes running lengthwise down the body. They may also appear to have red or blue spots along the sides and top of the body.

III. Geographic Distribution

Hylesia moths occur from Mexico to Argentina. The Douglas fir tussock moth is a serious pest in western North America. Some of the worst outbreaks have occurred in British

Figure 26.1   Hylesia moth (left) and white-marked tussock moth (right).

UGA0488035

Figure 26.4 White-marked tussock moth larva. (Photo courtesy John Ghent, Bugwood.org, used with permission)

Columbia, Idaho, Washington, Oregon, Nevada, California, Arizona, and New Mexico. The white-marked tussock moth occurs throughout most of North America. Gypsy moths occur in Europe and in the eastern United States.

IV. Biology and Behavior

Female *Hylesia* moths have barbed or spiny urticating setae that are called fleshettes. Fleshettes may break off and become airborne as large numbers of the moths emerge and gather around lights in towns or cities. *Hylesia* moths lay their eggs in clusters on the branches of trees, and the female coats the eggs with urticating spines. The life cycle is about 3 months in duration.

Tussock moths and their relatives are serious defoliators of forest, shade, and occasionally fruit trees. The Douglas fir tussock moth feeds primarily on fir trees. Usually, the first indication of attack appears in late spring. Larvae from newly hatched eggs feed on the foliage of the current year, causing it to turn brown. The small larvae are inconspicuous, but by mid-July, they are larger and more colorful. To feed in different areas of the tree, they lower themselves through the tree canopies by silken threads. By wind action, they often spread to nearby uninfested trees. By August, the larvae stop feeding and drop to the ground, crawling everywhere looking for a place to pupate. Female moths do not fly. Eggs are laid on tree trunks or branches, usually near the cocoon from which the female emerged.[8] Tussock moths overwinter in the egg stage.

Gypsy moths lay eggs on tree trunks during July and August in masses covered with froth and body hairs from the female. Eggs overwinter and the tiny first-stage larvae emerge the following spring, usually in late April or early May.[9] The females can fly, but only weakly. Dispersal of gypsy moth infestation is primarily by young larvae on silken threads traveling tree to tree or limb to limb by "ballooning" in the wind.

V. Treatment

Treatment mainly involves personal protection and avoidance of the offending species. During seasonal *Hylesia* (South America) moth emergence, it may help to turn off outdoor lights. In affected areas of the United States, air-conditioning and frequent changes of filters may help reduce airborne levels of scales or hairs. Sensitive persons should avoid walking in forests heavily infested with tussock moths. Acute urticarial lesions may respond satisfactorily to topical corticosteroid lotions and creams (desoximetasone gel has been frequently used), which reduce the intensity of the inflammatory reaction. Oral antihistamines may relieve itching and burning sensations. In more serious cases, oral steroids may be indicated. Rosen[3] reported that systemic administration of corticosteroids in the form of intramuscular triamcinolone acetonide has been remarkably effective in relieving severe itching due to gypsy moth dermatitis.

References

1. Delgado A. Venoms of Lepidoptera. In: Bettini S, ed., *Arthropod Venoms*. Berlin: Springer-Verlag; 1978:555–611.
2. Harwood RF, James MT. *Entomology in Human and Animal Health*. 7th ed. New York: Macmillan; 1979.
3. Rosen T. Caterpillar dermatitis. *Dermatol Clin*. 1990;8:245–252.
4. Gottschling S, Meyer S. An epidemic airborne disease caused by the oak processionary caterpillar. *Pediatr Dermatol*. 2006;23(1):64–66.
5. DeLong S. Mulberry tussock moth dermatitis. *J Epidemiol Comm Health*. 1981;35:1–4.
6. Perlman F, Press E, Googins GA, Malley A, Poarea H. Tussockosis: Reactions to Douglas fir tussock moth. *Ann Allergy*. 1976;36:302–306.
7. Alexander JO. *Arthropods and Human Skin*. Berlin: Springer-Verlag; 1984.
8. Borror DJ, Triplehorn CA, Johnson NF. *An Introduction to the Study of Insects*. 6th ed. Philadelphia: Saunders College Publishing; 1989.
9. Nichols JO. The gypsy moth in Pennsylvania – Its history and eradication. Pennsylvania Dept. Agri. Misc. Bull. No. 4404. 1962:13.

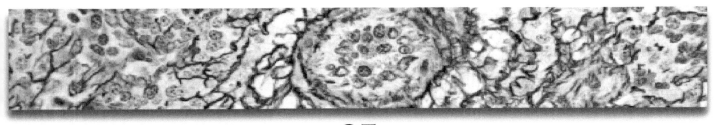

27

PENTASTOMES (TONGUE WORMS)

I. General and Medical Importance

Pentastomes (Phylum *Pentastomida*), sometimes classified as members of Arthropoda because of their chitinous exoskeleton, are wormlike parasites found mainly in the respiratory tracts of carnivorous mammals, reptiles, and birds.[1,2] There are approximately 130 species of pentastomes occurring worldwide. Pentastomiasis (human parasitism by pentastomes) may occur in the viscera, where nymphs develop in the peritoneal tissue, liver, spleen, lungs, and other organs,[3] or in the nasopharyngeal area, including the nasal passages, larynx, or eustachian tubes. Fortunately, human pentastomiasis is mostly asymptomatic and only found by radiography, surgery, or autopsy. Occasionally, human infestations may cause discomfort in the throat, paroxysmal coughing, and sneezing.

II. General Description of Some Species Involved

Pentastome worms vary in color and shape, depending on the species, but are generally colorless to yellow, ringed organisms with no apparent legs or body regions (Figure 27.1). The body may be cylindrical or flattened. Females may be up to 130 mm long and 10 mm wide; males are smaller, being up to 30 mm long. On either side of the mouth are two pairs of hollow, fanglike hooks which can be retracted into grooves like the claws of a cat.

III. Geographic Distribution

Two species of pentastomes account for 99% of the infections in humans.[4] The tongue worm, *Linguatula serrata* (Figure 27.2), which occurs worldwide, is often found in the nasal passages and frontal sinuses of several host animals,

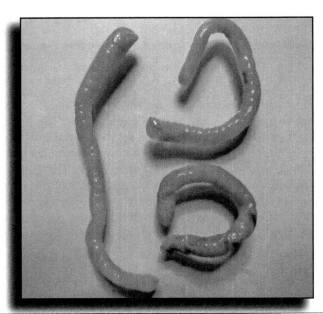

Figure 27.1 *Porocephalus* sp. (pentastomes) recovered from a rattlesnake in Kansas. (Photo courtesy Petra Jericke, Stuttgart, Germany, and Lawrence Bircham, Brandon, Mississippi, used with permission)

especially dogs. *Armillifer armillatus* (Figure 27.3) ordinarily inhabits the respiratory tract of certain snakes in central Africa but can be found in human visceral tissues.[3]

IV. Biology and Behavior

The life cycle of *Linguatula serrata* is provided here as a fairly typical example of pentastomid biology. Adults are often found in the nasal passages and frontal sinuses of canines and felines. Eggs produced pass out of their host in nasal discharges and are deposited in water or on vegetation. An intermediate host (e.g., rabbit or sheep) swallows contaminated water or vegetation, and the eggs hatch into primary

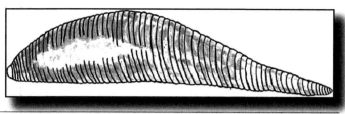

Figure 27.2   The tongue worm, *Linguatula serrata*.

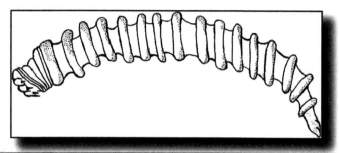

Figure 27.3   *Armillifer armillatus*, a common pentastomid found in snakes.

larvae, which penetrate the intestinal wall and become lodged in the liver, lungs, or mesenteric nodes. The larvae then pass through two molts resulting in a pupa-like stage. Up to seven more molts may occur before the nymphal stage develops (sometimes called an infective larva). The nymphs then migrate to the abdominal or pleural cavity of the inter-mediate host where they become encysted. When a defini-tive host (dog or cat) eats the intermediate host, the nymphs escape rapidly and migrate anteriorly, subsequently clinging to the lining within the host's mouth. From this point, they migrate to the nasal cavities, where they develop into adults. The adult stage may survive for up to 2 years.

V. Treatment

Treatment of pentostomiasis is usually not necessary, but in symptomatic cases, surgical removal of the parasites may be needed.[5]

References

1. Markell E, Voge M, John D. *Medical Parasitology.* 7th ed. Philadelphia: W.B. Saunders; 1992.
2. Lavrov DV, Brown WM, Boore JL. Phylogenetic position of the Pentastomida and (pan)crustacean relationships. *Proc Biol Sci.* 2004;271(1538):537–544.
3. Potters I, Desaive C, Van Den Broucke S, et al. Unexpected infection with *Armillifer* parasites. *Emerg Infect Dis.* 2017;23(12):2116–2118.
4. Binford CH, Conner DH. *Pathology of Tropical and Extraordinary Diseases.* Vol 2. Washington, DC: Armed Forces Institute of Pathology; 1976.
5. Hobmaier A, Hobmaier M. On the life cycle of *Linguatula rhinaria. Am J Trop Med* 1940;20:199–201.

28

SCORPIONS

I. General and Medical Importance

Scorpions are eight-legged arthropods that can inflict a painful sting. Over 1700 species occur worldwide on all major land masses except Antarctica.[1] Some species are more dangerous than others, depending on the type of venom. Size and appearance do not determine the medical importance of a scorpion; for example, scorpions in the genera Pandinus and Heterometrus (Old World) and Hadrurus (New World) may be huge with large pedipalps and appear menacing, but they constitute no serious health hazard. Most of these species can elicit local effects by their stings in a manner similar to that of bees and wasps; *Centruroides vittatus* is a common offender in the southwestern United States (see box). There is immediate sharp pain at the site of venom injection and often moderate local edema (which may be discolored). Dr. Scott Stockwell, formerly at the Walter Reed Biosystematics Unit, compared the sting to hitting one's thumb with a hammer. Regional lymph node enlargement, local itching, paresthesia, fever, and occasionally nausea and vomiting may also occur.[2] Signs and symptoms in a person stung by a scorpion with this type of venom usually subside in a few hours; however, it must be noted that a person with insect sting allergy could have a systemic reaction from this type of venom.

The other, more deadly, scorpion species are in the family Buthidae and have a venom that is more neurologic and hemolytic in activity. Systemic effects from such stings include drowsiness, abdominal cramps, blurred vision, spreading partial paralysis, muscle twitching, abnormal eye movements, profuse salivation, perspiration, priapism, hypertension, tachycardia, and convulsions.[3–5] Extreme restlessness, resembling a seizure, is a frequent presenting sign in children.[6] In fact, very small children may flail, writhe, and display roving eye movements. Death is probably due to respiratory paralysis, peripheral vascular failure, and/or myocarditis, and it

may occur at any interval between 1.5 and 42 hours after the sting.[2] Four cases of scorpion sting in the southwestern United States by our only dangerous species, *Centruroides sculpturatus*, caused severe clinical manifestations, including respiratory failure, metabolic acidosis, and severe multiorgan system disease.[7] Other than the sting of *C. sculpturatus* in Arizona and surrounding areas, scorpion stings in the United States are not usually life-threatening (barring allergic reaction), but scorpions constitute a significant health burden in other areas of the world, such as northern Africa and Central and South America. The number of severe envenomations in Mexico, for example, is estimated to be 250,000 per year,[8] in Morocco it is estimated to be 40,000 per year,[4] and in Brazil, it is approximately 6,000 per year.[9]

II. General Description

Scorpions are crablike in appearance, with pincers attached to their two front appendages (Figure 28.1). Their five-segmented tail terminates in a bulbous structure with a prominent curved stinger (Figure 28.2). Adult scorpions vary in size from 2 to 10 cm, depending on the species. Most American species are yellowish brown or brown in color (Figure 28.3). An identification key to the North American families and genera has been provided by Stockwell.[10] *Centruroides sculpturatus*, the dangerous U.S. species, is yellow to yellowish brown in color and is relatively small, usually 6 cm maximum (Figure 28.4B). It is a slender species with very narrow and elongate tail segments. Immatures have a diagnostic small spine or tubercle at the base of the stinger (Figure 28.5). A variant form of *C. sculpturatus* may also occur with two irregular black stripes down its back. At one time people thought this striped version of *C. sculpturatus* was another species named *C. gertschi.*

SCORPIONS

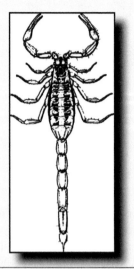

Centruroides vittatus, a common species.

IMPORTANCE

Painful sting; some species deadly.

DISTRIBUTION

Only one dangerous species in the United States; numerous species in other areas

LESION

Variable—sometimes local swelling and discoloration

DISEASE TRANSMISSION

None

KEY REFERENCES

Polis GA. *The Biology of Scorpions*. Stanford, CA: Stanford University Press; 1990

Stockmann R, Ythier E. *Scorpions of the World*. Verrieres-le-Buisson, France: N.A.P. Publishers; 2010

TREATMENT

Ice packs; anticonvulsants; vasodilators; sometimes antivenin serum for deadly species

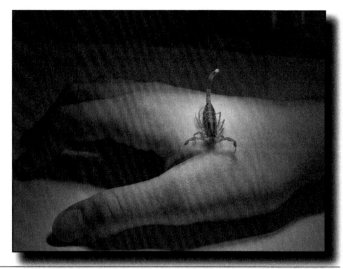

Figure 28.1   Striped scorpion on hand. (Photograph copyright 2008 by Jerome Goddard, Ph.D.)

Figure 28.2   Close-up of stinger of *Pandinus imperator*.

Figure 28.3   *Vejovis* spp. scorpions. (Photograph courtesy of Stoy Hedges and used with permission.)

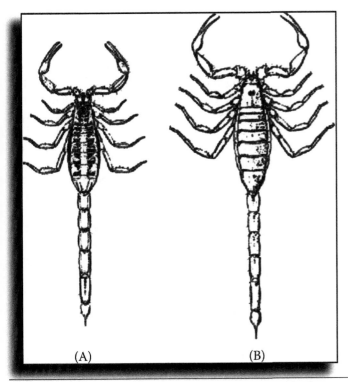

Figure 28.4 (A) *Centruroides vittatus*, a very common species in the southwestern United States, and (B) *Centruroides sculpturatus*, a dangerous U.S. species. (From Keegan HL. *Scorpions of Medical Importance*. Jackson, MS: University Press of Mississippi; 1980. With permission.)

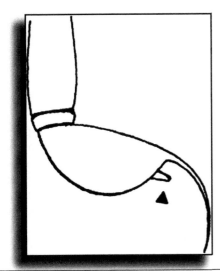

Figure 28.5 Side view of *Centruroides sculpturatus* stinger showing diagnostic small spine or tubercle. This structure is less obvious in adults.

III. Geographic Distribution

Table 28.1 lists some of the dangerous scorpions worldwide and their distributions. About 100 species occur in the United States, but only 4 of these naturally occur east of the Mississippi River. The most commonly encountered species in the southwestern United States is the common striped

Table 28.1 Geographic Distribution of Some Dangerously Venomous Scorpions

Region	Country	Species
Old World	Algeria	*Buthus occitanus*
		Androctonus australis
	Egypt	*Androctonus australis*
		A. amoreuxi
		Leiurus quinquestriatus
	Iraq	*Hemiscorpion lepturus*
		Androctonus crassicauda
	Israel	*Leiurus quinquestriatus*
		Androctonus crassicauda
		A. bicolor
	Jordan	*Buthus occitanus*
	Morocco	*Androctonus mauritanicus*
		A. australis
		A. amoreuxi
		Buthus occitanus
	South Africa	*Parabuthus triradulatus*
		P. transvaalensis
		P. villosus
	Sudan	*Buthus minax*
	Turkey	*Androctonus crassicauda*
		Leiurus quinquestriatus
		Mesobuthus gibbosus
New World	Argentina	*Tityus bahiensis*
	Brazil	*Tityus bahiensis*
		T. serrulatus
	Guyana	*Tityus cambridgei*
	Mexico	*Centruroides noxius*
		C. suffussus
		C. infamatus
		C. elegans
		C. limpidus
		C. exilicauda
	Trinidad	*Tityus trinitatis*
	United States	*Centruroides sculpturatus*[a]
	Venezuela	*Tityus trinitatis*
		Centruroides gracilis

Source: Adapted and updated from Keegan HL. *Scorpions of Medical Importance*. Jackson, MS: University of Mississippi Press; 1980, Chap. 3. With permission.

[a] Some authorities disagree on the taxonomic status of this species, calling it *C. exilicauda*, but consensus is that they are two distinct species.

scorpion, *Centruroides vittatus*, whose sting is as painful as a wasp or bee sting. The species is especially common in Texas and is responsible for numerous nonfatal stings each year. One study found more than 11,000 cases reported to Texas poison control centers over a five-year period.[11] The only dangerous scorpion (other than the allergic reaction) in the United States, *C. sculpturatus*, occurs primarily in Arizona (extending down into Sonora, Mexico) but also in southeastern California, Nevada, southern Utah, and southwestern

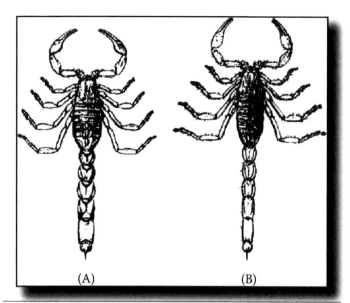

Figure 28.6 (A) *Androctonus australis*, and (B) *Buthus occitanus*. (From Keegan HL. Scorpions of Medical Importance. Jackson, MS: University Press of Mississippi; 1980. With permission.)

New Mexico. There has been controversy as to the taxonomic status of *C. sculpturatus*, with some authorities combining it with *C. exilicauda*,[12] although the consensus now is that they are two distinct species.[13]

Androctonus australis, the fat-tailed scorpion (Figure 28.6A), is a highly venomous, very aggressive scorpion found in North Africa in Egypt, Algeria, Tunisia, and Libya. Subspecies of *A. australis* may occur in the Middle East. The fat-tailed scorpion is responsible for 80% of all reported scorpion stings in Algeria; about one-third of them are fatal.[14] *Buthus occitanus* (Figure 28.6B) is a widely distributed, dangerous scorpion found in southern France, Spain, Italy, Greece, several Mediterranean islands, Israel, Jordan, and northern Africa. There are other *Buthus* species found in Europe as well, including *B. montanus*, *B. ibericus*, and *B. elongatus*.[15,16] The dangerously venomous scorpion, *Leiurus quinquestriatus*, is found in Turkey southward through Syria, Lebanon, Jordan, Israel, and down into northern Africa. In the Middle East, *Androctonus crassicauda* is responsible for many serious stinging incidents annually. *Tityus serrulatus* and *T. bahiensis* are probably the most dangerous species in Brazil. In South Africa, *Parabuthus triradulatus*, *P. transvaalensis*, and *P. villosus* are dangerous.

IV. Biology and Behavior

Female scorpions have pouches in which the young develop for a certain period, before being born alive; subsequently, they climb onto the mother's back and remain there until molting about 2 weeks later. After this molt, the young scorpions scatter to live solitary lives, molting six or seven more times before reaching maturity.

Scorpions are nocturnal feeders, feeding mostly on insects and spiders, although there are examples of the larger species being seen feeding on lizards, snakes, and other small vertebrates. They seize prey with their claws and paralyze it by their sting. During the day, scorpions remain concealed and are only encountered when people disturb them. Some of the most dangerous species are known as bark scorpions because they are often found under loose bark and in the crevices between the bases of palm tree leaves. They do not burrow but hide under boards, bricks, and other rubbish. Ground scorpions, on the other hand, burrow into loose sand or gravel and are capable of hibernating without food or water for up to 7 months.

Around homes, scorpions are usually found under the houses and in attics, which they enter by climbing between wall partitions. They will abandon attics as soon as the temperature gets too high, migrating downward into the walls and, if an opening is available, into the house itself. They favor attics with air-conditioning ducts that provide cool temperatures and a place to hide.

Because they will seek water, they are often found in bathrooms, kitchens, or laundry rooms at night. It is not uncommon to find them in bathtubs or sinks, which they entered seeking water and were unable to climb back out.

The amount of venom injected with a sting varies with the size of the scorpion and could even be none (a "dry sting"). Whittemore and Keegan[17] found that electrically stimulated specimens of a large species from the Middle East yielded an average of 0.48 mg of venom, whereas the average yield from a small, but highly dangerous species of *Centruroides* from Mexico was only 0.075 mg.

V. Treatment of Stings

Many species of scorpions are innocuous and produce stings followed by sharp pain or a burning sensation and a wheal, which usually disappears with no complications. For example, *Centruroides vittatus* stings in the southwestern U.S. are common, but rarely life-threatening. On the other hand, the venom of dangerous species is quite neurotoxic and can lead to a variety of systemic reactions, though rare (occurring in less than one-third of scorpion sting victims).[18] If this does happen, symptoms develop quickly, within the first few hours after a sting. A systemic reaction will commonly include leukocytosis, hyperglycemia, and lactic acidosis. Some patients will develop symptoms involving the sympathetic nervous system such as tachycardia, hypertension, hyperthermia, bladder dilatation, tachypnea, tremors, and convulsions. In contrast, other patients have been shown to have their parasympathetic nervous system involved, leading to symptoms that can include salivation, vomiting, diarrhea, miosis, bronchospasm, bradycardia, and hypotension. In particularly severe cases, patients will have manifestations of both these syndromes. After a sting by a dangerous species, it is important

Table 28.2 Clinical Grading System for Scorpion Stings (Adapted from Several Sources)

Clinical Grade	Approx. Proportion of Cases	Symptoms	Treatment	Remarks
1	67%	Local reaction only	No antivenom; aspirin, other analgesics, local anesthetics	Resolves rather quickly
2	23%	In addition to local reaction, pain and/or paresthesia distal from sting site; agitation and anxiety possible	Benzodiazepines	Cost-effectiveness analysis suggests antivenom not appropriate for this grade
3	*	Parasympathetic system dysfunction (i.e., bradycardia, hypotension, blurred vision, hypersalivation) OR sympathetic system dysfunction (i.e. tachycardia, hypertension, involuntary shaking)	Antivenom, benzodiazepine infusion, prazosin	Parasympathetic system dysfunction rarer than sympathetic system dysfunction
4	*	Both cranial nerve and somatic dysfunction, multiorgan failure, pulmonary edema, cardiogenic shock	Supportive care, benzodiazepine infusion, prazosin or dobutamine	

*The remaining 10% of cases are often lumped together into grades 3 and 4.

for patients to stay calm in order to minimize absorption of venom into the body. A pressure dressing over the sting site may be helpful, as well as application of ice packs. In adults, symptoms may be self-limited with resolution occurring in several hours. Children are at greatest risk of severe reactions.[6,19] In many countries, administration of a potent antivenom within a couple of hours of the sting may be needed for treatment of severe envenomation.[2,19] Antivenoms are commercially available for many (but not all) of the dangerous scorpions worldwide. Historically, there has been controversy concerning antivenom use in the United States, and there were no U.S. Food and Drug Administration (FDA)-approved products available. At one time, an antivenom for *Centruoides sculpturatus* envenomation was produced from goat serum at Arizona State University and was approved by the Arizona Board of Pharmacy for use within the state of Arizona (not FDA approved),[20] but there was a 58% rate of serum sickness in one study.[19] Newer products such as Anascorp® utilize purified fragments of IgG [F(ab')$_2$] which reduces negative side effects. One randomized, double-blind study involving 15 patients critically ill from scorpion stings and utilized this scorpion-specific F(ab')$_2$ antivenom showed significant resolution of the clinical syndrome within 4 hours, reduction of the need for sedation with midazolam, and reduction of levels of circulating unbound venom.[21] For this reason, Anascorp® was approved by the FDA.[8] However, it should be noted that while adverse reactions to Anascorp® are rare, hypersensitivity can occur and serum sickness may develop up to 3 weeks after treatment.[22]

Scorpion sting therapy other than the use of antivenom (see Table 28.2) may include mechanical ventilation to improve oxygenation and administration of CNS depressants such as midazolam (a benzodiazepine) and vasodilators such as selective α_1 blockers and calcium channel blockers.[4,20] Patients treated with midazolam should be carefully monitored for underventilation or apnea, which can lead to hypoxia or cardiac arrest.

References

1. Stockmann R, Ythier E. *Scorpions of the World.* Verrieres-le-Buisson, France: NAP Publishers; 2010.
2. Keegan HL. *Scorpions of Medical Importance.* Jackson, MS: University of Mississippi Press; 1980.
3. Bucherl W, Buckley E. *Venomous Animals and their Venoms.* Vol. 3. New York: Academic Press; 1971.
4. Ghalim N, El-Hafney B, Sebti F, et al. Scorpion envenomation and serotherapy in Morocco. *Am J Trop Med Hyg.* 2000;62:277–283.
5. Harwood RF, James MT. *Entomology in Human and Animal Health.* 7th ed. New York: Macmillan; 1979.
6. Polis GA. *The Biology of Scorpions.* Stanford: Stanford University Press; 1990.
7. Berg RA. Envenomation by the scorpion *Centruoides exilicauda* severe and unusual manifestations. *Pediatrics.* 1991;87:930–933.
8. Anonymous. FDA approves first scorpion antivenom. *Science (News Focus).* 2011;333:806.
9. Lourenco WR, Cloudsley-Thompson JL, Cuellar O, Von Eickstedt VRD, Barraviera B, Knox MB. The evolution of scorpionism in Brazil in recent years. *J. Venom. Anim. Tox.* 1996;2:32–42.
10. Stockwell S. Systematic observations on North American Scorpionida with a key and checklist of the families and genera. *J. Med. Entomol.* 1992;29:407–422.

11. Forrester MB, Stanley SK. Epidemiology of scorpion envenomations in Texas. *Vet. Hum. Tox.* 2004;46:219–221.

12. Williams S. Scorpions of Baja, California. *Occ Pap California Acad Sc.* 1980;135:1–67.

13. Valdez-Cruz NA, Davila S, Licea A, et al. Biochemical, genetic, and physiological characterization of venom components from two species of scorpions: *Centruroides exilicauda* and *Centruroides sculpturatus. Biochimie.* 2004;86:387–396.

14. Nichol J. *Bites and Stings – The World of Venomous Animals.* New York: Facts on File Inc.; 1989.

15. Lourenco WR, Vachon M. Considérations sur le genre *Buthus* Leach, 1815 en Espagne, et description de deux nouvelles espèces (Scorpiones, Buthidae). *Revista Ibérica de Aracnología.* 2004;9:81–94.

16. Rossi A. Notes on the distribution of the species of the genus *Buthus* (Leach, 1815) (Scorpiones, Buthidae) in Europe, with a description of a new species from Spain. *Bull. British Arachnol. Soc.* 2014;15:273–279.

17. Whittemore FW, Keegan HL. Medically important scorpions in the Pacific area. In: Keegan HL, MacFarlane WV, eds., *Venomous and Poisonous Animals and Noxious Plants of the Pacific Region.* Oxford: Pergamon Press; 1963:107–110.

18. Chippaux JP. Emerging options for the management of scorpion stings. *Drug. Des. Devel. Ther.* 2012;6:165–173.

19. Bond GR. Antivenin administration for *Centruroides* scorpion sting: risks and benefits. *Ann. Emerg. Med.* 1992;21:788–791.

20. Gilby R, Williams M, Walter FG, McNally J, Conroy C. Continuous intravenous midazolam infusion for *Centruroides exilicauda* envenomation. In: *North American Congress of Clinical Toxicology (Oral Presentation)*; Sept. 1998; Orlando, FL.

21. Boyer LV, Theodorou AA, Berg RA, et al. Antivenom for critically ill children with neurotoxicity from scorpion stings. *N. Engl. J. Med.* 2009;360(20):2090–2098.

22. Bope ET, Kellerman R. *Conn's Current Therapy.* Philadelphia: Elsevier Saunders; 2017.

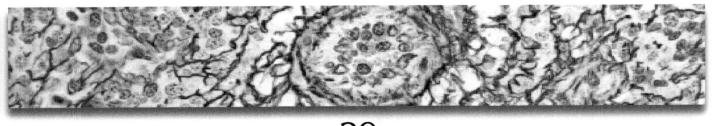

29

SPIDERS

I. Spiders in General

A. Medical Importance

There are almost 40,000 named species of spiders ranging in size (including legs) from a few millimeters to more than 17 cm (7 inches). Even some spiders that are not tarantulas are very large (Figures 29.1 and 29.2). All spiders, with the exception of the family Uloboridae, are venomous and use their venom to immobilize or kill prey; however, the chelicerae (fangs or pincerlike mouthparts) of many species are too short to penetrate human skin. The health burden of spider bites is significant. One study reported an average of 123,000 "spider bites" treated in emergency departments annually in the United States.[1] Of the species that can inflict human bites, the widow spiders (*Latrodectus* spp.), the violin spiders (*Loxosceles* spp.), and hobo spiders (*Tegenaria agrestis*) are the main causes of local or systemic effects from envenomization in the United States. In Australia, the Sydney funnel web spider (*Atrax robustus*) and related species can cause serious illness and death. In South America, the Brazilian Huntsman (*Phoneutria fera*) is a threat to humans, and Aranha Armadeira (*Phoneutria nigriventer*) has extremely potent venom.

It is important to note that at least 50 other species have been implicated in bites on humans. Bites from these species, although painful, generally are not considered as dangerous as the above-mentioned species. Sac spiders in the genus *Cheiracanthium* can sometimes pose health concerns.[2] In Europe, *Cheiracanthium punctorium* may cause painful burning and swelling at the bite site, while in North America *C. inclusum* and *C. mildei* bites may be associated with immediate local pain, erythema, and wheal.[3] The famous entomologist WC Reeves carefully documented a bite by *C. inclusum* wherein he described intense throbbing pain in the local area of the bite, subsiding completely within 12 hours.[4] As for other spider groups, Carpenter et al.[5] reported on a 15-year-old boy bitten by a *plectreurid* spider (*Plectreurys tristis*) in Kern County, California. Initially, the bite produced pain, edema, and slight pallor at the site of the wound, which persisted for 15–30 minutes; subsequently, the patient reported vague, diffuse numbness near the site. All these symptoms resolved within 2 hours. Because not all spider bites are life-threatening or lead to necrotic lesions, it is important for the patient to bring the offending spider into the clinic for identification. This can help physicians avoid the use of expensive treatments used for spider bites from dangerous species. In addition to the direct effects of spider venom, spider bites may produce secondary infections and allergic reactions (apparently rare in the United States).

B. Treatment of Bites

The treatment for most spider bites involves washing the bite site, applying cold compresses, avoiding exercise, treating with tetanus prophylaxis (if indicated), and taking analgesics for pain and antibiotics if secondary infection is suspected.[6] Additional measures are needed for bites by dangerous species, such as black widow, brown recluse, and funnel web spiders (see the text box).

II. Brown Recluse and Other Violin Spiders

A. General and Medical Importance

Violin spiders have venom that produces cutaneous lesions. The brown recluse spider (*Loxosceles reclusa*) is the most important of the violin spiders in the United States; however, several other species occur in the southwestern United States

Figure 29.1 Huge fish-eating spider collected in Mississippi—not a tarantula. (Photograph courtesy of Sheryl Hand and Dr. Sally Slavinski.)

Figure 29.2 Garden spider by hand. (Photograph copyright 2005 by Jerome Goddard, Ph.D., and courtesy of Gretchen Waggy.)

(*L. rufescens, L. deserta*, and *L. arizonica*) and the northeastern United States that can produce necrotic lesions. *L. laeta* is a notorious violin spider in South America.

Every year, the brown recluse spider (also called the fiddleback spider) is responsible for numerous reported cases of envenomization. Some of these "spider bites" are actually methicillin-resistant *Staphylococcus aureus* (MRSA) and difficult to distinguish from true brown recluse bites (see text box on facing page).[7] Brown recluse bites are usually localized and may produce considerable necrosis resulting in an unsightly scar.[8,9] Research on brown recluse venom in rabbits has indicated that it produces necrotic and hemolytic effects, but not neurologic (Figure 29.3). Brown recluse venom contains several fractions, but the primary dermonecrotic component appears to contain the phospholipase enzyme, sphingomyelinase D, which acts on tissues by activation of platelet aggregation, thrombosis, and massive neutrophil

OFTEN-ASKED QUESTION

HOW DO YOU KNOW IT REALLY IS A BROWN RECLUSE BITE?

More than 2000 brown recluse (BR) bites are reported to poison control centers each year, but if epidemiology and confirmed cases are any indication, most of them are due to something else.[1] Vetter[2] observed that the problem is that hundreds of BR bites are being diagnosed in areas where the spiders sparsely occur, or do not occur at all. I personally am aware of two correctional facilities in Mississippi where dozens of medically documented brown recluse spider bites have occurred, yet pest control professionals and Cooperative Extension Service entomologists have failed to find even one brown recluse. From the entomological perspective, it borders on the ridiculous to claim that dozens of people have been bitten by BR spiders when none can be found. They are not *that* reclusive.

Most spider experts agree that BR spiders are not aggressive. Bites typically occur defensively, only when the spider is accidentally trapped against human skin while a person is dressing or sleeping.[2] Often there are no biting incidents even when hundreds of the spiders are present in a dwelling. Vetter[3] reported finding over 2000 BR spiders in a home in Kansas occupied by a family of four with no bites recalled or reported by the family. Bite reactions from BR spiders are also probably exaggerated. We must keep things in perspective. Certainly, there is evidence that the venom may cause unsightly spots of necrosis on human skin; however, what is often not reported is that BR bite reactions can vary from no reaction, to a mild red wound, to a terrifying rotting flesh wound. In fact, Masters and King[4] reported that cutaneous necrosis usually does not develop after untreated BR bites. Even deaths reported from BR bites (from hemolytic anemia) may not be strongly supported by evidence.

Diagnosis of BR bites is based on clinical presentation and history (unless of course, the patient brings in the offending specimen). Even though at least one researcher has developed a diagnostic immunoassay,[5] there is currently no widely available laboratory test to confirm that a patient has been bitten by a BR. A very good mnemonic device to aid in brown recluse spider diagnosis has been published called NOT RECLUSE.[6] Each letter corresponds to features of "bites" that are not consistent with brown recluse envenomation, such as facts about the number, size, shape, and timing of the lesions.

The BR spider bite lesion is variable, but generally, it is a dry blue-gray or blue-white irregular sinking patch with ragged edges surrounded by erythema—the *red*, *white, and blue sign."*[1,4] The lesion often is asymmetric and gravity dependent due to the downward flow of venom through tissues. A perfectly symmetrical necrotic lesion is often not due to BR spider and may be a staph infection. Necrosis and sloughing of tissues may follow over a period of days or weeks, leading to an unsightly sunken scar. Many conditions may lead to spots of necrosis on human skin resembling BR spider bites and should be considered by physicians before assigning the diagnosis of BR spider bite. The most common cause of necrotic wounds misdiagnosed as BR spider bites are infections with *Staphylococcus* and *Streptococcus* species.

REFERENCES

1. Sandlin N. Convenient culprit: myths surround the brown recluse spider. *Amednews.com.* 2002; August 5, 2002. http://www.ama-assn.org/amednews/2002/08/05/hlsa0805.htm.
2. Vetter RS, Bush SP. The diagnosis of brown recluse spider bite is overused for dermonecrotic wounds of uncertain etiology. *Ann. Emerg. Med.* 2002;39:544.
3. Vetter RS, Barger DK. An infestation of 2055 brown recluse spiders and no envenomations in a Kansas home: implications for bite diagnoses in nonendemic areas. *J. Med. Entomol.* 2002;39:948
4. Masters EJ, King, Jr., LE. Differentiating loxoscelism from Lyme disease. *Emerg. Med.* 1994;36:47
5. Gomez HF, Krywko DM, Stoecker WV. A new assay for the detection of *Loxosceles* species spider venom. *Ann. Emerg. Med.* 2002;39:469
6. Stoecker WV, Vetter RS, Dyer JA. NOT RECLUSE – a mnemonic device to avoid false diagnoses of brown recluse spider bites. *JAMA Dermatol.* 2017;153:377–378.

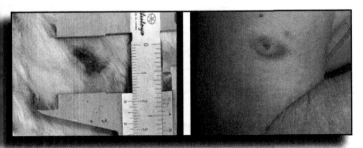

Figure 29.3 Brown recluse venom-induced lesion in a research animal (left) and on a human arm (right). (Photograph on left courtesy of Dr. Lane Foil, Louisiana State University; photograph on right courtesy of Jerome Goddard, Ph.D.)

infiltration.[6,10] The bite is usually painless until 3–8 hours later, when it may become red, swollen, and tender. Note: the central area of a brown recluse bite is generally pale or blue, not red. By 24 hours intense pain may develop. A study of 23 brown recluse spider bites revealed that 83% of the patients reported moderate to severe pain.[11] Clinical signs of brown recluse bites observed when patients are first seen in a clinical setting include erythema, cellulitis, rash, and blister.[12] Later, a black scab may develop, and eventually, an area around the site may decay and slough away, producing an ulcer from 1 to 25 cm in diameter (Figure 29.4). Finally, the edges of the wound thicken and become raised, whereas the central area is filled by scar tissue. Healing may require 1–3 months, and the end result may be a sunken scar. A helpful distinguishing sign of a brown recluse bite is the tendency of the lesion to extend downward in a gravitationally dependent manner (Figure 29.4). The direction of the lesion is dependent on the victim's position during the time the vessel damage was produced. Asymmetrical lesions (especially the "flowing-downhill" type) are rare from other arthropod bites. A very good mnemonic device to aid in brown recluse spider diagnosis has been published called NOT RECLUSE.[13] Each letter corresponds to features of "bites" that are not consistent with brown recluse envenomation, such as facts about the number, size, shape, and timing of the lesions.

Systemic reactions—called systemic loxoscelism—characterized by hematuria, anemia, fever, rash, nausea, vomiting, coma, and cyanosis have sometimes been reported from brown recluse bites. Systemic signs and symptoms are rare and, if they occur, usually begin 2–3 days following envenomization in contrast to the local cutaneous lesion that is evident by 12–24 hours.[14] Hemolysis, always heralded by rash and fever, has been reported occurring as early as 24 hours and as late as 3 days after the bite. It usually continues for 72 hours. There was a recent report of shock in a 13-year-old girl that began 9 hours after being bitten by a violin spider.[15]

B. General Description

The adult brown recluse is a medium-sized spider with a 20- to 40-mm leg span and a color range from tan to dark brown (see box and Figure 29.5). The most distinguishing characteristics are six eyes arranged in a semicircle of three pairs on top of the head and a violin-shaped marking (with base forward) extending from the area of the eyes to the beginning of the abdomen. Sometimes, however, the violin marking is not present in juveniles. Complicating matters, a violin-shaped marking is also found in common, harmless spiders such as the southern black house spider, *Kukulcania hibernalis.*

C. Geographic Distribution

Although its distribution may be more widespread, the brown recluse spider is most commonly found in the southcentral United States from about Oklahoma to Georgia and from Iowa down to the Mexican border (Figure 29.6). Related species, *Loxosceles deserta, L. rufescens, L. arizonica,* and *L. devia,* which mostly occur in the southwestern United States, may also cause necrotic arachnidism (see Figure 29.6). The Mediterranean recluse, *L. rufescens,* occurs from the southern

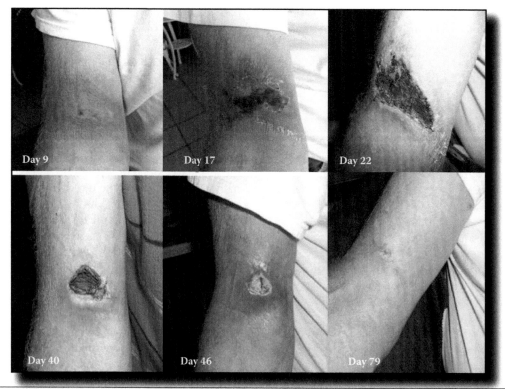

Figure 29.4   Brown recluse spider bite lesion development from 9 days post-bite to 79 days post-bite. (Photographs courtesy of Eugene Skiles and used with permission.)

Figure 29.5   Brown recluse spider. (Photograph courtesy of Dr. Barry Engber, North Carolina Department of Environment and Natural Resources, and used with permission.)

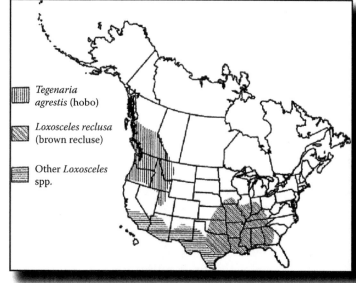

Tegenaria agrestis (hobo)

Loxosceles reclusa (brown recluse)

Other *Loxosceles* spp.

Figure 29.6   Approximate geographic distributions of spiders causing necrotic arachnidism in the United States and Canada. (From CDC, *MMWR* 1996;45:1.)

Mediterranean area and Middle East to at least as far north as the United Kingdom and western Russia. Interestingly, this spider has been found inhabiting federal government buildings in Washington, DC. In South America, there is a close relative of the brown recluse, *L. laeta*, that causes a severe necrotic lesion sometimes referred to as gangrenous spot. *L. laeta* has been introduced into the United States and is established in Los Angeles County, California, and possibly at Harvard University, Cambridge, Massachusetts.

D. Biology and Behavior

The brown recluse is well adapted to human dwellings, but in nature can be found underneath rocks, logs, and loose bark of dead trees.[16] They are nocturnal in feeding habits and most frequently are found inside houses in bathrooms, bedrooms, cellars, attics, the folds of clothing, cardboard boxes, and storage areas. Homes that are infested may easily contain hundreds of specimens. Brown recluse webs are sparse and irregular, usually constructed low to the ground outdoors. Indoors, they build their webs in out-of-the-way areas that are rarely disturbed.[17] The spiders are not aggressive but will bite if pressure is placed on them. At night, they may crawl into garments and bite people attempting to dress the next morning. Common sites are under the arms or on the lower parts of the abdomen. Brown recluse spiders are very resilient, being active in temperatures ranging from 40 to 110°F, and they may live as long as 2 years.

E. Treatment of Bites

The bite of a brown recluse spider is very shallow, so many times there is no immediate perception of the biting incident. If possible, persons bitten by a spider should make an effort to collect the spider and bring it into a clinic for identification. Positive identification of the spider by an expert can be helpful because bites of most spiders are temporarily painful but not dangerous. In the case of the brown recluse, prompt medical treatment may be needed to prevent more serious reactions.

A specific antidote (antivenin) has shown success in patients prior to the development of the necrotic lesion,[18] but currently, it is not widely available. Also, some brown recluse bites are unremarkable, not leading to necrosis; therefore, treatment is quite controversial. Some studies have indicated that application of ice is effective (the necrotic enzyme sphingomyelinase D increases in activity as temperature increases, so keep the temperature low.) There have been reports of successful treatment by early, total excision of the bite site followed by split-thickness skin grafting; however, more recent evidence argues against wound excision.[12] Systemic corticosteroids have been associated with slower healing of brown recluse bites.[19] Nitroglycerin apparently does not help.[20] Rees et al.[18] reported good success with the leukocyte inhibitor dapsone, which eliminated the need for surgery in many brown recluse bites. Others have also advocated dapsone.[21–23] Theoretical evidence supports its use—histological examination of bites in animals shows evidence of extensive neutrophil infiltration at the site of ultimate necrosis. Dapsone is contraindicated in persons with glucose-6-phosphate dehydrogenase (G6PD) deficiency due to potential massive hemolysis. It may produce side effects (even in non-G6PD patients) that could be confused with the systemic effects of brown recluse spider bite: malaise, nausea, and hemolysis. For this reason, before dapsone is administered in patients with spider bites, a baseline assessment of G6PD, a complete blood count, and a test of liver enzymes should be performed and be repeated weekly while the patient is receiving the drug.[24] On the other hand, there is some evidence that dapsone is completely ineffective. Randomized, blinded, controlled studies of venom effects in rabbits failed to show any benefit from the use of the drug.[25,26] At least one letter published in the

VIOLIN SPIDERS

Brown recluse spider. (Photograph courtesy of U.S. Department of Agriculture.)

IMPORTANCE

Several closely related species cause necrotic lesions; possibly hemolytic anemia

DISTRIBUTION

United States, South America, parts of Africa and Australia

LESION

Variable through time—papule with erythema, a large central zone of pallor, mottling, then blackening of the central zone, ulceration

DISEASE TRANSMISSION

None

KEY REFERENCES

Swanson D. and Vetter R. Bites of brown recluse spiders and suspected necrotic arachnidism. *New Engl. J. Med.* 2005;35:700

Diaz JH. The global epidemiology, syndromic classification, management, and prevention of spider bites. *Am. J. Trop. Med. Hyg.* 2004;71:239

TREATMENT

Controversial, no consensus—may include ice packs, antibiotics, corticosteroids.

New England Journal of Medicine, however, has cited clinical experience indicating that dapsone is the most suitable treatment and leads to a satisfactory resolution.[27] For systemic reactions to brown recluse spider bites, steroids may minimize hemolysis (especially in children) and may protect renal function.[28]

III. Black Widow and Other Widow Spiders

A. General and Medical Importance

There are more than 30 species of widow spiders occurring worldwide, but *Latrodectus mactans* is the one most generally associated with the name black widow spider. It is also sometimes referred to as the hourglass, shoe button, or pokomoo spider. Widow spiders in Europe are generally known as button spiders. The black widow spider injects a potent neurotoxin upon biting. One of the most famous and detailed accounts of bite effects from the black widow came from Baerg,[29] who allowed a specimen to bite him on the hand and then carefully documented the signs and symptoms over the course of several days. Initially, the bite itself produces a mild burning or stinging pain, but more than half of the patients in one study did not know they had been bitten.[30] Puncture marks may occur but are actually uncommon due to the spider's small fang size.[31] The venom exerts a specific effect on the central nervous system, causing depletion of acetylcholine at motor nerve endings and provoking the release of catecholamines at adrenergic nerve endings. Unlike the skin necrosis or systemic hemolysis associated with bites from the brown recluse spider, black widows produce pain in the regional lymph nodes (usually in the axilla or inguinal area), piloerection, increased blood pressure and white blood cell count, and profuse sweating and nausea. The bite site may appear as a bluish-red spot with a white areola and sometimes an urticarial rash. Significant envenomization may be accompanied by systemic symptoms of weakness, tremor, severe myalgia, muscular spasm, a rigid boardlike abdomen, and tightness in the chest. Bites on the torso are more likely to cause muscle cramps in the abdomen. Paralysis, stupor, and convulsions may occur in severe cases, and rarely death. The painful, rigid abdomen may be mistaken for appendicitis. Latrodectism can also be misdiagnosed as an alimentary toxic infection, acute psychosis, tabic crisis, pneumonia, tetanus, meningitis, acute renal failure, and various exanthematic diseases.[30] Black widow envenomation remains a significant public health issue in the U.S., with over 2500 calls annually to poison control centers.[32] Wingo[33] reported that the mortality rate from black widow bites is less than 1%, while Alexander[34] said it is 4–6%. Other, related widow spiders are perhaps more dangerous. The redback spider, *Latrodectus mactans hasselti*, ubiquitous in Australia, can occasionally cause death in humans, and its bite is the most common envenomization requiring antivenom in Australia, where at least 250 cases per year receive antivenom.[35] At least 17 deaths had been reported from bites by this species prior to the development of antivenom.[36]

Males and immature spiders also bite but usually produce milder symptoms. Some references indicate that the males are harmless owing to the lesser amount of venom injected because of their smaller size and proportionately shorter chelicerae.

WIDOW SPIDERS

Female black widow spider, *Latrodectus mactans*.

IMPORTANCE

Neurotoxic venom

DISTRIBUTION

Several species almost worldwide.

LESION

Minimal—two puncta

DISEASE TRANSMISSION

None

KEY REFERENCE

Diaz JH. The global epidemiology, syndromic classification, management, and prevention of spider bites. *Am. J. Trop. Med. Hyg.* 2004;71:239

TREATMENT

Ice packs; analgesics; muscle relaxants; sometimes antivenin; calcium gluconate.

B. General Description

Mature female black widow spiders are black with a leg span of 30–40 mm (see box). On the underside of the abdomen is a characteristic red or orange hourglass-shaped marking. There is considerable variation among species and even subspecies (see Figure 29.7). The adult brown widow varies in color from light tan to dark brown with banded legs (Figure 29.8). The hourglass marking on brown widows is sometimes light orange or even yellow. The red widow (central and southeastern Florida) has red legs and a red cephalothorax. Their hourglass is a single flattened triangle. Male black widow spiders are considerably smaller than females, averaging 16–20 mm in leg span, and have red and white marks on the dorsal side

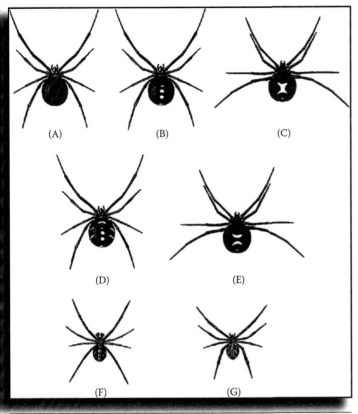

Figure 29.9 Black widow male. (Photograph copyright 2011 by Jerome Goddard, Ph.D.)

Figure 29.7 Widow spiders. Dorsal view of *Latrodectus mactans* adult female: (A) normal and (B) variant pattern. (C) Ventral view, (D) dorsal view, and (E) ventral view of *L. variolus*. (F) Dorsal view of *L. mactans* male. (G) *L. mactans* immature female.

of the abdomen (Figure 29.9). Immature widow spiders (spiderlings) are tan to grayish in color with very little or no black. They have orange and white markings on their abdomens that resemble "racing stripes."

C. Geographic Distribution

Five species of widow spiders occur in America, north of Mexico. Three (*Latrodectus mactans*, *L. variolus*, and *L. hesperus*) are similar in appearance. The black widow spider, *L. mactans*, is widely distributed in the United States, occurring from southern New England to Florida and west to California and Oregon; it is more common in the southern part of the range. The northern black widow spider, *L. variolus*, occurs in the New England area and adjacent Canada, south to Florida, and west to eastern Texas, Oklahoma, and Kansas; it is more common in the northern part of this range. The western black widow, *L. hesperus*, is found in western Texas, Oklahoma, and Kansas, north to the adjacent Canadian provinces, and west to the Pacific Coast states. The brown widow spider, *L. geometricus*, increasingly has been reported from various parts of the southern United States and southern California.[37] The red widow spider, *L. bishopi*, is limited to central and southeastern Florida.

Dangerous widow spiders in other parts of the world include *Latrodectus geometricus* (already mentioned occurring in the southern United States), occurring worldwide in the tropics and subtropics; the black widow subspecies *L. mactans tredecimquttatus*, in southern Europe; *L. mactans hasselti*, a problem in Australia and India; *L. mactans cinctus*, a dangerous widow spider in South Africa; and *L. mactans menavodi*, one of medical significance on the island of

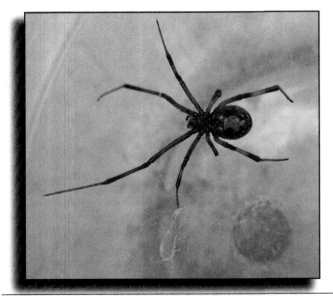

Figure 29.8 Brown widow, showing banded legs and egg sac with spicules or spines. (Photograph copyright 2008 by Jerome Goddard, Ph.D.)

CASE HISTORY

BROWN WIDOW SPIDER BITE

The brown widow, *Latrodectus geometricus*, has received a lot of attention over the past decade due to its rapid increase in numbers and geographic distribution across the entire southern United States.[1] Brown widows are often said to be less venomous than black widows (myth), although this may relate to their general lack of aggressiveness (not a myth). Usually, when disturbed, this species will draw up its legs and fall to the ground. This behavior rarely leads to envenomation. However, there are reports of bona fide bites from the brown widow from the southeastern United States,[2] and bites can occur in other places within the geographic range of the spider as well. In 2010, a woman living in Hawaii, where brown widows are common, was bitten by one of the spiders on her face near the ear. Within 2 hours, she was shaking and had a severe headache, body aches, and nausea. She said the bite hurt immediately and felt like a yellowjacket sting. She was taken to the hospital where her blood pressure was found to be 151/92 (high for her). Note that high blood pressure was a prominent symptom in a previously published bite case.[2] The woman was treated symptomatically and released in a few hours. By the next day, her symptoms had mostly resolved. Brown widows may seem reluctant to bite people, but when they do there can be intense pain, cramps, nausea/vomiting, and other discomfort for 24–48 hours.

REFERENCES

1. Brown KS, Necaise JS, Goddard J. Additions to the known distribution of the brown widow spider, *Latrodectus geometricus*. *J. Med. Entomol.* 2008;45:959–962
2. Goddard J, Upshaw S, Held D, Johnnson K. Severe reaction from envenomation by the brown widow spider, *Latrodectus geometricus*. *S. Med. J.* 2008;101:1269–1270

EFFECTS OF WIDOW SPIDER VENOM IN A HUMAN

In the early 1900s, Dr. EH Coleman, Los Altos, California, conducted a series of experiments on himself using black widow venom extracts (this account is described two places[1,2]). First, he dissected venom glands out of the spiders, made triturations of the glands and their contents, and then prepared a powdered form of the substance. He then divided the powder into "doses," took some of them, and made careful notes on the ensuing clinical effects. After taking 25 powders, his heart rate was reduced to 48 and his temperature was 99. He experienced a severe headache, clonic spasms of the thoracic and abdominal muscles, and "marked distress about the heart" with radiating pains extending to the left armpit and down to the elbow. He repeated the experiment twice with the same results.

Editorial note: Although the purity of Dr. Coleman's extracts cannot be ascertained, the clinical signs and symptoms he reported are (mostly) consistent with reports of black widow bite reactions today: intensely painful abdominal and leg cramps, chest pains mimicking heart attack, muscle weakness, stiffness, tremors, and loss of coordination. In addition, regional lymph nodes may become painful, especially axillary or inguinal lymph nodes.

REFERENCES

1. Kellogg VL. Spider poison. *J. Parasitol.* 1915;1:107–112
2. Pierce WD. *Sanitary Entomology: The Entomology of Disease, Hygiene, and Sanitation.* Boston, MA: Richard G Badger (Gorham Press); 1921

Madagascar. The redback spider, *L. mactans hasselti*, is now established in Japan (probably imported from Australia).[38]

D. Biology and Behavior

Widow spiders are found in various habitats in the wild: in protected places such as crawl spaces under buildings; in water meter housings; in holes in dirt embankments; in piles of rocks, boards, bricks, or firewood; and in shrubs or dense plant growth such as grain and cotton fields. Inside buildings, they avoid strong light and favor dark corners behind or underneath appliances or furniture that is seldom moved, garage corners, deep closets, damp cabinets, etc. With the advent of indoor plumbing, the most frequent site of encounter in previous decades, the outdoor privy, has become less common. Widow spiders will establish a web wherever conditions seem favorable, and if there is a suitable food source they may establish a long-term infestation.

Widow spiders are active primarily during the warm months of the year. Few adults survive cold winter weather, except indoors in heated places, but immature spiders may overwinter in large numbers. Mating takes place in the spring. Contrary to popular belief, which gave rise to the name black widow spider, the male is seldom killed and eaten by the female unless confined. If the female is well fed, the male usually leaves after mating and may mate again with other females. If the female is not well fed, she may consume the male for nutrients to produce the fertilized eggs, thus

CASE HISTORY

NOT ALL BLACK WIDOW SPIDERS LOOK THE SAME

An out-of-town emergency room (ER) physician called one night about a possible black widow (BW) spider bite in a young boy. The family had killed the spider and brought it in with the boy; the physician had the specimen in a vial and described it over the phone. The site of the bite was unremarkable, except for two small puncta. At the time of the call, the boy was not exhibiting any significant pain, nausea, sweating, or other symptoms usually associated with BW bites. The physician asked for identification advice and possible suggestions on what course of action to take.

Based on the description, the spider seemed to be an aberrant BW, a male BW, or possibly a closely related species, *Latrodectus variolus*. What seemed to be confusing the physician was the presence of several bright red hash marks on the dorsal side of the spider's abdomen (not present on a "normal" BW) in addition to the red hourglass-shaped marking on its underside. The physician was informed that in all likelihood the specimen was a BW and the case should be treated as such (based on the development of symptoms in that particular patient).

Comment: There are at least five species of widow spider in the United States. Physicians should be aware that not all widow spiders fit the typical black widow appearance (see figure). Some specimens have red markings on the dorsal side of the body (in addition to the red hourglass-shaped mark on the underside of the abdomen). In addition, in some species, the hourglass marking is "broken" and not complete.

Typical female black widow underside showing red hourglass marking (left) and related widow spider showing bright red markings on the dorsal side (right). (Photograph on left courtesy of Dr. James Jarratt, Mississippi Extension Service; photograph on right courtesy of Dr. Jerome Goddard II.)

Source: Adapted in part from Goddard, J. and Currier, M., Case histories of insect- or arachnid-caused illness, *J. Agromed.*, 2, 53–61, 1995. With permission.

contributing to the success of the species. The female lays 250–750 eggs at a time and spins a spherical sac of strong silk around them. There are usually three to nine egg sacs produced each summer.

The young hatch in 2–4 weeks. After emerging from the egg sac, they spin a strand of silk that enables them to be carried by the wind, dispersing them into the surrounding area. Male black widow spiders molt 4–7 times and females molt 7–9 times during a period of several weeks to several months, depending on the temperature, food availability, and other factors. More than one brood can be produced in a summer, and as many as 2000 offspring may be produced during a year. If they survive the winter, females may produce egg sacs the following year. Thus, they may live up to 3 years.

Widow spiders typically will be seen hanging upside down in their web, waiting for prey that includes many types of arthropods including insects and other spiders. The prey's contact with the web and subsequent struggle to free itself produces vibrations that the spider follows to locate the prey. The spider then seizes and bites the prey, injecting venom to paralyze and kill it so it can be consumed. Venom is produced in two glands located in the basal segment of the chelicerae and is injected through a tiny hole in each of the two fangs.

Widow spiders are reclusive; they usually will not bite humans unless provoked. When their web is disturbed, they will attempt to escape, but if teased or pursued they will bite in self-defense.

E. Treatment of Bites

A very good differential diagnosis table for widow spider bites has been published.[39] First aid for *Latrodectus* envenomization consists of cleansing the wound thoroughly and applying ice packs to slow absorption of the venom. Incision of the wound and suction to remove venom should not be attempted. Cramps may be helped with muscle relaxants. Older treatments for cramps such as calcium therapy have mostly been abandoned.[40] Pain may be relieved by administration of oral, or occasionally i.v., opioid analgesics such as morphine or fentanyl.[28] Tetanus vaccine status should be assessed. Physicians should monitor for shock and treat as necessary. Antivenin is available but rarely indicated.[41] It is sometimes used in severe cases[40] (especially Australian cases involving the redback spider) but is unnecessarily risky if given routinely for mild reactions. In fact, there have been at least two deaths from anaphylaxis resulting from administration of black widow antivenin.[42,43] Russell et al.[44] recommended the muscle relaxant methocarbamol for uncomplicated adult cases. Children, the very elderly, and persons with hypertensive disease or other medical conditions of risk need special attention, including hospitalization and possible antivenin. If antivenin is given, the content of one 2.5 ml vial (be sure to check the package insert for instructions) is usually administered by intravenous infusion after tests for serum sensitivity are made. Symptoms should start to subside in 0.5–3 hours after administration of the antivenin.

IV. Hobo Spiders

A. General and Medical Importance

Hobo spider bites have been reported to cause necrotic lesions in humans, although this claim has been challenged in recent years. Reports of necrotic spider bites in the states of Washington, Oregon, and Idaho steadily increased in frequency over the past 40 years. At first, these were attributed to the brown recluse spider, but intensive efforts to find brown recluses in or near patients' homes proved fruitless. Some scientists believe that local effects of hobo spider (formerly called aggressive house spider) envenomization may be similar to those of brown recluse bites and that the hobo spider may be the cause of necrotic arachnidism in the northwestern United States.[45,46] Akre and Myhre[47] recorded 52 serious bites from this species, *Tegenaria agrestis*, between 1989 and 1994. Epidemiological and laboratory toxicity studies, however, do not support common claims that hobo spiders cause necrotic lesions.[48,49] Further, bacterial causes of necrotic lesions after hobo spider bites have been ruled out.[50]

Hobo spider bites are usually painless at first. Induration may appear within 30 minutes, surrounded by an area of expanding erythema that may reach a diameter of 5–15 cm. Blisters develop within 15–35 hours, subsequently rupturing with a serous exudate encrusting the cratered wound. If any necrosis develops, an eschar may occur with underlying necrosis and eventual sloughing of affected tissues. Lesions may require up to 3 years to heal if the bite occurred in fatty tissue.[45] Systemic symptoms have been reported, including headache (often prolonged), nausea, weakness, fatigue, memory loss, and vision impairment. Protracted and severe systemic effects could possibly lead to death.[45]

Figure 29.10   Hobo spider or aggressive house spider. (Photograph copyright 2011 by Jerome Goddard, Ph.D.)

B. General Description

Hobo spiders are relatively large, very fast-running specimens, about 45 mm in length (including legs), with distinct chevron stripes on the abdomen (Figure 29.10). They have eight eyes with both the anterior and posterior eye rows in straight lines, and they have solid, light brown legs (not banded).

C. Geographic Distribution

Hobo spiders are native to Europe and were probably introduced into the Seattle area in the 1920s. They now occur from central Utah north to the Alaska Panhandle (see Figure 29.6). The hobo spider will probably continue to spread east in Montana and Wyoming and may or may not reach California.[51]

D. Biology and Behavior

Hobo spiders build funnel-shaped webs in or near houses. Common sites include rock walls, along house foundations, and in garages, piles of debris, and stacks of firewood. Both males and females build webs, and both may be in a single web in the late summer and fall. Mature males leave their webs in the fall to search for females. Male hobo spiders enter houses in large numbers in the fall but are usually restricted to basements or ground floor rooms because they are poor climbers. Males are more venomous than females and are responsible for most bites.

Figure 29.11 Sydney funnel web spider.

E. Treatment of Bites

Optimal treatment for necrotic spider bites is not well defined[45,52] (see previous discussion under violin spiders for current ideas).

V. Funnel Web Spiders

A. General and Medical Importance

The Sydney funnel web spider, *Atrax robustus*, is Australia's most dangerous spider, capable of causing death in as little as 15 minutes;[53] however, severe bite reactions are relatively rare, with only 5–10 cases occurring annually along a limited area of eastern Australia. There are several relatives of the Sydney funnel web that are also dangerous to humans such as *Hadronyche cerberea* and *H. formidabilis*.[54] In many cases, little or no venom is injected during the biting event and no symptoms develop. If envenomization does occur, the bite site becomes extremely painful. Systemic symptoms may develop within minutes owing to the toxin's direct effect on somatic and autonomic nerves, leading to the widespread release of neurotransmitter. Progressive hypotension and apnea may ensue.

B. General Description

The Sydney funnel web is a large, dark-colored aggressive spider with prominent fangs (Figure 29.11). The carapace is glossy dark brown to black, whereas the abdomen is usually dark plum to black. The spiders often range in size from 45 to 60 mm (including legs); body length alone ranges from 15 to 45 mm. Males are smaller than females. Spinnerets are obvious, fingerlike, and at the end of the abdomen. Interestingly, there are some North American trapdoor spiders that look very much like the dangerous funnel web spider and may lead to local panics (thinking they are now present in the United States).[55]

C. Geographic Distribution

Sydney funnel web spiders are only found within a 160-km radius of Sydney, Australia. Specifically, they occur from Newcastle to Nowra and west to Lithgow. Other related funnel webs occur all along the east coast of Australia.

D. Biology and Behavior

Sydney funnel webs live in burrows, rotting logs, tree holes, or crevices in rocks, where they build (as their name implies) a funnel-shaped web. *Hadronyche formidabilis* builds the funnel-shaped web in trees. The web characteristically contains irregular silken trap lines radiating out from the entrance. Colonies may consist of more than 100 spiders. Males may wander in search of females into houses during summer, especially during rainy weather.

E. Treatment of Bites

Immediate first aid should be administered for bites by any large black spider along the east coast of Australia, but especially in the Sydney area. A pressure bandage can be applied and the bitten limb immobilized using a splint. The spider should be captured, if possible, for identification. At the hospital, patients are carefully monitored for signs and symptoms for 4 hours. Little or no venom may have been injected during the bite. If signs and symptoms occur, such as mouth numbness, tongue spasms, nausea, vomiting, profuse sweating, salivation, and other muscle spasms, administration of antivenom is indicated (per package insert instructions). Patients should be monitored closely for the development of allergic reactions; however, severe allergic reactions to funnel web spider antivenom are uncommon.

VI. Tarantula Spiders

A. General and Medical Importance

Spiders in the family Theraphosidae are commonly called tarantulas in the United States (see box); however, in other parts of the world, this name is shared with other spiders, leading to confusion. Because of their great size and reputation, tarantulas are sometimes feared. This fear is mostly unfounded. Tarantulas attack only when roughly handled or deliberately provoked, and their bite is relatively minor (except in the case of a few tropical species that have more toxic venom). Bites of the North American species vary from being almost painless to a deep, throbbing pain that may last for hours. Baerg[56] allowed himself to be bitten twice and reported mild pain (like that of pinpricks) lasting 15–30 minutes; it was not accompanied by inflammation or swelling. Hypersensitive individuals could have more severe reactions. Tarantula venom consists mainly of hyaluronidase and a protein that is toxic to cockroaches and other arthropods for which it is intended. Interestingly, although there are no reported human deaths from tarantula bites, they may be fatal to domestic dogs.[57]

TARANTULA SPIDERS

Adult tarantula. (Adapted from Comstock, J.H. and Herrick, G.W., An Introduction to Entomology, Comstock, Ithaca, NY, 1940.)

IMPORTANCE

Painful bites; urticating hairs.

DISTRIBUTION

United States, Central and South America, Africa.

LESION

Bites—generally just pinprick type lesions.
Abdominal hairs—may cause itchy pruritic lesions.
Disease Transmission
None.

KEY REFERENCE

Diaz JH. The global epidemiology, syndromic classification, management, and prevention of spider bites. *Am. J. Trop. Med. Hyg.* 2004;71:239

TREATMENT

Bites—analgesics, tetanus prophylaxis.
Urticarial lesions—corticosteroids and antihistamines.

Many tarantulas occurring in the Western Hemisphere have urticarial hairs on the dorsal surface of the abdomen. These hairs may be "flicked off" or rubbed onto a predator by the spiders as a defensive mechanism. Urticarial hairs do not occur on African or Asian species. Contact with these urticarial hairs on the skin or mucous membranes may result in pruritic papular lesions that persist for weeks. If the hairs get into the eyes, complications can arise, with symptoms similar to ophthalmia nodosa.

B. General Description

Tarantulas are large hairy spiders. Their jaws (chelicerae) are attached in front of their head and can be moved up and down, opening parallel to the long axis of the body (Figure 29.12). Species native to the United States have a leg

Figure 29.12 Tarantula in hand. (Photograph copyright 2011 by Jerome Goddard, Ph.D.)

span of about 18 cm, but some South American jungle species have a span of 24 cm.

C. Geographic Distribution

About 30 species of tarantulas occur in the United States, mostly in the southwestern states. None occurs east of the Mississippi River. Since they are popular pets, they can be found in captivity all across the United States. The majority of tarantulas sold in pet stores in the United States are various species of *Aphonopelma*, *Grammostola*, and *Avicularia*, which are especially attractive as pets because of their bright colors.

D. Biology and Behavior

Tarantulas hatch from eggs into spiderlings that look like a miniature version of the adult. They mature in 10–12 years. A male tarantula usually does not live longer than a year after becoming sexually mature, but the female life span may exceed 15–20 years. In their native habitats, tarantulas dig burrows to rest in during the day. These burrows are often dug under large stones found in open hillside areas among mixed desert flora. Tarantulas are sluggish during daytime and may hibernate through the winter in colder areas. They emerge at night to hunt, usually ranging only a few yards from the burrow. During the mating season, males travel long distances from their burrows in search of females, but females and immatures remain near their burrow throughout their life. Tarantulas have extremely poor eyesight and detect their prey by vibrations. Their diet consists of arthropods and (rarely) small vertebrates.

E. Treatment of Bites

Treatment of tarantula bites consists of washing the bite site thoroughly, keeping the affected area elevated, and administering systemic analgesics if the wound is painful. Because tarantula mouthparts are dirty, tetanus vaccine status should be assessed. Treatment for urticaria produced by the abdominal hairs includes topical corticosteroids and oral antihistamines. If severe and widespread lesions occur, a short course of 1 or 2 weeks of oral corticosteroids may be considered.

References

1. O'Neil ME, Mack KA, Gilchrist J. Epidemiology of non-canine bite and sting injuries treated in U.S. emergency departments, 2001–2004. *Pub. Hlth. Rep.* 2007;122:764–775.

2. Peck WB, Whitcomb WH. Studies on the biology of the spider, Cheiracanthium inclusum. Univ. Arkansas Agri. Exp. Sta. Bull. No. 753, 1970; 76.

3. Mullen GR, Vetter RS. Spiders (Araneae). In: Mullen GR, Durden LA, eds., Medical and Veterinary Entomology. 2nd ed. New York: Elsevier; 2009:413–432.

4. Furman DP, Reeves WC. Toxic bite of a spider, Cheiracanthium inclusum Hentz. *Calif. Med.* 1957;87:114.

5. Carpenter TL, Bernacky BJ, Stabell EE. Human envenomation by Plectreurys tristis. *J. Med. Entomol.* 1991;28:477–478.

6. King LE, Jr. Spider bites. *Arch. Dermatol.* 1987;123:41–43.

7. CDC. Methicillin-resistant Staphylococcus aureus infections in correctional facilities – Georgia, California, and Texas, 2001–2003. *MMWR.* 2003;52(41): 992–996.

8. Atkins JA, Wingo CW, Sodeman WA. Probable cause of necrotic spider bite in the Midwest. Science. 1957;126:73–77.

9. Atkins JA, Wingo CW, Sodeman WA. Necrotic arachnidism. *Am. J. Trop. Med. Hyg.* 1958;7:165–169.

10. Masters E. Differentiating loxoscelism from cutaneous anthrax and Lyme erythema migrans. *Infect. Med.* 2008;25:29–35.

11. Payne KS, Schilli K, Meier K, et al. Extreme pain from brown recluse spider bites: Model for cytokine-driven pain. *JAMA Dermatol.* 2014;150(11):1205–1208.

12. DeLozier JB, Reaves L, King LE, Jr., Rees RS. Brown recluse spider bites of the upper extremity. *South Med. J.* 1988;81:181–184.

13. Stoecker WV, Vetter RS, Dyer JA. NOT RECLUSE – a mnemonic device to avoid false diagnoses of brown recluse spider bites. *JAMA Dermatol.* 2017;153:377–378.

14. Murray LM, Seger DL. Hemolytic anemia following a presumptive brown recluse spider bite. *Clin. Toxicol.* 1994;32:451–456.

15. Bey TA, Walter FG, Lober W, Schmidt J, Spark R, Schlievert PM. Loxosceles arizonica bite associated with shock. *Ann. Emerg. Med.* 1997;30:701–703.

16. Hite M. Notes on the natural habitat of the brown recluse spider. *Proceed. Ark. Acad. Sci.* 1964;18:10–12.

17. Sandidge JS, Hopwood JL. Brown recluse spiders: A review of biology, life history and pest management. *Trans. Kansas Acad. Sci.* 2005;108(3):99–108.

18. Rees R, Campbell D, Rieger E, King LE. The diagnosis and treatment of brown recluse spider bites. *Ann. Emerg. Med.* 1987;16:945–949.

19. Mold JW, Thompson DM. Management of brown recluse spider bites in primary care. *J. Am. Board. Fam. Pract.* 2004;17:347–352.

20. Lowry BP, Bradfiled JF, Carroll RG. A controlled trial of topical nitroglycerin in a New Zealand white rabbit model of brown recluse spider envenomation. *Ann. Emerg. Med.* 2001;37:161–165.

21. King LE, Jr., Rees RS. Dapsone treatment of a brown recluse bite. *J. Am. Med. Assoc.* 1983;250:648.

22. Nonavinakere VK, Stamm PL, Early J. A case study of brown recluse spider bite: Role of the community pharmacist in achieving a successful outcome. *J. Agromed.* 1996;3:37–43.

23. Rees RS, Altenbern DP, Lynch JB, King LE, Jr. Brown recluse spider bites: A comparison of early surgical excision versus dapsone and delayed surgical excision. *Ann. Surg.* 1985;202:659–663.

24. Swanson D, Vetter R. Bites of brown recluse spiders and suspected necrotic arachnidism. *N. Engl. J. Med.* 2005;352:700–707.

25. Elston DM, Miller SD, Young RJ, et al. Comparison of colchicine, dapsone, triamcinolone, and diphenhydramine therapy for the treatment of brown recluse spider envenomation: A double-blind, controlled study in a rabbit model. *Arch. Dermatol.* 2005;141(5):595–597.

26. Phillips S, Kohn M, Baker D, et al. Therapy of brown spider envenomation: A controlled trial of hyperbaric oxygen, dapsone, and cyproheptadine. *Ann. Emerg. Med.* 1995;25:363–368.

27. Masters EJ. Loxoscelism. *N. Engl. J. Med.* 1998;339:379.

28. Buescher LS. Spider bites and scorpion stings. In: Rakel RE, Bope ET, eds., *Conn's Current Therapy*. Philadelphia: Elsevier; 2005:1302–1304.

29. Baerg WJ. The effects of the bite of Latrodectus mactans Fabr. *J. Parasitol.* 1923;9:161–169.

30. Maretic Z. Latrodectism: Variations in clinical manifestations provoked by Latrodectus species of spiders. *Toxicon.* 1983;21:457–461.

31. Vetter RS, Ibister GK. Medical aspects of spider bites. *Ann. Rev. Entomol.* 2008;53:409–429.

32. Bronstein AC, Spyker DA, Cantilena LR, Green JL, Rumack BH, Giffin SL. Annual report. American Association of Poison Control Centers' National Poison Data System, 26th Annual Report. Clin *Toxicol.* 2008;47:911–1084.

33. Wingo CW. Poisonous Spiders. Missouri Agri. Exp. Sta. Bull. No. 738; 1960.

34. Alexander JO. *Arthropods and Human Skin*. Berlin: Springer-Verlag; 1984.

35. Sutherland SK. *Australian Animal Toxins: The Creatures, Their Toxins, and Care of the Poisoned Patient*. Melbourne: Oxford University Press; 1983.

36. Wiener S. Spider bite in Australia: An analysis of 167 cases. *Med. J. Australia.* 1961;48:44–47.

37. Brown KS, Necaise JS, Goddard J. Additions to the known distribution of the brown widow spider, Latrodectus geometricus. *J. Med. Entomol.* 2008;45:959–962.

38. Horton P. Redback spider is now established in Japan. *Br. Med. J.* 1997;314:1484.

39. Shackelford R, Veillon D, Maxwell N, et al. The black widow spider bite: Differential diagnosis, clinical manifestations, and treatment options. *J. Louisiana St. Med. Soc.* 2015;167:74–78.

40. Offerman SR, Daubert GP, Clark RF. The treatment of black widow spider envenomation with antivenin Latrodectus mactans: A case series. *Permanente J.* 2011;15:76–81.

41. Olson KR. Poisoning. In: Tierney LM, McPhee SJ, Papadakis MA, eds. *Current Medical Diagnosis and Treatment.* 41st ed. New York: Lange Medical Books/McGraw Hill; 2002.

42. Murphy CM, Hong JJ, Beuhler MC. Anaphylaxis with Latrodectus antivenin resulting in cardiac arrest. *J. Med. Toxicol.* 2011;7(4):317–321.

43. Clark RF. The safety and efficacy of antivenin Latrodectus mactans. *J. Toxicol. Clin. Toxicol.* 2001;39:125–127.

44. Russell FE, Wainsschel J, Gertsch WJ. Bites of spiders and other arthropods. In: Conn HF, ed. *Current Therapy.* Philadelphia: W.B. Saunders; 1973:868.

45. CDC. Necrotic arachnidism – Pacific Northwest, 1988–1996. *MMWR.* 1996;45:433–436.

46. Vest DK. Necrotic arachnidism in the northwest United States and its probable relationship to Tegenaria agrestis spiders. *Toxicon.* 1987;25:175–184.

47. Akre RD, Myhre EA. The great spider whodunit. *Pest Control Technol. Mag.* 1994;April issue:44–46.

48. Binford GJ. An analysis of geographic and intersexual chemical variation in the venom of the spider Tegenaria agrestis. *Toxicon.* 2001;39:955–968.

49. Vetter RS, Isbister GK. Do hobo spider bites cause dermonecrotic injuries. *Ann. Emerg. Med.* 2004;44:605–607.

50. Gaver-Wainwright MM, Zack RS, Foradori MJ, Lavine LC. Misdiagnosis of spider bites: Bacterial associates, mechanical pathogen transfer, and hemolytic potential of venom from the hobo spider, Tegenaria agrestis (Araneae: Agelenidae). *J. Med. Entomol.* 2011;48(2):382–388.

51. Vetter RS. Hobo spiders: Are they everywhere? *Pest Control Technol Mag.* 2003; December issue: 46–48.

52. Bope ET, Kellerman R. *Conn's Current Therapy.* Philadelphia: Elsevier Saunders; 2017.

53. Hawdon GM, Winkel KD. Spider bite: A rational approach. *Austr. Fam. Phys.* 1997;26:1380–1385.

54. Isbister GK, Gray MR, Balit CR, et al. Funnel-web spider bite: A systematic review of recorded clinical cases. *Med. J. Austr.* 2005;182(8):407–411.

55. Fox16.com. Man thinks he has discovered deadly Australian spider in Arkansas. *Fox16News.* 2008; Little Rock, Arkansas, 7/08, 9:34 am, http://www.fox16.com/news/story/Man-thinks-he-has-discovered-deadly-Australian/ZoOylotTcUSjsx3Ccg-emw.cspx.

56. Baerg WJ. Regarding the habits of tarantulas and the effects of their poison. *Sci. Mon.* 1922;14:482–486.

57. Diaz J. The global epidemiology, syndromic classification, management, and prevention of spider bites. *Am. J. Trop. Med. Hyg.* 2004;71:239–250.

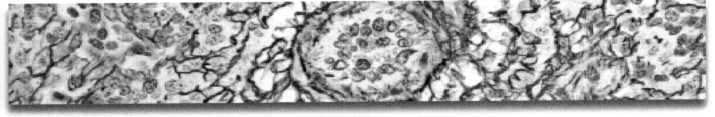

30

TICKS

I. General and Medical Importance

Ticks are bloodsucking ectoparasites that are efficient vectors of several different types of disease agents such as bacteria, protozoans, and viruses. In fact, they are second only to mosquitoes as arthropod vectors of human disease. In addition, tick bites cause a variety of acute and chronic skin lesions.[1] Listed in Table 30.1 are 16 of the common diseases produced by tick-transmitted agents, and a discussion of the major ones is provided in the following text. Disease transmission by arthropods is a complex phenomenon requiring a variety of unique factors to bring the pathogen, the host, and the vector into contact with each other. Even when this does occur, disease agent acquisition and subsequent transmission to another host may not always occur (these are concepts of host and vector competence). The reader is referred to other references[2-7] for more complete information on the natural history, epidemiology, and treatment of tick-borne diseases.

An important clinical consideration concerning transmission of tick-borne disease agents is how long the tick remains attached. Studies have shown that efficient transmission of Lyme disease spirochetes requires an infected tick to be attached at least 36–48 hours,[8,9] whereas rickettsial organisms (such as RMSF) may only require a few hours.[10] Worst-case scenario is the tick-borne viruses which may be transmitted in a matter of minutes.[11] Therefore, it is important to remind patients that prompt tick removal is critical for disease prevention (see Chapter 33 for more information).

A. Lyme Borreliosis

Lyme borreliosis (LB), caused by the bacterial spirochete *Borrelia burgdorferi* (there are several "genospecies" of this organism that are known to cause human disease), is a systemic tick-borne illness with many clinical manifestations that occurs over much of the world in temperate zones. Although rarely fatal,[12] the disease

may be long and debilitating, with cardiac, neurologic, and joint involvement. Initial symptoms include a flulike syndrome with headache, stiff neck, myalgias, arthralgias, malaise, and low-grade fever. Often, a more or less circular, painless, macular dermatitis is present at the bite site, called *erythema migrans* (EM). The EM lesion is characteristic for LB, although not all patients develop it. EM lesions may steadily increase in size with or without central clearing. Untreated EM and associated symptoms usually resolve in 3 to 4 weeks; however, the disease may disseminate within weeks or months, resulting in cardiac, neurologic, and joint manifestations. Symptoms may include Lyme carditis, cranial neuropathy, radiculopathy, diffuse peripheral neuropathy, meningitis, and asymmetric oligoarticular arthritis. Considerable concern exists in the medical community about people diagnosed with chronic LB who lack credible evidence of *B. burgdorferi* infection.[13,14] This may result in expensive or potentially dangerous treatments.[15,16] Sometimes referred to as the "Lyme Wars," concerned citizens and Lyme advocates likewise do not understand why mainstream medicine, based on the available evidence, has not embraced or sufficiently answered their questions regarding treatment of LB. Numbers of reported LB cases in Europe may be in the hundreds of thousands,[17] while in the United States officially there were 38,069 cases reported by the CDC in 2015,[18] although an indirect estimate is 300,000 per year.[19] In the United States, the vast majority of cases are from the northeastern and northcentral states (Figure 30.1). Although a number of microbiological and serological tests are available, diagnosis of LB is often based on clinical presentation. Serological diagnosis should be based on a two-step procedure: a sensitive enzyme-linked immunosorbent assay (ELISA) followed by immunoblot (IgM and IgG) if reactive. There have been recent proposals to change this recommended 2-tier algorithm to one consisting of two different types of ELISA tests (and eliminate the Western blot).[16] This approach would make the tests easier to perform, reduce subjectivity in interpreting Western blots, and be cheaper. Methods *not* recommended for diagnosis of LB include antigen

Table 30.1   Major Tick-Borne Diseases Worldwide

Disease	Causative Agent	Where Occurs	Tick Vectors
Lyme disease	Bacterium	United States, Europe, Japan, China, Australia	North America: *Ixodes scapularis, I. pacificus* Europe and Asia: *I. ricinus, I. persulcatus*
Rocky Mountain spotted fever	Bacterium	United States, Canada, Mexico, Central and South America	*Dermacentor variabilis, D. andersoni, Rhipicephalus sanguineus, Amblyomma cajennense*
Siberian tick typhus	Bacterium	Asiatic Russia, some islands in the Sea of Japan	Primarily *D. marginatus, D. silvarum, D. nuttalli*
Tick-borne encephalitis (TBE)	Virus	Europe	Western/Central Europe: *Ixodes ricinus* Central/Eastern Europe: *I. persulcatus*
Powassan (including "deer tick virus")	Virus	North America	*I. cookei* and *I. scapularis*
Boutonneuse fever	Bacterium	Southern Europe, India, Mediterranean area, Africa	*R. sanguineus, R. appendiculatus, Haemphysalis leachi*, others
American boutonneuse fever	Bacterium	Primarily southern half of United States, parts of South America	United States: *Amblyomma maculatum* South America: *A. triste*
Tularemia	Bacterium	United States, Europe, Russia, Japan, Canada, Mexico	United States: *A. americanum, D. variabilis* Europe: *D. nuttalli, I. ricinus*
Colorado tick fever	Virus	Mountain states of the western United States, British Columbia, Canada	*D. andersoni*
Ehrlichiosis	Bacterium	Mainly central and eastern half of the United States	*A. americanum, I. scapularis*
Anaplasmosis	Bacterium	North America (primarily where Lyme disease occurs), Europe and Asia	*I. scapularis, I. ricinus, I. persulcatus*, others
Babesiosis	Protozoan	The United States and Europe; foci in Massachusetts and New York areas, as well as Minnesota, Michigan, Washington, Oregon, and California	*I. scapularis, I. ricinus*, others
Relapsing fever	Bacterium	Almost worldwide; in the United States, primarily in Washington, Oregon, and California	*Ornithodoros* spp., U.S. cases mostly from *O. hermsi*
Crimean–Congo hemorrhagic fever	Virus	Europe, Asia, Africa	*Hyalomma marginatum, Hy. anatolicum*, others
Kyasanur forest disease	Virus	India, especially Karnataka State	*H. spinigera*
Tick paralysis	Salivary toxin	United States, particularly Montana, British Columbia border, Australia	*D. andersoni, D. variabilis, I. holocyclus*

tests of bodily fluids, polymerase chain reaction (PCR) of urine, and lymphocyte transformation tests.[20–22]

Lyme borreliosis is solely tick-borne. In the United States, *Ixodes scapularis* is the primary vector in the East and *Ixodes pacificus* in the West. In Europe, *I. ricinus* and *I. persulcatus* serve as principal vectors of the *Borrelia*. In Asia, vectors include *I. persulcatus, I. ovatus, I. granulatus, I. moschiferi*, as well as a few *Haemaphysalis* species.

Small animal host availability and diversity can affect transmission of zoonotic agents such as LB. If immature ticks feed on hosts that are refractory to infection with the LB spirochete, then overall prevalence of the disease agent in an area will decline. On the other hand, if an abundant host is available that also is able to be infected with *Borrelia burgdorferi* producing long and persistent spirochetemias, then prevalence of tick infection increases. This seems to be the case in the northeastern and upper midwestern United States. In those areas, the primary host for immature *Ixodes scapularis* is the white-footed mouse, which is capable of infecting nearly 100% of larval ticks during feeding. As infection can be transferred from tick stage to tick stage, this obviously leads to high numbers of infected nymphs and adults. In the western and southern United States, tick infection rates are much lower (and, hence, lower numbers of LB cases are reported). This is often attributed to the fact that immature stages of *I. scapularis* and *I. pacificus* feed primarily on lizards, which are incompetent as reservoirs and incapable of infecting ticks (Figure 30.2). Another factor affecting the dynamics of LB is the fact that nymphal *I. scapularis* is the stage primarily biting people and

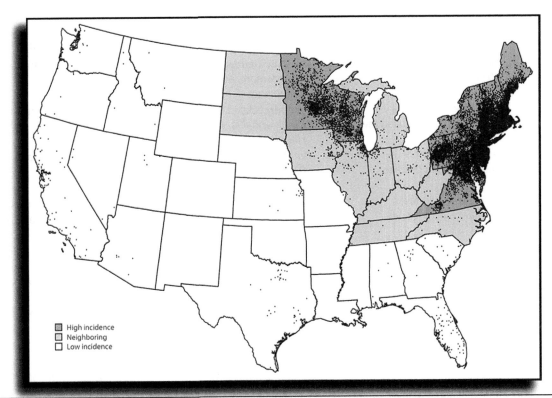

Figure 30.1 Lyme disease case distribution—United States. (Figure courtesy of U.S. Centers for Disease Control and Prevention, Atlanta, GA.)

Figure 30.2 Ticks may feed on a variety of vertebrate animals. (Photographs copyright 2012 by Jerome Goddard, Ph.D.)

transmitting the disease agent in the Northeast, whereas in the South nymphal *I. scapularis* rarely, if ever, bite humans.[23] Adult ticks are certainly capable of transmitting the LB agent in all areas—North, South, or West—but perhaps adult ticks are large enough to be easily seen and removed by people.[24] Nymphs, on the other hand, are about the size of a pinhead and may be easily overlooked or confused with a freckle.

Other tick species may be involved in the ecology of Lyme borreliosis in the United States. Also, there may be several, as yet undescribed, *Borrelia* species that cause Lyme-like illness, such as one recently described.[25] In the southern United States, there have been reports for years about an LB-like illness of unknown etiology which has associated EM lesions.[26]

The Centers for Disease Control and Prevention (CDC) often just classifies these cases as southern tick-associated rash illness (STARI) (see text box).

B. Spotted Fever Group Rickettsioses

Ticks may transmit a wide variety of rickettsial (bacterial) organisms to humans and other animals. The spotted fever group (SFG) contains rickettsial species related to the agent of Rocky Mountain spotted fever (RMSF), *Rickettsia rickettsii*, but there are many other rickettsial species in the spotted fever group; it contains at least 11 disease agents and 15 others with low or no pathogenicity to humans. Table 30.2 presents distributional and epidemiologic information on some of these human disease-causing SFG rickettsiae. The CDC reported 4,198 cases of spotted fever group rickettsiosis in 2015 (which includes RMSF), representing a 12% increase in incidence over the previous year.[18] Probably many more cases occur but go unreported. If an unusual febrile illness is treated successfully with doxycycline, there may be little interest in follow-up and reporting.

Rocky Mountain Spotted Fever Although named "Rocky Mountain" spotted fever (RMSF), cases of this SFG rickettsiosis actually occur more commonly elsewhere. RMSF has been reported from the United States, Mexico, Costa Rica, Panama, Colombia, Brazil, and Argentina. A fairly intense focus of RMSF occurs in a swath from North Carolina across Tennessee, Arkansas, and Oklahoma (Figure 30.3). At the

Table 30.2 Epidemiologic Information on Some Common Spotted Fever Group Rickettsiae

Rickettsia	Disease	Tick Vectors	Distribution
R. rickettsii	Rocky Mountain spotted fever (RMSF)	Primarily *Dermacentor variabilis, D. andersoni,* and *Rhipicephalus sanguineus*	Western Hemisphere
R. conorii	Boutonneuse fever or Mediterranean spotted fever	Primarily ticks in the genera *Rhipicephalus, Hyalomma,* and *Haemaphysalis*	Africa, Mediterranean Region, Middle East
R. parkeri	American boutonneuse fever or maculatum disease	*Amblyomma maculatum* (United States) and *A. triste* (South America)	Central and eastern United States, parts of South America
R. africae	African tick bite fever	*A. hebraeum* and *A. variegatum*	Sub-Saharan Africa
R. phillipi (formerly 364D agent)	Pacific Coast tick fever	*D. occidentalis*	Western United States (California)
R. slovaca	Tick-borne lymphadenopathy (TIBOLA)	*Dermacentor* species, especially *D. marginatus*	Europe
R. sibirica	North Asian tick typhus (Siberian tick typhus)	Primarily ticks in the genera *Dermacentor* and *Hyalomma*	Siberia, Central Asia, Mongolia
R. japonica	Japanese spotted fever	*Haemaphysalis flava, H. longicornis, Ixodes ovatus*	Japan and other parts of Orient
R. australis	Queensland tick typhus	*Ixodes holocyclus*	Australia
R. helvetica[a]	Not named[a]	*I. ricinus*	Europe
R. amblyommatis[a]	Not named[a]	*A. americanum*	Eastern United States

[a] There is some evidence that this agent may cause mild illness in some individuals.

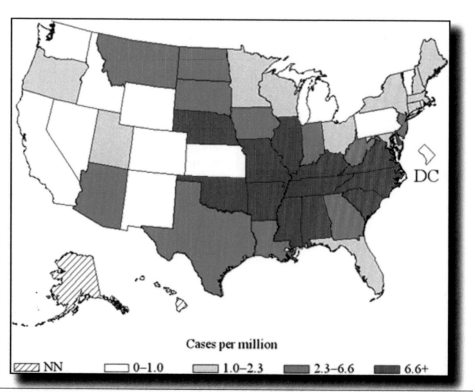

Figure 30.3 Case distribution of Rocky Mountain spotted fever. (Figure courtesy of U.S. Centers for Disease Control and Prevention, Atlanta, GA.)

LYME-LIKE ILLNESSES IN THE SOUTHERN UNITED STATES

There is controversy about the extent that Lyme disease occurs in the southern United States. Some states such as Florida have been declared endemic for Lyme disease (LD), although the American Lyme Disease Foundation states that LD-infected ticks in the southern states are "rare." Certainly, numerous cases are reported to state health departments and the Centers for Disease Control and Prevention each year. In Mississippi, for example, numerous physician reports of Lyme disease are submitted to the state health department annually, although these are almost never confirmed. Although confirmed cases of Lyme disease may be uncommon in the southern United States, the agent (*Borrelia burgdorferi*), vector (*Ixodes scapularis*), and reservoir hosts (white-footed mice and cotton rats/mice) have been reported from the South.[1,2] Interestingly, to date, true *B. burgdorferi* (in the "strictest sense") has not been isolated in culture from any human patient in the southern United States.[3]

STARI. Reports have persisted for years about the occurrence of a Lyme disease-like illness in the South.[4,5] The CDC often labels these southern Lyme disease-like illnesses as "southern tick-associated rash illness" (STARI) or Master's disease. Like Lyme disease, the characteristic feature is a skin lesion resembling EM, as well as flulike symptoms (see Figure 1). In contrast to Lyme disease, which is associated with *I. scapularis*, STARI is associated with the lone star tick, *Amblyomma americanum*. In addition, chronic sequelae from STARI have not been reported to date, although to our knowledge no specific studies to assess chronic STARI have been conducted. Some researchers first thought that a spirochete, *Borrelia lonestari*, occurring in *Amblyomma americanum*, was the cause of STARI, and the agent has been linked to at least one case of Lyme-like illness (see Figure 2).[6] *B. lonestari* can readily be visualized by fluorescent antibody staining in a small percentage of field-caught lone star ticks and can be detected by polymerase chain reaction (PCR) assays.[7] However, in a study by Wormser and colleagues,[8] *B. lonestari* was not detected by PCR or culture from skin biopsies of patients in Missouri with erythema migrans (EM). Patients also were seronegative to *B. burgdorferi*, suggesting another cause of the lesions. Also, Philipp and colleagues[9] found that STARI patients from Missouri were negative on the C6 Lyme enzyme-linked immunosorbent assay to detect antibodies to *B. burgdorferi*. Thus, evidence of a borreliosis in STARI patients is lacking based on typical assays for *B. burgdorferi*.

Allergic reactions to tick bite. Further complicating our understanding of Lyme disease in the South is an apparent hypersensitivity reaction to saliva of the lone star tick that sometimes occurs 1 to 3 days following a bite. I have seen such a reaction several times (unpublished data). This hypersensitivity reaction may resemble EM and is often up to 8 cm in diameter, ring-like, raised, and vesicular. Studies of such lesions are lacking; nevertheless, the lesions are probably not true EM because there is little or no incubation period, lesions often fade in a few days, and they are raised (often vesicular). In southern states where physicians do not see many cases of true Lyme disease, these hypersensitivity reactions could be misdiagnosed as Lyme disease. Recent evidence of rapid development of IgE antibodies to tick salivary proteins (after tick bite) in people residing in areas where *Amblyomma americanum* occurs supports the idea of hypersensitivity reactions.[10]

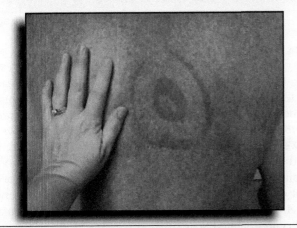

Figure 1 Erythema migrans on patient's back after tick bite. (Photograph courtesy of Dr. Ashley Lovell, USDA Wildlife Services, and used with permission.)

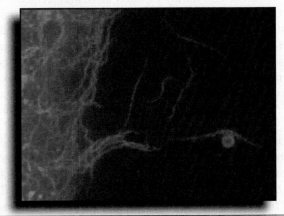

Figure 2 Borrelia lonestari, associated with the lone star tick, *Amblyomma americanum*. (Photograph courtesy of Dr. Andrea Varela-Stokes, Mississippi State University.)

REFERENCES

1. Oliver JH, Gao L, Lin T. Comparison of the spirochete *Borrelia burgdorferi* S.L. isolated from the tick *Ixodes scapularis* in southeastern and northeastern United States. *J. Parasitol.* 2008;94(6):1351–1356

2. Rudenko N, Golovchenko M, Grubhoffer L, Oliver, Jr., JH. *Borrelia carolinensis* sp. nov., a new (14th) member of the *Borrelia burgdorferi* sensu lato complex from the southeastern region of the United States. *J. Clin. Microbiol.* 2009;47(1):134–141

3. Dennis DT. Rash decisions: Lyme disease or not? *Clin. Infect. Dis.* 2005;41:966–968

4. Felz MW, et al. Solitary erythema migrans in Georgia and South Carolina. *Arch Dermatol.* 1999;135(11):1317–1326

5. Masters EJ, Donnell HD. Lyme and/or Lyme-like disease in Missouri. *Missouri Med.* 2005;92(7):346–353

6. James AM, Liveris D, Wormser G, Schwartz I, Montecalvo MA, Johnson B. *Borrelia lonestari* infection after a bite by an *Amblyomma americanum* (L.). *J. Infect. Dis.* 2001;183:1810–1814

7. Varel AS, Moore VA, Little SE. Disease agents in *Amblyomma americanum* from northeastern Georgia. *J. Med. Entomol.* 2004;41(4):753–759

8. Wormser G, Masters EJ, Liveris D, Nowakowaski J, Nadelman R, Holmgren D, Bittker S, Cooper D, Wang G, Schwartz I. Microbiologic evaluation of patients from Missouri with erythema migrans. *Clin. Infect. Dis.* 2005;40:423–428

9. Philipp MT, Masters E, Wormser GP, Hogrefe W, Martin D. Serologic evaluation of patients from Missouri with erythema migrans-like skin lesions with the C6 Lyme test. *Clin Vaccine Immunol.* 2006;13(10):1170–1171

10. Commins SP, et al. The relevance of tick bites to the production of IgE antibodies to the mammalian oligosaccharide galactose-alpha-1,3-galactose. *J. Allergy. Clin. Immunol.* 2001;127(5):1286–1293

time of initial presentation, patients often have the classic triad of RMSF: fever, rash, and history of tick bite. Other characteristics are malaise, severe headache, chills, and myalgias. Sometimes, gastrointestinal symptoms such as abdominal pain, vomiting, and diarrhea are reported, especially early in the illness. More than one member of the family may be infected.[27,28] The rash, appearing on the third day or after, usually begins on the extremities and then spreads to the rest of the body; however, there have been confirmed cases without rash. Mental confusion, coma, and death may occur in severe cases. Untreated, the mortality rate is about 20%; even with treatment, the rate is approximately 5%.

Laboratory findings may include hyponatremia (20 to 50% of cases), thrombocytopenia (30 to 50% of cases), anemia (5 to 25% of cases), and mildly elevated aminotransferase levels (40 to 60% of cases). Although PCR (on blood and tissue) and immunohistochemistry (tissue) can be used to confirm RMSF, these tests are not widely available. Serological tests such as immunofluorescent assay (IFA) on acute and convalescent sera may help confirm the disease later (although there may be cross-reactivity with other SFG rickettsiae such as *Rickettsia parkeri*, see next section). *It is vital to understand that diagnosis and treatment decisions for RMSF are based on clinical and epidemiologic clues and should never be delayed pending laboratory confirmation or even further development of signs and symptoms such as a rash.*[29,30] *RMSF can be fulminating and result in death rapidly if untreated, so prompt treatment with doxycycline is warranted (or chloramphenicol if allergic).*[31]

Rocky Mountain spotted fever is usually transmitted by the bite of an infected tick (Figure 30.4). Not all tick species are effective vectors of the rickettsia and, even in the

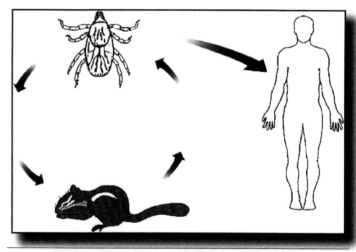

Figure 30.4   Life cycle of Rocky Mountain spotted fever agent.

vector species, not all ticks are infected; therefore, the presence of an infected tick in an area is like a needle in a haystack. Generally, only 1% or less of vector ticks in an area are infected. Several tick vectors may transmit RMSF organisms, but the primary ones are the American dog tick, *Dermacentor variabilis*, in the eastern United States, and *D. andersoni* in the West. The brown dog tick, *Rhipicephalus sanguineus*, was found to be the vector in an outbreak of RMSF in Arizona.[32] Certain *Amblyomma* species are also known vectors of the agent of RMSF in Mexico through parts of South America[33] and *A. americanum* may be involved in the southern and southcentral United States.[34,35]

American Boutonneuse Fever   Another rapidly emerging SFG rickettsiae closely related to the agent of RMSF is

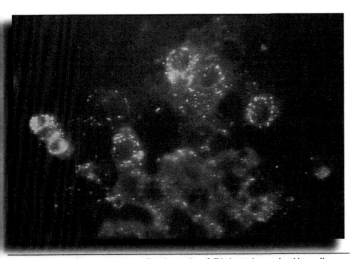

Figure 30.5 Fluorescent antibody stain of *Rickettsia parkeri* in cell culture. (Photograph courtesy of Dr. Andrea Varela-Stokes, Mississippi State University.)

Rickettsia parkeri (Figure 30.5), which was first found in cattle and Gulf Coast ticks, *Amblyomma maculatum*, in 1939.[36] For nearly a century following the discovery of the agent of RMSF, *R. rickettsii* was considered the sole tick-borne rickettsia unequivocally associated with human disease in the United States. Soon after the initial isolation of *R. parkeri* and during subsequent decades, investigators speculated about a possible role for this bacterium as an agent of disease in humans.[37,38] Parker and others identified similarity among the "maculatum agent" (*R. parkeri*) and the agents of other SFG rickettsioses, including boutonneuse fever, South African tick bite fever, and Siberian tick typhus, by comparing clinical and serological characteristics of the infections in guinea pigs.[36,39–41] These observations have been validated by contemporary phylogenetic analyses showing relationships among SFG rickettsiae that position *R. parkeri* most closely with Old World, eschar-producing SFG pathogens such as *R. africae*, *R. conorii*, *R. mongolotimonae*, and *R. sibirica*.[38,42,43] It naturally follows that *R. parkeri* should also produce eschars. Goddard[44] reported eschar formation in guinea pigs infected with *R. parkeri*, and the first human case of this eschar-characterized disease was described by Paddock et al.[45] Since then, at least 50 other cases have been identified, primarily in the southeastern and mid-Atlantic states. Seasonally, cases have occurred between April and October, with a peak from July through September. Because of clinical similarities between *R. parkeri* infection and boutonneuse fever, one descriptive moniker for this newly recognized rickettsiosis is American boutonneuse fever (ABF).[46] In addition, due to cross-reactivity in serological tests between *R. parkeri* and several other SFG rickettsiae, including *R. rickettsii*, many reported cases of RMSF may actually be ABF. Clinical differences between ABF and RMSF are primarily an inoculation eschar and a vesicular rash in ABF, and the fact that RMSF is a much more severe illness.[47,48] Interestingly, thus far there have been no cases of ABF in children.

Boutonneuse Fever Boutonneuse fever (BF), or Mediterranean spotted fever, caused by *Rickettsia conorii*, is widely distributed in Africa, areas surrounding the Mediterranean, southern Europe, and India. The name is derived from the manifestation of a papular rather than a macular rash. Typically, there is also development of an inoculation eschar at the tick bite site. BF often resembles a mild form of RMSF, characterized by mild to moderately severe fever, headache, and a rash usually involving the palms and soles; however, a few cases of severe illness have been associated with BF. Several tick species serve as vectors of the agent to humans, but especially *Rhipicephalus sanguineus*, *R. appendiculatus*, and *Amblyomma hebraeum*.

African Tick Bite Fever African tick bite fever (ATBF), caused by the spotted fever group rickettsia, *Rickettsia africae*, is clinically similar to BF with the exception that there is usually an absence of rash in ATBF patients.[49,50] ATBF also produces an eschar. The disease has been recognized in sub-Saharan Africa (especially South Africa) and the French West Indies and is transmitted by *Amblyomma hebraeum* and *A. variegatum* ticks.[51,52] ATBF appears to be an emerging cause of influenza-like illness in travelers.[53]

Tick-Borne Lymphadenopathy Tick-borne lymphadenopathy (TIBOLA), sometimes also known as *Dermacentor*-borne necrosis erythema lymphadenopathy (DEBONEL), is a relatively recently described tick-borne illness occurring in Europe during the wintertime.[54] The causative agent is *Rickettsia slovaca*, and vectors are ticks in the genus *Dermacentor*. Bites from these ticks usually appear on the head,[55] and there is often eschar development at bite sites, along with painful cervical lymphadenopathy. Alopecia may also develop at the eschar site.

Siberian Tick Typhus Siberian tick typhus (STT), or North Asian tick typhus, caused by *Rickettsia sibirica*, is very similar clinically to RMSF with fever, headache, and rash. The disease can be mild to severe but is seldom fatal. STT was first recognized in the Siberian forests and steppes in the 1930s but now is known to occur in many areas of Asiatic Russia and on islands in the Sea of Japan. Various hard ticks are vectors of the agent, but especially *Dermacentor marginatus*, *D. silvarum*, *D. nuttalli*, and *Haemaphysalis concinna*.

Queensland Tick Typhus Queensland tick typhus (QTT), caused by *Rickettsia australis*, occurs along the east coast of Australia and is named after the state of Queensland. It is primarily restricted to dense forests interspersed with grassy savanna or secondary scrub. Most patients have fever, headache, and rash that may be vesicular and petechial—even pustular. Commonly, there is an eschar at the site of the tick bite. The agent of QTT is transmitted to humans by the bite of an infected *Ixodes holocyclus* tick.

Pacific Coast Tick Fever   Cases of Pacific Coast tick fever (PCTF), first recognized in California in 2010, are caused by *Rickettsia phillipi* (formerly known as the 364D agent) and are associated with an eschar at the bite site.[56] Much is yet unknown about the clinical presentation as only a few cases have been described, but symptoms may include fever, headache, malaise, and lymphadenopathy.[57] The only known vector is *Dermacentor occidentalis*. Cases have been reported from July through September.

C. Ehrlichiosis

*Ehrlichia a*re organisms in the family Anaplasmataceae that primarily infect circulating leukocytes. Much of the knowledge gained concerning ehrlichiae has come from the veterinary sciences, with intensive studies on *Ehrlichia* (*Cowdria*) *ruminantium* (cattle, sheep, goats), and *Ehrlichia equi* (horses). Canine ehrlichiosis, caused by *Ehrlichia canis*, wiped out 200 to 300 military working dogs during the Vietnam War.[58] In the United States, human cases of ehrlichiosis were unknown until a report in March 1986 of a 51-year-old man who had been bitten by a tick in Arkansas and was sick for 5 days before being admitted to a hospital in Detroit.[59] He had malaise, fever, headache, myalgia, pancytopenia,

abnormal liver function, renal failure, and high titers of *E. canis* antibodies that fell sharply during convalescence. The patient was thought to have the dog disease. It turned out not to be the case; he had an infection with *E. chaffeensis* (a hitherto unknown agent). For this reason, in the literature, there are several reports of human infection in the United States with *E. canis*, when, in fact, human ehrlichiosis is usually caused by one to three closely related *Ehrlichia* organisms as follows. One of these, *Ehrlichia chaffeensis*, the most frequently reported, is the causative agent of human monocytic ehrlichiosis (HME), which occurs mostly in the southern and southcentral United States (sporadic cases of HME have also been reported in Europe) and infects mononuclear phagocytes in blood and tissues.[60] HME is not an insignificant disease—1288 cases were reported to the CDC in 2015 (Figure 30.6).[18] The second, *E. ewingii*, mostly a dog and deer pathogen, infects granulocytes and causes a clinical illness similar to HME but thus far has only been identified in a few patients, most of whom were immune-compromised. The disease became reportable in 2008; however, a recent study suggests that cases of *E. ewingii* infection are underreported.[61] The third ehrlichial agent, the *E. muris*-like *agent* (sometimes called EMLA or *E. eauclairensis*), causes fever, malaise, headache, lymphopenia, and elevated liver enzymes.[62] Thus far, it

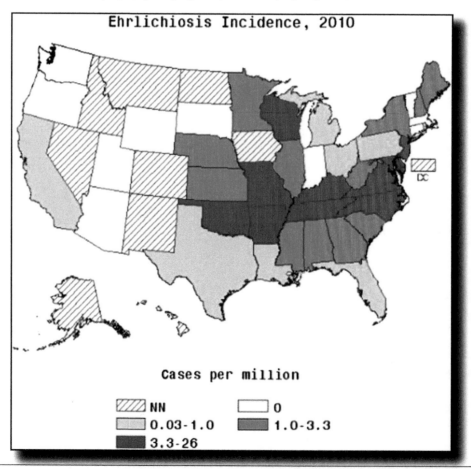

Figure 30.6   Distribution of reported cases of *Ehrlichia chaffeensis*. (Figure courtesy of U.S. Centers for Disease Control and Prevention, Atlanta, GA.)

has only been reported in a couple hundred patients from the upper midwestern United States, and the agent is transmitted by the deer tick, *Ixodes scapularis.*[63]

Clinical and laboratory manifestations of infection with the various ehrlichial agents are similar.[64] The patient usually presents with fever, headache, myalgia, progressive leukopenia (often with a left shift), thrombocytopenia, and anemia. In addition, there may be moderate elevations in levels of hepatic transaminases. Sometimes there is a cough, gastroenteritis, or meningitis. Only about 2–40% of the time is there a rash (more common with HME). Diagnosis depends mainly on clinical findings, although IFA tests may be used to detect antibodies against the respective *ehrlichial* agent. The gold standard serologic test is IFA performed on paired acute and convalescent sera demonstrating a fourfold rise in antibody titers. Antibodies may not be detectable in the first week of illness, so a negative test during that time does not rule out infection. At one time, human anaplasmosis (see next section) was classified as "human granulocytic ehrlichiosis," and due to this confusion over HGE/HGA terminology, physicians should carefully check what tests they are ordering. Further, due to cross-reactivity, tests for ehrlichiosis alone in anaplasmosis-endemic areas can result in an inaccurately high ehrlichiosis incidence and underrecognition of actual anaplasmosis cases.[65]

Ehrlichiosis is transmitted to humans almost exclusively by tick bite, but the different ehrlichial agents have different tick vectors. *Ehrlichia chaffeensis* and *E. ewingii* are both transmitted by the lone star tick (LST), while the EMLA has only been detected in *Ixodes scapularis*. LSTs generally occur from central Texas east to the Atlantic Coast and north to approximately Iowa and New York. WT deer, possibly along with dogs or other mammals, serve as reservoir hosts for the agent, and LSTs are the most important vectors; however, detection of the HME agent in other tick vectors and a few cases outside the distribution of LST may indicate that additional vectors occur.

D. Anaplasmosis

Anaplasma phagocytophilum infects granulocytes and causes human granulocytic anaplasmosis (HGA). For many years, this disease was called human granulocytic ehrlichiosis (HGE) and is often included in the older medical literature under that label. Complicating matters further, sometimes commercial laboratories may still refer to tests for HGA as human granulocytic ehrlichiosis tests. HGA is mostly reported from the upper midwestern and northeastern United States. There were 3656 cases of HGA reported to the CDC in 2015, a 30% increase over 2014.[18] The case fatality rate is 0.3% but can be higher in older patients.[31]

Clinical and laboratory manifestations of infection with HGA include fever, headache, myalgia, progressive leukopenia (often with a left shift), thrombocytopenia, and anemia. In addition, there may be moderate elevations in levels of hepatic transaminases. Sometimes there is a cough, gastroenteritis, or meningitis. Illness due to HGA is somewhat milder than with HME; reported fatality rates are about 1 and 2% for HGA

and HME, respectively. Some research has indicated that both agents alter the patient's immune system, allowing opportunistic infections such as fungal pneumonia to occur. Diagnosis depends mainly on clinical findings, although IFA tests may be used to detect antibodies against the anaplasmal agent. The gold standard serologic test is IFA performed on paired acute and convalescent sera demonstrating a fourfold rise in antibody titers. Antibodies may not be detectable in the first week of illness, so a negative test during that time does not rule out infection. Due to confusion over HGE/HGA terminology, physicians should carefully check what tests they are ordering. If HGA is suspected, physicians should make sure the test detects antibodies to *Anaplasma phagocytophilum*.

Anaplasmosis (HGA) is transmitted to humans almost exclusively by bites from the deer tick, *Ixodes scapularis*. The ecology of HGA is not well understood at this time. It has been diagnosed mostly in patients from the upper Midwest, the northeastern United States, and the Pacific Coast area, although cases have also been reported sporadically elsewhere, including Europe and Asia. The tick vector in the United States is *Ixodes scapularis*, the same species that transmits the agent of Lyme borreliosis; thus, there is the possibility of co-infection with Lyme borreliosis and HGA (and even babesiosis).[66] In Europe and Asia, the vectors are *I. ricinus* and *I. persulcatus*, respectively. Animal reservoirs of the HGA agent include small rodents and possibly deer.

E. Babesiosis

Human babesiosis is a tick-borne disease primarily associated with two protozoa of the family Piroplasmordia: *Babesia microti* and *Babesia divergens*, although other newly recognized species may also cause human infection (Figure 30.7). The disease is a malaria-like syndrome characterized by fever, fatigue, and hemolytic anemia lasting from several days to a few months. In terms of clinical manifestations, babesiosis may vary widely, from asymptomatic infection to a severe, rapidly fatal disease.

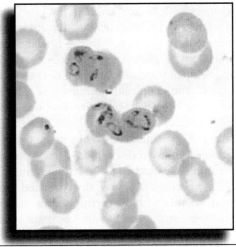

Figure 30.7 Babesia organisms in blood smear. (Photograph courtesy of U.S. Centers for Disease Control and Prevention, Atlanta, GA.)

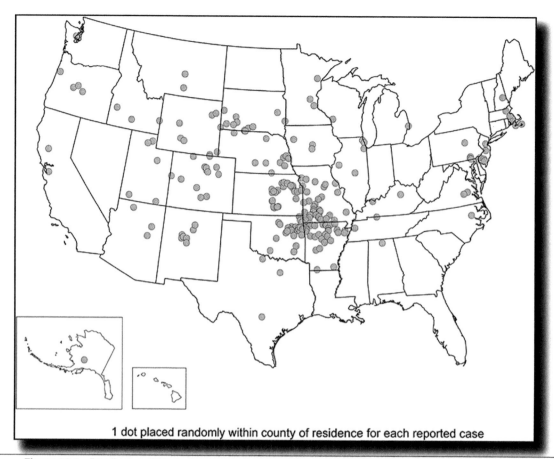

Figure 30.8 Distribution of tularemia cases in the United States, 2016. (Figure courtesy of U.S. Centers for Disease Control and Prevention, Atlanta, GA.)

The first demonstrated case of human babesiosis in the world was reported in Europe in 1957. Since then, there have been at least 28 additional cases in Europe.[67] Most European cases occurred in asplenic individuals and were caused by *Babesia divergens*, a cattle parasite. In the United States there have been thousands of cases of babesiosis (most with intact spleens), mainly caused by *Babesia microti*, mostly from southern New England, and specifically Nantucket, Martha's Vineyard, Shelter Island, Long Island, and Connecticut.[68,69] In 2015, there were 2074 cases reported by the CDC.[18] In the United States, cases of human infection by *Babesia microti* are caused by bites from the same tick that transmits the agent of Lyme borreliosis, *Ixodes scapularis*. The tick vector in Europe is believed to be the European castor bean tick, *Ixodes ricinus*, one of the most commonly encountered ticks in Central and Western Europe.

Babesiosis is very similar clinically to malaria; in fact, confusion between the two diseases is often reported in the scientific literature. Headache, fever, chills, nausea, vomiting, myalgia, altered mental status, disseminated intravascular coagulation, anemia with dyserythropoiesis, hypotension, respiratory distress, and renal insufficiency are common to both diseases. The symptoms of babesiosis do not show periodicity. The incubation period varies from 1 to 4 weeks. Physical examination of patients is generally unremarkable,

although the spleen and liver may be palpable. Diagnosis of babesiosis is based on recognition of the organism within erythrocytes in Giemsa-stained blood smears.

Babesial parasites, along with members of the genus *Theileria*, are called *piroplasms* because of their pear-shaped intraerythrocytic stages. There are at least 100 species of tick-transmitted *Babesia* parasitizing a wide variety of vertebrate animals. Some notorious ones are as follows: *Babesia bigemina*, the causative agent of Texas cattle fever; *B. canis* and *B. gibsoni*, canine pathogens; *B. equi*, a horse pathogen that occasionally infects humans; *B. divergens*, a cattle parasite that infects humans; and *B. microti*, a rodent parasite that infects humans. New *Babesia* species have been recovered from sick humans in the United States and Europe and have tentatively been variously designated as the WA1 agent, the CA1 agent, the *Babesia divergens*-like/MO1 agent, and the EU1 agent.[67,70,71] The WA1 agent (now named *B. duncani*), isolated from a patient in Washington State, was particularly interesting because the man was only 41 years old, had an intact spleen, and was immunocompetent.[67] There have been several reported cases of babesiosis in Washington State due to *B. duncani* and also a case from another, entirely different *Babesia*.[72] Obviously, there is much to be learned about the many and varied *Babesia* species and their complex interactions in nature.

F. Tularemia

Tularemia, sometimes called *rabbit fever* or *deer fly fever*, is a bacterial zoonosis that occurs throughout temperate climates of the Northern Hemisphere. Historically, approximately 150 to 300 cases occur in the United States each year, with most cases usually occurring in Arkansas, Missouri, and Oklahoma (Figure 30.8).[73,74] There were 314 cases reported to the CDC during 2015, a 74% increase from 2014.[18] The causative organism, *Francisella tularensis*, is a small, Gram-negative, nonmotile coccobacillus named after Sir Edward Francis (who did the classical early studies on the organism) and Tulare, CA (where it was first isolated). The disease may be contracted in a variety of ways: food, water, mud, articles of clothing, and (particularly) arthropod bites. Arthropods involved in transmission of tularemia include ticks, biting flies, and possibly even mosquitoes. Ticks account for a high percentage of human cases; one study of 78 clinical cases in Missouri with a known exposure source showed that 72% were associated with tick bite.[73] Tularemia may present as several different clinical syndromes, including glandular, ulceroglandular, oculoglandular, oropharangeal, pneumonic, and typhoidal. In general, the clinical course is characterized by an influenza-like attack with severe initial fever, temporary remission, and a subsequent febrile period of at least 2 weeks. Later, a local lesion with or without glandular involvement may occur. Additional symptoms vary depending on the method of transmission and form of the disease (see the following discussion). Untreated, the mortality rate for tularemia is about 8%; early diagnosis and treatment can reduce that to 1 to 2%.

Depending on the route of entry of the causative organism, tularemia may be classified in several ways. The most common is ulceroglandular, resulting from cutaneous inoculation and characterized by an ulcer with sharp undetermined borders and a flat base. Location of the ulcers may help identify the mode of transmission. Ulcers on the upper extremities are often a result of exposure to infected animals, whereas ulcers on the lower extremities, back, or abdomen most often reflect arthropod transmission. When there is lymphadenopathy without an ulcerative lesion, the classification glandular tularemia is used. If the tularemia bacterium enters via the conjunctivae, oculoglandular tularemia may result. Oropharyngeal tularemia results from ingestion of contaminated food or water. If airborne transmission of the agent is involved, the pneumonic form occurs. These patients often present with fever, a nonproductive cough, dyspnea, and chest pain. Finally, tularemia may be classified as typhoidal, characterized by disseminated infection mimicking typhoid fever, brucellosis, tuberculosis, or some of the RMSF-type infections.

In ticks, tularemia infection occurs in both the gut and body tissues and hemolymph fluid (tick blood). Infection is known to persist for many months and even years in some species. Tularemia organisms may be passed from tick stage to tick stage and to the offspring of infected female ticks. The three major North American ticks involved in transmission of tularemia

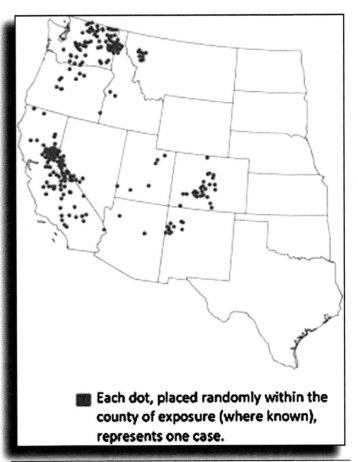

Each dot, placed randomly within the county of exposure (where known), represents one case.

Figure 30.9 Distribution of relapsing fever cases in the United States. (Figure courtesy of U.S. Centers for Disease Control and Prevention, Atlanta, GA.)

organisms are the lone star tick, *Amblyomma americanum*; the Rocky Mountain wood tick, *Dermacentor andersoni*; and the American dog tick, *D. variabilis*.[75] Both the lone star tick and the American dog tick occur over much of the eastern United States; the Rocky Mountain wood tick occurs in the West. All three tick species are avid human biters. In Central and Western Europe, *Ixodes ricinus* is probably a vector. *Dermacentor nuttalli* may be a vector in Russia. *D. dagestanicus* (= *niveus*) and *Rhipicephalus pumilio* may be vectors in Kazakhstan.

G. Relapsing Fever

Tick-borne (endemic) relapsing fever (TBRF) is a systemic spirochetal disease characterized by periods of fever lasting 3 to 5 days alternating with afebrile periods of 5 to 7 days. The total number of relapses can vary from 1 to 10 or more, lasting 2 or 3 weeks. Transitory petechial rashes are common during the initial febrile period. Untreated, the mortality rate is between 2 and 10%. Several hundred cases are reported worldwide each year, with approximately 20 to 50 of those being diagnosed in the United States (primarily from California, Washington, and Colorado) (Figure 30.9). TBRF is caused by a variety of tick-adapted *Borrelia* species (some

Figure 30.10 Distribution of Old World tick-borne relapsing fever.

authors maintain that all of the tick-adapted strains are really just one species). The spirochetes are transmitted to humans by several species of soft ticks in the genus *Ornithodoros*. Interestingly, in experimental animals, a single spirochete is sufficient to cause infection. As soft ticks generally feed for only a short period of time (30 minutes or so), the victim may be unaware of any recent tick bites. Rodents serve as a natural source of infection for ticks, and transmission is by tick bite (saliva) and also sometimes through contamination of the bite wound with infective coxal fluid produced by feeding ticks just before they detach. Transstadial and transovarial transmission of the agent occurs readily; thus, the ticks may be reservoirs of infection. In the Old World, the disease is endemic across central Asia, northern Africa, tropical Africa, and parts of the Middle East (Figure 30.10). Foci of infection are restricted to *Ornithodoros*-infested areas such as huts, caves, log cabins, cattle barns, and uninhabited

houses. Outbreaks in the United States have most often been associated with *O. hermsi* and *O. turicata* ticks in mountain cabins or rented state or federal park cabins (Figure 30.11).[76,77] One outbreak of 11 cases resulted from a 1-day family gathering in a remote New Mexico cabin,[78] while another cluster of cases occurred in high school students attending an outdoor education camp.[79]

H. Colorado Tick Fever

Colorado tick fever (CTF) is a generally moderate, acute, self-limited, febrile illness caused by a *Coltivirus* in the Reoviridae. Typically, the onset of CTF is sudden, with chilly sensations, high fever, headache, photophobia, mild conjunctivitis, lethargy, myalgias, and arthralgias. The body temperature pattern may be biphasic, with a 2- to 3-day febrile period, a remission lasting 1 to 2 days, then another 2 to 3 days of

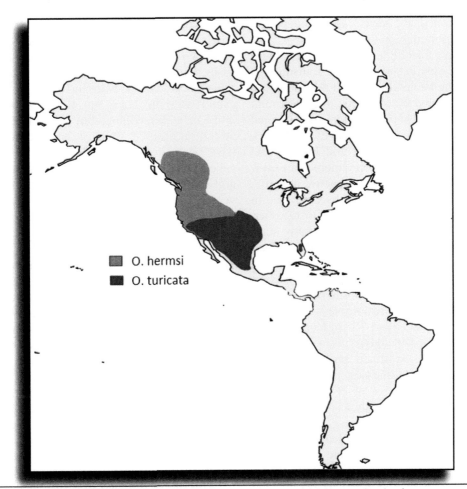

Figure 30.11 Distribution of *Ornithodoros hermsi* and *Ornithodoros turicata*, vectors of tick-borne relapsing fever.

fever, sometimes with more severe symptoms.[80] Rarely, the disease may be worse in children with encephalitis, myocarditis, or tendency to bleed. Infrequently, a transient rash may accompany infection. Recovery is usually prompt, but patients may be hospitalized or even (rarely) die. In one study of 91 cases, 18% were hospitalized.[81] CTF occurs in areas above 4000 feet in at least 11 western states (South Dakota, Montana, Wyoming, Colorado, New Mexico, Utah, Idaho, Nevada, Washington, Oregon, and California) and in British Columbia and Alberta, Canada. Exact case numbers are difficult to ascertain, as many cases may be mild and sick persons fail to seek medical care, but approximately 200 to 400 cases are reported in the United States annually. Peak incidence is during April and May at lower elevations and during June and July at higher elevations. The virus is maintained in nature by cycles of infection among various small mammals and the ticks that parasitize them. Infection in humans is by the bite of an infected tick. Several tick species have been found infected with the virus, but *Dermacentor andersoni* is by far the most common. This tick is especially prevalent where there is brushy vegetation to provide good protection for small mammalian hosts of immature ticks and yet with sufficient forage to attract large hosts required for the adults.

I. Tick-borne Encephalitis

The term tick-borne encephalitis (TBE) is generally used to describe disease entities caused by at least three subtypes of a flavivirus: European tick-borne encephalitis (TBEV-Eur), Siberian (TBEV-Sib), and Far Eastern (TBEV-FE). A couple of these agents, however, occur in the Western Hemisphere (see below). The three Old World TBE diseases differ in severity, with the Far Eastern form, sometimes known as Russian spring–summer encephalitis (RSSE), being the worst. In Central Europe, the typical case shows a biphasic course with an early, viremic, flulike stage, followed about a week later by the appearance of signs of meningoencephalitis.[82] CNS disease is relatively mild, but occasional severe motor dysfunction and permanent disability occur. The case fatality rate is 1 to 5%.[83] On the other hand, TBEV-FE is characterized by violent headache, high fever, nausea, and vomiting. Delirium, coma, paralysis, and death may follow; the mortality rate is about 25 to 30%. A recent report showed that new variants of TBE virus in Russia may produce a hemorrhagic syndrome.[84,85] Another member of the TBE serocomplex is called *louping ill*—named after a Scottish sheep disease— which, in humans, also displays a biphasic pattern and is generally mild. As mentioned, the virus infects sheep; few cases are actually ever reported in humans.

Reported case numbers for TBE (excluding the few louping ill cases) are approximately 10,000 annually.[7] Transmission to humans is mostly by the bite of an infected tick; however, infection may also be acquired by consuming infected milk and uncooked milk products. The distribution and seasonal incidence of TBE is closely related to the activity of the tick vectors *Ixodes ricinus* in western and central Europe and *I. persulcatus* in central and eastern Europe (there is an overlap of the two species). *Ixodes ricinus* is most active in spring and autumn. Two peaks of activity may be observed: one in late March to early June, and one from August to October. *Ixodes persulcatus* is usually active in spring and early summer. Apparently, *I. persulcatus* is more cold-hardy than *I. ricinus*, thus inhabiting harsher, more northern areas.

Powassan encephalitis (POW)—also in the TBE serocomplex—is a rare infection of humans that mostly occurs in the northeastern United States, adjacent regions of Canada, and parts of Russia. Characteristically, there is a sudden onset of fever with temperature up to 40°C along with convulsions. Also, accompanying encephalitis is usually severe, characterized by vomiting, respiratory distress, and prolonged, sustained fever. Only a few dozen cases of POW have been reported in North America,[86,87] although its reported incidence is increasing.[18] Recognized cases have occurred in children and adults, with a case fatality rate of approximately 50%. POW is transmitted in an enzootic cycle among ticks (primarily *Ixodes cookei*) and rodents and carnivores. *Ixodes cookei* only occasionally bites people; this may explain the low case numbers. Antibody prevalence to POW in residents of affected areas is less than 1%, indicating that human exposure to the virus life cycle is a rare event.

Deer tick encephalitis, closely related to POW, is another clinical entity in the TBE complex which was first discovered in North America in the late 1990s.[88] Few clinical cases have ever been described, although at least one death has been attributed to this virus.[89] The agent has been found along the Atlantic Coast and in Wisconsin and is primarily associated with the deer tick, *Ixodes scapularis*.

J. Other Tick-Borne Viruses

Kyasanur forest disease (KFD), caused by a flavivirus, is a severe hemorrhagic disease of humans and other primates in the Karnataka State of India, where it causes hundreds of cases and dozens of deaths every year. A close relative of the agent of KFD, Alkahurma virus, occurs in Saudia Arabia.[90] KFD is transmitted to humans by the tick, *Haemaphysalis spinigera*, primarily as people venture into forest fringes collecting firewood. In the winter of 2012–2013, the virus expanded over 400 km geographically and is now considered to be emerging.[91]

In the last decade, several new tick-borne viruses have been identified. Severe fever with thrombocytopenia syndrome (SFTS) was first described in rural areas of China, Japan, and Korea in 2011–2012. This disease, caused by a Phlebovirus, has thus far produced at least 2,500 human cases,[92] and is believed to be transmitted by *Haemaphysalis longicornis* ticks.[93] Heartland virus (a *Phlebovirus*) is associated with the Lone Star tick, *Amblyomma americanum* and has been recognized in Missouri, Oklahoma, Kentucky, and Tennessee.[94,95] Only about 30 cases of heartland virus have been identified. A few cases of a new *Thogotovirus* called bourbon virus have been identified in the Midwest and southern United States, also with *A. americanum* as the vector.[96,98]

K. Tick Paralysis

Tick paralysis is characterized by an acute, ascending, flaccid motor paralysis that may terminate fatally if the tick is not located and removed. The causative agent is believed to be a salivary toxin produced by ticks when they feed. In the strictest sense, tick paralysis is not a zoonosis; however, many contend that zoonoses should include not only infections that humans acquire from animals but also diseases induced by noninfective agents such as toxins and poisons.[99] The disease is more common than one might think. In North America, hundreds of cases have been documented from the Montana–British Columbia region.[100,101] It occurs in the southeastern United States as well. Tick paralysis is also especially common in Australia. Sporadic cases may occur in Europe, Africa, and South America.

The site of tick bite in a case of tick paralysis looks no different from that in cases without paralysis. There is a latent period of 4 to 6 days before the patient becomes restless and irritable. Within 24 hours there is an acute ascending lower motor neuron paralysis of the Landry type. It usually begins with weakness of the lower limbs, progressing in a matter of hours to falling down and obvious incoordination, which is principally due to muscle weakness, although rarely there may also be true ataxia. Finally, cranial nerve weakness with dysarthria and dysphagia leads to bulbar paralysis, respiratory failure, and death. In children, presenting features may include restlessness, irritability, malaise, and sometimes anorexia and vomiting. A tick may usually be found attached to the patient, usually on the head or neck. Some controversy exists over whether or not severity of symptoms is related to the proximity of the attached tick to the patient's brain. In one study, the case fatality rate in patients with ticks attached to the head or neck was higher than that in patients with ticks attached elsewhere; however, the difference was not statistically significant. Although ticks causing paralysis are often attached to the head or neck, it must be noted that cases of paralysis may occur from tick bites anywhere on the body (published examples: external ear, breast, groin, and back[102]). Once the tick is found and removed, all symptoms usually disappear rapidly (there are exceptions to this, especially in Australia with *Ixodes holocyclus*).

CASE HISTORY

TICK PARALYSIS

A 4-year-old female presented to a Mississippi hospital emergency department on July 3 with a 24-hour history of worsening weakness that began as difficulty in walking. Upon admission, the child was unable to stand, sit up, or even raise herself from bed. The physical exam revealed normal vital signs, mild lethargy, total body weakness, and some drooping of the eyelids. Tick paralysis was suspected, and upon closer examination a large tick was found attached on her scalp. The tick was removed and identified as a partially engorged female American dog tick (see figure). This tick is known to cause paralysis in humans and animals in the eastern United States. The patient remained stable but continued to exhibit total body motor weakness and difficulty swallowing, resulting in drooling when she attempted to drink liquids. When the child failed to improve after 4 hours, she was transferred to a regional pediatric intensive care unit. After arrival in the ICU, she quickly began regaining motor strength over the next 2 to 3 hours. Residual motor weakness lasted for a total of about 7 to 8 hours after tick removal. One interesting aspect of this case is how long it took the patient to recover. In most cases of tick paralysis in North America (this is not the case in Australia), the patient quickly recovers after tick removal.

Female American dog tick, *Dermacentor variabilis*, involved in a case of tick paralysis.

TICKS

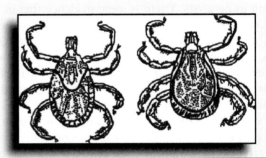

Female and male hard ticks. (From Goddard J. Ticks and Tick-borne Diseases Affecting Military Personnel, USAFSAM-SR-89-2. San Antonio, TX: School of Aerospace Medicine, U.S. Air Force; 1989.)

IMPORTANCE

Annoyance; disease transmission; tick paralysis; red meat allergy

DISTRIBUTION

Numerous species worldwide

LESION

Red papule with erythema; sometimes nodular; occasionally erythematous ring-like lesions due to hypersensitivity to tick saliva

DISEASE TRANSMISSION

Lyme disease, Rocky Mountain spotted fever, tularemia, relapsing fever, tick-borne encephalitis, powassan, heartland, and bourbon viruses, ehrlichiosis, anaplasmosis, and others

KEY REFERENCES

Arthur DR. *Ticks and Disease*. New York: Pergamon Press; 1962: 445

Gugliemone AA, et al. The Argasidae, Ixodidae, and Nuttalliellidae (Acari: Ixodida) of the world: a list of valid species names. *Zootaxa*. 2010;2528:1–28

TREATMENT

Generally, none needed for bites other than palliatives after tick removal; watch for development of tick-borne disease

As many as 43 tick species in 10 genera have been incriminated in tick paralysis in humans, other mammals, and birds (see Figure 30.2 for a photograph of a tick feeding on a bird).[103] However, human cases of the malady mostly occur in only a few geographic regions, caused by three main tick species. In the northwestern U.S. and British Columbia region of North America, the Rocky Mountain wood tick, *Dermacentor andersoni*, is the principal tick involved. In the southeastern United States, a related species, *Dermacentor variabilis*, known as the American dog tick, is the main cause of tick paralysis. Human cases in Australia are due to the Australian paralysis tick, *Ixodes holocyclus*. Interestingly, not all feeding female ticks—even of the species known to cause paralysis—produce paralysis. Why, out of hundreds of tick bites, does one result in paralysis? There is some evidence that in cattle, sheep, and dogs, numerous ticks feeding simultaneously (to reach a minimum dose) is necessary to elicit paralysis. In humans, however, one tick is usually involved. Most researchers believe that tick paralysis is caused by a toxin, but its nature is not well characterized. Generally, it is thought that the toxin is produced in the salivary glands of the female tick as she feeds. One alternative view would be that the toxin is produced in tick ovaries and subsequently passes to the salivary glands during later stages of tick engorgement. Although the vast majority of cases are due to female ticks, there are reports of male ticks causing limited paralysis. This fact seems to argue against the ovary toxin theory. There are other theories for the cause of the paralysis such as host reactions to components of the tick saliva or possibly symbiotic rickettsial organisms commonly found in tick salivary glands.

II. General Biology and Ecology

There are three families of ticks recognized today: (1) Ixodidae (hard ticks), (2) Argasidae (soft ticks), and (3) Nuttalliellidae (a small, curious, little-known family represented by a single species that possesses some characteristics of both hard and soft ticks). The terms *hard* and *soft* refer to the presence in the Ixodidae of a dorsal scutum or "plate" that is absent in the Argasidae.

Hard ticks display sexual dimorphism; males and females look conspicuously different (see box), and the blood-fed females are capable of enormous expansion (Figure 30.12). Their mouthparts are anteriorly attached and visible in dorsal view (Figures 30.13 and 30.14B). There is no true head, but their mouthparts appear as one (Figure 30.15). When eyes are present, they are located dorsally on the sides of the scutum.

Soft ticks are leathery and nonscutate, without obvious sexual dimorphism. The integument may be wrinkled or granulated (genus *Otobius*), mammillate (*Ornithodoros*) (Figure 30.16 and 30.17), or tuberculate (*Antricola*). Soft tick mouthparts are subterminally attached (in adult and nymphal stages) and not visible from dorsal view (Figure 30.14A). Eyes, when present, are located laterally in folds above the legs.

There are major differences in the biology of hard and soft ticks. Some hard tick species have a one-host life cycle, wherein engorged larvae and nymphs remain on the host after feeding; they then molt, and the subsequent stages reattach and feed. The adults mate on the host, and only engorged females drop off to lay eggs on the ground. Although some hard ticks complete their development on only one or two hosts, most commonly encountered ixodids have a three-host life cycle (Figure 30.18). In this case, adults mate on a host (except for some *Ixodes* spp.), and the fully fed female drops from a host animal to the ground and lays from 2000 to 18,000 eggs, after which she dies. The eggs hatch in about 30 days

Figure 30.12   Engorged female tick. (Photograph courtesy of Dr. Blake Layton, Mississippi Cooperative Extension Service, Mississippi State University.)

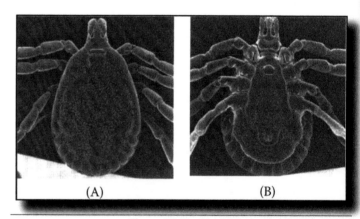

(A)   (B)

Figure 30.13   Hard tick showing (A) dorsal and (B) ventral aspects. (National Institutes of Health photograph courtesy of Dr. Jim Keirans.)

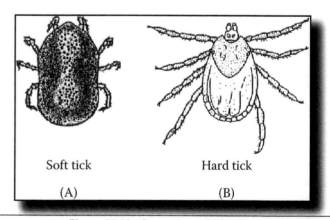

Soft tick   Hard tick

(A)   (B)

Figure 30.14   Comparison of (A) soft and (B) hard ticks. (From Goddard J. Ticks and Tick-borne Diseases Affecting Military Personnel, USAFSAM-SR-89-2. San Antonio, TX: School of Aerospace Medicine, U.S. Air Force; 1989.)

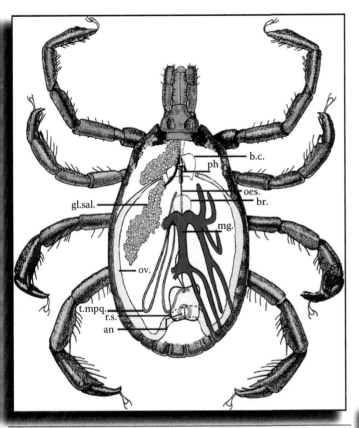

Figure 30.15 Drawing of a hard tick, showing internal and external structures. Note that the salivary glands (gl. sal.) are not located in the "head." (Figure courtesy of Sylvia Burnett, Mississippi State Department of Health.)

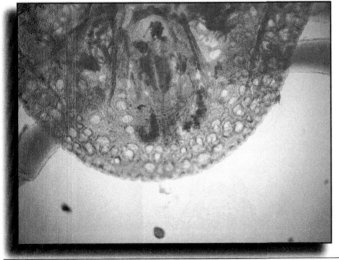

Figure 30.16 Anterior of a soft tick showing mammillated surface.

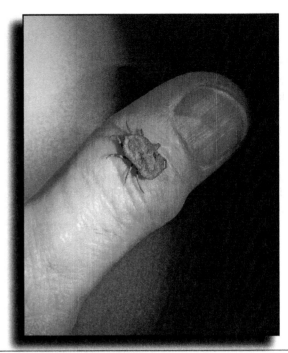

Figure 30.17 A soft tick, *Ornithodoros* sp. (Photo copyright 2018 by Jerome Goddard, Ph.D.)

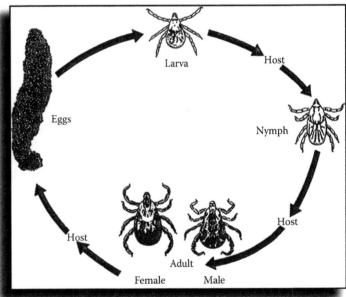

Figure 30.18 Tick life cycle. (Figure courtesy of Tennessee Valley Authority, Knoxville.)

into a six-legged seed tick (larval) stage, which feeds predominantly on small animals. The fully fed seed ticks drop to the ground and transform into eight-legged nymphs. These nymphs seek an animal host, feed, and drop to the ground. They then molt into adult ticks, thus completing the life cycle. Figure 30.19A shows all the motile life stages of hard ticks.

Many hard tick species *quest* for hosts, whereby they climb blades of grass or weeds and remain attached, forelegs outstretched, awaiting a passing host. For this reason, dragging a white flannel cloth through brushy areas works well for collecting ticks (Figure 30.20). Questing ticks may travel up a blade of grass (to quest) and back down to the leaf litter where humidity is high (to rehydrate) several times a day. Also, hard ticks will travel toward a carbon dioxide source. Adult ticks are more adept at traveling through vegetation

OFTEN-ASKED QUESTION

HOW DO TICKS GET ON PEOPLE?

There is considerable confusion among the public as to how ticks get on people. Folklore has it that they live in pine thickets, hide under bark, and/or fall out of trees on unsuspecting passersby. Frequently, park rangers or other outdoor workers erroneously recommend a hat for protection against ticks (presumably to guard against falling ticks), but ticks cannot fly, jump, or swim. They quest for host animals by climbing vegetation and passively waiting. The height of questing varies by species and life stage, but usually is 3 feet or less (often just a few inches). If environmental conditions are favorable, they may stay atop the vegetation for hours or even days. When a potential host approaches, signaled by vibrations or carbon dioxide, questing ticks may wave their legs (see figure) or move around on the plant trying to get on the animal. Once on the host, they crawl around for some time looking for a suitable place to feed. As ticks often attach to the scalp of humans, it is falsely assumed that they fall out of trees.

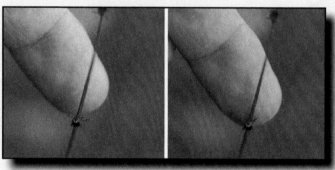

Ticks crawl up blades of grass to "quest" for passing hosts. This tick immediately latched onto the author's finger when touched. (Photograph copyright 2010 by Jerome Goddard, Ph.D.)

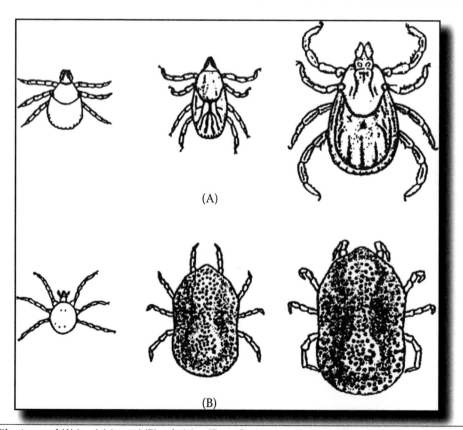

(A)

(B)

Figure 30.19 Motile life stages of (A) hard ticks and (B) soft ticks. (From Goddard, J. Ticks and Tick-borne Diseases Affecting Military Personnel, USAFSAM-SR-89-2. San Antonio, TX: School of Aerospace Medicine, U.S. Air Force; 1989.)

Figure 30.20 Tick collection by dragging or flagging with a white flannel cloth.

than the minute larvae. Studies have shown that adult lone star ticks may travel up to 10 m (33 feet) to a carbon dioxide source, but other species such as *Ixodes scapularis* will only travel distances of 1 to 2 meters (3.3 to 6.6 feet) toward a carbon dioxide source. Goddard[104] demonstrated minimal (less than 2 feet) lateral movement by questing adult *I. scapularis* in a mark release recapture study.

Ticks feed by cutting a small hole into the host epidermis with their chelicerae and inserting the hypostome into the cut, thereby attaching to the host. Blood flow is presumably maintained with the aid of an anticoagulant from the salivary glands. Some hard ticks secure their attachment to the host by forming a cement cone around the mouthparts and surrounding skin. Two phases are recognized in the feeding of nymphal and female hard ticks: (1) a growth feeding stage characterized by slow continuous blood uptake, and (2) a rapid engorgement phase occurring during the last 24 hours or so of attachment.

The biology of soft ticks differs from hard ticks in a number of ways (see Figure 30.19B for life stages). Adult female soft ticks feed and lay eggs several times during their lifetime. Soft tick species may also undergo more than one nymphal molt before reaching the adult stage. With the exception of larval stages of some species, soft ticks do not firmly attach to their hosts for several days like the Ixodidae. They are adapted to feeding rapidly and leaving the host promptly.

The expansion capability of hard ticks sometimes causes confusion among nonspecialists. I have often been sent fully engorged hard ticks removed from dogs with instructions to "identify enclosed soft tick"; this misconception arises because engorged ixodids do sometimes appear "soft." Another common misconception is that flat, unengorged hard ticks and engorged hard ticks represent different species. Ranchers

often speak of two "species" on their cattle; "the large swollen species" and the "small flat, brown species."

Hard ticks and soft ticks occur in different habitats. In general, hard ticks occur in brushy, wooded, or weedy areas containing numerous deer, cattle, dogs, small mammals, or other hosts. Soft ticks are generally found in animal burrows or dens, bat caves, dilapidated or poor-quality human dwellings (huts, cabins, etc.), or animal-rearing shelters. Many soft tick species thrive in hot and dry conditions, whereas ixodids are more sensitive to desiccation (the genus *Hyalomma* is an exception) and are, therefore, usually found in areas protected from high temperatures, low humidities, and constant breezes.

Most hard ticks, being sensitive to desiccation, must practice water conservation and uptake. Their epicuticle contains a wax layer that prevents water movement through the cuticle. Water can be lost through the spiracles; therefore, resting ticks keep their spiracles closed most of the time (opening them only once or twice an hour). Tick movement and the resultant rise in carbon dioxide production cause the spiracles to open about 15 times an hour with a corresponding water loss.

The development, activity, and survival of hard ticks are influenced greatly by temperature and humidity within the tick microhabitat. Lancaster[105] found that lone star tick eggs reared in an environment of less than 75% humidity would not hatch. Lees[106] demonstrated that *Ixodes ricinus* died within 24 hours if kept in a container at 0% relative humidity (RH) but survived 2 to 3 months at 90% RH. Because of their temperature and humidity requirements, as well as host availability, hard ticks tend to congregate in areas providing these factors. Ecotonal areas (ecological interface areas such as between forests and fields) are excellent habitats for hard ticks. Open meadows and prairies, along with climax forest areas, support the fewest lone star ticks. Ecotone areas and small openings in the woods are usually heavily infested. In a study by Semtner et al.,[107] lone star tick populations decreased with an increase in distance from the ecotone. Studies in Virginia demonstrated that American dog ticks tend to be especially abundant along trails, roadsides, and forest boundaries surrounding old fields or other clearings.[108110]

Deer and small mammals thrive in ecotonal areas, thus providing blood meals for ticks. In fact, deer are often heavily infested with hard ticks in the spring and summer months. The optimal habitat of white-tailed deer has been reported to be the forest ecotone, as this area supplies a wide variety of browse and frequently offers the greatest protection from natural enemies. Many favorite deer foods are also found in the low trees of an ecotone, including greenbrier, sassafras, grape, oaks, and winged sumac.

Ticks are not evenly distributed in the wild; instead, they are localized in areas providing their necessary temperature, humidity, and host requirements. Such biological characteristics of ticks, when known, may enable humans to avoid these parasites.

III. Tick Identification

Ticks are large mites, both having eight legs and one (apparent) disk-shaped body region. One exception is that the larval stage only has six legs (Figure 30.21) and should not be confused with insects that are six-legged. However, ticks differ from insects by having the one obvious body region (insects have three), and from other mites in having a toothed hypostome and no claws on their palps. Like those of many other arthropod groups, tick morphological structures have been assigned rather long and somewhat confusing names. Figures 30.22 and 30.23 are provided to acquaint the reader with these terms. Certain common tick species are variously colored (Figure 30.24) (see the paper by Schachat[111] for explanations for tick color patterns) and can sometimes be identified by comparisons with good photographs, but for the most

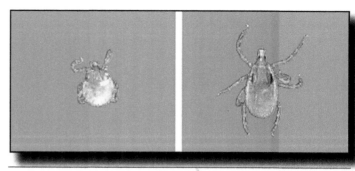

Figure 30.21 Tick larva (left) and nymph (right). (Photographs courtesy of Dr. Blake Layton, Mississippi State University Extension Service.)

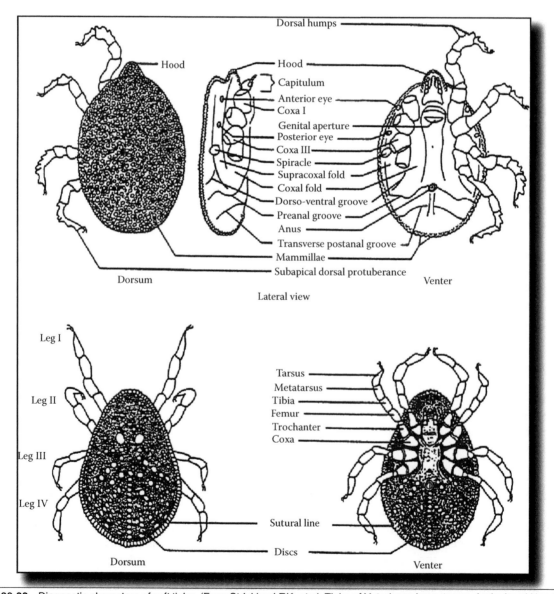

Figure 30.22 Diagnostic characters of soft ticks. (From Strickland RK, et al. Ticks of Veterinary Importance, Agricultural Handbook No. 485. Washington, DC: U.S. Department of Agriculture; 1976.)

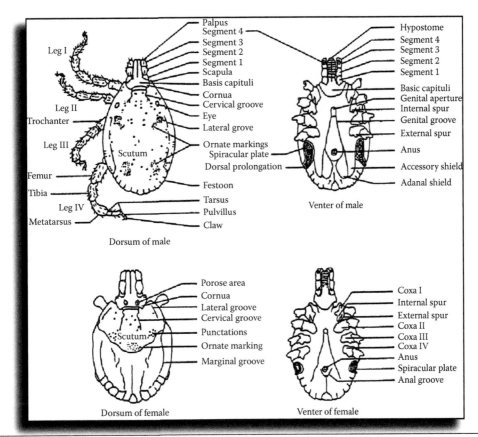

Figure 30.23 Diagnostic characters of hard ticks. (From Strickland RK, et al. Ticks of Veterinary Importance, Agricultural Handbook No. 485. Washington, DC: U.S. Department of Agriculture; 1976.)

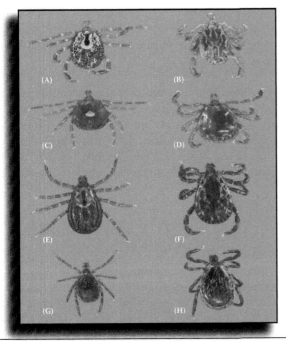

Figure 30.24 Some common hard ticks occurring in the eastern United States: (A,B) Female and male Gulf Coast tick; (C,D) female and male lone star tick; (E,F) female and male American dog tick; (G,H) female and male black-legged tick. (Photographs from Goddard J and Layton B. A Guide to the Ticks of Mississippi, Mississippi Agricultural and Forestry Station Bulletin 1150. Mississippi State University; 2006.)

part identification keys are required. Some identification keys are well written and relatively easy to use (a good example is the pictorial key to adult hard ticks east of the Mississippi River[112]); however, tick identification to the species level is probably best accomplished by an expert to avoid misinformation. In a "study" that the author was asked to review, more than 30% of the ticks collected and identified from Mississippi did not even occur in this country.

IV. Discussion of Some of the Common U.S. Species*

A. *Ornithodoros hermsi* Wheeler, Herms, and Meyer

Medical Importance Primary vector of tick-borne relapsing fever (TBRF) spirochetes in the Rocky Mountain and Pacific Coast states; implicated in several TBRF outbreaks.[79]

---

* The descriptions for each tick species are only general comments about their macroscopic appearance. Tick identification to the species level is quite difficult, utilizing a number of microscopic characteristics. Specific identifications should be performed by specialists at institutions that routinely handle such requests (universities, extension services, state health departments, the military, etc.).

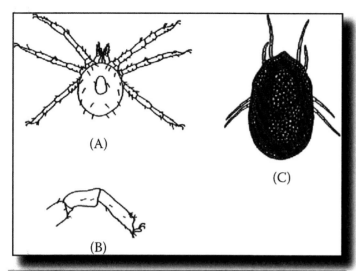

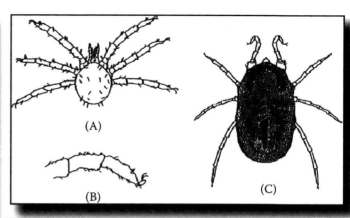

Figure 30.25 (A) Larva, (B) foreleg, and (C) adult of *Ornithodoros hermsi*. (From Goddard J. Ticks and Tick-borne Diseases Affecting Military Personnel, USAFSAM-SR-89-2. an Antonio, TX: School of Aerospace Medicine, U.S. Air Force; S, 1989.)

Description Typical-looking soft tick about 10 mm long, gray in color, and covered with numerous bumplike projections (mammillae); as with all soft ticks, head (mouthparts in this case) is not visible in dorsal view. The foreleg is depicted in Figure 30.25.

Distribution California, Nevada, Idaho, Oregon, Utah, Arizona, Washington, and Colorado, as well as in British Columbia, Canada (Figure 30.11).

Hosts Rodents and humans.

Seasonality Varies with geographic location, hosts, and habitat.

Remarks Often found infesting corners and crevices of vacation or summer cabins; they remain attached to a host for only about 15 to 30 minutes. There are usually four nymphal molts; the cycle from egg to egg takes about 4.5 months. They are often found in coniferous forests at elevations above 1000 meters.

B. Relapsing Fever Tick, *Ornithodoros turicata* (Dugès)

Medical Importance May produce an intense irritation and edema at the bite site in humans; serves as a vector of relapsing fever spirochetes in portions of Kansas, Oklahoma, Texas, and other southwestern states.

Description About 10 mm long, gray in color, and covered with bumplike projections (mammillae); the foreleg is depicted in Figure 30.26.

Figure 30.26 (A) Larva, (B) foreleg, and (C) adult of *Ornithodoros turicata*. (From Goddard J. Ticks and Tick-borne Diseases Affecting Military Personnel, USAFSAM-SR-89-2. San Antonio, TX: School of Aerospace Medicine, U.S. Air Force; 1989.)

Distribution Texas, New Mexico, Oklahoma, Kansas, California, Colorado, Arizona, Florida, and Utah; reported from Mexico in the states of Aguascalientes, Coahuila, Guanajuato, Morelos, Queretaro, San Luis Potosi, and Sinaloa (Figure 30.11); also reportedly found in Venezuela, Honduras, Bolivia, Chile, and Argentina. Records from Central and South America are perhaps incorrect (see Guglielmone[113] for an extensive discussion about argasid distribution records).

Hosts Collected from rattlesnakes, turtles, birds, rodents, rabbits, sheep, cattle, horses, pigs, and humans.

Seasonality Varies with geographic location, hosts, and habitat; may be active in warmer geographic areas throughout the year.

Remarks Often found in burrows used by rodents or burrowing owls. The bite is painless but may be followed in a few hours by intense local irritation and swelling; subsequently, subcutaneous nodules may form and persist for months. There are three to five nymphal stages; the time required for development from larva to adult is approximately 6 months.

C. Lone Star Tick, *Amblyomma americanum* (Linnaeus)

Medical Importance Goddard and Varela-Stokes[114] provided a detailed review of the disease potential of this species. It transmits the pathogen of tularemia to humans; it is a known vector of the agent of human ehrlichiosis (*Ehrlichia chaffensis*) and is reported to rarely transmit the agent of Rocky Mountain spotted fever (RMSF). *Borrelia* species spirochetes (but not true Lyme disease agent) have been recovered from this species.[115] This tick is also found naturally infected with heartland virus (pathogenic), *Rickettsia amblyommatis* (mostly nonpathogenic),[116,117] *Rickettsia parkeri* (pathogenic),[118,119] and

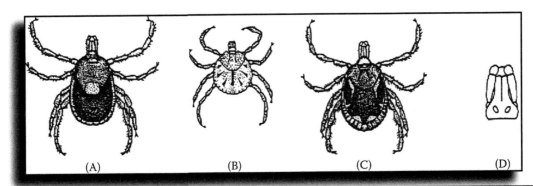

Figure 30.27 (A) Adult female, (B) nymph, (C) adult male, and (D) dorsal view of capitulum of *Amblyomma americanum*. (From Goddard J. Ticks and Tick-borne Diseases Affecting Military Personnel, USAFSAM-SR-89-2. San Antonio, TX: School of Aerospace Medicine, U.S. Air Force; 1989.)

Ehrlichia ewingii (pathogenic).[120] Indirectly, this species may cause red meat allergy in humans.[121]

Description Palps are long, with segment 2 at least twice as long as segment 3. Mouthparts are visible from above (in contrast to the soft ticks); eyes are present, but not in sockets. It is a reddish-brown tick species. Adult females have a distinct single white spot on their back (scutum); males have no single spot and instead have inverted horseshoe-shaped markings at the posterior edge of their dorsal side (Figure 30.27).

Distribution Central Texas east to the Atlantic Coast and north to approximately Iowa, New York, and southern New England; Expanding northward; reported from Mexico in several northern states (Figure 30.28);[122] also occasionally reported from Panama, Venezuela, Argentina, Guatemala, Guayana, and Brazil (although many Central and South American records of this species may not be valid).

Hosts Extremely aggressive and nonspecific in its feeding habits; all three motile life stages will feed on a wide variety of mammals, including humans, and ground-feeding birds.

Seasonality Varies by location, but adults and nymphs are generally active from early spring through midsummer, with larvae being active from late summer into early fall.

Remarks Probably the most annoying and commonly encountered tick occurring in the southern United States; in some rural areas, almost every person has been bitten by these ticks at one time or another. The "seed ticks" occurring in late summer in the southern United States are most often this species (Figure 30.29) and produce itchy red lesions on persons bitten by them.[123] Lone star ticks are especially found in ecotonal zones between forested and open (meadow) areas, usually where there is an abundance of deer or other hosts; they seldom occur in high numbers in the middle of pastures or meadows because of low humidities and high daytime temperatures present in those areas. Larvae may survive

Figure 30.28 Approximate geographic distribution of *Amblyomma americanum*.

from 2 to 9 months; nymphs and adults 4 to 15 months each. Females usually deposit 3000 to 8000 eggs. Females may be falsely referred to as the "spotted fever tick" because of the single white spot visible on their back; however, this spot has nothing to do with the presence or absence of RMSF organisms. Adults have very long mouthparts and can produce painful bites.

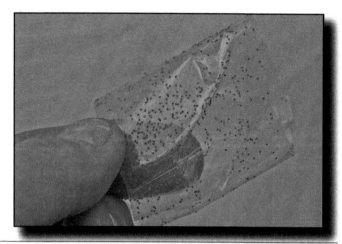

Figure 30.29 Larval lone star ticks on a piece of tape, showing extremely small size.

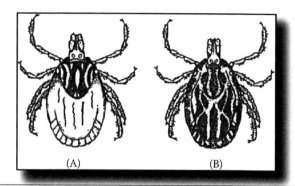

Figure 30.30 (A) Adult female and (B) male *Amblyomma maculatum*.

D. Gulf Coast Tick, *Amblyomma maculatum* Koch

Medical Importance Known vector of American boutonneuse fever (infection with *Rickettsia parkeri*)[48]; may cause tick paralysis; nuisance effects due to painful bites.

Description Macroscopically, somewhat similar to the American dog tick, except that it has metallic markings (instead of white) and long mouthparts typical of all *Amblyomma*. It is a large tick species with long mouthparts visible from above. Adult females have metallic white or gold markings on the scutum (Figure 30.30); males have numerous, mostly connected, linear spots of golden white.

Distribution *Amblyomma maculatum* in the narrowest sense has historically been found along of Atlantic and Gulf coast areas (generally 100 to 200 miles inland) and south into parts of Mexico (Figure 30.31); Currently expanding inland and northward in the U.S. The *A. maculatum* complex (broader sense of the species) extends into South America (Figure 30.31).

Hosts Adults on large animals, including deer, cattle, sheep, and humans; larvae and nymphs on small mammals and ground-feeding birds such as cotton rats, meadow larks, and bobwhite quail.

Seasonality Variable depending on geographic location; larvae are generally active from July through November, and nymphs may be active almost year-round, but especially in the spring (March). Adults can be found from March through September, usually peaking in activity during August.

Remarks Increasingly a pest in the southern United States; large ticks which are often found in the ears of cattle, producing great irritation, destruction of cartilage, and drooping,

Figure 30.31 Approximate geographic distribution of *Amblyomma maculatum*.

or *gotched* ears.[124] Produce painful bites in humans; interestingly, can survive in the middle of open fields and pastures in direct sunlight.

E. Rocky Mountain Wood Tick, *Dermacentor andersoni* Stiles

Medical Importance Primary vector of RMSF in the Rocky Mountain states and also known to transmit the causative agents of Colorado tick fever and tularemia; it produces cases of tick paralysis in the United States and Canada each year.[125]

Description   Adults have shorter mouthparts than the *Amblyomma* species and are usually dark brown or black with bright white markings on the scutum (see Figures 30.32 and 30.33 for pattern); basis capituli are rectangular when viewed from above. A pair of medially directed spurs are on the first pair of coxae.

Distribution   Found from the western counties of Nebraska and the Black Hills of South Dakota to the Cascade and Sierra Nevada Mountains; also reported from northern Arizona and northern New Mexico to British Columbia, Alberta, and Saskatchewan, Canada (Figure 30.34).[126,127]

Hosts   Immatures prefer many species of small mammals such as chipmunks and ground squirrels, whereas adults feed mostly on cattle, sheep, deer, humans, and other large mammals.

Seasonality   Larvae feed throughout the summer, and adults usually appear in March, disappearing by July; nymphs

may continue to be present, although in diminishing numbers, until late summer.

Remarks   Especially prevalent where there is brushy vegetation to provide good protection for small mammalian hosts of immatures and with sufficient forage to attract large hosts required by adults; unfed larvae may live for 1 to 4 months,

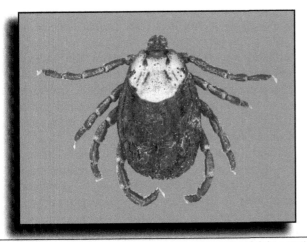

Figure 30.32   Adult female *Dermacentor andersoni*, which is a vector of RMSF, Colorado tick fever, and tularemia in the western United States. (Photograph courtesy of Dr. Blake Layton, Mississippi State University Extension Service.)

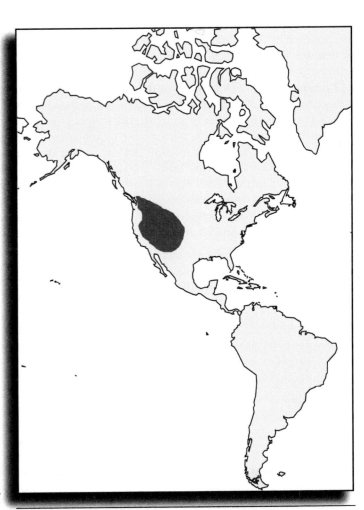

Figure 30.34   Approximate geographic distribution of *Dermacentor andersoni*.

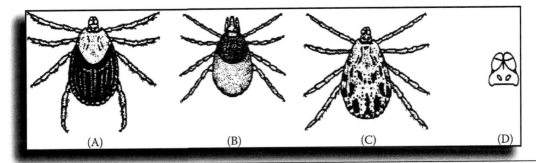

Figure 30.33   (A) Adult female, (B) nymph, (C) adult male, and (D) dorsal view of capitulum of *Dermacentor andersoni*. (From Goddard J. Ticks and Tick-borne Diseases Affecting Military Personnel, USAFSAM-SR-89-2. San Antonio, TX: School of Aerospace Medicine, U.S. Air Force; 1989.)

CASE HISTORY

TICK ON MAN'S EARDRUM

One week after sleeping on the ground near Mexican Hat, Utah, a man began to have severe pain in his right ear. After a few days, lymphadenitis developed in the neck region surrounding the ear. He was seen by a family physician approximately 7 days later, who referred him to an ear, nose, and throat specialist (ENT). After 4 days (and 18 days after camping), the ENT examined the patient's eardrum and found a small, "crablike" creature attached, which he removed (see figure). As none of the physicians in the medical group knew exactly what the creature was, the patient—with specimen in hand—was referred to a parasitologist at the University of Mississippi Medical Center. The parasitologist identified the specimen as a tick and referred the patient to the Mississippi Department of Health for more specific identification. The tick was identified as a nymphal spinose ear tick, *Otobius megnini*. The patient recovered uneventfully.

Comment: This tick belongs to the soft tick family. Most ticks seen by people are hard ticks—the ones that attach to a vertebrate host such as a dog or deer and remain attached for a week or so. Soft ticks are classified in another entire tick family as they differ from hard ticks in appearance and behavior. Soft ticks have little sexual dimorphism and have a wrinkled, granulated integument; the mouthparts are generally not visible when the specimen is viewed from above. Soft ticks are especially adapted to dry climates or dry conditions within wet climates, and generally only feed on their vertebrate hosts for a short time (1 hour or less), not remaining attached for days. Most soft tick species in the United States occur in the West. The tick involved in this particular case differs from other soft ticks in that the larval and nymphal stages invade ears of cattle, horses, sheep, deer, and other wild animals and remain attached for long periods of time. There have been records of this species remaining in the ears of animals for as long as 121 days. The scientific literature contains several records of this species being found in the ears of humans.

Tick removed from patient's eardrum.

This case is interesting and illuminates several issues. For one thing, people travel and may bring back with them all sorts of parasites or microbes. Species that normally do not occur in an area may be seen from time to time by practicing physicians. Second, none of the physicians knew that this creature was a tick. To their credit, the physicians knew that the specimen was some sort of parasite and referred the man to a parasitologist. The parasitologist—a broadly trained, organismal-level scientist knowing a wide range of internal and external parasites on sight—immediately knew it was a tick. The medical community needs such scientists, especially in light of rapid, modern travel methods and immigration (both legal and illegal). In the last 20 years or so, there has been a drift away from organismal-level training, with increasing emphasis on molecular biology, but there will always be a need for scientists who can "eyeball" a specimen and place it in its appropriate taxonomic group. A key to treatment strategies in such cases is knowledge of the biology and behavior of the parasite and what diseases, if any, it transmits.

---

nymphs for 10 months or more, and adults 14 months or longer. Females deposit about 4000 eggs.

F. American Dog Tick, *Dermacentor variabilis* Say

Medical Importance A medically important tick in the United States; it is the primary vector of RMSF in the East and also transmits tularemia and causes tick paralysis.[125,128]

Description Adults are dark brown or black with short, thick mouthparts; this tick has dull or bright white markings on the scutum (see Figure 30.35 for pattern). Basis capituli is rectangular when viewed from above; a pair of medially directed spurs are on the first pair of coxae.

Distribution Throughout the United States except in parts of the Rocky Mountain region; also established in Nova Scotia, Manitoba, and Saskatchewan, Canada. It has been reported in Mexico from Chiapas, Guanajuato, Hidalgo, Oaxaca, Puebla, San Luis Potosi, Sonora, Tamaulipas, and Yucatan (Figure 30.36).

Hosts Immatures feed primarily on small mammals (particularly rodents); adults prefer the domestic dog, but will readily bite humans.

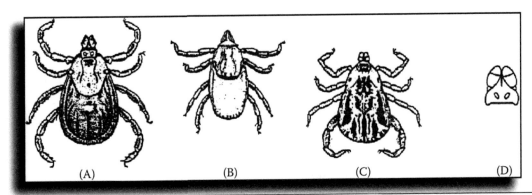

Figure 30.35 (A) Adult female, (B) nymph, (C) adult male, and (D) dorsal view of capitulum of *Dermacentor variabilis*. (From Goddard J. Ticks and Tick-borne Diseases Affecting Military Personnel, USAFSAM-SR-89-2. San Antonio, TX: School of Aerospace Medicine, U.S. Air Force; 1989.)

Seasonality Adults active from about mid-April to early September, nymphs predominate from June to early September, and larvae are active from about late March through July.

Remarks Principal vector of RMSF in the central and eastern United States; deticking dogs is an important mode of RMSF transmission that may be overlooked (handpicking *Dermacentor variabilis* from dogs is dangerous because infected tick secretions on the hands may be transmitted through contact with the eyes, mucous membranes, etc.). Unfed larvae may live up to 15 months, nymphs 20 months, and adults up to 30 months or longer. Females deposit 4000 to 6500 eggs.

G. Western Black-Legged Tick, *Ixodes pacificus* Cooley and Kohls

Medical Importance Known to be a vector of Lyme borreliosis spirochetes; most, if not all, cases of Lyme borreliosis occurring in California are transmitted by this tick, which also transmits the agent of human granulocytic anaplasmosis (HGA). There are reports of Type I (IgE-mediated) hypersensitivity reactions in humans as a result of bites by this species.

Description No white markings on dorsal side, no eyes or festoons (Figure 30.37) (like other members of genus). Anal groove encircles the anus anteriorly. Males have sclerotized ventral plates. Adults are generally dark brown in color. Females have moderately long mouthparts. This tick looks almost identical to *Ixodes scapularis*.

Distribution Along the Pacific coastal margins of British Columbia, Canada, and the United States, possibly extending into Baja, California, and other parts of Mexico (Figure 30.38).[129] It also has been reported from at least one area in Arizona.

Hosts Immatures feed on numerous species of small mammals, birds, and lizards; in certain areas of California, the

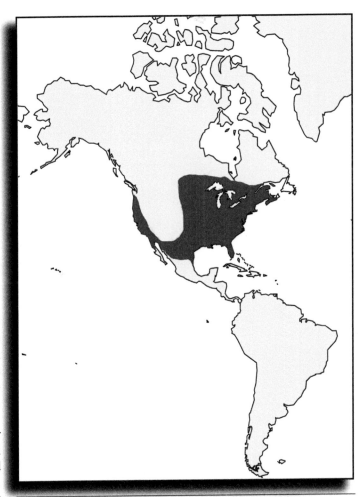

Figure 30.36 Approximate geographic distribution of *Dermacentor variabilis*.

predominance of feeding is on lizards. Adults feed primarily on Columbian black-tailed deer.

Seasonality Adults primarily active from fall to late spring, with immatures being active in the spring and summer.

Remarks Adult females, like *Ixodes scapularis*, have long mouthparts, enabling them to be especially painful parasites

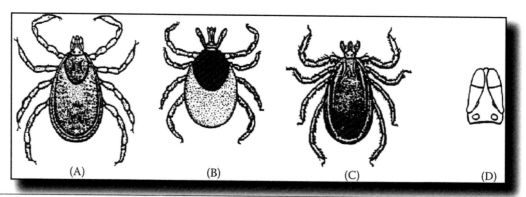

Figure 30.37 (A) Adult female, (B) nymph, (C) adult male, and (D) dorsal view of capitulum of *Ixodes pacificus*. (From Goddard J. Ticks and Tick-borne Diseases Affecting Military Personnel, USAFSAM-SR-89-2. San Antonio, TX: School of Aerospace Medicine, U.S. Air Force; 1989.)

Figure 30.38 Approximate geographic distribution of *Ixodes pacificus*.

of humans. Adults are most abundant in the early spring. Infection rates with Lyme borreliosis spirochetes are usually in the range of 1 to 5% compared to rates of 25 to 75% in *I. scapularis* (effect may be related to vector competence or host preferences of the immatures). Immatures will bite people.

H. Black-Legged Tick, *Ixodes scapularis* Say

Taxonomic note: The name *Ixodes dammini* is often seen in older scientific literature associated with Lyme disease and an explanation is needed. For several years, a northern form of *Ixodes scapularis* was thought to be a distinct species named

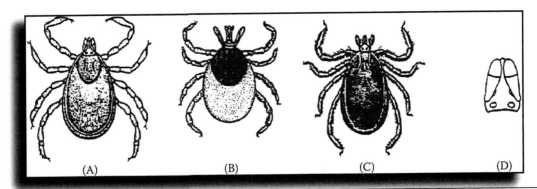

Figure 30.39 (A) Adult female, (B) nymph, (C) adult male, and (D) dorsal view of capitulum, of *Ixodes scapularis*. (From Goddard J. Ticks and Tick-borne Diseases Affecting Military Personnel, USAFSAM-SR-89-2. San Antonio, TX: School of Aerospace Medicine, U.S. Air Force; 1989.)

Ixodes dammini (see Goddard[130] for a discussion of this issue). Subsequently, evidence was produced indicating that the two are one species.[131] Morphologically, they are almost identical; however, there are important behavioral differences, especially in the immature stages.

Medical Importance   Primary vector of the causative agent of Lyme borreliosis, especially in the northeastern and upper midwestern areas of the United States. This tick is a vector of the protozoan *Babesia microti* in the Northeast and Upper Midwest and is a vector of the agent of human granulocytic anaplasmosis (HGA) and also a new *Ehrlichia* called the *Ehrlichia muris*-like agent or *E. eauclairensis*.[63] It is also vector of "deer tick virus" (closely related to POW) and is sometimes said to be a vector of *Bartonella henselae*, but this claim is thus far not well documented.[132]

Description   Adults have no eyes, festoons, or white markings on their dorsal side (Figure 30.39); they have an anal groove that encircles the anus anteriorly. Males have sclerotized ventral plates. It is dark brown in color (occasionally the abdomen from dorsal view is light brown or orangish).

Distribution   New England states and New York, also Upper Midwest and Ontario Canada, south to Florida, and westward to Texas and Oklahoma; also reported from several Mexican states (Figure 30.40).

Hosts   Immatures feed on lizards, small mammals, and birds. Adults prefer deer but will bite people; in Mexico, additional hosts include dogs, cattle, deer, and rabbits.

Seasonality   In the United States, adults are active in fall, winter, and spring (Figure 30.41); this seasonal activity pattern is interesting, as most people do not think of ticks as being active in the dead of winter. Immatures are active in spring and summer.

Figure 30.40 Approximate geographic distribution of *Ixodes scapularis*.

Remarks   Congregates along paths, trails, and roadways in various types of forested areas such as those with mature pine hardwoods with dogwood, wild blueberry, privet, blackberry, huckleberry, and sweetgum. This tick inflicts a painful bite. Adult males rarely bite. Nymphs of northern *Ixodes scapularis*

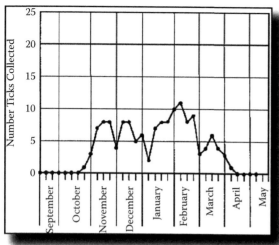

Figure 30.41 Seasonal activity of adult *Ixodes scapularis* in Mississippi, as determined by dragging vegetation with a white flannel cloth.

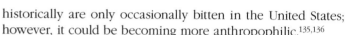

Figure 30.42 Adult female *Rhipicephalus sanguineus*. (Photo courtesy Dr. Blake Layton, Mississippi State University Extension Service).

bite people aggressively during summer months and can be collected with a drag cloth. Nymphs of southern *I. scapularis* rarely, if ever, bite people and can rarely be collected with drag cloths. Most hard ticks acquired by persons in the south-central and southeastern states in the winter months are of this species; in one study,[133] they were most often collected questing at around 20°C but were collected on days as cold as 6.9°C.

I. Brown Dog Tick, *Rhipicephalus sanguineus* Latreille

Medical Importance Now known to transmit the agent of RMSF in the western U.S.; in southern Europe and Africa, this tick is a known vector of *Rickettsia conorii*, the causative agent of boutonneuse fever.

Description Light to dark brown in color, with no white markings on the dorsum (Figure 30.42); it has hexagonal-shaped basis capituli. Festoons and eyes are both present.

Distribution Probably the most widely distributed of all ticks, being found almost worldwide (Figure 30.43).[134] In the Western Hemisphere, it has been reported from most of the United States and southeastern and southwestern parts of Canada; it has also been reported from most of Mexico, Argentina, Venezuela, Colombia, Brazil, Nicaragua, Panama, Uruguay, Paraguay, Galapagos Islands, Surinam, British Guiana, French Guiana, Peru, Costa Rica, and the Caribbean islands of Cuba, Jamaica, and the Bahamas. It is also widely distributed throughout Eurasia, Africa, and Australia.

Hosts The dog is the principal host, although in immature stages it sometimes attacks numerous other animals. Humans

historically are only occasionally bitten in the United States; however, it could be becoming more anthropophilic.[135,136]

Seasonality May be active in the warmer parts of its range year-round; however, in temperate zones, adults and immatures are primarily active from late spring to early fall.

Remarks Most often found indoors in and around pet bedding areas; it has a strong tendency to crawl upward and can often be seen climbing the walls of infested houses. It is associated with homes and yards of pet owners and is seldom found out in the middle of a forest or uninhabited area. Unfed larvae may survive as long as 8.5 months, nymphs 6 months, and adults 19 months. Females usually lay 2000 to 4000 eggs.

V. Discussion of Some Major Pest Species in Other Areas of the World

A. Eyeless Tampan, *Ornithodoros moubata* Murray

Medical Importance Known vector of African tick-borne relapsing fever spirochetes in eastern, central, and southern Africa.

Description Like many other soft ticks, this tick is about 9 to 12 mm long in the adult stage. It has a bumpy integument (mammillated) and protuberances on the tarsi (Figure 30.44).

Distribution Throughout eastern Africa and the northern portions of southern Africa, extending into the drier parts of central Africa (Figure 30.45).

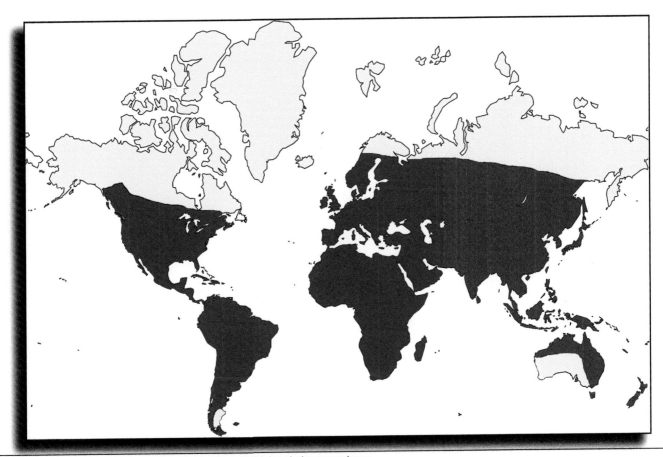

Figure 30.43 Approximate geographic distribution of *Rhipicephalus sanguineus*.

Hosts Humans, warthogs, domestic pigs, antbears, and porcupines.

Seasonality Varies with geographic location, hosts, and habitat.

Remarks Often found in cracks in walls and in earthen floors of huts. Female usually lays six to seven batches of eggs (several hundred per batch) during her lifetime. Larvae do not feed; nymphs engorge in about 20 to 25 minutes. Usually, there are four nymphal molts for males and five for females. This tick is able to live up to 5 years without feeding.

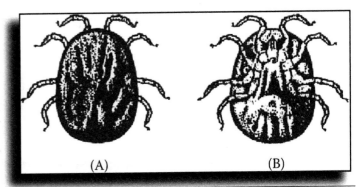

Figure 30.44 (A) Dorsal and (B) ventral view of *Ornithodoros moubata*. (From Goddard J. Ticks and Tick-borne Diseases Affecting Military Personnel, USAFSAM-SR-89-2. San Antonio, TX: School of Aerospace Medicine, U.S. Air Force; 1989.)

B. *Ornithodoros rudis* (includes *venezuelensis*) Karsch

Sometimes this species is listed as *Carios rudis*,[137] although not all tick taxonomists agree.[138]

Medical Importance Most important vector of relapsing fever spirochetes in Panama, Colombia, Venezuela, and Ecuador.

Description Unlike some of the other soft ticks, this tick has no dorsal humps on legs (see Figure 30.46).

Distribution Panama, Paraguay, Colombia, Venezuela, Peru, and Ecuador (Figure 30.47).

Hosts Domestic birds and humans.

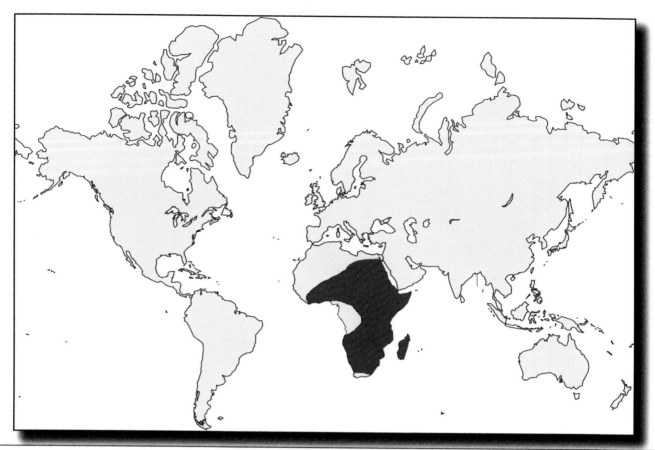

Figure 30.45 Approximate geographic distribution of *Ornithodoros moubata*.

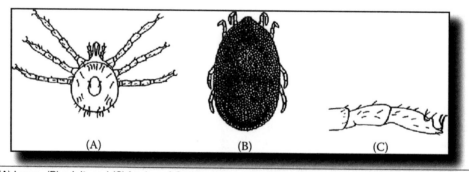

Figure 30.46 (A) Larva, (B) adult, and (C) foreleg of *Ornithodoros rudis*. (From Goddard J. Ticks and Tick-borne Diseases Affecting Military Personnel, USAFSAM-SR-89-2. San Antonio, TX: School of Aerospace Medicine, U.S. Air Force; 1989.)

Seasonality Varies with geographic location, hosts, and habitat; it may be active in warmer areas throughout the year.

Remarks Appears especially adapted as a parasite of humans but feeds on other animals; it is a night feeder, with the larval stages engorging rapidly. It has three to four nymphal stages, and the developmental time from larvae to adult is about 3 months.

C. *Ornithodoros talaje* Guérin-Méneville

Medical Importance Transmits the agent of relapsing fever to humans in Guatemala, Panama, and Colombia.

Description Typical soft tick with large disks (round spots mostly on the dorsal side of the integument) and no humps on tarsi (Figure 30.48).

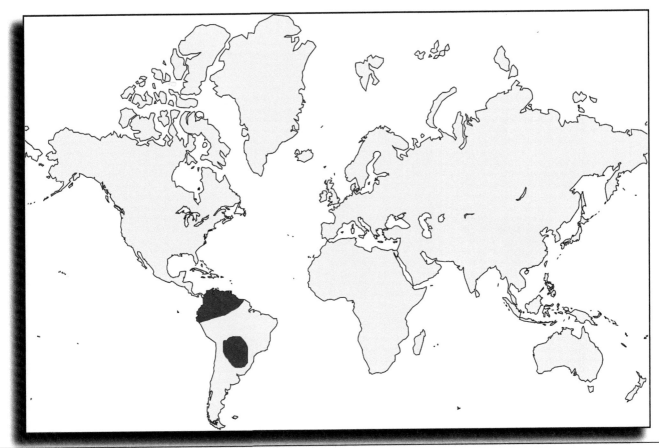

Figure 30.47 Approximate geographic distribution of *Ornithodoros rudis*.

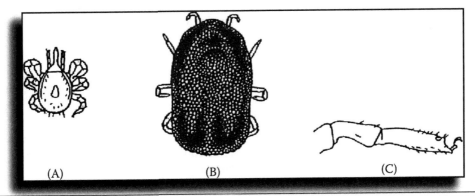

Figure 30.48 (A) Larva, (B) adult, and (C) foreleg of *Ornithodoros talaje*. (From Goddard J. Ticks and Tick-borne Diseases Affecting Military Personnel, USAFSAM-SR-89-2. San Antonio, TX: School of Aerospace Medicine, U.S. Air Force; 1989.)

Distribution Reported in Florida, Texas, Arizona, Nevada, Kansas, New Mexico, and California; however, Hoogstraal[139] maintained that in the United States it has only been reported from Kansas and California. It also occurs in Mexico in the states of Baja California, Chiapas, Guerrero, Morelos, Oaxaca, Puebla, Sinaloa, Sonora, Veracruz, and Yucatan. It has also been reported from Venezuela, Uruguay, Brazil, French Guiana, Panama, Ecuador, and Chile. Hoffman[140] noted that this species has also been reported from Guatemala, Colombia, Argentina, and Galapagos Islands, although, according to Keirans et al.,[141] the *Ornithodoros talaje* reported from Galapagos Islands actually is *O. galapagensis* (Figure 30.49).

Hosts Rodents (principally) and humans, as well as birds, bats, pigs, cattle, horses, opossums, and snakes.

Figure 30.49 Approximate geographic distribution of *Ornithodoros talaje*.

Seasonality Varies with geographic location, hosts, and habitat; it may be active in warmer geographic areas throughout the year.

Remarks Adults are seldom observed in dwellings and are not avid parasites of humans. Larvae remain attached to a host for several days. It has three to four nymphal stages, and developmental time from larva to adult is about 8 months.

D. Cayenne Tick, *Amblyomma cajennense* Fabricius

Taxonomic note: The cayenne tick is actually a complex of several species that look almost identical.[142] Here we just treat them as one.

Medical Importance Probably the most commonly encountered and aggressive of all Central and South American ticks; it is considered a vector of RMSF rickettsiae in Mexico, Panama, Colombia, and Brazil.

Description Long mouthparts, eyes, and festoons similar to that of *Amblyomma americanum*. Males have weblike ornamentation radiating from the center of the scutum, and

females also have extensive ornamentation and festoons with tubercules at the posterior edge (Figure 30.50).

Distribution Extreme southern Texas, south throughout most of Mexico and Central America and into parts of South America; several Caribbean islands, including Cuba and Jamaica; Brazil, Honduras, Venezuela, Costa Rica, Uruguay, Ecuador, Nicaragua, and Bolivia (Figure 30.51). Hoffman[140] reported that it also occurs in Guatemala, Colombia, Guyana, Paraguay, and Argentina.

Hosts All active stages commonly attack people, domestic and wild animals, and ground-frequenting birds.

Seasonality May be active in tropical areas year-round; activity may diminish in midwinter in the cooler areas at the northernmost and southernmost extent of distribution.

Remarks Very similar to *Amblyomma americanum* in aggressiveness and nonspecific feeding habits; basically, where the southernmost distribution of *A. americanum* stops, *A. cajennense* picks up and continues southward throughout Central and South America. The longevity of larvae, nymphs, and adults and the numbers of eggs laid by engorged females

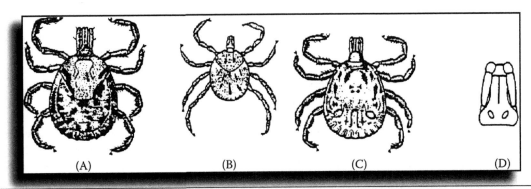

Figure 30.50   (A) Adult female, (B) nymph, (C) adult male, and (D) dorsal view of capitulum of *Amblyomma cajennense*. (From Goddard J. Ticks and Tick-borne Diseases Affecting Military Personnel, USAFSAM-SR-89-2. San Antonio, TX: School of Aerospace Medicine, U.S. Air Force; 1989.)

Figure 30.51   Approximate geographic distribution of *Amblyomma cajennense*.

are similar to those of *A. americanum*; as with *A. americanum*, *A. cajennense* has long mouthparts and produces painful bites.

E. Bont Tick, *Amblyomma hebraeum* Koch

Medical Importance   One of several ixodid vectors of *Rickettsia conorii*, the agent of boutonneuse fever and also a vector of African tick bite fever organisms, *Rickettsia africae*.

Description   Long mouthparts, eyes, and festoons typical of other *Amblyomma* spp. (Figure 30.52). Males have a pattern of pale stripes and spots on a dark brown background, and

females have central and lateral pale patches against a dark background.

Distribution   Distributed throughout southern Africa, specifically eastern Botswana, South Africa, southern and central Zimbabwe, and southern Mozambique;[143] it has been accidentally introduced into the United States on several occasions (primarily on rhinoceroses) but each time has been successfully eliminated (Figure 30.53).

Hosts   Immatures feed on many medium- and large-sized mammals such as dogs, cats, and cattle, but particularly on wild hares. Adults parasitize a variety of domestic and wild mammals but seem to prefer cattle and antelopes. All life stages will bite people.

Seasonality Active in spring, summer, and fall months; in South Africa, adults are most abundant on hosts during the late summer and autumn.

Remarks Larvae, like their U.S. cousins, the lone star ticks, are troublesome pests of people; they attach themselves in large numbers on the legs and about the waist, causing intense irritation, rashlike lesions, and occasional pustules.

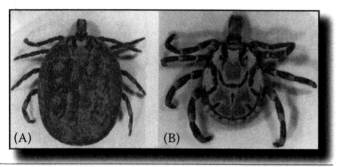

Figure 30.52 (A) Female and (B) male *Amblyomma hebraeum*. (From Goddard J. Ticks and Tick-borne Diseases Affecting Military Personnel, USAFSAM-SR-89-2. San Antonio, TX: School of Aerospace Medicine, U.S. Air Force; 1989.)

Unfed larvae may live for up to 11 months, nymphs for 8 months or more, and adults for 22 months or longer. Females deposit about 15,000 eggs.

F. *Dermacentor marginatus* Sulzer

Medical Importance Primary vector of Siberian tick typhus rickettsia in Eurasia; this tick is a vector of *Rickettsia slovaca* (tick-borne lymphadenopathy) in Europe,[144] a vector of viruses in the tick-borne encephalitis complex, and a possible vector of Omsk hemorrhagic fever virus.

Description Usually ornate specimens with both eyes and festoons present (Figure 30.54) and with rectangular basis capituli dorsally. Males have a weakly defined spur on the posterodorsal margin of the second palpal segment and a mixture of both large and small punctations over the scutum; females have small ventral spurs on trochanters II and III.

Distribution Many areas of western and central Europe; specific countries include Afghanistan, Albania, Bulgaria, Czech Republic, Portugal, Slovakia, Spain, France, Germany, Greece, Hungary, Romania, Switzerland, Iran, Iraq, Poland, Italy, Turkey, Yugoslavia, Russia, Belarus, Ukraine, Kazakhstan, Uzbekistan, Kyrgyzstan, Georgia, Azerbaijan, and Armenia (Figure 30.55).

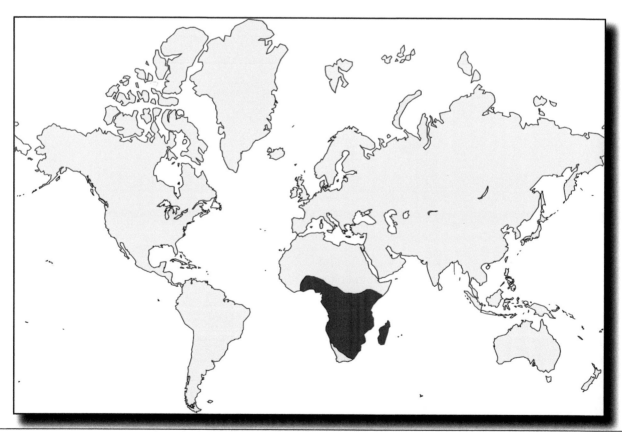

Figure 30.53 Approximate geographic distribution of *Amblyomma hebraeum*.

Figure 30.54 Male *Dermacentor marginatus*. (Photo courtesy Dr. Lorenza Beati and the U.S. National Tick Collection)

Hosts Adults parasitize horses, cattle, sheep, people, dogs, buffalo, swine, camels, and hedgehogs; immatures feed most frequently on small mammals, especially rodents.

Seasonality Adults are generally active in spring and again in autumn. Larvae usually peak in activity in June and July; nymphs in July and August.

Remarks This tick inhabits shrubby areas, low forests, marshes, lowlands, alpine steppes, and semidesert areas; in southeastern France, it is found in close association with woods where oaks, *Quercus pubescens*, predominate.[145]

G. *Dermacentor nuttalli* Olenev

Medical Importance One of several known vectors of Siberian tick typhus; also vector of agent of tularemia in northern Eurasia.

Description Characteristics common to all members of genus *Dermacentor*; females have no internal spurs on coxa IV and no cornua (Figure 30.56).

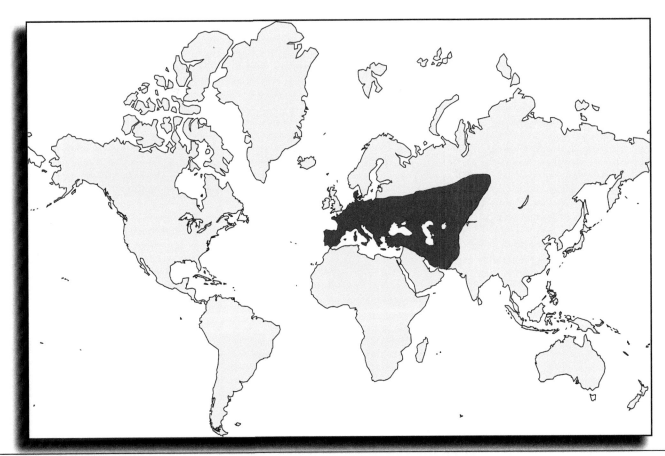

Figure 30.55 Approximate geographic distribution of *Dermacentor marginatus*.

Distribution   Through central and eastern Siberia, Asiatic Russia, northern Mongolia, and China (Figure 30.57); occasionally reported from Ukraine and Kazakhstan.

Hosts   Immatures generally parasitize small mammals such as field mice, rats, marmots, hamsters, hares, cats, and dogs. Adults feed predominantly on larger hosts, such as horses, cattle, camels, sheep, dogs, and humans.

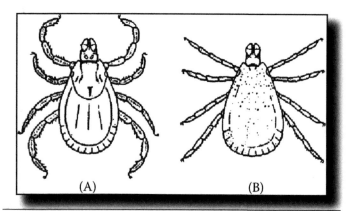

Figure 30.56   (A) Female and (B) male *Dermacentor nuttalli*. (From Goddard J. Ticks and Tick-borne Diseases Affecting Military Personnel, USAFSAM-SR-89-2. San Antonio, TX: School of Aerospace Medicine, U.S. Air Force; 1989.)

Seasonality   Larvae and nymphs active from mid-June to mid-August with adults active primarily in the spring (peaking in mid-May); Splisteser and Tyron[146] reported high population numbers of adult *Dermacentor nuttalli* in steppe regions of Mongolia from mid-March to late May.

Remarks   Seems to be especially associated with high grasslands; generally is not found in dense forests, river lowlands, or hilly wooded country. Unfed adults usually overwinter in cracks in soil and occasionally in burrows of rodents. They cease questing and become inactive at temperatures below 10°C (50°F).

H. *Dermacentor silvarum* Olenev

Medical Importance   Vector of Siberian tick typhus rickettsia in Eurasia and Asia, as well as viruses in the tick-borne encephalitis complex; may be a vector of Lyme borreliosis spirochetes in Asia.

Description   Males have ornamentation similar to that in Figure 30.58B, with a prominent dorsal spur on trochanter I. Females have coxa IV without internal spurs. Cornua are present and ventral spurs lacking on basis capituli.

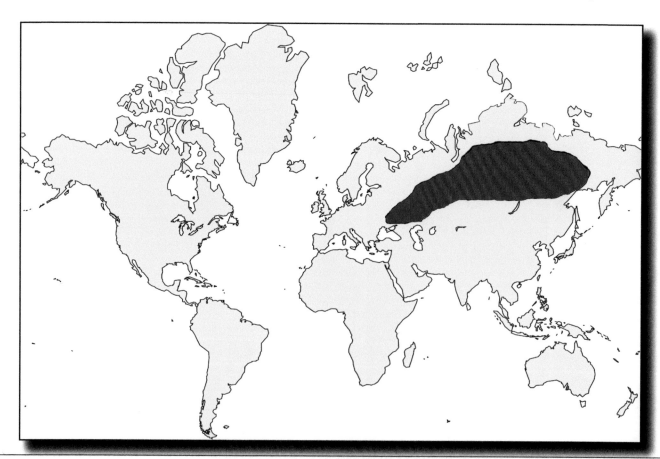

Figure 30.57   Approximate geographic distribution of *Dermacentor nuttalli*.

Distribution Primarily in eastern and far eastern Russia and northern Mongolia; also reported from Belarus, Ukraine, Lithuania, Latvia, Estonia, Georgia, Kazakhstan, Uzbekistan, Turkmenistan, Kyrgyzstan, Romania, and Yugoslavia (Figure 30.59).

Hosts Adults have been collected from people, horses, cattle, sheep, dogs, fox, and deer. Larvae and nymphs feed on numerous species of small mammals.

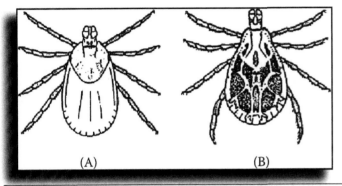

(A) (B)

Figure 30.58 (A) Female and (B) male *Dermacentor silvarum*. (From Goddard J. Ticks and Tick-borne Diseases Affecting Military Personnel, USAFSAM-SR-89-2. San Antonio, TX: School of Aerospace Medicine, U.S. Air Force; 1989.)

Seasonality Bimodal pattern of seasonal activity of adults with one peak in early June and another in early September;[147] larvae are most active in June and July and nymphs from June to mid-August.

Remarks Inhabitants of forest steppe zones; they are most numerous in birch–aspen marshes, glades in mixed forests, cultivated areas in taiga forests, and other localized dense shrub areas and secondary-growth forest.

l. *Haemaphysalis concinna* Koch

Medical Importance Vector of Siberian tick typhus rickettsia and viruses in the tick-borne encephalitis complex; may be a vector of Lyme borreliosis spirochetes in Asia.

Description As a group, these are small inornate ticks with festoons but without eyes (Figure 30.60); the second palpal segment projects beyond the lateral margin of the basis capituli (gives palpi appearance of being triangular—should not be confused with basis capituli being angular, as in the *Rhipicephalus* ticks). Males have pincerlike palps; females have a hypostome dentition of 6/6, and scutum is broadest in the middle.

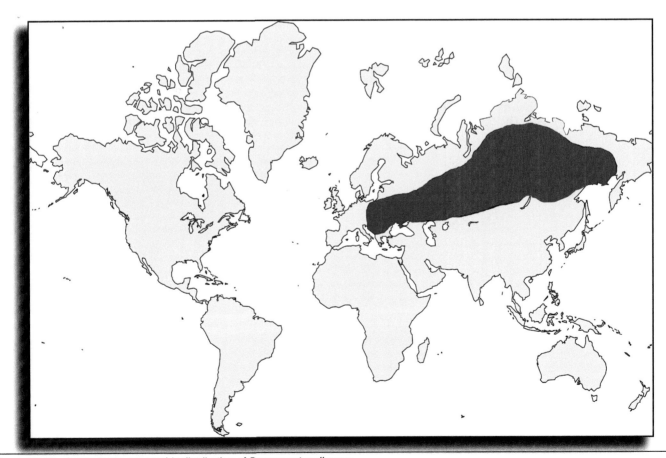

Figure 30.59 Approximate geographic distribution of *Dermacentor silvarum*.

Distribution Widely distributed in forests of temperate Eurasia, including most of Central Europe, Estonia, Latvia, Belarus, Ukraine, Russia, Kazakhstan, Uzbekistan, Turkmenistan, China, Japan, Korea, and Vietnam (Figure 30.61).

Hosts Adults feed on large wild and domestic mammals; immatures infest smaller mammals and birds, sometimes even reptiles. Both adults and nymphs bite people.

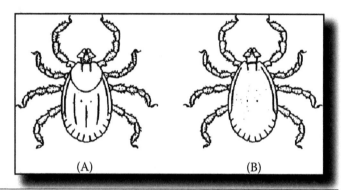

Figure 30.60 (A) Female and (B) male *Haemaphysalis concinna*. (From Goddard J. Ticks and Tick-borne Diseases Affecting Military Personnel, USAFSAM-SR-89-2. San Antonio, TX: School of Aerospace Medicine, U.S. Air Force; 1989.)

Seasonality All stages active from spring to autumn; peak adult activity is in June.

Remarks This tick is found chiefly in deciduous and mixed forests, grass tussock swamps, birch–aspen groves, and alpine taiga forests; it is reported to be abundant in low-lying areas with high humidities.

J. Yellow Dog Tick, *Haemaphysalis leachi* (Audouin) (including *H. l. muhsami*)

Medical Importance Vector of boutonneuse fever rickettsia (human infection with rickettsia may also be acquired by contamination of skin and eyes with infectious fluids from crushing ticks).

Description Typical *Haemaphysalis* sp. (Figure 30.62) characteristically appears to have large, wedge-shaped mouthparts as viewed from above (persons identifying these specimens should be careful not to confuse them with the brown dog tick, *Rhipicephalus sanguineus*).

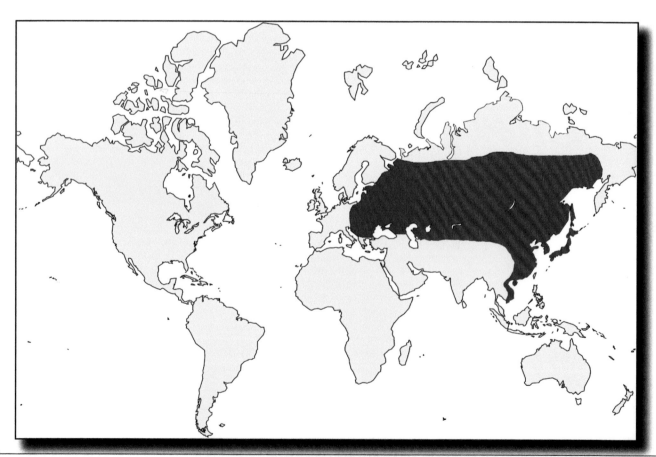

Figure 30.61 Approximate geographic distribution of *Haemaphysalis concinna*.

Distribution Primarily in tropical and southern Africa (although there are records from Algeria, Libya, and Egypt); records from India and Southeast Asia probably represent related but distinct species (Figure 30.63).

Hosts Immatures usually parasitize field rodents; adults commonly bite domestic dogs but will also bite people readily; subspecies *H. leachi muhsami* prefers small carnivores (mongooses, wildcats, etc.) instead of canines.

Seasonality Most active from late spring to early fall.

Remarks Very common on dogs; in some areas, it is more prevalent on dogs than *Rhipicephalus sanguineus*. Usually, two generations are produced each year. Unfed larvae may survive at least 169 days, nymphs 52 days, and adults 210 days. Females lay up to 5000 eggs.

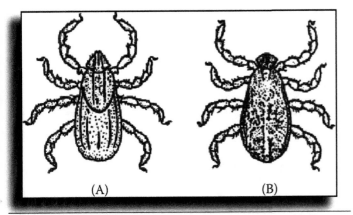

Figure 30.62 (A) Female and (B) male *Haemaphysalis leachi*. (From Goddard J. Ticks and Tick-borne Diseases Affecting Military Personnel, USAFSAM-SR-89-2. San Antonio, TX: School of Aerospace Medicine, U.S. Air Force; 1989.)

K. *Haemaphysalis spinigera* Neumann

Medical Importance Primary vector of the virus of Kyasanur forest disease (KFD) in India.

Description Apart from usual generic characteristics of *Haemaphysalis* (inornate, no eyes, festoons present, triangular-shaped second palpal segment), males have long spurs on coxae I and IV and poorly developed spurs on coxae II and III (Figure 30.64). Females have no spurs on the trochanters.

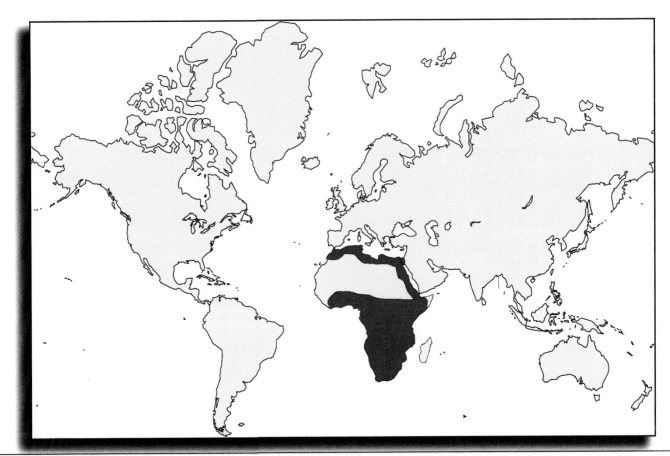

Figure 30.63 Approximate geographic distribution of *Haemaphysalis leachi*.

Distribution Widely distributed in central and southern India; also reported from Southeast Asia and Indonesia (Figure 30.65).

Hosts Immatures parasitize a wide range of small mammals and birds. Adults prefer large mammals such as cattle, monkeys, bears, and tigers. Nymphs avidly bite humans.

Seasonality Generally active in the spring, summer, and fall; immatures peak in numbers from September to November.

Remarks People in the KFD endemic areas (Karnataka, India) turn from agricultural pursuits to wood gathering in forests during the season of peak immature activity, and this activity greatly increases human–tick contact. Hoogstraal[148] reported tick population increases in India because of recent increased cattle-grazing practices in and beside forests. Immatures thrive on numerous small vertebrates hiding in dense lantana thickets where sections of forests were cleared.

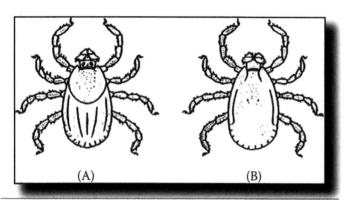

Figure 30.64 (A) Female and (B) male *Haemaphysalis spinigera*. (From Goddard J. Ticks and Tick-borne Diseases Affecting Military Personnel, USAFSAM-SR-89-2. San Antonio, TX: School of Aerospace Medicine, U.S. Air Force; 1989.)

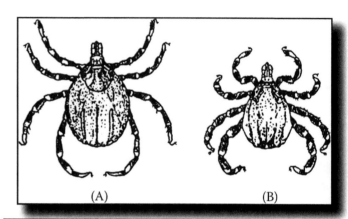

Figure 30.66 (A) Female and (B) male *Hyalomma asiaticum*. (From Goddard J. Ticks and Tick-borne Diseases Affecting Military Personnel, USAFSAM-SR-89-2. San Antonio, TX: School of Aerospace Medicine, U.S. Air Force; 1989.)

Figure 30.65 Approximate geographic distribution of *Haemaphysalis spingera*.

L. Asiatic *Hyalomma, Hyalomma asiaticum* Schulze and Schlottke

Medical Importance   Vector of agent of Siberian tick typhus.

Description   In general, this tick has scant ornamentation, long mouthparts, eyes present in sockets, and festoons (although the festoons are not always clearly delineated);

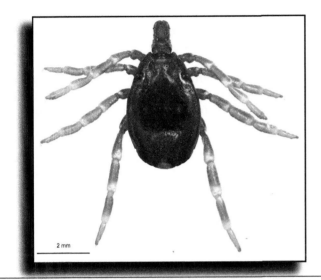

Figure 30.67 *Hyalomma ticks are large and fast moving. (Photo courtesy Dr. Lorenza Beati and the U.S. National Tick Collection).*

often the legs appear banded (Figure 30.66). Females have a scutum longer than wide. Both sexes have white or yellowish bands on legs (Figure 30.67).

Distribution   Widely distributed in Asia, from Syria in the West to eastern China in the East (Figure 30.68).[149]

Hosts   Adults parasitize all domestic animals, especially camels, cattle, horses, and sheep; people, hares, boars, and hedgehogs are less frequently attacked. Immatures feed on hedgehogs, rodents, hares, cats, and dogs.

Seasonality   Most active in spring and summer throughout range.

Remarks   In southwestern Kirghiz (former U.S.S.R.), the foci of Siberian tick typhus occur where southern steppes give way to foothill semidesert zone; in these areas, it is associated with red-tailed birds along dry waterways overgrown with shrubs and along irrigation canals.

M. Small Anatolian *Hyalomma, Hyalomma anatolicum* Koch

Taxonomic note: Hyalomma excavatum, a closely related species, was previously considered to be the same as *H.*

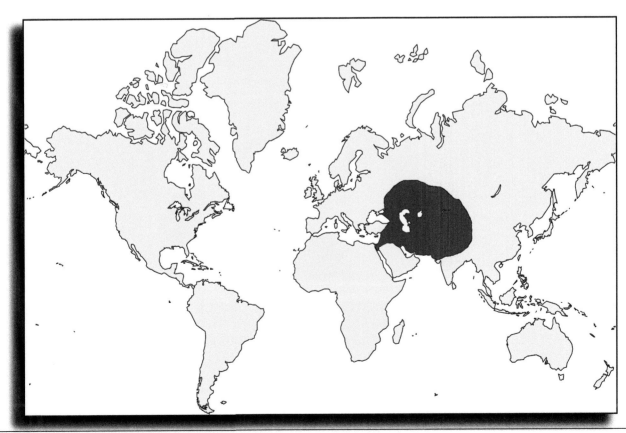

Figure 30.68   Approximate geographic distribution of *Hyalomma asiaticum*.

anatolicum, but now it is a valid species known simply as the "small *Hyalomma*."[150,151]

Medical Importance Vector of virus of Crimean–Congo hemorrhagic fever.

Description Usual characteristics of all *Hyalomma* ticks (scant ornamentation, eyes present in sockets, long

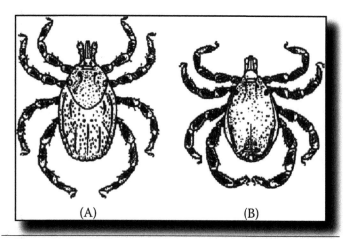

Figure 30.69 (A) Female and (B) male *Hyalomma anatolicum*. (From Goddard J. Ticks and Tick-borne Diseases Affecting Military Personnel, USAFSAM-SR-89-2. San Antonio, TX: School of Aerospace Medicine, U.S. Air Force; 1989.)

mouthparts, banded legs); small and similar in appearance to *H. asiaticum* (Figure 30.69).

Distribution Throughout northern Africa, portions of the Near East, Asia Minor, southern Europe, Russia, Ukraine, Georgia, Azerbaijan, Armenia, Kazakhstan, Uzbekistan, Turkmenistan, and India (Figure 30.70).

Hosts All stages have been observed feeding on hares in forests near Casablanca; it is also an avid parasite of humans and many domestic animals.

Seasonality Varies with latitude throughout range; in general, adults infest domestic animals from March to October, and larvae and nymphs from July to September. All stages are most abundant in early August.

Remarks Engorged larvae and unfed adults are the usual overwintering stages; they hibernate in cracks and crevices in wooden animal shelters in Russian climate and in rodent burrows in African desert conditions. Larvae may survive up to 241 days, nymphs up to 246 days, and adults over 1 year.

N. *Hyalomma marginatum* Koch

Medical Importance Efficient vector of the Crimean–Congo hemorrhagic fever virus.

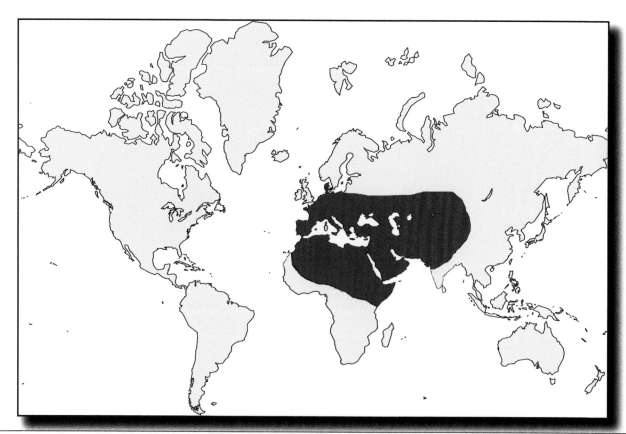

Figure 30.70 Approximate geographic distribution of *Hyalomma anatolicum*.

Description Large ticks with banded legs (Figure 30.71).

Distribution Most common in southern Europe, Arabia, and Asia Minor, as well as central, southern, and southeastern Asia and westward into Africa (Figure 30.72).[152]

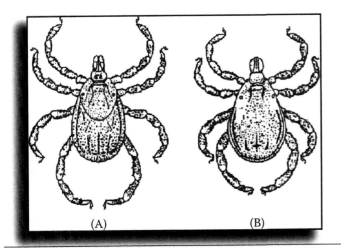

Figure 30.71 (A) Female and (B) male *Hyalomma marginatum*. (From Goddard J. Ticks and Tick-borne Diseases Affecting Military Personnel, USAFSAM-SR-89-2. San Antonio, TX: School of Aerospace Medicine, U.S. Air Force; 1989.)

Hosts Adults attack humans and wild and domestic ungulates, especially cattle and horses; immatures are specific to certain birds and hares.

Seasonality According to Hoogstraal,[153] this tick is rarely seen in winter throughout much of its distribution, but they begin to appear in March and continue until October. Maximum densities are reached in April, May, and June. Nymphs are active throughout summer.

Remarks Extremely hardy, often existing under varied conditions of cold, heat, and aridity; it often occurs in high numbers. It is an aggressive human parasite and may act as either a two-host or three-host tick. Unfed adults can survive over 2 years. Females deposit between 4000 and 15,000 eggs.

O. Australian Paralysis Tick, *Ixodes holocyclus* Neumann

Medical Importance Primary cause of tick paralysis cases in Australia; the bite is also known to cause Type I hypersensitivity reactions in humans.

Description Inornate with no eyes or festoons (Figure 30.73). Legs I and IV of males are reddish, legs II and III of

Figure 30.72 Approximate geographic distribution of *Hyalomma marginatum*.

males are yellowish. The scutum of the female is broadest posterior to middle, with numerous punctations of varying sizes. Coxae are large and trapezoid shaped.

Distribution   Primarily in New Guinea and along eastern coastal areas of Australia (Figure 30.74).

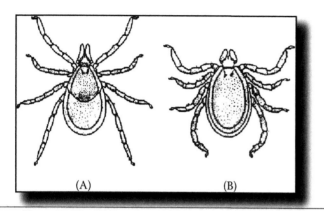

Figure 30.73   (A) Female and (B) male *Ixodes holocyclus*. (From Goddard J. Ticks and Tick-borne Diseases Affecting Military Personnel, USAFSAM-SR-89-2. San Antonio, TX: School of Aerospace Medicine, U.S. Air Force; 1989.)

Hosts   Parasitizes humans, other mammals, and birds; seems to especially prefer sheep, cattle, dogs, cats, and bandicoots.

Seasonality
Active in warmer months of year.

Remarks   Primarily in heavily vegetated rainforest areas of eastern coastal Australia; bandicoot is a natural host (bandicoot populations are increasing near urban areas owing to control campaigns against dingoes and foxes[154]).

P. Taiga Tick, *Ixodes persulcatus* Schulze

Medical Importance   Vector of virus of Russian spring–summer encephalitis and Lyme borreliosis spirochete in Europe and Asia.

Description   Characteristics common to all species in genus *Ixodes*; it is very similar in appearance to commonly encountered European castor bean tick, *I. ricinus* (Figure 30.75).

Figure 30.74   Approximate geographic distribution of *Ixodes holocyclus*.

Distribution Central and northeastern Europe, Russia, Ukraine, Belarus, Kazakhstan, Uzbekistan, Kyrgyzstan, China, and Japan (Figure 30.76).

Hosts Larvae and nymphs feed on a wide variety of small forest mammals and birds; adults parasitize larger wild and domestic mammals. It readily bites people.

Seasonality This tick is found on hosts in late spring and summer months; Zemskaya[155] found that, after overwintering, adults usually resumed activity in the eastern part of the Russian plains during the last 10 days of April, when upper soil layers warmed up to 5 to 10°C (41 to 50°F); they remained active for 65 to 95 days.

Remarks The most common tick species in northeastern Europe and northern Asia, it apparently is more cold-hardy than *I. ricinus*, thus inhabiting harsher, more northern areas. It inhabits small-leaved forests near primary coniferous forests, such as spruce–basswood combinations (commonly referred to as *taiga*).

Figure 30.75 Female *Ixodes persulcatus*. (Photo courtesy Dr. Lorenza Beati and the U.S. National Tick Collection.)

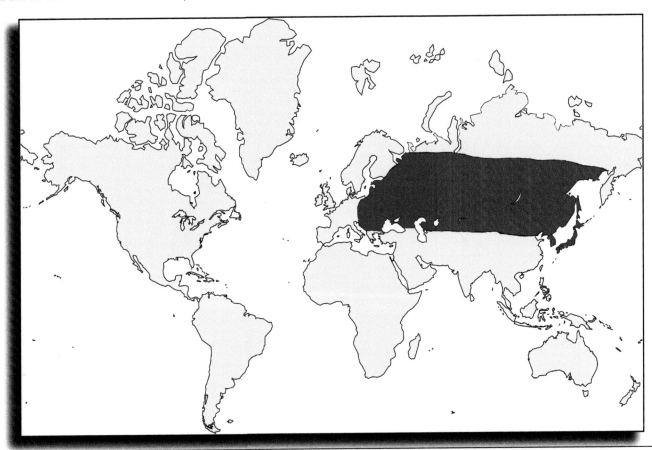

Figure 30.76 Approximate geographic distribution of *Ixodes persulcatus*.

Figure 30.77 Female *Ixodes ricinus*. (Photo courtesy Dr. Lorenza Beati and the U.S. National Tick Collection.)

Q. European Castor Bean Tick, *Ixodes ricinus* Linnaeus

Medical Importance Primary vector of Lyme borreliosis spirochetes in Europe; also known to transmit viruses of tick-borne encephalitis complex.

Description Long mouthparts and dark brown to black in color; females may have a portion posterior to the scutum that is orangish (Figure 30.77). It resembles and is closely related to the American species *Ixodes scapularis*.

Distribution Common throughout most of Europe, including the British Isles (Figure 30.78); it is also found in scattered locations in northern Africa and parts of Asia.

Hosts Immatures recorded from lizards, small mammals, and birds; adults feed mostly on sheep, cattle, dogs, horses, and deer and are avid parasites of humans.

Seasonality In temperate regions of range, it is most active in spring and autumn. Two peaks of activity may be observed: late March to early June and August to October. In northern Africa, it is most active in winter.

Remarks This is probably the most commonly encountered tick in Central and Western Europe; more than 90% of tick

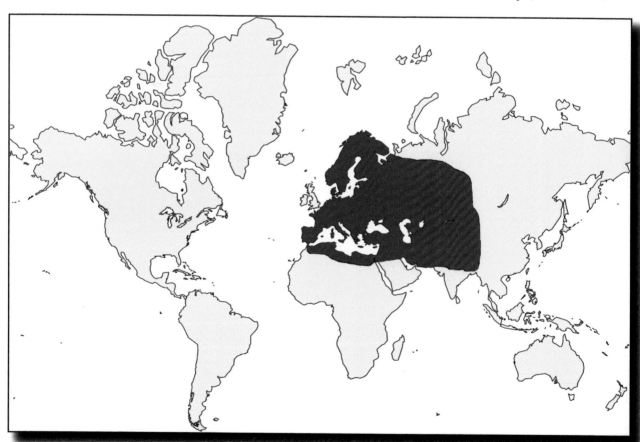

Figure 30.78 Approximate geographic distribution of *Ixodes ricinus*.

Figure 30.79 Male *Rhipicephalus appendiculatus*. (Specimen provided by Dr. K.Y. Mumcuoglu, Hebrew University, Jerusalem.)

bites in England and Ireland are from *Ixodes ricinus* nymphs. As long as 3 years is usually required to complete the life cycle; larvae feed the first year, nymphs the second year, and adults the third year. Females deposit 2000 to 3000 eggs.

R. Brown Ear Tick, *Rhipicephalus appendiculatus* Neumann

Medical Importance   One of several vectors of boutonneuse fever *rickettsia*.

Description   Inornate with both eyes and festoons present (Figure 30.79); it looks similar to cosmopolitan *Rhipicephalus sanguineus*. Males have a distinctly pointed dorsal projection on coxa I; females have flat eyes (not obviously convex).

Distribution   Throughout southern Africa up to about 10°N latitude (Figure 30.80).

Figure 30.80   Approximate geographic distribution of *Rhipicephalus appendiculatus*.

Hosts Cattle primarily (for adult ticks as well as immatures), in addition to humans, domestic animals, and wild game such as antelope and buffalo.

Seasonality Larvae are most abundant on hosts from May to July, nymphs occur on hosts from June to September, and adults are most active from November to March.

Remarks One to three generations occur annually, depending on the length and number of rainy seasons in its range. It seems especially sensitive to desiccation, as during consecutive dry years it tends to die out. Unfed larvae may survive as long as 10 months, nymphs 15 months, and adults 24 months. Females lay up to 6000 eggs.

References

1. McGinley-Smith DE, Tsao SS. Dermatoses from ticks. *J. Am. Acad. Dermatol.* 2003;49:363–392.

2. Burgdorfer W. A review of Rocky Mountain spotted fever: Its agent, and its vectors in the U.S. *J. Med. Entomol.* 1975;12:269–278.

3. Goddard J. *Ticks and Tick-borne Diseases Affecting Military Personnel*. San Antonio, TX: USAF, School of Aerospace Medicine; 1989.

4. Heymann DL, (Ed.). *Control of Communicable Diseases Manual*. 20th ed. Washington, DC: American Public Health Association; 2015.

5. Hoogstraal H. Ticks in relation to human diseases caused by viruses. *Ann. Rev. Entomol.* 1966;11:261–308.

6. Hoogstraal H. Ticks in relation to human diseases caused by *Rickettsia* species. *Ann. Rev. Entomol.* 1967;12:377–420.

7. Goodman JL, Dennis DT, Sonenshine DE. *Tick-borne Diseases of Humans*. Washington, DC: ASM Press; 2005.

8. Crippa M, Rais O, Gern L. Investigations on the mode and dynamics of transmission and infectivity of *Borrelia burgdorferi* sensu stricto and *Borrelia afzelii* in Ixodes ricinus ticks. *Vector Borne Zoonotic Dis.* 2002;2(1):3–9.

9. Piesman J. Transmission of Lyme disease spirochetes (*Borrelia burgdorferi*). *Exp. Appl. Acarol.* 1989;7:71–80.

10. Saraiva DG, Soares HS, Soares JF, Labruna MB. Feeding period required by *Amblyomma aureolatum* ticks for transmission of *Rickettsia rickettsii* to vertebrate hosts. *Emerg. Infect. Dis.* 2014;20(9):1504–1510.

11. Ebel GD, Foppa I, Spielman A, Telford SR, III. A focus of deer tick virus transmission in the northcentral United States. *Emerg. Infect. Dis.* 1999;5:570–573.

12. Kugeler KJ, Griffith KS, Gould LH, et al. A review of death certificates listing Lyme disease as a cause of death in the United States. *Clin. Infect. Dis.* 2011;52(3):364–367.

13. Callahan P, Tsouderous T. Chronic Lyme disease: A dubious diagnosis. *Chicago Tribune, Health Section.* 2010; December 8 issue.

14. O'Connell S. Lyme borreliosis. In: Palmer SR, Soulsby L, Torgerson PR, Brown DWG, eds. *Oxford Textbook of Zoonoses: Biology, Clinical Practice, and Public Health Control.* 2nd ed. Oxford, UK: Oxford University Press; 2011:82–91.

15. Dennis DT. Rash decisions: Lyme disease or not? *Clin. Infect. Dis.* 2005;41:966–968.

16. Shapiro ED. Lyme disease in 2018: What is new (and what is not). *JAMA.* 2018;Electronic publication ahead of print, August 2, 2018.

17. Ginsberg HS, Faulde MK. Ticks. In: Bonnefoy X, Kampen H, Sweeney K, eds. *Public Health Significance of Urban Pests.* Copenhagen: WHO Regional Office for Europe; 2008:303–345.

18. CDC. Summary of notifiable infectious diseases and conditions – United States, 2015. *MMWR.* 2017;64(53):1–144.

19. Kuehn BM. CDC estimates 300,000 U.S. cases of Lyme disease annually. *JAMA.* 2013;310:1110.

20. CDC. Caution regarding testing for Lyme disease. *MMWR.* 2005;54:125.

21. FDA. Beware of ticks and Lyme disease. U.S. Food and Drug Administration Consumer Health Information, update on Lyme disease, www.fda.gov/consumer/updates/lymedisease 062707.html; 2007.

22. Wilske B, Fingerle V, Schulte-Spechtel U. Microbiological and serological diagnosis of Lyme borreliosis. *FEMS Immunol. Med. Microbiol.* 2007;49(1):13–21.

23. Piesman J. Ecology of *Borrelia burgdorferi* sensu lato in North America. In: Gray JS, Kahl O, Lane RS, Stanek G, eds., *Lyme Borreliosis – Biology, Epidemiology, and Control.* New York: CABI Publishing;2002:223–249.

24. Falco RC, Fish D, Piesman J. Duration of tick bites in a Lyme disease-endemic area. *Am. J. Epidemiol.* 1996;143:187–192.

25. Pritt BS, Mead PS, Johnson DK, et al. Identification of a novel pathogenic *Borrelia* species causing Lyme borreliosis with unusually high spirochaetaemia: A descriptive study. *Lancet Infect. Dis.* 2016;3099(15):00464–00468.

26. Goddard J. Not all erythema migrans lesions are Lyme disease. *Am. J. Med.* 2016;epub ahead of print, doi:10.1016/j.amjmed.

27. CDC. Fatal cases of Rocky Mountain spotted fever in family clusters – Three states, 2003. *MMWR.* 2004;53:407–410.

28. Conwill DE, Oakes T, Brackin BT. Rocky Mountain spotted fever: Fatalities in Mississippi, 1986. *Miss. Morb. Rep.* 1987;5:1–2.

29. Dumler JS, Walker DH. Rocky Mountain spotted fever – Changing ecology and persisting virulence. *N. Engl. J. Med.* 2005;353:551–553.

30. O'Reilly M, Paddock CD, Elchos BN, Goddard J, Childs J, Currier M. Physician knowledge of the diagnosis and management of Rocky Mountain spotted fever. *Ann. N. Y. Acad. Sci.* 2003;990:295–301.

31. Biggs H, Behravesh CB, Bradley KK, et al. Diagnosis and management of tickborne rickettsial diseases: Rocky Mountain spotted fever and other spotted fever group rickettsioses, ehrlichioses, and anaplasmosis – United States. *MMWR, R&R.* 2016; 65:1–45.

32. Demma LJ, Traeger MS, Nicholson WL, et al. Rocky Mountain spotted fever from an unexpected tick vector in Arizona. *N. Engl. J. Med.* 2005;353(6):587–594.

33. Sexton DJ, Walker DH. Spotted fever group rickettsioses. In: Guerrant RL, Walker DH, Weller PF, eds. *Tropical Infectious Diseases: Principles, Pathogens, and Practice.* 3rd ed. New York: Saunders Elsevier; 2011:323–329.

34. Berrada ZL, Goethert HK, Cunningham J, Telford SR, 3rd. *Rickettsia rickettsii* (Rickettsiales: Rickettsiaceae) in *Amblyomma americanum* (Acari: Ixodidae) from Kansas. *J. Med. Entomol.* 2011;48(2):461–467.

35. Breitschwerdt EB, Hegarty BC, Maggi RG, Lantos PM, Aslett DM, Bradley JM. *Rickettsia rickettsii* transmission by a lone star tick, North Carolina. *Emerg. Infect. Dis.* 2011;17(5):873–875.

36. Parker RR. A pathogenic rickettsia from the Gulf Coast tick, *Amblyomma maculatum.* Proceedings of the Third International Congress on Microbiology. New York, 1940: 390–391.

37. Anigstein L, Bader MN. Investigations on rickettsial diseases in Texas, Part I. *Texas Rep. Biol. Med.* 1943;1:105–116.

38. Stothard DR, Fuerst PA. Evolutionary analysis of the spotted fever and typhus groups of *Rickettsia* using 16sRNA gene sequences. *Syst. Appl. Microbiol.* 1995;18:52–61.

39. Bell EJ, Stoenner HC. Immunologic relationships among the spotted fever group of rickettsias determined by toxin neutralization tests in mice with convalescent animal serums. *J. Immunol.* 1960;84:171–182.

40. Lackman DB, Parker RR, Gerloff RK. Serological characteristics of a pathogenic rickettsia occurring in *Amblyomma maculatum. Public Health Rep.* 1949;64:1342–1349.

41. Parker RR, Kohls GM, Cox GW, Davis GE. Observations on an infectious agent from *Amblyomma maculatum. Public Health Rep.* 1939;54:1482–1484.

42. Goddard J. Historical and recent evidence for close relationships among *Rickettsia parkeri, R. conorii, R. africae,* and *R. siberica:* Implications for rickettsial taxonomy. *J. Vector Ecol.* 2009;34:238–242.

43. Xu W, Raoult D. Taxonomic relationships among spotted fever group rickettsias as revealed by antigenic analysis with monoclonal antibodies. *J. Clin. Microbiol.* 1998;36:887–896.

44. Goddard J. Experimental infection of lone star ticks, *Amblyomma americanum* (L.), with *Rickettsia parkeri* and exposure of guinea pigs to the agent. *J. Med. Entomol.* 2003;40:686–689.

45. Paddock CD, Sumner JW, Comer JA, et al. *Rickettsia parkeri* – A newly recognized cause of spotted fever rickettsiosis in the United States. *Clin. Infect. Dis.* 2004;38:805–811.

46. Goddard J. American Boutonneuse Fever – A new spotted fever rickettsiosis. *Infect. Med.* 2004;21:207–210.

47. Ekenna O, Paddock CD, Goddard J. Gulf Coast tick rash illness caused by *Rickettsia parkeri. J. Miss. State Med. Assoc.* 2014;55:216–219.

48. Paddock CD, Goddard J. The evolving medical and veterinary importance of the Gulf Coast tick. *J. Med. Entomol.* 2015;52:230–252.

49. Kelly PJ, Beati L, Mason PR, Matthewman LA, Roux V, Raoult D. *Rickettsia africae* sp. nov.: The etiological agent of African tick bite fever. *Int. J. Syst. Bacteriol.* 1996; 46:611–614.

50. Kelly PJ, Beati L, Matthewman LA, Mason PR, Dasch GA, Raoult D. A new pathogenic spotted fever group rickettsia from Africa. *J. Trop. Med. Hyg.* 1994;97:129–137.

51. Jensenius M, Fournier PE, Kelly PJ, Myrvang B, Raoult D. African tick bite fever. *Lancet Infect. Dis.* 2003;3:557–564.

52. Kelly PJ. *Rickettsia africae* in the West Indies. *Emerg. Infect. Dis.* 2006;12:224–225.

53. Bohaty BR, Hebert AA. African tick-bite fever after a game hunting expedition. *N. Engl. J. Med.* 2015;372:e14.

54. Walker DH. Rickettsiae and rickettsial infections: The current state of knowledge. *Clin. Infect. Dis.* 2007;45:539–544.

55. Ibarra V, Oteo JA, Portillo A, et al. *Rickettsia slovaca* infection: DEBONEL/TIBOLA. *Ann. N. Y. Acad. Sci.* 2006;1078:206–214.

56. Shapiro MR, Fritz CL, Tait K, et al. Rickettsia 364D: A newly recognized cause of eschar-associated illness in California. *Clin. Infect. Dis.* 2010;50(4):541–548.

57. Johnston SH, Glaser CA, Padgett KA, et al. *Rickettsia* spp. 364D causing a cluster of eschar-associated illness, California. *Ped. Infect. Dis. J.* 2013;32(9):1036–1039.

58. Walker DH, Dumler JS. Emergence of the ehrlichioses as human health problems. *Emerg. Infect. Dis.* 1996;2:18–28.

59. Maeda K, Markowitz N, Hawley RC, Ristic M, Cox D, McDade JE. Human infection with *Ehrlichia canis* a leukocytic rickettsia. *N. Engl. J. Med.* 1987;316:853–856.

60. Dumler JS, Bakken JS. Ehrlichial diseases of humans: Emerging tick-borne infections. *Clin. Infect. Dis.* 1995;20:1102–1110.

61. Harris RM, Couturier BA, Sample SC, Coulter KS, Casey KK, Schlaberg R. Expanded geographic distribution and clinical characteristics of *Ehrlichia ewingii* infections, United States. *Emerg. Infect. Dis.* 2016;22:862–865.

62. Pritt BS, Sloan LM, Johnson DKH, et al. Emergence of a new pathogenic Ehrlichia species, Wisconsin and Minnesota, 2009. *N. Engl. J. Med.* 2011;365(5):422–429.

63. Wormser GP, Pritt B. Update and commentary on four emerging tick-borne infections: *Ehrlichia muris*-like agent, *Borrelia miyamotoi,* deer tick virus, Heartland virus, and whether ticks play a role in transmission of *Bartonella henselae. Infect. Dis. Clin. NA.* 2015;29(2):371–381.

64. Dumler JS. Ehrlichiosis and anaplasmosis. In: Guerrant RL, Walker DH, Weller PF, eds. *Tropical Infectious Diseases: Principles, Pathogens, and Practice.* 3rd ed. New York: Saunders Elsevier; 2011:339–344.

65. CDC. Anaplasmosis and ehrlichiosis – Maine, 2008. *MMWR.* 2009;58(37):1033–1036.

66. Holman MS, Caporale DA, Goldberg J, et al. *Anaplasma phagocytophilum, Babesia microti,* and *Borrelia burgdorferi* in *Ixodes scapularis* in southern coastal Maine. *Emerg. Infect. Dis.* 2004;10:744–746.

67. Gorenflot A, Moubri K, Precigout E, Carcy B, Schetters TP. Human babesiosis. *Ann. Trop. Med. Parasitol.* 1998;92:489–501.

68. CDC. Babesiosis – Connecticut. *MMWR.* 1989;38:649–650.

69. Markell E, Voge M, John D. *Medical Parasitology.* 7th ed. Philadelphia, PA: WB Saunders; 1992.

70. Lempereur L, De Cat A, Caron Y, et al. Fist molecular evidence of potentially zoonotic *Babesia mcroti* and *Babesia* sp. EU1 in *Ixodes ricinus* ticks in Belgium. *Vector Borne Zoonotic Dis.* 2010;11:125–130.

71. Thomford JW, Conrad PA, Telford SR, III, et al. Cultivation and phylogenetic characterization of a newly recognized human pathogenic protozoan. *J. Infect. Dis.* 1994;169:1050–1056.

72. Herwaldt BL, de Bruyn G, Pieniazek NJ, et al. *Babesia divergens*-like infection, Washington State. *Emerg. Infect. Dis.* 2004;10(4):622–629.

73. CDC. Tularemia – Missouri, 2000–2007. *MMWR.* 2009;58:744–748.

74. Spach DH, Liles WC, Campbell GL, Quick RE, Anderson DEJ, Fritsche TR. Tick-borne diseases in the United States. *N. Engl. J. Med.* 1993;329:936–947.

75. Farlow J, Wagner DM, Dukerich M, et al. *Francisella tularensis* in the United States. *Emerg. Infect. Dis.* 2005;11:1835–1841.

76. CDC. Outbreak of relapsing fever – Grand Canyon National Park, Arizona. *MMWR.* 1991;40:296–297.

77. Thompson RS, Burgdorfer W, Russell R, Francis BJ. Outbreak of tick-borne relapsing fever in Spokane County, Washington. *J. Am. Med. Assoc.* 1969;210:1045–1049.

78. CDC. Tickborne relapsing fever outbreak after a family gathering – New Mexico, August 2002. *MMWR.* 2003;52:809–812.

79. CDC. Tickborne relapsing fever outbreak at an outdoor education camp – Arizona, 2014. *MMWR.* 2015;64:651–652.

80. Emmons R. Ecology of Colorado tick fever. *Ann. Rev. Microbiol.* 1988;42:49–64.

81. Brackney MM, Marfin AA, Staples JE, et al. Epidemiology of Colorado tick fever in Montana, Utah, and Wyoming, 1995–2003. *Vector Borne Zoonotic Dis.* 2010;10(4):381–385.

82. Monath TP, Johnson KM. Diseases transmitted primarily by arthropod vectors. In: Last JM, Wallace RB, eds. *Public Health and Preventive Medicine.* 13th ed. Norwalk, CT: Appleton and Lange; 1992.

83. Gresikova M, Calisher CH. Tick-borne encephalitis. In: Monath TP, ed., *The Arboviruses: Epidemiology and Ecology.* Vol 4. Boca Raton, FL: CRC Press; 1989:177–184.

84. Loktev VB, Ternovoy VA, Kurgukov GP, et al. New variants of tick-borne encephalitis discovered by retrospective investigation of fatal cases of tick-borne encephalitis with hemorrhagic syndrome occurring in Novosibirsk Region (Russia) during summer of 1999. (Program containing abstracts), International conference on Emerging Infectious Diseases, Atlanta, March 24–27. 2002;(supplement):11.

85. Ternovoi VA, Protopopova EV, Chausov EV, et al. Novel variant of tickborne encephalitis, Russia. *Emerg. Infect. Dis.* 2007;13:1574–1578.

86. Hinten SR, Beckett GA, Gensheimer KF, et al. Increased recognition of Powassan encephalitis in the United States, 1999–2005. *Vector Borne Zoonotic Dis.* 2008;8(6):733–740.

87. Nuttall PA, Labuda M. Tick-borne encephalitis subgroup. In: Sonenshine DE, Mather TN, eds. *Ecological Dynamics of Tick-borne Zoonoses.* New York: Oxford University Press; 1994:351.

88. Telford SR, III, Armstrong PM, Katavolos P, et al. A new tick-borne encephalitis-like virus infecting New England deer ticks, *Ixodes dammini. Emerg. Infect. Dis.* 1997;3:165–170.

89. Tavakoli NP, Wang H, Dupuis M, et al. Fatal case of deer tick virus encephalitis. *N. Engl. J. Med.* 2009;360(20):2099–2107.

90. Tesh RB, Solomon T. Japanese encephalitis, West Nile, and other flavivirus infections. In: Guerrant RL, Walker DH, Weller PF, eds. *Tropical Infectious Diseases.* 3rd ed. New York: Saunders Elsevier Publishing; 2011.

91. Stone R. Monkey fever unbound. *Science (News Focus).* 2014;345:128–133.

92. Liu Q, He B, Huang SY, Wei F, Zhu X-Q. Severe fever with thrombocytopenia syndrome, an emerging tick-borne zoonosis. *Lancet Infect. Dis.* 2014;14:763–772.

93. Yun SM, Song BG, Choi W, et al. First isolation of severe fever with thrombocytopenia syndrome virus from *Haemaphysalis longicornis* ticks collected in SFTS outbreak areas in the Republic of Korea. *Vector Borne Zoonotic Dis.* 2016;16:66–70.

94. McMullan LK, Folk SM, Kelly AJ, et al. A new Phlebovirus associated with severe febrile illness in Missouri. *N. Engl. J. Med.* 2012;367(9):834–841.

95. Pastula DM, Turabelidze G, Yates KF, et al. Heartland virus disease – United States, 2012–2013. *MMWR.* 2014;63:270–271.

96. Kosoy OI, Lambert AJ, Hawkinson DJ, et al. Novel thogotovirus associated with febrile illness and death, United States, 2014. *Emerg. Infect. Dis.* 2015;21(5):760–764.

97. Savage HM, Godsey MS, Panella NA, et al. Surveillance for tick-borne viruses near the location of a fatal human case of Bourbon virus in eastern Kansas. *J. Med. Entomol.* 2018;55:701–705.

98. Savage HM, Burkhalter KL, Godsey MSJ, et al. Bourbon virus in field-collected ticks, Missouri, USA. *Emerg. Infect. Dis.* 2017;23(12):2017–2022.

99. Kocan AA. Tick paralysis. *J. Am. Vet. Med. Assoc.* 1988;192:1498–1500.

100. Gregson JD. Tick paralysis: An appraisal of natural and experimental data. Canada Dept. Agri. Monograph No. 9; 1973:48.

101. Schmitt N, Bowmer EJ, Gregson JD. Tick paralysis in British Columbia. *Can. Med. Assoc. J.* 1969;100:417–421.

102. Stanbury JB, Huyck JH. Tick paralysis: A critical review. *Medicine.* 1945;24:219–242.

103. Gothe R, Kunze K, Hoogstraal H. The mechanisms of pathogenicity in the tick paralysis. *J. Med. Entomol.* 1979;16:357–369.

104. Goddard J. Ecological studies of *Ixodes scapularis* in Mississippi: Lateral movement of adult ticks. *J. Med. Entomol.* 1993;30:824–826.

105. Lancaster JL, Jr. Control of the lone star tick. University of Arkansas Agr. Exp. Sta. Rep. Ser. No. 67. 1957:39.

106. Lees AD. The water balance in *Ixodes ricinus* and certain other species of ticks. *Parasitology* 1946;37:1–20.

107. Semtner PJ, Howell DE, Hair JA. The ecology and behavior of the lone star tick (Acarina: Ixodidae). I. The relationship between vegetative habitat type and tick abundance and distribution in Cherokee Co., Oklahoma. *J. Med. Entomol.* 1971;8:329–335.

108. Sonenshine DE. The ticks of Virginia. Virginia Polytechnic Institute and State University Res. Div. Bull. No. 139. 1979:42.

109. Sonenshine DE, Atwood EL, Lamb JT. The ecology of ticks transmitting Rocky Mountain spotted fever in a study area in Virginia. *Ann. Entomol. Soc. Am.* 1966;59:1234–1262.

110. Sonenshine DE, Levy GF. Ecology of the American dog tick, *Dermacentor variabilis* in a study area in Virginia. II. Distribution in relation to vegetative types *Ann. Entomol. Soc. Am.* 1972;65:1175–1182.

111. Schachat SR, Robbins RG, Goddard J. Color patterning in hard ticks. *J. Med. Entomol.* 2018;55:1–13.

112. Keirans JE, Litwak TR. Pictorial key to the adults of hard ticks, Family Ixodidiae, East of the Mississippi River. *J. Med. Entomol.* 1989;26:435–448.

113. Guglielmone AA, Estrada-Pena A, Keirans JE, Robbins RG. Ticks of the Neotropical Zoogeographic Region. Special Publication, International Consortium on Ticks and Tick-borne Diseases, Houten, The Netherlands; 2003: 1–173.

114. Goddard J, Varela-Stokes AS. Role of the lone star tick, *Amblyomma americanum*, in human and animal diseases. *Vet. Parasitol.* 2009;160:1–12.

115. Barbour AG, Maupin GO, Teltow GJ, Carter CJ, Piesman J. Identification of an uncultivable *Borrelia* species in the hard tick *Amblyomma americanum*: Possible agent of a Lyme disease-like illness. *J. Infect. Dis.* 1996;173:403–409.

116. Apperson CS, Engber B, Nicholson WL, et al. Tick-borne diseases in North Carolina: Is *Rickettsia amblyommii* a possible cause of rickettsiosis reported as Rocky Mountain spotted fever? *Vector Borne Zoonotic Dis.* 2008;8:597–606.

117. Karpathy SE, Slater KS, Goldsmith CS, Nicholson WL, Paddock CD. *Rickettsia amblyommatis* sp. nov., a spotted fever group Rickettsia associated with multiple species of *Amblyomma* ticks in North, Central and South America. *Int. J. Syst. Evol. Microbiol.* 2016;66(12):5236–5243.

118. Cohen SB, Yabsley MJ, Garrison LE, et al. *Rickettsia parkeri* in *Amblyomma americanum* ticks, Tennessee and Georgia, USA. *Emerg. Infect. Dis.* 2009;15:1471–1472.

119. Goddard J, Norment BR. Spotted fever group rickettsiae in the lone star tick. *J. Med. Entomol.* 1986;23:465–472.

120. Long SW, Pound JM, Yu X. *Ehrlichia* prevalence in *Amblyomma americanum* in central Texas. *Emerg. Infect. Dis.* 2004;10:1342–1343.

121. Ramey K, Stewart PH. Top ten facts you should know about "Alpha-gal," the newly described delayed red meat allergy. *J. Miss. State Med. Assoc.* 2016;57:279–281.

122. Guzman-Cornejo C, Robbins RG, Guglielmone AA, Montiel-Parra G, Perez TM. The *Amblyomma* of Mexico: Identification keys, distribution, and hosts. *Zootaxa.* 2011;2998:16–38.

123. Goddard J, Portugal JS. Cutaneous lesions due to bites by larval *Amblyomma americanum*. *JAMA Dermatol.* 2015;151:1373–1375.

124. Edwards KT. Gotch ear: A poorly-described, local, pathologic condition of livestock associated primarily with the Gulf Coast tick, *Amblyomma maculatum*. *Vet. Parasitol.* 2011;183:1–7.

125. Goddard J. Tick paralysis. *Infect. Med.* 1998;15:28–31.

126. Cooley RA. The genera *Dermacentor* and *Otocentor* in the United States, with studies in variation. U.S. National Institutes of Health Bulletin No. 171: 1–89; 1938.

127. Yunker CE, Keirans JE, Clifford CM, Easton ER. *Dermacentor* ticks (Acari: Ixodoidea: Ixodidae) of the new world: A scanning electron microscope atlas. *Proc. Entomol. Soc. Wash.* 1986;88:609–627.

128. Goddard J. Arthropod transmission of tularemia. *Infect. Med.* 1998;15:306–308.

129. Guzman-Cornejo C, Robbins RG. The genus *Ixodes* in Mexico: adult identification keys, diagnoses, hosts, and distribution. *Revista Mexicana de Biodiversidad* 2010;81(2):289–298.

130. Goddard J. Ticks and Lyme disease. *Infect. Med.* 1997;14:698–700,702.

131. Oliver JH, Owsley MR, Hutcheson HJ, et al. Conspecificity of the ticks *Ixodes scapularis* and *Ixodes dammini*. *J. Med. Entomol.* 1993;30:54–63.

132. Telford SRI, Wormser GP. *Bartonella* spp. transmission by ticks not established. *Emerg. Infect. Dis.* 2010;16(3):379–384.

133. Goddard J. Ecological studies of adult *Ixodes scapularis* in central Mississippi: Questing activity in relation to time of year, vegetation type, and meteorologic conditions. *J. Med. Entomol.* 1992;29:501–506.

134. Walker JB, Keirans JE, Horak IG. *The Genus Rhipicephalus (Acari, Ixodidae): A Guide to the Brown Ticks of the World.* New York: Cambridge University Press; 2000.

135. Carpenter TL, McMeans MC, McHugh CP. Additional instances of human parasitism by the brown dog tick. *J. Med. Entomol.* 1990;27:1065–1066.

136. Goddard J. Focus of human parasitism by the brown dog tick, *Rhipicephalus sanguineus*. *J. Med. Entomol.* 1989;26: 628–629.

137. Klompen JSH, Oliver JH, Jr. Systematic relationships in the soft ticks. *Syst. Entomol.* 1993;18:313–331.

138. Gugliemone AA, Robbins RG, Apanaskevich DA, et al. The Argasidae, Ixodidae, and Nuttalliellidae of the world: A list of valid species names. *Zootaxa.* 2010;2528:1–28.

139. Hoogstraal H. Argasid and nuttalliellid ticks as parasites and vectors. *Adv. Parasitol.* 1985;24:135–238.

140. Hoffman A. Monografia de los Ixodoidea de Mexico, I Parte. *Rev. Soc. Mexicana Hist. Nat.* 1962;23:191–305.

141. Keirans JE, Clifford CM, Hoogstraal H. *Ornithodoros yunkeri*, new species from seabirds and nesting sites in the Galapagos Islands. *J. Med. Entomol.* 1984;21:344–350.

142. Nava S, Beati L, Labruna M, Cáceres A, Mangold A, Guglielmone A. Reassessment of the taxonomic status of *Amblyomma cajennense* (Fabricius, 1787) with the description of three new species, *Amblyomma tonelliae* n. sp., *Amblyomma interandinum* n. sp. and *Amblyomma patinoi* n. sp. *Ticks Tick-borne Dis.* 2014;5:252–276.

143. Voltzit OV, Keirans JE. A review of the African *Amblyomma* species. *Acarina.* 2003;11(2):135–214.

144. Selmi M, Bertolotti L, Tomassone L, Mannelli A. *Rickettsia slovaca* in *Dermacentor marginatus* and tick-borne lymphadenopathy, Tuscany, Italy. *Emerg. Infect. Dis.* 2008;14(5):817–820.

145. Gilot B, Pautou G. Distribution and ecology of *Dermacentor marginatus* in the french alps and their piedmont. *Acarologia.* 1983;24:261–273.

146. Splisteser H, Tyron U. Studies on the ecology and behavior of *Dermacentor nuttalli* in the Mongolian Republic. *Monat. Veter.* 1986;41:126–128.

147. Pomerantzev BI, Serdyukova GV. Ecological observations of ticks of the family Ixodidae, vectors of spring-summer encephalitis in the Far East. *Parazit. Sborn. Zool. Inst. Akad. Nauk. USSR.* 1947;9:47–67.

148. Hoogstraal H. Changing patterns of tick-borne diseases in modern society. *Ann. Rev. Entomol.* 1981;26:75–99.

149. Apanaskevich DA, Horak IG. The genus *Hyalomma*. XI. Redescription of all parasitic stages of *H. asiaticum* with notes on its biology. *Exp. Appl. Acarol.* 2010;52(2):207–220.

150. Apanaskevich DA, Horak IG. The genus *Hyalomma* Koch, 1844. II. Taxonomic status of *H. anatolicum* Koch, 1844 and *H. excavatum* Koch 1844, with redescriptions of all stages. *Acarina.* 2005;13(2):181–197.

151. Robbins RG, Carpenter TL. An annotated list of the tick common names authored by Harry Hoogstraal (1917–1986). *Syst. Appl. Acarol.* 2011;16(2):99–132.

152. Apanaskevich DA, Horak IG. The genus *Hyalomma* Koch, 1844 V. Re-evaluation of the taxonomic rank of taxa comprising the *H. marginatum* Koch complex of species with redescription of all parasitic stages and notes on biology. *Int. J. Acarol.* 2008;34(1):13–42.

153. Hoogstraal H. African Ixodoidea. I. Ticks of the Sudan. U.S. Navy Bur. Med. Sur., Washington, DC, Res. Rep. NM 005-050, 29.07, 89 pp.; 1956.

154. Bagnall BG, Doube BM. The Australian paralysis tick, *Ixodes holocyclus. Aust. Vet. J.* 1975;51:151–160.

155. Zemskaya AA. Seasonal activity of adult ticks *Ixodes persulcatus* in the eastern part of the Russian plain. *Folia Parasitol.* 1984;31:269–276.

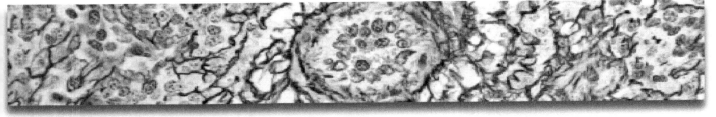

31

WASPS (YELLOWJACKETS, HORNETS, AND PAPER WASPS)

I. Yellowjackets

A. General and Medical Importance

Venomous wasps in the genera *Vespula* and *Dolichovespula* are called yellowjackets (see box and Figure 31.1) and comprise about 26 species. Some species nest near (in the ground) or in human dwellings and can be a nuisance and health threat to people. Yellowjackets produce painful stings and may cause death due to sting allergy (see Chapter 2). Cellulitis may occur following stings by some scavenger yellowjackets. Foraging yellowjackets are particularly numerous around recreation areas and refuse collection sites, where they are attracted to meats, sweet carbohydrates, soda pop, etc.

B. General Description

The name "yellowjacket" refers to the typical yellow and black bands on the abdomen (Figure 31.2), although some species are actually black and white, such as the bald-faced hornet. The pattern of markings on the yellowjacket gaster (most prominent portion of the abdomen) is often diagnostic as to species (Figure 31.3). Most species are smaller than paper wasps (1.5–2.0 cm) and more robust in appearance like honey bees. Wasps, yellowjackets, and hornets have inconspicuous hairs on their bodies that are not feathered (when observed under magnification). An outstanding online color identification guide to yellowjackets and wasps is provided by Buck et al.[1]

C. Geographic Distribution

Vespula maculifrons, the eastern yellowjacket, occurs from Minnesota to Texas and eastward. It is probably the most troublesome yellowjacket in the eastern and southeastern United States. *Vespula vulgaris*, the common yellowjacket, is holarctic, being widely distributed across Europe, Asia, and North America, although recent research suggests that *V. vulgaris* occurring in the United States may be a different species called *V. alascensis*.[2] *Vespula squamosa*, the southern yellowjacket, occurs from Wisconsin to Texas and southeastward to the Atlantic Coast. Yellowjackets that have been introduced into nonnative habitats seem to be especially troublesome pests. The introduced German yellowjacket, *V. germanica*, is rapidly becoming a problem in the northeastern United States as well as the Midwest. Also, the western yellowjacket, *V. pennsylvanica*, is a severe pest in the western United States. In certain years, this species appears in such great numbers that it surpasses mosquitoes and flies as an annoyance to campers.[3] In addition, this species has been introduced into Hawaii, where it builds perennial colonies and is a pest in sugarcane fields.

D. Biology and Behavior

Yellowjackets of the genus *Dolichovespula* build aerial nests. This is the case for the large black and white species called the bald-faced hornet (*D. maculata*). Actually, this species is not a hornet and should be called an aerial yellowjacket (see Section II). The other genus of yellowjacket, *Vespula*, usually builds nests in underground sites. Yellowjacket nests are multicombed with a surrounding paper envelope. There may be multiple entrances to the underground nest. Mature size of yellowjacket colonies in temperate regions ranges from 500 cells in 2 combs in some *Vespula* spp. to 15,000 cells in 8–10 combs in some *Paravespula* spp. Mature colonies may have from as few as 75 to over 5000 worker yellowjackets. Most yellowjackets typically have annual nesting cycles, but in

YELLOWJACKETS

Worker yellowjacket. (From CDC, *Pictorial Keys to Arthropods, Reptiles, Birds, and Mammals of Public Health Significance*, U.S. Centers for Disease Control and Prevention, Atlanta, GA, 1963.)

IMPORTANCE

Painful stings; allergic reactions

DISTRIBUTION

Numerous species almost worldwide

LESION

Central white spot with erythematous halo; amount of local swelling is variable

DISEASE TRANSMISSION

None

KEY REFERENCE

Akre RD, Greene A, MacDonald JF, Landolt PJ, Davis HG. Yellowjackets of America North of Mexico, Agriculture Handbook No. 552. Washington, DC: U.S. Department of Agriculture; 1981: 102

TREATMENT

Pain relievers, antipruritic lotions for local reactions; systemic reactions may require antihistamines, epinephrine, and other supportive measures

Figure 31.1 Yellowjacket. (Photograph courtesy of Armed Forces Pest Management Board.)

Figure 31.2 Yellowjacket workers tending nest. (Photograph courtesy of Dr. James Jarratt, Entomology Department, Mississippi State University.)

some of the southernmost areas of their distribution, perennial colonies of *V. squamosa* and *V. maculifrons* have been reported. This has led to the discovery in Florida of nests that are 6–9 feet tall and contain over 100,000 cells. A 4-foot-tall yellowjacket nest has been found in Mississippi (Figure 31.4). A single overwintered queen yellowjacket initiates and builds a nest without aid from other queens. After the first brood of workers emerge, the queen ceases foraging and building activities and becomes the egg layer. At that point, she rarely leaves the colony. The queen is much larger than her workers

(1.5–3 times the workers' size). A yellowjacket colony usually survives until late fall (September to November), depending on the species and locality. There is rapid colony growth during late summer. In the fall, hundreds of new queens emerge from the colony and mate with males (which were also produced in the fall), and each inseminated queen hibernates in a protected place until the following spring when the cycle is repeated. Some species scavenge for decaying protein and carbohydrates at carrion, garbage cans, rotting fruit, picnic areas, and meat-processing plants. This seems to be a problem, especially in the fall, resulting in a serious stinging hazard to people.

E. Treatment of Stings

The treatment recommendations given for yellowjacket stings are generally those recommended for all stinging wasps. Local treatment of yellowjacket stings involves using ice packs and pain relievers to minimize pain, washing the wound to lessen the chances of secondary infection, and administering oral

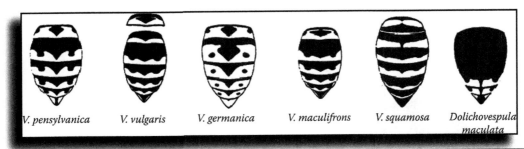

Figure 31.3 Gaster patterns of some yellowjackets (not drawn to scale). (From Akre RD, et al. **Yellowjackets of America North of Mexico**, Agriculture Handbook No. 552. Washington, DC: U.S. Department of Agriculture; 1981.)

OFTEN-ASKED QUESTION

HOW DO YELLOWJACKETS BUILD A NEST IN THE GROUND?

People often encounter yellowjackets when cutting grass, hiking, playing golf, or participating in other outdoor activities during late summer. Unfortunately, they may fly up the pants legs when emerging from their holes in the ground. Stinging events can be quite severe with multiple stings possible. Why do these little pests come out of a hole in the ground? How do they construct a nest underground? First of all, the hole in the ground is just an entrance to a nest underground. Even though the nests are underground, they are still paper nests with the familiar combs and a paper envelope surrounding them. Raw materials gathered by yellowjackets to construct the paper include wood or vegetable fibers mixed with salivary secretions. They may even utilize human products such as blankets, newspapers, or cardboard. The queen yellowjacket selects a site (usually already a hole, crack, or cavern in the ground) and builds the first few nest cells. Almost simultaneously, she constructs a paper envelope around the developing nest (see Figures 1 and 2). When the first batch of workers emerge, the queen then switches from provisioning or building the nest to solely egg laying. Workers then take over the jobs of foraging for food and nest building. Nest building is continuous from the time the first workers emerge until the colony declines (usually fall). Accordingly, underground nests may get quite large in late summer, containing thousands of workers. As the nest grows, yellowjackets excavate dirt to enlarge the underground cavity. Most of the evacuated earth is carried outside the burrow and dropped. As summer ends and fall approaches, yellowjackets generally (there are exceptions) produce new queens and the colony dies. Newly formed queens spend the winter in hollow trees, stumps, wall voids of houses, etc., waiting to emerge in the spring and start the process all over again.

Figure 1 Yellowjacket nest in early stages of development (with an individual block of cells removed from inside).

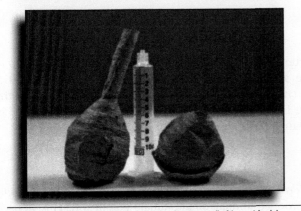

Figure 2 Early stages of yellowjacket nest (left) and bald-faced hornet nest (right).

antihistamines to counteract the direct release of histamine (not IgE-mediated). In the case of a large local reaction, rest and elevation of the affected arm or leg may also be needed. For allergic reactions, administration of epinephrine, antihistamines, and other supportive treatment may be required (see Chapter 2).

II. Hornets

A. General and Medical Importance

In the United States, the term hornet is often misapplied to the yellowjacket, *Dolichovespula maculata*, because it is large with

HORNETS

Worker hornet, *Vespa crabro.*

IMPORTANCE

Painful stings; allergic reactions

DISTRIBUTION

Several species, almost worldwide; only one U.S. species: *Vespa crabro*

LESION

Central white spot with erythematous halo; amount of local swelling is variable

DISEASE TRANSMISSION

None

KEY REFERENCE

Akre RD, Davis HG. Biology and pest status of venomous wasps. *Annu. Rev. Entomol.* 1978;23:215–238

TREATMENT

Pain relievers, antipruritic lotions for local reactions; systemic reactions may require antihistamines, epinephrine, and other supportive measures

Figure 31.4 Huge yellowjacket nest built in a roadside ditch which was over 4 feet tall. (Photograph courtesy of Dr. James Jarratt, Mississippi State University Extension Service.)

behind the eyes, and the ocelli (small, simple eyes) are remote from the margin of the head.

C. Geographic Distribution

The true hornets belong to the genus Vespa and occur in Europe and Asia. The only species in the United States is *Vespa crabro*, which was accidentally introduced into the eastern United States around 1850. It now occurs sparsely in the Atlantic Seaboard states into the South.

D. Biology and Behavior

Vespa crabro is very similar in its biology to yellowjackets. They produce annual, single-female colonies usually housed in a multiple-combed nest with a thick brown envelope in sheltered aerial locations such as hollow trees, attics, and wall voids of houses. This brown paper nest is easily distinguishable from the typical gray paper nests of other wasps and bald-faced hornets (Figure 31.6). The nests are generally large because of the large individual cells and contain about 1500–3000 cells in 6–9 combs with 200–400 workers. Hornets hunt insects, and several species attack honey bee colonies.[4] Although not usually scavengers, they do occasionally puncture the skin of ripe fruits and feed on them.

white and black markings and it builds large aerial nests. True hornets are represented in North America by only one species, *Vespa crabro*, the brown hornet or European hornet (see box). This large hornet produces a very painful sting and may produce allergic reactions in sensitive individuals (see also Chapter 2).

B. General Description

Again, it is important to mention that the commonly called bald-faced hornet that produces a large, egg-shaped aerial nest (Figure 31.5) is not a true hornet—it is a yellowjacket. *Vespa crabro* (the only true U.S. hornet) is a distinctive wasp with a large, robust body (2.5–3.5 cm long) and characteristic brown, orange, and red coloration. The head is swollen

Figure 31.5 Hornet's nest (left) and cut-away (right). Actually, this is an aerial yellowjacket nest. (Photograph by Jerome Goddard, Ph.D.)

Figure 31.6 European hornet's nest with attending workers. (Photograph courtesy of Dr. James Jarratt, Mississippi State University Extension Service.)

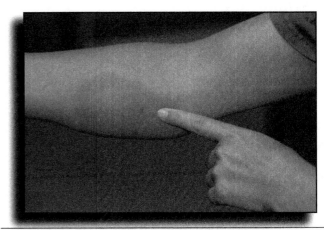

Figure 31.7 Large local sting reaction to a red wasp sting. (Photograph copyright 2012 by Jerome Goddard, Ph.D., and courtesy of Audrey Sheridan.)

E. Treatment of Stings

Hornet stings are treated the same as yellowjacket stings.

III. Paper Wasps

A. General and Medical Importance

This large and diverse wasp group is usually divided into two major divisions depending on lifestyle: solitary or social. Of the 15,000 or so species of stinging wasps worldwide, 95% are solitary species that are not aggressive toward people. The social wasps, on the other hand, form large colonies and can pose serious threats to humans by their stings (Figure 31.7). The most common social groups in temperate regions are the yellowjackets and hornets (subfamily Vespinae) and the paper wasps (*Polistes* and *Mischocyttarus*). Paper wasps have a great affinity for building their nests on or around buildings; therefore, human encounters and stinging incidents are common. Wasp stings may lead to large local and systemic reactions (see Chapter 2 for in-depth coverage), as well as unusual manifestations such as acute renal failure and polyradiculoneuropathy.[5,6]

B. General Description

Paper wasps have elongated, slim bodies (2–2.5 cm long) and are variously colored yellow, black, brown, and red depending on the species (see box and Figure 31.8). Some are striped, yellowish or white and brown, and are occasionally confused with yellowjackets. The introduced pest species, *Polistes dominula*, especially looks like a yellowjacket.[7] Males of many paper wasps have more antennal segments causing them to curl at the tip, more yellow on their faces (frons and clypeus), and distinctly longer faces than females of the same species. An outstanding online color identification guide to wasps is provided by Buck et al.[1]

PAPER WASPS

Paper wasp. (From CDC, Pictorial Keys to Arthropods, Reptiles, Birds, and Mammals of Public Health Significance. Atlanta, GA: U.S. Centers for Disease Control and Prevention; 1963.)

IMPORTANCE

Painful stings; allergic reactions

DISTRIBUTION

Numerous species almost worldwide

LESION

Central white spot with erythematous halo; amount of local swelling variable

DISEASE TRANSMISSION

None

KEY REFERENCE

Akre RD, Davis HG. Biology and pest status of venomous wasps, *Annu. Rev. Entomol.* 1978;23:215–238

TREATMENT

Pain relievers, antipruritic lotions for local reactions; systemic reactions may require antihistamines, epinephrine, and other supportive measures

C. Geographic Distribution

Paper wasps in the genus *Polistes* include about 200 species distributed worldwide in temperate and tropical regions. *P. rubiginosa* (now divided into two species, *P. carolina* and *P. perplexus*) is a bright red-orange wasp that is common throughout the southern United States (Figure 31.8A). *P. exclamans* (Figure 31.8B), also commonly found in the southern United States, is the striped species most often mislabeled as yellowjackets. *P. annularis* (Figure 31.8C) is a large, dark red paper wasp that nests near permanent bodies of water. It is common in the southern United States. In the western United States and California, *P. aurifer* and *P. apachus* are significant pests. *P. fuscatus* is a commonly encountered wasp found throughout Canada and the United States (Figure 31.8D). *P. dominula* is a particularly troubling introduced Palearctic paper wasp species that now infests a large swath of the northern United States (Figure 31.9).[8,9] *P. gallicus* is a species occurring widely throughout southern France and Italy. *Polistes* species common in Japan include *P. smelleni* and *P. jadwigae*. Species of *Mischocyttarus* occurring in the southern and southwestern states include *M. flavitarsis*, *M. mexicanus*, *M. navajo*, and *M. phthisicus*.

Figure 31.9 Worker European paper wasp, *Polistes dominula*.

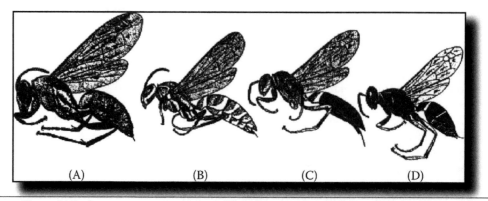

Figure 31.8 Some common paper wasps: (A) *Polistes perplexus*, (B) *P. exclamans*, (C) *P. annularis*, and (D) *P. fuscatus*. (From CDC, **Pictorial Keys to Arthropods, Reptiles, Birds, and Mammals of Public Health Significance**. Atlanta, GA: U.S. Centers for Disease Control and Prevention; 1963.)

Figure 31.10 European paper wasp nests in eaves.

D. Biology and Behavior

Paper wasps build a nest consisting of a single, open-faced comb of gray paper or other fibers. The nest is attached to various substrates (buildings, trees, shrubs) by a single petiole. Certain species such as *Polistes dominula* are especially adept at nesting in anthropogenic structures and are well adapted to the urban and suburban environment (Figure 31.10).[10] In general, *Polistes* wasps feed on any extrafloral juices, plant sap, sweets, and various arthropods. They catch caterpillars to feed their brood. In temperate regions, each inseminated queen hibernates in a protected site and emerges in the spring to initiate nest building. Usually, a single female, called a foundress, initiates the nest, but in some species other individual females, called cofoundresses, may join the original foundress and help build and provision the nest. These multiple foundresses (usually two to six) compete for reproductive dominance in this small colony. Eventually, a dominance hierarchy or pecking order develops with one foundress (i.e., queen), and it lays more of the eggs than the other foundresses. After the emergence of the first brood of workers, the other foundresses lose their reproductive status and the colony usually develops into a single queen (monogynous) society. The workers perform most colony tasks, especially nest building and foraging; however, reproduction remains the primary function of the queen. Although the queen is distinct in behavior from the other colony members, she is not usually discernibly different in appearance from the workers.

The colony grows rapidly through the summer, and mature colony size is usually reached in August or September. Usurpation, wherein wasps take over or use other wasp nests, is quite common in social wasps such as yellowjackets and paper wasps. The practice is exacerbated when nesting sites in the immediate vicinity are already occupied. Typical mature *Polistes* colonies probably contain 30–75 adults and 100–200 cells, although very large colonies of 150–200 adults and nests of 1000–1900 cells have been reported. In late summer and early fall, males and reproductive females (future queens) are produced. The colony declines as workers die and the reproductives leave the nest to mate. After mating, males soon die or do not usually survive the winter, whereas inseminated females seek out and survive in winter hibernation sites. Large aggregations of males and females are sometimes seen near hibernation sites or in tall trees, towers, and buildings during late fall, on warm days of winter, or in early spring. This "house-dwelling" hibernation behavior of *Polistes* may lead to human at stings any time of the year.[11] The following spring, the annual cycle is repeated as females emerge from hibernation to initiate a new nest.

E. Treatment of Stings

Paper wasp stings are treated in the same way as yellowjacket stings.

References

1. Buck M, Marshall SA, Cheung DKB. Identification atlas of the Vespidae (Hymenoptera, Aculeata) of the northeastern Nearctic region. *Can. J. Arthr. Ident.* 2008;5:1–492.
2. Carpenter JM, Glare TR. Mis-identification of *Vespula alascensis* as *V. vulgaris* in North America (Hymenoptera: Vespidae; Vespinae). *Am. Mus. Novitates.* 2010;3690:1–7.
3. Ebeling W. *Urban Entomology.* Berkeley, CA: University of California Press; 1978.
4. Harwood RF, James MT. *Entomology in Human and Animal Health.* 7th ed. New York: Macmillan; 1979.
5. Ridolo E, Albertini R, Borghi L, Meschi T, Montanari E, Dall'Aglio PP. Acute polyradiculoneuropathy occurring after hymenoptera stings: A clinical case study. *Int. J. Immunopathol. Pharmacol.* 2005;18(2):385–390.
6. Vikrant S, Pandey D, Machhan P, Gupta D, Kaushal SS, Grover N. Wasp envenomation-induced acute renal failure: A report of three cases. *Nephrology (Carlton).* 2005;10(6):548–552.
7. White G. Hunting *Polistes dominulus. Pest Control Technology Magazine.* 2004; September issue: 58–64.
8. Hathaway MA. *Polistes gallicus* in Massachusetts (Hymenoptera: Vespidae). *Psyche.* 1981;88:169–173.
9. Liebert AE, Gamboa GJ, Stamp NE, et al. Genetics, behavior and ecology of a paper wasp invasion: *Polistes dominulus* in North America. *Ann. Zool. Fenn.* 2006;43:595–624.
10. Silagi SA, Gamboa GJ, Klein CR, Noble MA. Behavioral differences between two recently sympatric paper wasps, the native *Polistes fuscatus* and the invasive *Polistes dominulus. Great Lakes Entomol.* 2003;36:99–104.
11. Alexander JO. *Arthropods and Human Skin.* Berlin: Springer-Verlag; 1984.

Part IV

Personal Protection Measures against Arthropods

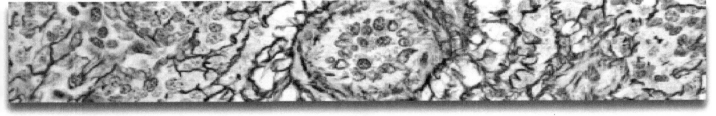

32
PROS AND CONS OF INSECT REPELLENTS

I. Introduction

Insect repellents are chemicals that cause insects to make directed, oriented movements away from the source of the repellent. In light of disease transmission by insects and other arthropods, chemical substances that have repellent effects or interfere with biting are wonderful because they enable us to go places and do things in insect- or disease-infested areas. Undoubtedly, repellents have prevented thousands of cases of malaria, dengue fever, encephalitis, and other mosquito-borne diseases; for example, a recent study demonstrated that persons practicing two or more personal protective behaviors against mosquito bites, including repellents, reduced the risk of West Nile virus (WNV) infection by half.[1] In recent years, however, concerns about the potential adverse health effects of insect repellents have increased, especially for those containing the active ingredient DEET. N,N-diethyl-3-methylbenzamide (DEET) is one of the most effective and widely used insect repellents available.[2] It repels a variety of mosquitoes, chiggers, ticks, fleas, and biting flies, and an estimated 50 to 100 million people in the United States use it each year.[3,4] This chapter discusses various chemical repellents, their modes of action, possible side effects, and precautions necessary to prevent adverse reactions.

II. Mosquito Repellents

A. DEET Products

Previously called N,N-diethyl-m-toluamide, DEET remains the gold standard of currently available insect repellents. The chemical was discovered by U.S. Department of Agriculture (USDA) scientists and patented by the U.S. Army in 1946. It was registered for use by the public in 1957. Twenty years of empirical testing of more than 20,000 other chemical compounds have not resulted in another marketed product with the duration of protection and broad-spectrum effectiveness of DEET.[2] DEET is sold under numerous brand names and is formulated in various ways and concentrations—creams, lotions, sprays, extended-release formulations, etc. (Figure 32.1). Concentrations of DEET range from about 5 to 100%, and, generally, products with higher concentrations of DEET have longer repellence times.[5] At some point, however, the direct correlation between concentration and repellency breaks down. For example, in one study, 50% DEET provided about 4 hours of protection against *Aedes aegypti* mosquitoes, but increasing the concentration to 100% provided only 1 additional hour of protection.[6] DEET is absorbed through the skin into the systemic circulation; one study showed that about 10 to 15% of each dose can be recovered from the urine. Other studies have shown lower skin absorption values in the range of 5.6 to 8.4%. It is possible that the solvent used in a product affects absorption. One study showed that ethanol may increase permeation of DEET.[7] Regardless, the lowest concentration of DEET providing the longest repellency should be chosen for use. Products containing 10 to 35% DEET will provide adequate protection from biting insects under most circumstances (Figure 32.2). The U.S. military developed a polymer-based, extended-release formulation containing 35% DEET, which is available to the general public through the 3M Corporation under the brand name Ultrathon®. This long-acting formulation has been shown to protect against mosquitoes for up to 12 hours.[8] The American Academy of Pediatrics recommends that children should probably not be exposed to DEET concentrations higher than 30% and the CDC says it should not be used on babies less than 2 months old.

B. Picaridin

Picaridin (Figure 32.3), also known as Bayrepel®, is an effective alternative to DEET products that provides long-lasting

Figure 32.2 Repellency of DEET against mosquitoes; hand on the right has no DEET on it. (Photograph copyright 2004 by Jerome Goddard, Ph.D.)

Figure 32.1 Several commercially available insect repellents. Natrapel® and lemon eucalyptus are examples of natural products without the active ingredient DEET.

protection against mosquito bites.[9] This relatively new repellent has been used worldwide since 1998. As opposed to DEET, picaridin is nearly odorless, does not cause skin irritation, and has no adverse effect on plastics; however, even though the product is long lasting and effective against mosquitoes, in some cases it does not provide protection for as long as DEET.[10,11] One field study demonstrated 5-hour protection against *Culex annulirostris* mosquitoes with picaridin vs. 7-hour protection with DEET.[10] The American Academy of Pediatrics recommends that children should probably not be exposed to picaridin concentrations higher than 10%, and the CDC says it should not be used on babies less than 2 months old.

C. Plant-Derived Substances

Plant-derived substances that provide some repellency against mosquitoes include citronella, cedar, verbena, lemon eucalyptus, pennyroyal, geranium, lavender, pine, cajeput, cinnamon, rosemary, basil, thyme, allspice, garlic, and peppermint; however, some of these products provide only temporary protection, if any at all. One study testing DEET-based products against Buzz Away® (containing citronella, cedarwood, eucalyptus, lemongrass, alcohol, and water) and Green Ban® (containing citronella, cajuput, lavender, safrole-free sassafras, peppermint, bergaptene-free bergamot, calendula, soy, and tea tree oils) demonstrated essentially no repellency against *Aedes aegypti*.[5] Other studies with Buzz Away®, however, indicated that the product does have repellency, for about 2 hours.[12] A plant-based repellent that was released in the United States in 1997, Bite Blocker® (containing soybean oil, geranium oil, and coconut oil) has shown good repellency against certain mosquitoes for up to

7 hours.[13] Also in that study, Natrapel® (containing 10% citronella) prevented mosquito biting for about 1–2 hours.[13] Oil of lemon eucalyptus (OLE) has performed well in a number of scientific studies and is now listed on the CDC website as one possible alternative to DEET. OLE has been shown to provide up to 6 hours of protection against mosquito bites. The CDC does not recommend the use of OLE on children less than 3 years old.

Citronella candles have been marketed as backyard mosquito repellents for years. One study compared the ability of commercially available 3% citronella candles, 5% citronella incense, and plain candles to prevent bites by *Aedes* mosquitoes under field conditions.[14] Persons near the citronella candles had 42% fewer bites than controls, but ordinary candles provided a 23% reduction. The efficacy of plain candles and citronella incense did not differ. The ability of plain candles to decrease biting may be because they serve as a decoy source of warmth, moisture, and carbon dioxide.

D. Permethrin

Permethrin, actually a pesticide rather than a repellent, is a synthetic pyrethroid available for use against mosquitoes, but it can only be used on clothing. The product is sold in lawn, garden, or sporting goods stores as an aerosol under the name Permanone®, Repel Gear and Clothing Spray®, or something similar (Figure 32.3). It is nonstaining, nearly odorless, and resistant to degradation by light, heat, or immersion in water. Interestingly, it can maintain its potency for at least 2 weeks, even through several launderings.[15] Permethrin is highly effective. In one study, persons wearing permethrin-treated socks and sneakers received almost 75% fewer bites than persons wearing untreated items (see section below for more information about permethrin and protection from ticks).[16] Permethrin can be applied to clothing, tent walls, and mosquito nets; in fact, sleeping under permethrin-impregnated mosquito nets has been tried extensively in malaria prevention campaigns in Africa, New Guinea, Pakistan, and Malaysia.[17] Factory-based impregnation methods for incorporating permethrin into clothing have been developed that soak the fabric or use a new polymer-coating technique. The polymer-coating methodology is safe, and permethrin impregnation lasts the life of the clothing.[18] For personal protection, the combination of permethrin-treated clothing and DEET-treated skin creates almost complete protection against mosquito bites. In field trials conducted in Alaska, persons wearing permethrin-treated uniforms and 35% DEET (on exposed skin) had more than 99.9% protection (1 bite/hour) over 8 hours, whereas unprotected persons received an average of 1188 bites/hour.[19]

E. Skin-So-Soft®

Avon's Skin-So-Soft® bath oil is often used as a mosquito repellent and is discussed here because of its widespread use. Apparently, Skin-So-Soft does have some transient repellency for mosquitoes. Rutledge et al.[20] reported that the bath oil exhibits repellency for *Aedes aegypti*, but the effect is short lived. They reported that Skin-So-Soft is not nearly as effective as DEET (gram for gram). In another study, the Avon product provided 0.64 hours of protection from *A. albopictus* mosquito bites compared to greater than 10 hours of protection provided by 35% DEET.[15] Avon now markets products under the Skin-So-Soft label that contain an EPA-recognized repellent.

III. Tick Repellents

A. DEET Products

Use of repellents, along with other personal protection methods, can greatly reduce tick biting (Figure 32.4). One study[21] demonstrated that DEET on military uniforms provided between 10 and 87.5% protection against ticks, depending on species and life stage of the tick. There was an average of 59.8% protection against all species of ticks. Obviously, protection levels in the 50% range are less than desirable, considering the fact that just one tick can transmit a tick-borne disease. In a U.S. Army repellent rating system, DEET is assigned a 2× value, whereas permethrin is given a 3× rating.[22] DEET products are simply not as effective in protecting

Figure 32.3   Permethrin (for clothing) and picaridin (for skin and clothing) insect repellents.

Figure 32.4   Repellents and personal protective equipment can help prevent tick biting. (Photograph copyright 2011 by Jerome Goddard, Ph.D.)

from tick infestation as permethrin products; however, the advantage of DEET is that it can be applied to human skin in places likely to be encountered by ticks: ankles, legs, and arms.

B. Permethrin

By far the most effective tick repellent is permethrin, which is sold under various brand names, but especially Permanone® or Repel Gear and Clothing Spray® (Figure 32.3), a synthetic pyrethroid pesticide with very low mammalian toxicity. It is for use on clothing only and is not to be applied directly to human skin. In one study, a pressurized spray of 0.5% permethrin was compared to 20 and 30% DEET products on military uniforms in a highly infested tick area. A 1-minute application of permethrin provided 100% protection, compared to 86 and 92% protection with the two DEET products, respectively.[23] Additionally, permethrin has been shown to remain in clothing, providing 100% protection against ticks after several washings. Generally, application of permethrin to clothing is done by a slow, sweeping application of the aerosol spray until clothing is slightly wet. Label instructions should be followed. Some people looking for protection against ticks hang the clothing to be worn on a clothesline, spray it, and put it on after drying. When ticks subsequently crawl on the clothing treated with permethrin they are either killed or repelled. Permethrin is extremely effective against New World ticks, but there is some evidence that not all tick species are equally repelled by permethrin. One Old World species, the camel tick, actually showed a high tolerance to permethrin and an increased biting response when exposed to the product.[24]

IV. Health Concerns Associated with DEET Products

A. Background

DEET has been used for over 60 years by millions of people worldwide. Although it has an excellent safety record, there have been sporadic reports of adverse reactions associated with its use. Most of these have resulted from accidental exposure, such as swallowing, spraying into the eye, or repeated application, although at least one case occurred in an 18-month-old boy following brief exposure to low-strength (17.6%) DEET.[25] Although most complaints have involved transient minor skin or eye irritation, rare cases of toxic encephalopathy have been reported, especially in children. Adverse reactions have included headache, nausea, behavioral changes, disorientation, muscle incoordination, irritability, confusion, difficulty sleeping, respiratory distress, and even convulsions and death. In one report, six girls, ranging in age from 17 months to 8 years, developed behavioral changes, ataxia, encephalopathy, seizures, and coma

after repeated cutaneous exposure to DEET; three later died. However, if DEET products (preferably not the high-concentration products such as 50 to 100%) are properly applied and used according to their label directions, they are generally considered safe. Use of DEET products according to U.S. Environmental Protection Agency guidelines (see next section) will greatly reduce the possibility of toxicity.

B. Safe Application Rates and Methods

Except under extraordinary conditions, high concentrations of DEET should not be used. Products with 10 to 35% DEET will provide adequate protection under most conditions. For children, even lower concentrations may be warranted. The American Academy of Pediatrics recommends that repellents used on children contain no more than 30% DEET or 10% picaridin. The following guidelines will help ensure the safe use of DEET-based repellents. Remember, repellents should only be applied to clothing and exposed skin according to the product label directions.

DO

Use aerosols or pump sprays for skin and for treating clothing; these products provide even application.
Use liquids, creams, lotions, or sticks for more precise application on exposed skin.
After outdoor activity, wash DEET-covered skin with soap and water.
Always keep insect repellents out of the reach of small children.

DON'T

Apply to eyes, lips, or mouth or over cuts, wounds, or irritated skin.
Overapply or saturate skin or clothing.
Apply to skin under clothing.
Apply more often than directed on the product label.

References

1. Loeb M, Elliott SJ, Gibson B, et al. Protective behavior and West Nile virus risk. *Emerg. Infect. Dis.* 2005;11:1433–1436.
2. Fradin MS. Mosquitoes and mosquito repellents: A clinician's guide. *Ann. Intern. Med.* 1998;128:931–940.
3. Abramowicz M, Zucotti G, Pflomm JM. Insect repellents. *Med. Lett. Drugs Ther.* 2016;58(1498):83–85.
4. CDC. Seizures temporarily associated with use of DEET insect repellents – New York and Connecticut. *MMWR.* 1989;38:678–680.
5. Chou JT, Rossignol PA, Ayres JW. Evaluation of commercial insect repellents on human skin against Aedes aegypti. *J. Med. Entomol.* 1997;34:624–630.

6. Buescher MD, Rutledge LC, Wirtz RA, Nelson JH. The dose-persistence relationship of DEET against Aedes aegypti. *Mosq. News.* 1983;43:364–366.

7. Stinecipher J, Shah J. Percutaneous permeation of N,N-diethyl-m-toluamide from commercial mosquito repellents and the effect of solvent. *J. Toxicol. Environ. Health.* 1997;52:119–135.

8. Anonymous. Insect repellents. *Med. Lett. Drugs Ther.* 2016;58:83–84.

9. CDC. Updated Information Regarding Mosquito Repellents. Atlanta, GA: U.S. Centers for Disease Control and Prevention; 2009. http://www.cdc.gov/ncidod/dvbid/westnile/repellent updates.htm.

10. Frances SP, Waterson DG, Beebe NW, Cooper RD. Field evaluation of repellent formulations containing deet and picaridin against mosquitoes in Northern Territory, Australia. *J. Med. Entomol.* 2004;41(3):414–417.

11. Klun JA, Khrimian A, Margaryan A, Kramer M, Debboun M. Synthesis and repellent efficacy of a new chiral piperidine analog: Comparison with Deet and Bayrepel activity in human-volunteer laboratory assays against *Aedes aegypti* and *Anopheles stephensi. J. Med. Entomol.* 2003;40(3):293–299.

12. Fradin MS, Day JF. Comparative efficacy of insect repellents against mosquito bites. *N. Engl. J. Med.* 2002;347:13–18.

13. Barnard DR, Xue RD. Laboratory evaluation of mosquito repellents against Aedes albopictus, Culex nigripalpus, and Ochierotatus triseriatus (Diptera: Culicidae). *J. Med. Entomol.* 2004;41(4):726–730.

14. Lindsay RL, Surgeoner GA, Heal JD, Gallivan GJ. Evaluation of the efficacy of 3% citronella candles and 5% citronella incense for protection against field populations of *Aedes* mosquitoes. *J. Am. Mosq. Control Assoc.* 1996;12:293–294.

15. Schreck CE, McGovern TP. Repellents and other personal protection strategies against *Aedes albopictus. J. Am. Mosq. Control Assoc.* 1989;5:247–250.

16. Miller NJ, Rainone EE, Dyer MC, Gonzalez ML, Mather TN. Tick bite protection with permethrin-treated summer-weight clothing. *J. Med. Entomol.* 2011;48(2):327–333.

17. Service MW. Mosquitoes. In: Lane RP, Crosskey RW, eds., *Medical Insects and Arachnids.* London: Chapman and Hall; 1996:120–240.

18. Bonnefoy X, Kampen H, Sweeney K. Public Health Significance of Urban Pests. Copenhagen: World Health Organization Europe; 2008: 569.

19. Lillie TH, Schreck CE, Rahe AJ. Effectiveness of personal protection against mosquitoes in Alaska. *J. Med. Entomol.* 1988;25:475–478.

20. Rutledge LC, Wirtz RA, Buescher MD. Repellent activity of a proprietary bath oil, Skin-So-Soft. *Mosq. News.* 1982;42:557–560.

21. Evans SR, Korch GW, Jr., Lawson MA. Comparative field evaluation of permethrin and DEET-treated military uniforms for personal protection against ticks. *J. Med. Entomol.* 1990;27:829–834.

22. Evans SR. Personal Protective Techniques Against Insects And Other Arthropods Of Military Significance. U.S. Army Environmental Hygiene Agency, Aberdeen Proving Ground, Maryland, TG No. 174; 1991:90.

23. Schreck CE, Snoddy EL, Spielman A. Pressurized sprays of permethrin or deet on military clothing for personal protection against *Ixodes dammini. J. Med. Entomol.* 1986;23:396–399.

24. Fryauff DJ, Shoukry MA, Schreck CE. Stimulation of attachment in a camel tick, *Hyalomma dromedarii:* The unintended result of sublethal exposure to permethrin-treated fabric. *J. Med. Entomol.* 1994;31:23–29.

25. Briassoulis G, Narlioglo M, Hatzis T. Toxic encephalopathy associated with use of DEET insect repellents: A case analysis of its toxicity in children. *Human Exp. Toxicol.* 2001;20:8–13.

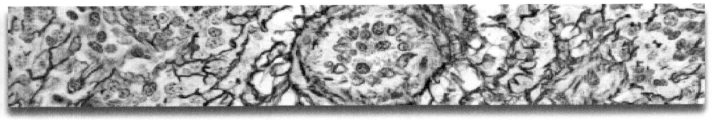

33

ARTHROPOD-SPECIFIC PERSONAL
PROTECTION TECHNIQUES

I. Protection from Mosquitoes

A. Avoidance

Practical nonchemical measures for mosquito avoidance include limiting outdoor activity after dark and avoiding known mosquito-infested areas (e.g., swamps, marshes) during the peak mosquito season. In addition, people who have to be outdoors after dark in the mosquito season should wear long sleeves and long pants.

B. Screening

Probably one of the most basic and effective sanitation measures to limit arthropod–human contact is that of screen wire windows and doors. Screens are constructed of various metals or plastic and are ordinarily 16 × 16 × 20 mesh. They should be tight fitting over window openings. Screen doors should be hung so that they open outward. Being such a basic protection measure, screens are sometimes overlooked; however, their importance cannot be overemphasized. Dr. Goddard personally investigated a fatality due to eastern equine encephalitis in which the family had no screens on the house. By their own testimony, the family members stated, "Mosquitoes eat us up every night!"

C. Netting

During the mosquito season, people may choose to use protective headgear or jackets made of netting when outdoors in heavily infested areas (Figures 33.1 and 33.2). Also, those camping may sleep under mosquito nets for protection from mosquitoes. This becomes mandatory on safaris or other trips to the tropics or subtropics for protection against disease-carrying mosquitoes. Ordinarily, mosquito netting is made of cotton or nylon with 23 to 26 meshes per inch. Netting should not be allowed to lie loosely on the head or body, because mosquitoes can feed through the net wherever it touches the skin; therefore, construction of a crude frame may help keep the net away from the body (Figure 33.3). Insecticide-treated nets (ITNs) are a mainstay for malaria and filariasis prevention in Africa, Southeast Asia, and South America, and they are widely promoted by the World Health Organization and charitable groups such as the Bill and Melinda Gates Foundation. These nets are usually treated with pyrethroids such as permethrin or deltamethrin and may remain effective for months or even a year or two if not rinsed or washed. One study in Papua New Guinea showed that bites from mosquitoes ranged from 6.4 to 61.3 bites per person per day before bed net distribution, and from 1.1 to 9.4 bites for 11 months after distribution (P<0.001).[1] A recent worrisome trend is development of insecticide resistance to the pyrethroids in ITNs, leading to loss of efficacy,[2] although their use is still very effective.[3]

II. Protection from Ticks

A. Boots, Trousers, Tape, and Repellents

Personal protection techniques for tick bites include avoiding tick-infested woods, tucking pant legs into boots or socks, and using repellents on pant legs and socks (Figures 33.4 through 33.6) (also see Online Supplemental Material). Tucking trousers into boots or socks forces ticks to crawl up the outside of one's pants, thus making them easier to spot and remove. Wide masking tape may be used to tape down pants legs at the ankles, or tape can even be placed around the ankles or thighs with the sticky side exposed to protect against

Figure 33.1 Head net for protection against mosquito bites. (Photograph courtesy of Joseph Goddard.)

Figure 33.3 Bed net used for protection against mosquitoes at night. (Photograph courtesy of Rachel Freeman Ford, RN.)

Figure 33.2 Protective jacket made of netting.

Figure 33.4 Tucking pant legs into socks for tick protection.

ticks crawling up the legs (Figure 33.7). One research study showed that significantly fewer nymphal deer ticks were picked up on 5-minute walks in the woods when boots were worn with ankles taped than when sneakers were worn with socks exposed.[4] As for repellents (see also previous chapter), a laboratory study with formulations of DEET, IR3535, and picaridin showed that those products (actually any of the three) containing at least 20% active ingredient were highly effective in repelling lone star tick nymphs for 12 hours.[5] In addition, permethrin products formulated for clothing application are extremely effective for tick protection.[6]

B. Unorthodox Methods

Many unorthodox (or at least, questionable) methods of tick protection are commonly used by people. Flea collars worn around the ankles, Avon's Skin-So-Soft®, vitamin C, garlic, sulfur powder, pantyhose, and others have been reported as "great tick repellents." Although these efforts may possibly provide some degree of protection, tucking the trousers with proper use of permethrin (Permanone® Repel Gear and Clothing Spray® and other brand names) is much more effective.[7] Besides, using pet products such as flea collars may be dangerous due to absorption of the pesticide through the skin (Figure 33.8); they have not been approved by the Food and Drug Administration (FDA) for human wear.

C. Tick Removal

What is the best way to remove a tick after it gets on you? The answers are varied, depending on whom you ask and what

Figure 33.5 Tucking pants legs into rubber boots for tick protection.

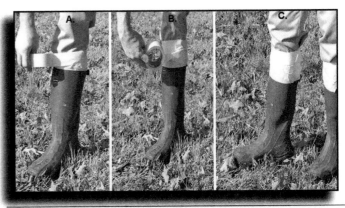

Figure 33.7 "Taping up" with masking tape for tick protection. The last layer or two should be with the sticky side out.

Figure 33.6 The use of both boots and repellents can be very effective against ticks.

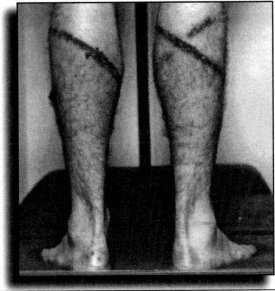

Figure 33.8 Skin lesions from human use of animal flea collars for tick protection. (Photograph courtesy of S Evans and R Fitzsimmons, U.S. Army Environmental Hygiene Agency.)

part of the country you are in because many folklore methods are available. Hard ticks attach themselves firmly to a host for a feeding period of several days and are especially difficult to remove. Methods such as touching attached ticks with a hot match; coating them with mineral oil, petroleum jelly, or some other substance; or "unscrewing" them are but a few of the home remedies that supposedly induce them to back out. The theory behind coating a tick with fingernail polish, petroleum jelly, or mineral oil is that covering the spiracles with a substance will interfere with their breathing and make the ticks back out; however, ticks are able to shut off their spiracles (at least temporarily) and certainly do not breathe via the mouthparts. Even if coating a tick with a substance causes it to back out, the time required to accomplish this is unacceptable (usually 1 to 4 hours).

Because the lengthy feeding period is an important factor in disease transmission by ticks, it is crucial that a tick is removed as soon as possible to reduce chances of infection by disease organisms. During many years of field research with ticks, the authors have had to remove ticks, and pulling them straight off with blunt forceps (tweezers) seemed to work best (Figure 33.9) (see also Online Supplemental Material for a video). There has been some research in this area. A professor at Ohio State University did a very good study of this problem.[8] He evaluated five methods commonly used for tick removal: (1) petroleum jelly, (2) fingernail polish, (3) 70% isopropyl alcohol, (4) hot kitchen match, and (5) forcible removal with forceps. Results showed that that the commonly advocated methods are either ineffective or, worse, actually created greater problems. If petroleum jelly or some other substance causes the tick to back out on its own (and most often it does not), the cement surrounding the mouthparts used for attachment remains in the skin, where it continues to cause

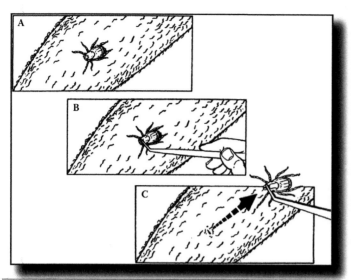

Figure 33.9 Recommended method for tick removal: Grab tick with forceps as close to the skin as possible and pull straight off. (From Goddard J. *Ticks and Tick-borne Diseases Affecting Military Personnel*, USAFSAM-SR-89-2. San Antonio, TX: School of Aerospace Medicine, U.S. Air Force; 1989.)

irritation. Touching the tick with a hot match may cause it to burst, increasing the risk of disease pathogen exposure. Furthermore, hot objects may induce ticks to salivate or regurgitate infected fluids into the wound. Unscrewing a tick is likely to leave broken mouthparts in the host's skin.

The Ohio State paper offered the following recommendations for tick removal: (1) use blunt forceps or tweezers; (2) grasp the tick as close to the skin surface as possible, and pull upward with steady, even pressure; (3) take care not to squeeze, crush, or puncture the tick; (4) do not handle the tick with bare hands because infectious agents may enter via mucous membranes or breaks in the skin; and (5) after removing the tick, disinfect the bite site and wash hands thoroughly with soap and water.

Many tick-borne diseases can be successfully treated with antibiotics in their initial stages; therefore, early diagnosis is imperative. For this reason, marking the day of a tick bite on a calendar is a good idea. If unexplained disease symptoms occur within 2 weeks from this day, persons should be reminded to see their physicians and specifically inform him or her of the tick bite. This method has proved to be very helpful in the diagnosis of tick-borne disease. Although there are a number of well-known tick removal methods (mostly folklore), the best one seems to be the simplest: pull them straight off with blunt forceps and disinfect the bite site.

III. Protection from Other Arthropods

A. Biting Midges

Biting midges (no-see-ums), the tiny slender gnats that are extremely common along the Atlantic and Gulf coasts, are small enough to pass through ordinary screen wire used to cover windows and doors. Accordingly, a finer-mesh screen wire must be used to prevent entry of these flies into dwellings. Outdoors, DEET repellents and long-sleeved shirts and long pants can provide relief in infested areas. Avon's Skin-So-Soft is often used as a repellent for biting midges, and controlled studies indicate that the product provides some protection[9]; however, the effectiveness is probably because the oiliness of Skin-So-Soft traps the tiny midges on the skin surface.

B. Sand Flies

Sand flies are small, delicate, mosquito-like insects (usually less than 5 mm long) that inflict painful bites. Although the most serious problems from sand flies occur in tropical countries, there are many species in temperate zones, as well. The authors have collected them in mosquito light traps in Mississippi.[10] Because sand flies do not bite through clothing, long sleeves, trousers, and socks should be worn in areas where sand flies are active.[11] In heavily infested areas, head nets, gloves, and repellent-treated net jackets and hoods can provide additional protection. Campsites should be chosen that are high, breezy, open, and dry. Fine-mesh bed nets should be used in areas where sand flies are present.

C. Chiggers (Red Bugs)

Chigger mites (sometimes called red bugs) occur in grass, weeds, or leaves and get onto passing vertebrate hosts; therefore, personal protection measures for chiggers are similar to those for ticks. Tucking pants legs into socks or boots and spraying clothing with DEET-based repellents provide fairly good protection (Figure 33.4). Treating clothing with permethrin is very effective against chiggers. One study showed a 74.2% increase in protection from chiggers compared to untreated clothing and use of repellents.[12] In addition, after exposure to infested outdoor areas, hot soapy baths or showers will help remove any chiggers, attached or unattached.

D. Kissing Bugs

Kissing bugs (*Triatoma* species), the vectors of Chagas' disease in Mexico and Central and South America, are nocturnal insects that seek refuge by day in the cracks and crevices of poorly constructed houses or in the loose roof thatching of huts. Personal protection from the bugs involves avoidance (if possible)—not sleeping in thatched-roof huts in endemic areas—and exclusion methods such as bed nets. Prevention and control of domestic species of triatomines can be accomplished by proper construction of houses, wise choice of building materials, sealing cracks and crevices, and precision-targeting with insecticides in the home.

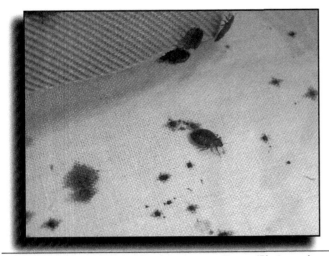

Figure 33.10 Bed bugs and fecal spots on mattress. (Photograph courtesy Armed Forces Pest Management Board and Dr. Harold Harlan)

E. Bed Bugs

Bed bugs are increasingly being reported worldwide in hotels and hostels, apartments, and even single-family homes. Upon initial introduction into a dwelling, bed bugs are generally found in/near the bed, recliners, and couches where people spend a lot of time. However, older, more established infestations may result in bugs all over the place, on the ceiling, behind pictures on the wall, under edges of carpet, etc. As for personal protection, repellents are largely ineffective, so inspection of hotel or hostel rooms before spending the night is highly recommended. In fact, we recommend placing suitcases in the bathroom and (then) carefully inspecting the room before unpacking belongings. Mattresses, box springs, and behind the headboard (if possible) should be carefully inspected upon initial entry into a rented room. If bed bugs or their fecal spots are found (Figure 33.10) (see Chapter 13 for more details), another room should be obtained immediately. Pest control and environmental sanitation to prevent/control bed bug infestations are the most effective preventive measures for bed bugs. For one thing, after traveling, be cautious about unpacking suitcases on the bed or even inside the house. Suitcases can be stored in the garage or outbuildings. Clothes should be washed and dried on the hottest settings on washers and dryers. Since all stages (including eggs) of bed bugs are killed at temperatures of 104–122°F (40–50°C), clothes should be dried in a tumble dryer at the HOT setting for at least 30 minutes.[13]

F. Fire Ants

Fire ants are extremely aggressive ants that currently inhabit much of the southern United States. They were accidentally imported into the United States between 1918 and 1940 and continue to spread. Outdoors, fire ants can be recognized by their mounds, which are elevated earthen mounds 8 to 90 cm high, surrounded by relatively undisturbed vegetation. When disturbed, the ants characteristically boil out of their mounds in great numbers, stinging their victims. Personal protection from fire ants primarily involves recognition and avoidance of their mounds. Rubber boots are effective in preventing attacks when outdoors (see Online Supplemental Material). Insect repellents are apparently of no value against fire ant attacks; however, there is apparently some benefit in wearing socks, at least in delaying ant attacks.[14]

References

1. Reimer IJ, Thomsen EK, Tisch DJ, et al. Insecticidal bed nets and filariasis transmission in Papua New Guinea. *N. Engl. J. Med.* 2013;369(8):745–753.
2. N'Guessan R, Corbet V, Akogbeto M, Rowland ME. Reduced efficacy of insecticide-treated nets and indoor residual spraying for malaria control in pyrethroid resistance area, Benin. *Emerg. Infect. Dis.* 2007;13:199–206.
3. Tokponnon FT, Ogouyémi AH, Sissinto Y, et al. Impact of long-lasting, insecticidal nets on anaemia and prevalence of *Plasmodium falciparum* among children under five years in areas with highly resistant malaria vectors. *Malaria J.* 2014;13(76).
4. Carroll JF, Kramer M. Different activities and footwear influence exposure to host-seeking nymphs of *Ixodes scapularis* and *Amblyomma americanum. J. Med. Entomol.* 2001;38:596–600.
5. Carroll JF, Benante JP, Kramer M, Lohmeyer KH, Lawrence KL. Formulations of Deet, picaridin, and IR3535 applied to skin repel nymphs of the lone star tick for 12 hours. *J. Med. Entomol.* 2010;47:699–704.
6. Schreck CE, Snoddy EL, Spielman A. Pressurized sprays of permethrin or deet on military clothing for personal protection against *Ixodes dammini. J. Med. Entomol.* 1986;23:396–399.
7. Evans SR, Korch GW, Jr., Lawson MA. Comparative field evaluation of permethrin and DEET-treated military uniforms for personal protection against ticks. *J. Med. Entomol.* 1990;27:829–834.
8. Needham GR. Evaluation of five popular methods for tick removal. *Pediatrics.* 1985;75:997–1002.
9. Schreck CE, Kline DL. Repellency determinations of four commercial products against six species of ceratopogonid biting midges. *Mosq. News.* 1981;41:7–10.
10. Goddard J. New records for the phlebotomine sand fly *Lutzomyia shannoni* (Dyar) in Mississippi. *J. Miss. Acad. Sci.* 2005;50:195–196.
11. Rutledge LC, Gupta RK. Moth flies and sand flies (Psychodidae). In: Mullen GR, Durden LA, eds. *Medical and Veterinary Entomology, Second Edition.* New York: Elsevier; 2009:153–168.
12. Breeden GC, Schreck CE, Sorensen AL. Permethrin as a clothing treatment for personal protection against chigger mites. *Am. J. Trop. Med. Hyg.* 1982;33:589–592.
13. Pinto L. How to kill bed bugs by laundering. Pest Control Technology Magazine, March issue, p. 66; 2017.
14. Goddard J. Personal protection measures against fire ant attacks. *Ann. Allergy Asthma Immunol.* 2005;95:344–349.

GLOSSARY

Abdomen: The hindmost of the three body divisions in insects.

Acaracide: A substance poisonous to ticks or mites.

Aculeate: Pertaining to the members of the order Hymenoptera, which sting—the bees, wasps, and ants.

Alate: Winged form; in ants, alates are the winged reproductive forms.

Anal: The posterior wing veins.

Antennae (sing. Antenna): A pair of sensory segmented appendages located on the head above the mouthparts.

Anterior: Toward the front end.

Anthropophilic: Describes any bloodsucking arthropod that prefers humans to other animals as its food source.

Apiary: Any place where honey bees are kept.

Apical: At the tip or end.

Apterous: Without wings.

Arbovirus: An arthropod-borne virus.

Arista: A large bristle located on the last antennal segment of some flies.

Atopy: IgE-dependent allergy often arising from an unknown exposure to an antigen.

Babesiosis: Infection with a protozoan organism in the genus Babesia; often a malaria-like illness.

Bacon therapy: Method of removing embedded bot fly larvae by occlusion of the punctum (breathing hole in the skin made by the larvae) with raw meat or pork.

Basis capitulum: Basal portion of the capitulum on which the tick mouthparts are attached. Various shapes (hexagonal, rectangular, subtriangular, etc.) are possible in hard ticks.

Beak: The proboscis of a sucking insect.

Bifid: Clearly divided into two parts.

Brood: All the immature members of a colony, including eggs, larvae, and pupae.

Brood cell: A cell made by a worker bee in which to lay an egg.

Bug: Term loosely used to denote any arthropod; technically, meaning only members of the insect order Hemiptera.

Cantharidin: A chemical produced by certain beetles (especially Family Meloidae) that causes blistering on the skin of humans.

Calypter: A basal lobe or lobes on the posterobasal portion of the axillary membrane of the wings of some Diptera.

Canine scabies: A condition caused by *Sarcoptes scabiei* which can be temporarily transferred to humans from dogs, causing itching and papular or vesicular lesions primarily on the waist, chest, or forearms.

Carapace: The sclerotized (hardened) plate forming the dorsal surface of an arachnid cephalothorax.

Carnivorous: Feeding on flesh (animals).

Caste: A group of morphologically distinct individuals within a colony often having distinctive behavior (e.g., workers, queens, males).

Caterpillar: The larva of a butterfly or moth having a cylindrical body, a well-developed head, thoracic legs, and abdominal prolegs.

Cell: Terminology used in describing areas of an insect wing; an area of the wing enclosed by veins.

Cephalothorax: Head and thorax combined; characteristic of arachnids.

Cerci (sing. Cercus): Paired appendages at the posterior end of the abdomen, as in a cockroach.

Chelae (sing. Chela): The second pair of appendages (pedipalps) of scorpions and pseudoscorpions; modified into pincers.

Chelicerae: Paired appendages of an arachnid, highly variable in shape and size. In scorpions, they are short, chelate, and lacking a poison gland; in spiders, they terminate in a sharp tip with a venom duct; and, in ticks, they lie dorsally to the hypostome, completing the cylindrical mouthparts that are inserted when a tick feeds.

Chiggers: Common name given to the larvae of mites in the Family Trombiculidae.

Chitin: A complex nitrogenous carbohydrate forming the main skeletal substance of arthropods.

Class: A grouping used in classification; a division of a phylum.

Classification: The arrangement of species in a hierarchical system of categories and taxa.

Clypeus: That part of the insect head below the front to which the labrum is attached.

Cocoon: A silken enclosure secreted by a larva just before pupation.

Colony: Individuals, other than a single mated pair, that cooperate to build a nest or rear offspring.

Complete metamorphosis: Type of insect development in which there are egg, larva (caterpillar), pupa (resting stage), and adult stages.

Compound eye: Insect eye composed of many individual elements represented externally by hexagonal facets.

Contiguous: Touching one another.

Cornua: Small projections extending from the dorsal, posterolateral angles of basis capituli in ticks.

Copra itch: Condition involving dermatitis caused by epidermal infestation of grain or flour mites.

Costa: The thickened anterior vein of the insect wing.

Coxae (sing. Coxa): Basal segments of the leg; in ticks, small sclerotized plates on the venter representing the first segment of the leg to which the trochanters are movably attached. From anterior to posterior, the coxae are designated by Roman numerals I, II, III, and IV. Bifid coxae are those that are cleft, divided, or forked.

Dentition: Refers to the presence and arrangement of denticles (teeth). In ticks, the numerical arrangement of the rows of denticles on the ventral side of the hypostome is expressed by dentition formulas. Thus, dentition 3/3 means that there are three longitudinal rows of denticles on each side of the median line of the hypostome.

Desensitization (also called Immunotherapy): Elimination or reduction of allergic sensitivity; usually accomplished through a programmed course of antigen treatment.

Deutonymph: The third stage of a mite.

Diapause: Delay in development of insects in response to environmental conditions.

Dichotomous: Divided into two parts. Insect identification keys are often dichotomous, giving the reader two choices after each question.

Direct effects: Medical or health effects produced directly by arthropods, including bites, stings, and blisters.

Diurnal: Active in the daytime.

Dried fruit dermatitis: Dermatitis caused by grain or flour mites found on dried fruit or other stored food products.

Dorsal: Pertaining to the back or top of the body.

Ecdysis: The process of shedding the skin or exoskeleton.

Ectoparasite: A parasite that feeds on the outer surface of the body.

Ehrlichiosis: Disease caused by one of several rickettsia-like bacterial organisms in the genus *Ehrlichia*.

Elytra: Thick or leathery front wing of beetles.

Envenomization (also called Envenomation): The poisonous effects caused by the bites, stings, secretions, stinging hairs of insects, other arthropods, certain other invertebrate animals, or the bites of reptiles.

Enzootic: Disease of animals constantly present in an area.

Epidemic typhus: Disease transmitted by lice which has historically wiped out large numbers of people during natural disasters or wars.

Epizootic: Describes any disease of animals, the number of cases of which exceeds that normally expected.

Erucism: Urtication caused by moth or butterfly larvae.

Exuvium: The cast exoskeleton of an arthropod.

Eyes: In insects, either simple (singular) or compound, the eyes are variously arranged on the head. In ticks, eyes, when present, are located on the edges of the scutum. They are about even with the site of leg I attachment in hard ticks. Soft ticks may have eyes on their lateral margins near coxae I and II.

Facet: The external surface of one of the individual units of a compound eye, as in the fly.

Family: A category in the hierarchy of classification; a division of an order.

Femur: The third leg segment outward from the insect body.

Festoons: Uniform rectangular areas, separated by distinct grooves, located on the posterior margin of most genera of hard ticks.

Fiddleback spider: Alternative common name for the brown recluse spider, *Loxosceles reclusa*.

Filariasis: Disease caused by filarial worms (nematodes), which invade lymphatic tissues and are transmitted by mosquitoes.

Filth Flies: Members of the fly families Muscidae, Sarcophagidae, and Calliphoridae, which are domestic non-biting flies commonly seen in and around human dwellings.

Flagellum: in arthropods, the third and succeeding segments of most antennae.

Flagellomere: A division of the flagellum of the antennae.

Foci (sing. Focus): With reference to a disease, specific areas in which the disease is prevalent.

Follicle mites: Minute, wormlike mites living in hair follicles or sebaceous glands of mammals.

Foundress: An individual, usually a fertilized female, that founds a new colony. All subsequent offspring are her daughters and sons.

Galea: Portion of some insect mouthparts; specifically, the outer lobe of the maxilla.

Gaster: The prominent part of wasp or ant abdomen, separated from the other body parts by a thin connecting segment called a petiole or pedicel.

Genera (sing. Genus): Categories in the classification hierarchy to which species are assigned.

Gradual metamorphosis: Type of insect development in which there are egg, nymph, and adult stages; no wormlike larval stage present.

Gravid: Full of mature eggs; ready to lay eggs.

Grocer's itch: Dermatitis caused by exposure to mite-infested food products, especially grain and flour mites.

Grub: Term used for a thick-bodied, sluggish, often white insect larva.

Goblets: Small, round structures located in the spiracular plate of ticks.

Grooves: On ticks, linear depressions or furrows, primarily on the ventral surface.

Halteres (sing. Haltere): Small knoblike structures on each side of the thorax of a fly immediately behind each wing.

Harvest mites: Common name given to mites in the Family Trombiculidae.

Haustellate: Having mouthparts adapted for sucking blood.

Head: The anterior body region of an insect bearing eyes, antennae, and mouthparts. Ticks and mites have no true head.

Hemelytron: The forewing of the true bugs (order Hemiptera).

Hemimetabolous: Simple metamorphosis.

Hemocoel: The major body cavity of insects containing blood.

Hemolymph: Arthropod blood.

Hemolytic anemia: Shortage of red blood cells due to their premature destruction; sometimes a complication as a result of brown recluse spider bite.

Herbivorous: Feeding on plants.

Hibernation: A period of lethargy or suspension of most bodily activities with a greatly reduced respiration rate, occurring mostly during periods of low temperature.

Histamine: Organic substance released from tissues during an allergic reaction to injury or invasion by an antigen, causing dilation of local blood vessels.

Holometabolous: Complete metamorphosis.

Holoptic: In flies, the eyes touching above.

Hood: The anterior projection of the integument in soft ticks above and covering the mouthparts.

Hypognathus: Head and mouthparts situated ventrally (pointed down).

Hypopharynx: Mouthpart structure, located medially (that is, anterior to the labium). In many sucking insects, this structure contains the salivary channel.

Hypostome: In ticks, the median ventral structure of the mouthparts that lies parallel to and between the palps. It bears the recurved teeth or denticles.

Imago: The adult stage of an insect.

Inornate: In ticks, the absence of a color pattern on the scutum.

Instar: An insect between successive molts.

Joint: An articulation; the area of flexion between sections of an appendage.

Labellum: Insect mouthpart; the tip of the labium.

Labial palpi (sing. Labial palpus): Segmented sensory appendages of the labium of insects.

Labium: "Lower lip" of insect mouthparts.

Labrum: "Upper lip" of insect mouthparts.

Laciniae (sing. Lacinia): Insect mouthparts; the inner lobe of the maxilla.

Large local reaction: Reaction to sting or bite that is exaggerated but still contiguous with the sting site.

Larvae (sing. Larva): An immature stage of an insect having complete metamorphosis but excluding the egg or pupal stage. Also, a six-legged first instar mite or tick.

Larviparous: Method of reproduction by bringing forth larvae that have already hatched in the female's body.

Lepidopterism: Urtication caused by hairs, scales, or spines of adult moths or butterflies.

Lesion: An injury to body tissue; often a spot or mark.

Maggot: A legless larva (usually limited in usage to certain families of Diptera) that has no well-developed head region.

Malphigian tubules: Long and slender blind tubes in the hemocoel that open into the beginning of the hind intestine of insets; excretory in function.

Mammillae (sing. Mammilla): Elevations of various forms found on the integument of *Ornithodoros* tick species.

Mandibles: The most anterior pair of two pairs of insect mouthpart structures.

Maxillae (sing. Maxilla): The pair of mouthpart structures lying behind the mandibles.

Maxillary palpi (sing. Maxillary palpus): Segmented sensory structures located on the maxillae.

Mesosoma: The seven abdominal segments of scorpions.

Metamorphosis: The series of changes through which an insect passes in developing from egg to adult.

Metasoma: The tail of scorpions.

Molt: The process of shedding the exoskeleton.

Musciform: Resembling a fly.

Myiasis: The invasion of human or other animal tissues by dipterous larvae.

Nearctic: One of the zoogeographical regions of the Earth that includes Canada, Alaska, Greenland, the United States, and the temperate northern part of Mexico.

Neotropical: South America, Central America (including Mexico), and the Antilles.

Nits (sing. Nit): Eggs of lice.

Nocturnal: Being active during night.

Nomenclature: The scientific names of living organisms and the application of these names.

Norwegian scabies (also called crusted scabies): Serious scabies infestation in which millions of mites inhabit thick crusts over the skin.

Nymph: An immature insect that does not have a pupal stage (e.g., cockroach, grasshopper, bed bug). Also, an eight-legged immature tick or mite.

Obligate parasite: A parasitic association in which the parasite cannot complete its life cycle without a suitable host.

Ocelli (sing. Ocellus): The simple eyes of arthropods.

Oothecae (sing. Ootheca): Egg case in cockroaches.

Opisthosoma: The entire abdomen in arachnids.

Opisthognathus: Mouthparts situated and directed toward the posterior (backward).

Order: A division of a class in the hierarchy of categories.

Ornate: Definite enamel-like color pattern superimposed on the base color of the integument in hard ticks.

Ovipositor: Tubular structure used by many insects to lay eggs; modified into a stinger in the ants, wasps, and bees.

Palpi (sing. Palpus or Palp): In insects, a segmented process on the maxillae or labium. In ticks, paired articulated appendages located on the front and sides of the basis capituli and lying parallel to the hypostome.

Papules (sing. Papule): Small elevations of the skin that are usually inflamed.

Parasite: Any organism that lives in or on, and at the expense of, another organism.

Parthenogenesis: Condition in which egg development can occur without fertilization.

Pectines: Feathery sensory organs of scorpions; believed to sense ground vibrations.

Pedicel: In spiders, the petiole (waist) between the cephalothorax and abdomen; in ants and other Hymenoptera, the stalk between thorax and abdomen.

Pedipalps: The second pair of appendages of an arachnid.

Petechiae (sing. Petechia): Pinpoint, flat, round, purplish red spots.

Petiole: The narrowed portion of the abdomen of certain Hymenoptera.

Pharynx: In insects, the anterior region of the foregut, located between the mouth and the esophagus.

Pheromone: A chemical produced by one individual that causes a specific reaction by other members of the same species.

Phylum: A major division of a taxonomic kingdom; a category.

Plumose: As in antennae, meaning featherlike.

Prepupa: The quiescent stage immediately before the pupal stage in insects.

Proboscis: A beaklike projection containing various arrangements of mouthparts and their modifications.

Prognathous: With the head horizontal and the jaws directed forward.

Pronotum: Dorsal shield over the anterior segment of the thorax; in cockroaches, the pronotum looks like the head.

Protonymph: In mites, the second instar.

Pseudopustule: An elevated dome-like skin lesion filled with necrotic material, resembling a pustule, developing after the sting of a fire ant.

Pubescent: Covered with short, fine hairs; appearing hairy.

Pupae (sing. Pupa): A nonfeeding and inactive stage (except mosquitoes) between the larvae and adults.

Puparium: A shell or case produced by the hardening of the last larval skin.

Quinones: Caustic, highly volatile hydrocarbons secreted by some arthropods as a defensive measure.

Red bugs: Common name given to the larvae of mites in the Family Trombiculidae. Also known as chiggers.

Red mite of poultry: Another name for the chicken mite, Dermanyssus gallinae, commonly found on domestic fowl, pigeons, English sparrows, starlings, and other birds.

Rickettsiae (sing. Rickettsia): Single-celled, very small, intracellular bacteria. Notorious members of this group include the causative agents of Rocky Mountain spotted fever, Rickettsia rickettsii, and murine typhus, Rickettsia typhi.

Rickettsialpox: Disease caused by a bacterium in the spotted fever group of rickettsiae and transmitted by the house mouse mite.

Segment: A ring or subdivision of the body, or of an appendage, between areas of flexibility, with muscles attached for movement.

Sclerite: Hardened plate or portion of an insect body.

Screwworms: Larvae of a particular group of calliphorid flies which are obligate parasites of living flesh.

Scutum: The sclerotized dorsal plate posterior to the capitulum in hard ticks. It covers almost the entire dorsal surface in the male and about one half of the dorsal surface in the unengorged female.

Sensillae (sing. Sensilla): Setae, bristles, or hairs having a sensing function.

Sexual dimorphism: Morphological differences between males and females of a species.

Skin scraping: Sample of skin taken for analysis for the presence of mites such as scabies.

Species complex: A group of closely related species, the taxonomic relationships of which are sometimes unclear, making precise identification difficult.

Spiracle: A breathing pore that marks the external opening of the tracheal system.

Spurs: In ticks, coxal spurs are projections from the posterior surface of the posterior margin of the coxae; they may be rounded or pointed, large or small.

Stylet: A needlelike structure, especially the elongated portion of the piercing–sucking type of insect mouthparts.

Stylostome: A straw-like feeding tube produced by chiggers (Family Trombiculidae) when attached to their host, through which semidigested material is drawn to the mouth.

Sylvatic: Describes any disease usually acquired in a forest or other uncultivated, unoccupied areas, rather than in an urban environment or other areas developed by humans.

Synonym: Another name used for a species, or other taxon, invalid because it is either of a more recent date or invalidly proposed.

Systemic: Affecting the entire body.

Systemic loxoscelism: Whole-body reaction to brown recluse spider bite characterized by hematuria, anemia, fever, rash, nausea, vomiting, coma, and cyanosis.

Tarsi (sing. Tarsus): The terminal leg segments.

Taxa (sing. Taxon): Groups of organisms classified together.

Taxonomy: The naming and arranging of species and groups into a system of classification.

Telson: In scorpions, the last segment of the "tail"; it bears the stinger.

Thorax: A body region in insects, located behind the head, that bears the legs and wings.

Thousand leggers: Common name for millipedes.

Tibia: The fourth segment (from the body outward) of an insect leg.

Trachea: An internal respiratory tube.

Transovarial transmission: The production (by an infected vector) of infected eggs that hatch into individuals likewise infected and capable of transmitting the infecting organism.

Transstadial transmission: The survival of parasites or pathogens through successive stages (larva, nymph, and adult).

Trench fever: Disease caused by the bacterium Bartonellaquintana, transmitted to humans by the body louse.

Trochanter: The portion of the leg between the coxa and the femur.

Trombiculosis or Trombidiosis: Infestation with trombiculid mite larvae.

Trophallaxis: Exchange of alimentary canal liquid among colony members of social insects.

Vanillism: Condition (dermatitis) from exposure to stored products mites such as those in dried vanilla.

Vein: A tube running through the membrane of the wings of insects.

Wheat pollard itch: Condition (dermatitis) from exposure to mites in stored food products.

Wigglers: Common name for mosquito larvae.

Worker: A member of a nonreproductive or sterile caste that contributes to a colony welfare by rearing offspring of reproductives. In Hymenoptera, workers are the ones that sting.

Yaws: Tropical infection of the skin, bones, and joints caused by a spirochete (bacterium) and transmitted by either direct contact with infected people or by flies.

Zoogeographic regions: The six divisions of the Earth's surface distinguished by major differences in the organisms present.

Zoonoses (sing. Zoonosis): Diseases of other animals that may be transmitted to humans.

INDEX

Taylor & Francis Group
an **informa** business

Taylor & Francis eBooks

www.taylorfrancis.com

A single destination for eBooks from Taylor & Francis
with increased functionality and an improved user
experience to meet the needs of our customers.

90,000+ eBooks of award-winning academic content in
Humanities, Social Science, Science, Technology, Engineering,
and Medical written by a global network of editors and authors.

TAYLOR & FRANCIS EBOOKS OFFERS:

A streamlined
experience for
our library
customers

A single point
of discovery
for all of our
eBook content

Improved
search and
discovery of
content at both
book and
chapter level

REQUEST A FREE TRIAL
support@taylorfrancis.com

 Routledge
Taylor & Francis Group

 CRC Press
Taylor & Francis Group

9781032338521